Lecture Notes in Networks and Systems

Volume 540

The series "Lecture Notes in Networks and Systems" publishes the latest developments in Networks and Systems—quickly, informally and with high quality. Original research reported in proceedings and post-proceedings represents the core of LNNS.

Volumes published in LNNS embrace all aspects and subfields of, as well as new challenges in, Networks and Systems.

The series contains proceedings and edited volumes in systems and networks, spanning the areas of Cyber-Physical Systems, Autonomous Systems, Sensor Networks, Control Systems, Energy Systems, Automotive Systems, Biological Systems, Vehicular Networking and Connected Vehicles, Aerospace Systems, Automation, Manufacturing, Smart Grids, Nonlinear Systems, Power Systems, Robotics, Social Systems, Economic Systems and other. Of particular value to both the contributors and the readership are the short publication timeframe and the world-wide distribution and exposure which enable both a wide and rapid dissemination of research output.

The series covers the theory, applications, and perspectives on the state of the art and future developments relevant to systems and networks, decision making, control, complex processes and related areas, as embedded in the fields of interdisciplinary and applied sciences, engineering, computer science, physics, economics, social, and life sciences, as well as the paradigms and methodologies behind them.

Indexed by SCOPUS, INSPEC, WTI Frankfurt eG, zbMATH, SCImago.

All books published in the series are submitted for consideration in Web of Science.

For proposals from Asia please contact Aninda Bose (aninda.bose@springer.com).

Vinit Kumar Gunjan · Jacek M. Zurada
Editors

Proceedings of 3rd International Conference on Recent Trends in Machine Learning, IoT, Smart Cities and Applications

ICMISC 2022

Springer

Editors
Vinit Kumar Gunjan
Department of Computer Science and Engineering
CMR Institute of Technology
Hyderabad, India

Jacek M. Zurada
Department of Electrical and Computer Engineering
University of Louisville
Louisville, KY, USA

ISSN 2367-3370 ISSN 2367-3389 (electronic)
Lecture Notes in Networks and Systems
ISBN 978-981-19-6087-1 ISBN 978-981-19-6088-8 (eBook)
https://doi.org/10.1007/978-981-19-6088-8

This Springer imprint is published by the registered company Springer Nature Singapore Pte Ltd.
The registered company address is: 152 Beach Road, #21-01/04 Gateway East, Singapore 189721, Singapore

Preface

The rising technologies of artificial intelligence (AI) and machine learning (ML) are advancing businesses more quickly than ever. Success in this era of digital transformation depends on leveraging analytics to uncover vast amounts of data with detailed insights. In the past, these insights were uncovered manually through extensive data analysis, but data complexity and data complexity both continue to rise. The most recent technologies for data scientists are AI and ML, which allow them to quickly modify the data to verify its usefulness.

The word "smart cities" first gained popularity in the 1990s as a means of describing the application of technology and creativity in urban planning. Since that time, urbanization has increased dramatically. The percentage of people who live in urban regions is predicted to rise to around 70% by 2050, from about 55% in 2020. Rapid urbanization will increase the demand for intelligent and sustainable settings while also putting more strain on available resources. The demands of citizens for a greater standard of living will rise.

A new phase in computing history known as the "Internet of Things" is about to begin (IoT). No matter what you want to call it—machine to machine, machine to infrastructure, machine to environment, Internet of Everything, Internet of Intelligent Things, and intelligent system—it is occurring, and it has enormous promise.

Selected papers from the International Conference on Recent Trends in Machine Learning, IoT, Smart Cities and Applications, 2022 are collected in this volume. It comprises chosen papers arranged according to the approaches they take and the contributions they make to the conference's topic. The chapters in this book provide an overview of the key concepts and theories underlying the technologies and applications discussed, with a focus on face recognition, evolutionary algorithms like genetic algorithms, automotive applications, and automation devices using artificial neural networks, business management systems, the Internet of Things (IoT), machine learning, data science, and contemporary speech processing systems. Additionally covered in this book are contemporary developments in sensor networks, VLSI systems, and medical diagnostic systems. Where appropriate, a discussion of learning and software modules in deep learning algorithms is added. In a nutshell, the book will shed light on the IoT and ML-based smart city innovations that are now being used.

Hyderabad, India
Louisville, USA

Vinit Kumar Gunjan
Jacek M. Zurada

Contents

Editors and Contributors

About the Editors

Dr. Vinit Kumar Gunjan is Associate Professor in Department of Computer Science and Engineering at CMR Institute of Technology Hyderabad (Affiliated to Jawaharlal Nehru Technological University, Hyderabad). An active researcher; published research papers in IEEE, Elsevier and Springer Conferences, authored several books and edited volumes of Springer series, most of which are indexed in SCOPUS database. Awarded with the prestigious Early Career Research Award in the year 2016 by Science Engineering Research Board, Department of Science and Technology Government of India. Senior Member of IEEE, Additional Secretary, Vice Chair Computational Intelligence Society and An active Volunteer of IEEE Hyderabad section; volunteered in the capacity of Treasurer, Secretary and Chairman of IEEE Young Professionals Affinity Group and IEEE Computer Society. Was involved as organizer in many technical and non-technical workshops, seminars and conferences of IEEE and Springer. During the tenure He had an honour of working with top leaders of IEEE and awarded with best IEEE Young Professional award in 2017 by IEEE Hyderabad Section.

Dr. Jacek M. Zurada (SM'85, F'96, LF'14) is Professor of Electrical and Computer Engineering and Director of the Computational Intelligence Laboratory at the University of Louisville, Louisville, Kentucky, USA, where he served as Department Chair and a Distinguished University Scholar. He was Professor at Princeton University, Northeastern University, Auburn University, the National University of Singapore, Nanyang Technological University in Singapore, Chinese University of Hong Kong, the University of Chile, Santiago, Toyohashi University of Technology, Japan, the University of Stellenbosch, South Africa, the University of Marie-Curie, Paris, France, and was Postdoctoral Fellow at the Swiss Federal Institute of Technology, Zurich, Switzerland.

Contributors

M. A. Abitha Department of Electrical and Electronics Engineering, Government Engineering College Thrissur, Affiliated to APJ Abdul Kalam Technological University, Thrissur, Kerala, India

C. G. Abraham Mangalam College of Engineering, Ettumanoor, India

Sudhir D. Agashe Department of Instrumentation and Control Engineering, College of Engineering, Pune, India

Ruchika Aggarwal GGSIPU, Delhi, India

Sanskar Aggarwal Department of Software Engineering, Delhi Technological University, Delhi, India

Shruti Aggarwal Department of Computer Science and Engineering, Chandigarh University, Gharuan, Punjab, India

Upanshu Aggarwal CSED, Thapar Institute of Engineering and Technology, Patiala, India

Abhay Aggrawal Department of Computer Science, PES University, Bengaluru, Karnataka, India

Sumit Ahire Department of Information Technology, Pimpri Chinchwad College of Engineering, Pune, Maharashtra, India

D. P. Akarsha Nitte Meenakshi Institute of Technology, BSF Campus, Yelahanka, Bengaluru, Karnataka, India

S. Akshatha Department of Computer Science, PES University, Bengaluru, Karnataka, India

B. Amogha Reva University, Bangalore, India

Gadiraju Aravind Department of Computer Science and Engineering, Amrita Vishwa Vidyapeetham, Coimbatore, India

J. Arockiya Jerson Christ University, Pune, Lavasa, India

Isha Arora Department of Computer Science, PES University, Bengaluru, Karnataka, India

J. Arun Pandian School of Information Technology & Engineering, Vellore Institute of Technology, Vellore, Tamil Nadu, India

R. Aruna Computer Science and Engineering, Vel Tech Rangarajan Dr. Sagunthala R&D Institute of Science and Technology, Chennai, Tamil Nadu, India

V. Ashwin Department of Computer Science and Engineering, Amrita Vishwa Vidyapeetham, Amritapuri, India

Randheer Bagi Parul University, Vadodara, Gujarat, India

Aarushi Batta SCIT, Manipal University, Jaipur, Rajasthan, India

Sonal Bera Delhi Technological University, New Delhi, India

Rajiv Ramesh Bhandari Department of Computer Science and Engineering, K L Education Foundation, Vaddeswaram, Guntur Dist., AP, India

Sohel A. Bhura Jhulelal Institute of Technology, Nagpur, India

Abha Bodas Department of Information Technology, Pimpri Chinchwad College of Engineering, Pune, India

Gautam Borkar IT, RAIT, Mumbai University, Mumbai, India

Amit R. Bramhecha Department of Computer Engineering, Late Sau. K. B. Jain College of Engineering, Chandwad, India

S. Brunda Nitte Meenakshi Institute of Technology, BSF Campus, Yelahanka, Bengaluru, Karnataka, India

Harshavardhana Chadaram Department of Computer Science and Engineering, Amrita Vishwa Vidyapeetham, Amritapuri, India

Mukul Chaudhari Department of Information Technology, Pimpri Chinchwad College of Engineering, Pune, Maharashtra, India

Anamika Chauhan Delhi Technological University, New Delhi, India

Brig Anuj Chawla National Defence College, Delhi, India

Aditi Chougule Department of Information Technology, Pimpri Chinchwad College of Engineering, Pune, India

Prashant Dahiwale Department of Computer Engineering, Government Polytechnic Daman, Daman, India

Manjula S. Dalabanjan Department of Mathematics, Don Bosco Institute of Technology, Bengaluru, India;
Department of Research and Development, Don Bosco Institute of Technology, Bengaluru, India

Darsh Dave Parul University, Vadodara, Gujarat, India

Rohini Deshpande Reva University, Bangalore, India

A. M. Devagopal Department of Computer Science and Engineering, Amrita Vishwa Vidyapeetham, Amritapuri, India

M. Shyamala Devi Computer Science and Engineering, Vel Tech Rangarajan Dr. Sagunthala R&D Institute of Science and Technology, Chennai, Tamil Nadu, India

Gulbakshee Dharmale Department of Information Technology, Pimpri Chinchwad College of Engineering, Pune, Maharashtra, India

Gulbakshi Dharmale Department of Information Technology, Pimpri Chinchwad College of Engineering, Pune, India

T. V. K. Dheeraj CSE, Vel Tech Rangarajan Dr Sagunthala R&D Institute of Science and Technology, Chennai, Tamilnadu, India

Satyavathi Divadari Micro Focus India Pvt Ltd, PES University, Bangalore, India

J. Divya Udayan Department of Computer Science and Engineering, Amrita Vishwa Vidyapeetham, Amritapuri, Coimbatore, India

K. Ezhilarasi KCG College of Technology, Chennai, India

D. Ganesha Srinivas Department of Computer Science and Engineering, Amrita Vishwa Vidyapeetham, Amritapuri, India

Nikita Ganvkar Department of Computer Science, PES University, Bengaluru, Karnataka, India

Ritik Garg Delhi Technological University, New Delhi, India

Kamlesh Gautam Poornima College of Engineering, Jaipur, Rajasthan, India

Nitesh Ghodichor CSE, SRK University, Bhopal, India

Payal Ghutke G H Raisoni College of Engineering, Nagpur, India

Animesh Giri Department of Computer Science, PES University, Bengaluru, Karnataka, India

Parth Gite Department of Information Technology, Pimpri Chinchwad College of Engineering, Pune, Maharashtra, India

Saikat Gochhait Symbiosis Institute of Digital and Telecom Management, Constituent of Symbiosis International University, Pune, Maharashtra, India

Susmitha Valli Gogula Department of CSE, Gokaraju Rangaraju Institute of Engineering and Technology, Hyderabad, Telangana, India

Somya Goyal Manipal University Jaipur, Jaipur, Rajasthan, India

Ayush Gupta Manipal University Jaipur, Jaipur, Rajasthan, India

Esha Gupta SCIT Manipal University Jaipur, Jaipur, Rajasthan, India

Tanuj Gupta SCIT, Manipal University, Jaipur, Rajasthan, India

A. Hari Kishan Reddy Department of Computer Science, PES University, Bengaluru, Karnataka, India

A. Harshith Varma Amrita Vishwa Vidyapeetham, Amritapuri, India

Anil Hiwale MIT World Peace University, Pune, India

Prasad Honnavalli PES University, Bangalore, India

Anuja Jadhav Department of Information Technology, Pimpri Chinchwad College of Engineering, Savitribai Phule Pune University, Pune, India

Prem Jadhav Department of Information Technology, Pimpri Chinchwad College of Engineering, Pune, India

Apurva Jagtap Department of Information Technology, PCCoE, Pune, Maharashtra, India

Neha Jagwani BMS College of Engineering, Bangalore, Karnataka, India

S. Jahnavi Computer Science and Engineering Department, Dayananda Sagar Academy of Technology and Management, Visvesvaraya Technological University, Bangalore, Karnataka, India

V. K. Jain Sage University Bhopal, Bhopal, India

Swati Jaiswal Department of Information Technology, Pimpri Chinchwad College of Engineering, Savitribai Phule Pune University, Pune, India

Harshit Jha Manipal University Jaipur, Jaipur, Rajasthan, India

Raviraj Joshi Indian Institute of Technology Madras, L3Cube Pune, Pune, India

Sonali Joshi G H Raisoni College of Engineering, Nagpur, India

Athreayasa Kalyan Jupudi CSE, Vel Tech Rangarajan Dr Sagunthala R&D Institute of Science and Technology, Chennai, Tamilnadu, India

E. V. N. Jyothi Department of Information Technology, Keshav Memorial Institute of Technology, Hyderabad, India

Shubham Kadam Department of Information Technology, PCCoE, Pune, Maharashtra, India

Pratibha Kamal Guru Gobind Singh Indraprastha University, Delhi, India

Shevgoor Adithya Kamath Department of Computer Science, PES University, Bengaluru, Karnataka, India

P. Kanakaraja Department of Electronics and Communication Engineering, Koneru Lakshmaiah Education Foundation, Greenfields, Vaddeswaram, A.P., India

K. Kanchanadevi Computer Science and Engineering, Vel Tech Rangarajan Dr. Sagunthala R&D Institute of Science and Technology, Chennai, Tamil Nadu, India

Sunil Karamchandani Department of Electronics and Telecommunication, D.J. Sanghvi College of Engineering, Mumbai, India

Shreenidhi Karjagi Department of Information Technology, Pimpri Chinchwad College of Engineering, Pune, Maharashtra, India

Atul Kathole Department of Information Technology, Pimpri Chinchwad College of Engineering, Savitribai Phule Pune University, Pune, India

Komalpreet Kaur CSED, Thapar Institute of Engineering and Technology, Patiala, India

Karan Kavatage Department of Information Technology, Pimpri Chinchwad College of Engineering, Pune, Maharashtra, India

M. Keertish Kumar Christ University, Pune, Lavasa, India

Khushbu Khandait Babasaheb Naik College of Engineering, Pusad, India

Meera Khandekar Department of Instrumentation and Control Engineering, College of Engineering, Pune, India

Shakyadeep Khobragade Department of Information Technology, Pimpri Chinchwad College of Engineering, Pune, Maharashtra, India

Uday Kocherla Computer Science and Engineering, Vel Tech Rangarajan Dr, Sagunthala R&D Institute of Science and Technology, Chennai, Tamil Nadu, India

Harshal Kotalwar Department of Information Technology, Pimpri Chinchwad College of Engineering, Pune, Maharashtra, India

Priyank Koul Department of Computer Science, PES University, Bengaluru, Karnataka, India

Maddala Kranthi Department of Computer Science and Engineering, PACE Institute of Technology and Sciences, Ongole, India

Pranav Kul Department of Information Technology, Pimpri Chinchwad College of Engineering, Pune, Maharashtra, India

Mugdha Kulkarni Symbiosis Centre for Information Technology (SCIT), Symbiosis International (Deemed University) (SIU), Pune, India

Nupur Kulkarni Department of Information Technology, PCCoE, Pune, Maharashtra, India

Pavuluri Kumar Jaswanth Department of Computer Science and Engineering, Amrita Vishwa Vidyapeetham, Coimbatore, India

P. Keshav Kumar Computer Science and Engineering, Vel Tech Rangarajan Dr. Sagunthala R&D Institute of Science and Technology, Chennai, Tamil Nadu, India

Parteek Kumar CSED, Thapar Institute of Engineering and Technology, Patiala, India

Samuel Roushan Kumar Nitte Meenakshi Institute of Technology, BSF Campus, Yelahanka, Bengaluru, Karnataka, India

Sachin Kushwaha Delhi Technological University, Delhi, India

Onkar Litake SCTR's Pune Institute of Computer Technology, L3Cube, Pune, Pune, India

N. Malarvizhi CSE, Vel Tech Rangarajan Dr Sagunthala R&D Institute of Science and Technology, Chennai, Tamilnadu, India

Ruchika Malhotra Department of Software Engineering, Delhi Technological University, Delhi, India

C. Manjunatha Nitte Meenakshi Institute of Technology, BSF Campus, Yelahanka, Bengaluru, Karnataka, India

G. V. Manoj Kumar Department of Computer Science and Systems Engineering, Andhra University College of Engineering, Visakhapatnam, AP, India

Ashutosh Manorkar Pimpri Chinchwad College of Engineering, Pune, India

G. Maria Kalavathy CSE, St. Joseph's College of Engineering, Chennai, India

Kaja Masthan Department of Computer Science and Engineering, Sphoorthy Engineering College, Hyderabad, India

Sanjay Mate Department of Computer Science Engineering, Sangam University, Bhilwara, India

Shweta Meena Delhi Technological University, Delhi, India

Vishal Menon Department of Computer Science and Engineering, Amrita Vishwa Vidyapeetham, Amritapuri, India

Lavitra Misal Pimpri Chinchwad College of Engineering, Pune, India

M. Monisha Department of ECE, VISTAS, Chennai, India

Md. Mubasheer Uddin Department of CSE, Gokaraju Rangaraju Institute of Engineering and Technology, Hyderabad, Telangana, India

Arjuna Muduli Department of Electronics and Communication Engineering, Koneru Lakshmaiah Education Foundation, Greenfields, Vaddeswaram, A.P., India

Chinmay Mukhedker Department of Information Technology, Pimpri Chinchwad College of Engineering, Pune, Maharashtra, India

Naresh Babu Muppalaneni National Institute of Technology Silchar, Silchar, Assam, India

M. Nagaraju Department of CSE, Gokaraju Rangaraju Institute of Engineering and Technology, Hyderabad, Telangana, India

Rathinn Nair Melath Department of Computer Science and Engineering, Amrita Vishwa Vidyapeetham, Amritapuri, India

Varsha Namdeo CSE, SRK University, Bhopal, India

C. Nandini Computer Science and Engineering Department, Dayananda Sagar Academy of Technology and Management, Visvesvaraya Technological University, Bangalore, Karnataka, India

Kaustubh Narkhede Pimpri Chinchwad College of Engineering, Pune, India

K. R. Nataraj Department of Mathematics, Don Bosco Institute of Technology, Bengaluru, India

G. B. Nijaguna Nitte Meenakshi Institute of Technology, BSF Campus, Yelahanka, Bengaluru, Karnataka, India

Devika Nikam Department IT, PCCOE, Savitribai Phule University, Pune, India

C. V. Nikshep Nitte Meenakshi Institute of Technology, BSF Campus, Yelahanka, Bengaluru, Karnataka, India

Khyati R. Nirmal Department of Computer Engineering, Late Sau. K. B. Jain College of Engineering, Chandwad, India

P. A. Nived Department of Computer Science and Engineering, Amrita Vishwa Vidyapeetham, Amritapuri, India

Suhas Palukuri Computer Science and Engineering, Vel Tech Rangarajan Dr, Sagunthala R&D Institute of Science and Technology, Chennai, Tamil Nadu, India

J. Arun Pandian School of Information Technology & Engineering, Vellore Institute of Technology, Vellore, India

Jaspreet Singh Pannu Department of Information Technology, Pimpri Chinchwad College of Engineering, Pune, Maharashtra, India

Chandni Patel Einfochips, Ahmedabad, Gujarat, India

Rashmi Pathak National Institute of Technology Silchar, Silchar, Assam, India

Parth Patil SCTR's Pune Institute of Computer Technology, L3Cube, Pune, Pune, India

Sonali Patil Department of Information Technology, Pimpri Chinchwad College of Engineering, Savitribai Phule Pune University, Pune, Maharashtra, India

Wani Patil G H Raisoni College of Engineering, Nagpur, India

Raunak Pawar Department of Instrumentation and Control Engineering, College of Engineering, Pune, India

Rohini Pise Department of Information Technology, PCCoE, Pune, Maharashtra, India

G. Poornima BMS College of Engineering, Bangalore, Karnataka, India

A. Prakash CMR Institute of Technology, Hyderabad, India

B. Prasad Department of Electronics and Communication Engineering, Koneru Lakshmaiah Education Foundation, Greenfields, Vaddeswaram, A.P., India

Varun Prasannan Department of Information Technology, Pimpri Chinchwad College of Engineering, Pune, Maharashtra, India

N. Preethi Christ University, Pune, Lavasa, India

Srikar Rachabattuni Department of Computer Science and Engineering, Amrita Vishwa Vidyapeetham, Amritapuri, India

R. Radeep Krishna Mangalam College of Engineering, Ettumanoor, India

K. Raghavendra Reddy Department of Electronics and Communication Engineering, Koneru Lakshmaiah Education Foundation, Greenfields, Vaddeswaram, A.P., India

V. Raj Thaneeghavl CSE, SRK University, Bhopal, India

D. S. Rajnor Department of Computer Engineering, Late Sau. K. B. Jain College of Engineering, Chandwad, India

Divya Rajput Department IT, PCCOE, Savitribai Phule University, Pune, India

Aparna Ranade SCTR's Pune Institute of Computer Technology, L3Cube, Pune, Pune, India

Archana Ranade Deenanath Mangeshkar Hospital and Research Centre, Pune, India

Hiral Rane Department IT, PCCOE, Savitribai Phule University, Pune, India

Sujata Rani CSED, Thapar Institute of Engineering and Technology, Patiala, India

Nilesh Ranpura Einfochips, Ahmedabad, Gujarat, India

Kunjam Nageswara Rao Department of Computer Science and Systems Engineering, Andhra University College of Engineering, Visakhapatnam, AP, India

Roshani Raut Department of Information Technology, Pimpri Chinchwad College of Engineering, Savitribai Phule Pune University, Pune, India

Sudarsi Namrata Ravindra Department of CSE, GRIET, Hyderabad, Telangana, India

Maithili Sabane SCTR's Pune Institute of Computer Technology, L3Cube, Pune, Pune, India

N. Sai Ganesh Amrita Vishwa Vidyapeetham, Amritapuri, India

P. Sai Kishore Reddy Department of Computer Science, PES University, Bengaluru, Karnataka, India

Kasindula Sai Padmakar Department of Computer Science and Engineering, Amrita Vishwa Vidyapeetham, Coimbatore, India

M. Sai Santosh Anirudh Department of Electronics and Communication Engineering, Koneru Lakshmaiah Education Foundation, Greenfields, Vaddeswaram, A.P., India

N. Charan Sai Computer Science and Engineering, Vel Tech Rangarajan Dr. Sagunthala R&D Institute of Science and Technology, Chennai, Tamil Nadu, India

Hitika Saini CSED, Thapar Institute of Engineering and Technology, Patiala, India

Abdul Saleem Department of Electrical and Electronics Engineering, Government Engineering College Thrissur, Affiliated to APJ Abdul Kalam Technological University, Thrissur, Kerala, India

Mahesh R. Sanghavi Department of Computer Engineering, Late Sau. K. B. Jain College of Engineering, Chandwad, India

Neha Sathe MIT World Peace University, Pune, India

Samineni Sathvik Chowdary Department of Computer Science and Engineering, Amrita Vishwa Vidyapeetham, Coimbatore, India

H. Satya Akash Department of CSE, Gokaraju Rangaraju Institute of Engineering and Technology, Hyderabad, Telangana, India

Salunke Savitha Department of CSE, GRIET, Hyderabad, Telangana, India

Vishwajeet Sawant Department of Information Technology, Pimpri Chinchwad College of Engineering, Pune, India

Yosi Shacham Electrical Engineering, Tel Aviv University, Tel Aviv, Israel

Kashish Shah Department of Electronics and Telecommunication, D.J. Sanghvi College of Engineering, Mumbai, India

Moksha Shah Department of Electronics and Telecommunication, D.J. Sanghvi College of Engineering, Mumbai, India

Riyanshi Shah Department of Electronics and Telecommunication, D.J. Sanghvi College of Engineering, Mumbai, India

Bhuvika Sharma CSED, Thapar Institute of Engineering and Technology, Patiala, India

Riya Sharma SCIT Manipal University Jaipur, Jaipur, Rajasthan, India

Sanyam Sharma Department of Software Engineering, Delhi Technological University, Delhi, India

Abhishek Shinde Department of Information Technology, Pimpri Chinchwad College of Engineering, Pune, India

Pratiksha Shirsath Department of Information Technology, Pimpri Chinchwad College of Engineering, Pune, India

Snigdha Shrivastav SCIT Manipal University Jaipur, Jaipur, Rajasthan, India

Aditi Shukla SCIT, Manipal University, Jaipur, Rajasthan, India

M. Shyamala Devi Computer Science and Engineering, Vel Tech Rangarajan Dr, Sagunthala R&D Institute of Science and Technology, Chennai, Tamil Nadu, India

Sai Siddharth Cilamkoti Department of Computer Science and Engineering, Amrita Vishwa Vidyapeetham, Amritapuri, India

Siddhartha Singh Bhadauriya Delhi Technological University, Delhi, India

Abhishek Singh Amrita Vishwa Vidyapeetham, Amritapuri, India

Jaspreet Singh Department of Computer Science and Engineering, Chandigarh University, Gharuan, Punjab, India

Krishna Kumar Singh Symbiosis Centre for Information Technology (SCIT), Symbiosis International (Deemed University) (SIU), Pune, India

Nanhay Singh Netaji Subhas University of Technology, Delhi, India

Rishabh Sinha Delhi Technological University, New Delhi, India

G. Sita Ratnam Lendi Institute of Engineering and Technology, Jonnada, India

P. Sivakumar Kalasalingam Academy of Research and Education, Krishnankoil, India

Vijender Kumar Solanki CMR Institute of Technology, Hyderabad, India

Vikas Somani Department of Computer Science Engineering, Sangam University, Bhilwara, India

Badal Soni National Institute of Technology Silchar, Silchar, Assam, India

Riya Soni Department of Electronics and Communication, Dharmsinh Desai University, Nadiad, Gujarat, India

Swati Sorte G H Raisoni College of Engineering, Nagpur, India

C. S. Soumya Nitte Meenakshi Institute of Technology, BSF Campus, Yelahanka, Bengaluru, Karnataka, India

K. M. Sreedivya Carmel College of Engineering and Technology, Punnapra, India

Amola Srivastava Symbiosis Institute of Digital and Telecom Management, Constituent of Symbiosis International University, Pune, Maharashtra, India

Devesh Kumar Srivastava SCIT Manipal University Jaipur, Jaipur, Rajasthan, India

Mandeddu Sudhakar Reddy Department of ECE, VISTAS, Chennai, India

Lekha Suresh Department of Computer Science, PES University, Bengaluru, Karnataka, India

J. Surya Prasad PES University, Bangalore, India

B. Sai Tarun Computer Science and Engineering, Vel Tech Rangarajan Dr. Sagunthala R&D Institute of Science and Technology, Chennai, Tamil Nadu, India

Meera Arun Thorat Pimpri Chinchwad College of Engineering, Pune, India

D. Thrinesh Department of CSE, Gokaraju Rangaraju Institute of Engineering and Technology, Hyderabad, Telangana, India

Juhi Tikyani Symbiosis Centre for Information Technology (SCIT), Symbiosis International (Deemed University) (SIU), Pune, India

J. Udayan Divya Department of Computer Science and Engineering, Amrita Vishwa Vidyapeetham, Amritapuri, India

K. Vamshi Krishna Reddy Department of Computer Science, PES University, Bengaluru, Karnataka, India

G. Vamsidhar Reddy Amrita Vishwa Vidyapeetham, Amritapuri, India

M. Ananya Varma Department of CSE, GRIET, Hyderabad, Telangana, India

Manikanta Velpuri CSE, Vel Tech Rangarajan Dr Sagunthala R&D Institute of Science and Technology, Chennai, Tamilnadu, India

Amit Verma University Centre for Research and Development, Chandigarh University, Gharuan, Punjab, India

Karun Verma CSED, Thapar Institute of Engineering and Technology, Patiala, India

Sourabh Singh Verma Manipal University Jaipur, Jaipur, India

Y. Vijayalata Department of Computer Science and Engineering, KG Reddy College of Engineering & Technology, Hyderabad, India

A. Vishal Chandra Amrita Vishwa Vidyapeetham, Amritapuri, India

Sonali Vyas University of Petroleum and Energy Studies, Dehradun, India

Janhavi Wagh Department IT, PCCOE, Savitribai Phule University, Pune, India

Omkar Walhekar Department of Information Technology, PCCoE, Pune, Maharashtra, India

Vandana Yalla Department of CSE, GRIET, Hyderabad, Telangana, India

Prognostic Investigation into Melancholic Maladies in Hinterlands

Sunil Karamchandani, Kashish Shah, Moksha Shah, and Riyanshi Shah

Abstract Depression is one of the clinically mental health ailments which results in a person regularly feeling sad, low or lack of interest and distorts the natural way of living. According to the statistics provided by the World Health Organization, yearly around one-fourth of inhabitants in India sustain from Major Depressive Disorder or MDD (commonly known as depression) untreated. The proportion of psychiatrist aid ranges from 0.05 for every 100,000 people in Madhya Pradesh to 1.2 in the south-eastern region of Kerala. National Health Statistics reflect a higher weighted prevalence of mental morbidity in urban metros relative to the hinterlands; however, this is due to their anosognosia. The proposed work aims to accurately predict depression based on variables related to the immediate surroundings of the individual as well as the distribution of their finances using supervised learning algorithms. Out of predictions made using four algorithms on data from a survey conducted on a demographic similar to India's rural population, extreme Gradient Boosting showed the most promising results. The purpose of the proposed research is to have a prognosis of Major Depressive Disorder in the initial stages through a rapid test by professionals and avoid tragic happenings like self-destruction and suicide.

Keywords Feature engineering · Hyperparameter tuning · Machine learning · Major Depressive Disorder · Mental health

1 Introduction

Major depressive disorder is a severe mental disorder that profoundly affects an individual's quality of life [1]. It is a complex disease whose exact causes are still varied [2]. Depression may be seen in people having serious illnesses, people going through

S. Karamchandani · K. Shah (✉) · M. Shah · R. Shah
Department of Electronics and Telecommunication, D.J. Sanghvi College of Engineering, Vile Parle (W), Mumbai 400056, India
e-mail: kashishatitshah@gmail.com

S. Karamchandani
e-mail: Sunil.Karamchandani@djsce.ac.in

V. K. Gunjan and J. M. Zurada (eds.), *Proceedings of 3rd International Conference on Recent Trends in Machine Learning, IoT, Smart Cities and Applications*, Lecture Notes in Networks and Systems 540, https://doi.org/10.1007/978-981-19-6088-8_1

significant life changes or the death of near and dear ones, while some have a family history of depression. Depression is deemed to be the leading cause of disability in the entire world. On 27 February 2017, the health agency of the United Nations reported estimates stating that depression affects more than 300 million people worldwide [3]. A report released on the same day by the UN WHO confirms that an estimated 4.4% of the entire world's population suffers from depression and an increase of 18% in the number of people living with depression between 2005 and 2015 [4]. Undoubtedly, these numbers have been on the rise since 2017. In 2019, the onset of the COVID-19 pandemic led to increased concerns about mental health, including concerns about suicide ideation. In January of 2021, a surprising 41% of adults identified anxiety and Major Depressive Disorder symptoms. This statistic has been majorly stable since the spring of 2020 [5]. The United Nations reported that higher-than-usual levels of symptoms of depression and anxiety were recorded in various countries worldwide [6]. This is because causes of depression have increased as people are under physical, emotional, and psychological stress. In a recent study carried out by Catherine K. Ettman et al., survey data from 1,441 participants was analysed by researchers [7]. The results of the survey showed that depression symptoms were three times higher than ever. During the lockdown, numbers reported were 27.8%, whereas, before the lockdown, numbers were 8.5%.

Kenya is an agriculture-dependent country with an estimated 80% of the population dependent on agriculture for livelihood and employment [8]. It also has a rural population of 72.49% as of 2019 [9]. Hence, despite having a massive difference in the population, Kenya's demographic is not very different from India's. Agriculture is the backbone of the Indian economy as well. India has one of the largest and institutionally most complex agricultural research systems globally [10]. More than 70% of the Indian population resides in rural areas. In rural communities, the individuals are primarily involved in the agricultural sector and farming practices to meet their livelihood requirements [11].

In India, an estimated 56 million people suffer from depression. A maximum of only 12% seeks aid for mental health concerns out of 20% of India's population that suffers from mental illness [12]. In a national survey conducted by the Ministry of Health and Family Welfare amongst adults, it was found that 5.25% of the survey-takers had Major Depressive Disorder during their lifetime, whereas 2.68% of the survey-takers had it at the time. It was also found that approximately 2.5% of the rural survey-takers had depression. Out of the total percentage of the survey-takers who had Major Depressive Disorder, 3.5% and 3.6%, respectively, belonged to the low-income and not literate social demographic. The study also showed that 1.3% of survey-takers who showed High Suicide Risk was from the Lowest Income Quintile. A study conducted in 51 districts of Madhya Pradesh, India, to know about mental health systems showed that only 13.7% of the districts had been covered under the District Mental Health Programme, and around 11.8% of district/general hospitals were involved in providing mental health services. The availability of psychiatrists was 0.05 per Lakh population. This study conducted in 2015–16 concluded that there was an urgent need to increase and update mental health facilities and mental health-care staff [13]. However, these limited services are not as accessible to rural India as to the standards found in urban areas. People are oblivious to the mental disorders faced by them and go to religious healers to find a solution, further intensifying the stigma around psychiatry. On the other hand, villages have become centres to experiment with community-oriented techniques to cure mental disorders. A strong sense of community can be constructive in addressing psychological issues at the village level [14].

This study aims to explore the reasons for developing Major Depressive Disorder amongst the rural population of India. With adequate results, the approach can assess depression in a similar demographic like rural India. The main area of focus is the interdependencies of factors like an agro-based livelihood, varied assets, family-related statistics, and the consequent financial stresses affecting an individual's mental health. Depression prediction based on these causes has been discussed in this paper.

2 Literature Survey

The scientific report on 'Predictive modelling of depression and anxiety using electronic health records and a novel machine learning approach with artificial intelligence' presents a study on 4184 undergraduates conducted at the University of Nice-Sophia-Antipolis the students also had to undergo a basic medical examination. Features ranged from basic information like nationality, age, gender, the field of study, learning disabilities, satisfaction with living conditions, family details, financial difficulties, health conditions, etc. After feature engineering using biomedical domain-specific knowledge, the data was trained using a fivefold validation technique and classified using six algorithms—a neural network using Bayesian hyperparameter optimization, XGBoost, support vector machine, K-nearest neighbours, and random forest. All models excluding the XGBoost were used to make predictions. These predictions were then used to train XGBoost. Predictions from each model become features of the final XGBoost classifier in the form of a prediction matrix [15].

The paper 'Machine Learning-based Approach for Depression Detection in Twitter Using Content and Activity Features' deals with identifying whether a person is depressed based on their tweets and Twitter activity. A study to train and test various machine learning algorithms was conducted to determine whether a user on the social networking site of Twitter is depressed by analysing Twitter activity, including tweets. User activity features were 13 in number. Account measures were deduced from user activity, and the features mentioned above were merged. Predictions were made using decision tree, linear kernel SVM, radial kernel SVM, and Naive Bayes [16].

In this paper on 'Prediction of Depression amongst Senior Citizens using Machine Learning Classifiers', WEKA, a software tool developed by the University of Waikato in New Zealand that can train and predict decisions using different classifiers like a tree-based classifier, rule-based classifier, meta classifier, multi-layer perceptron, support vector machine, BayesNet classifiers, etc. For this research, five classifiers were compared to choose the best one for depression prediction amongst senior citizens. BayesNet classifier, logistic regression, multilayer perceptron, sequential minimal optimization in SVM (SMO), decision trees classifier were used to make predictions [17].

3 Methodology

3.1 Data Set Description and Cleaning

This study survey involves 75 constraints based on age, gender, marital status, household size, financial and health condition, and alcohol consumption. The research candidates were also screened for depression using a verified instrument. The tool used to screen depression was an epidemiological measure (examining and interpreting the arrangement, paradigm and parameters of health and disorder conditions in a specific set of inhabitants) rather than a clinical (based on medicine) measure. The data set contains 1430 sample cases collected on different days of the week. Directly feeding the data before cleaning can reduce the prediction model's performance and harm the accuracy. Hence, the data was cleaned before feeding to the prediction algorithm. It involved taking care of missing values, removing outliers, duplicate, incorrect formatting due to file encodings, and normalizing the data set. After cleaning the data, 62 columns were left after dropping corrupted and almost empty columns.

3.2 Feature Engineering

The process of feature engineering is cardinal in nature amongst machine learning techniques. Numerical and categorical imputation based on the column types was applied in this study. Outliers have been detected by percentile, and standard deviation method and two columns have been dropped with the observation of boxplots. One-hot-encoding has been implemented on categorical columns. Finally, specific numerical columns have undergone the scaling process. The numerical columns have been scaled, bypassing the minimum and maximum values. Feature selection was performed with the help of the correlation matrix. The higher the value of correlation will be, the stronger is the relation between them. A Pearson correlation was also performed, which returns values between -1 and 1. After analysing the correlation matrix, one of each of the highly correlated columns was dropped. The features that were highly correlated to the output column were selected. Finally, Principal Component Analysis was performed to narrow down further the dimensionality of the data being fed into the models. PCA is a method used to improve baseline models.

The original features in a data set can be approximated to fewer variables. Using this method, the cumulative explained variance ratio of the original features can be studied to decide which features account for the maximum variance seen in the features of the data set. The PCA function is initiated, and the number of features to be considered was set to 20 and the features from the training data fit the PCA function. However, after PCA, some features were still positively skewed, which were normalized using log transform. Following all these steps, the resulting 20 features were fed into the models [18].

3.3 Training, Testing, and Hyperparameter Tuning

After the data cleaning and feature selection, the final input factors are reduced to 20 to avoid overfitting. The objective of this study is to determine whether the candidate is depressed or not. The final output from the data set is a binary classification with two possible outputs, i.e. Yes for depressed and No for not depressed. After studying the data set and the requirements, four prediction algorithms were selected to determine whether the inhabitant is depressed or not with the 20 features obtained by exploratory data analysis, feature selection with filtering and Principal Component Analysis. These algorithms were—decision trees, logistic regression, k-neighbours classifier, and extreme Gradient Boosting. Hyperparameter tuning involves choosing a set of optimal hyperparameters for a machine learning algorithm. This study combines the four algorithms above with hyperparameter tuning to obtain the best results. Since finding the right combination of hyperparameters to tune can be a difficult feat manually, this study opted for Grid Search to optimize the performance of the models. Grid Search uses the traditional method for hyperparameter optimization, and it searches through a specified subset of hyperparameters. Grid Search was used to set the values of model-specific hyperparameters for logistic regression, k-neighbours, and decision trees. For Gradient-Boosted decision trees, there are at least five hyperparameters that can be tuned. Here, this was done using Grid Search as well as manually. The improvement in the metrics for XGBoost before and after applying hyperparameter tuning has been shown in the results.

3.4 XGBoost Based Depression Prediction

XGBoost is an ensemble model amongst other machine learning techniques. It uses boosting to improve weak classification models sequentially and improve their performance to match strong classification models. XGBoost is a model built on decision trees [17]. XGBoost algorithm minimizes the overall loss function. It is used to booster up trees. It picks up a tree or a F base learner that minimizes the loss while using new models, as shown in Fig. 1. The loss function is shown in (1), where g represents the first-order derivative of loss at previous iteration denoted by $\partial L(y,f)/\partial f$, h represents the second-order derivative of loss at previous iteration denoted by $(\partial\text{^}2\, L(y,f))/(\partial f\text{^}2)$, 'm' represents the number of trees, and 'n' represents the number of values in the training set.

$$L^m = \sum_{i=1}^{n} (g_i f_m(x_i) + h_i f_m(x_i)) + \Omega(f_m) \tag{1}$$

XGBoost offered the best results, as shown in the results section. Hyperparameter tuning was applied in this algorithm to obtain the best results. To do the same, 'n_estimators', 'max_depth' and the number of steps were tuned. The first two were adjusted using Grid Search, whereas the number of steps was decided manually. Through Grid Search, the best values of n_estimators and max_depth were obtained as 100 and 10, respectively. This result can be further fortified by observing and comparing the values of these hyperparameters with the Log Loss as shown in Figs. 2 and 3.

We have obtained the trend of Log Loss versus max_depth and the trend of Log Loss versus n_estimators. From the trends, it can be confirmed that the values obtained from Grid Search are the best as they give an optimum Log Loss of−0.669900. Now, with the two hyperparameters tuned, the number of steps can be decided as well. This was done by observing the change in error with each step for which the algorithm trained. The number of steps varied from 0 to 60 with an increment of 10 each time. This process was done before and after hyperparameter tuning to show a comparison between the two cases.

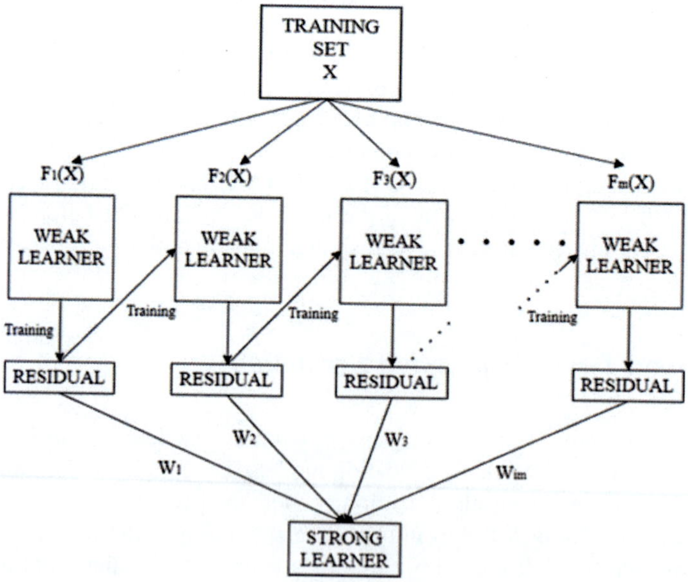

Fig. 1 XGBoost model process based on Gradient Descent

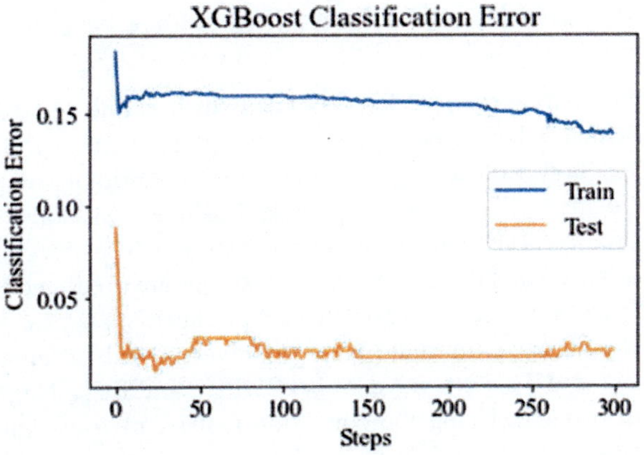

Fig. 2 Classification error versus steps before hyperparameter tuning

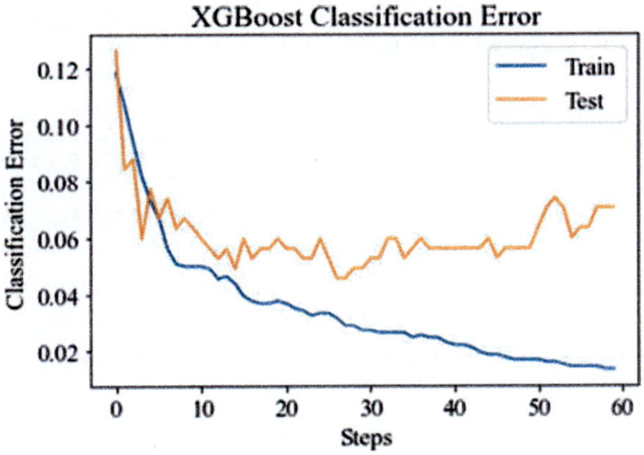

Fig. 3 Classification Error versus steps after hyperparameter tuning

Figures 2 and 3 show the error trend vs the number of steps before and after hyperparameter tuning, respectively. Before tuning, the difference between the training and testing error is enormous compared to the same after tuning. Therefore, it can be said that the model performance improved due to hyperparameter tuning. The number of steps was decided based on the trend after hyperparameter tuning. From Fig. 3, it can conclude that around step number 20, the model shows an ideal combination of training and testing error. The training error is only slightly less than the testing error, which should be the case as the model is more familiar with training data and would show lesser error and higher accuracy for training data. To select a value of the number of steps for which the classification error is not almost the same for both training and testing sets, but the difference is also not too vast, the value of 20 steps was selected. Therefore, for the number of steps = 20, the confusion matrix was calculated before and after hyperparameter tuning. The model performance before and after hyperparameter tuning helps in quantifying the impact of hyperparameter tuning on the XGBoost algorithm, as discussed in the results section.

4 Results and Discussion

Evaluating each model to estimate the best one was done using the confusion matrix, which is best used to describe the performance of classification models. It quantifies how well an algorithm classifies data points by providing the number of True Positives, True Negatives, False Positives, and False Negatives. True Positives are predicted yes and were depressed; True Negatives are cases that are expected no and were not depressed; False Positives are predicted yes but were not depressed and False Negatives are cases that are predicted no but were depressed. An optimal model should have minimum False Positive and False Negative and Maximum True Positive and True Negative. False Positive and False Negative should have been evenly distributed. Using the confusion matrix, metrics that help better compare various classifiers can be derived. These include—accuracy, precision, recall, F-measure, area under ROC Curve, Kappa and Gini coefficient. Accuracy is the measure of correctly predicted values with respect to total predicted values. It is given by the formula shown in (2).

$$\text{Accuracy} = \frac{\text{TP} + \text{TN}}{\text{TP} + \text{TN} + \text{FP} + \text{FN}} \tag{2}$$

Precision is the ratio of correct positive predictions to the total predicted positives. For all the algorithms, accuracy was twofold—training accuracy and testing accuracy to determine how well the model was fitting. Recall is the ratio of correct positive predictions to the total positive's examples. The F-measure is a metric that conveys the balance between precision and recall, which is calculated as in (3).

$$F\text{-measure} = \frac{2 \times \text{Recall} \times \text{Precision}}{\text{Recall} + \text{Precision}} \tag{3}$$

The area under the ROC curve graphically represents the performance of a classification model by plotting the True Positive rate, which is the same as recall against the False Positive Rate given in (4).

$$\text{FPR} = \frac{\text{FP}}{\text{FP} + \text{TN}} \tag{4}$$

Based on the metrics, the results of these models can be categorized into two parts—one where the training accuracy exceeds the testing accuracy and another where the testing accuracy exceeds the training accuracy (Table 1).

The models that pertain to the first category are XGBoost and k-neighbours classifier. Ideally, for any classification problem, since the model learns from the training data itself, the training accuracy should be undoubtedly higher than the testing accuracy, which is the case for these two algorithms. Hence, these algorithms can be considered as the better performing set amongst the four. At first glance, it may seem like k-neighbours perform better due to a higher training accuracy (99.388%) than

Table 1 Comparison of metrics

Algorithm	Training Acc	Test metrics						
		Test Acc	Precision	Recall	F-Measure	Gini Coeff	Kappa Coeff	AUC
XGBoost	0.98162	0.95762	0.06250	1	0.11765	0.94737	0.11180	0.97368
K-neighbours	0.99388	0.87413	0.02703	1	0.05263	0.87368	0.04614	0.93684
Decision tree	0.85039	0.90909	0.03704	1	0.07143	0.90877	0.06512	0.95439
LR	0.83555	0.95454	0.07143	1	0.13333	0.95438	0.12764	0.97719

XGBoost (98.162%), but that is a false perception of the actual results. The test accuracy, on the contrary, is lesser for k-neighbours (87.413%) out of the two, increasing the gap between the training and testing accuracies. For a good-performance model, the testing accuracy should only be slightly lesser than the training accuracy. Suppose the testing accuracy is significantly less than the training accuracy as seen in k-neighbours classifier. In that case, it implies that the model is trying to accommodate any combination of features encountered while learning on the training set, i.e. the model is overfitting on the training set.

Comparatively, XGBoost not only shows a high training accuracy but also testing accuracy (95.762%) as well. The gap between the training and testing accuracy is less, which means that relative to the training data, the algorithm does not do much worse while making predictions on data that it hasn't previously encountered which makes XGBoost the superior algorithm. Other metrics like the F-measure, recall, and precision can bolster the conclusion drawn above. The precision for XGBoost (0.06250) is almost three times as that of k-neighbours classifier (0.02703), which confirms that the former handles True Positives and True Negatives, people who are depressed and were predicted as depressed and those who were not depressed and were predicted as not depressed, in a better manner. The F-measure, which combines the precision and recall results for XGBoost (0.11765), is better than k-neighbours (0.05263), almost two times. The Kappa coefficient also follows a similar trend. A greater value of area under ROC curve and Gini coefficient for XGBoost (0.97368 and 0.94737, respectively) than k-neighbours (0.97368 and 0.94737, respectively) means that it more accurately classifies True Positives than True Negatives. The closer these values are to 1, the closer their predictions are to being 100% correct. This is important as it is more important to classify a depressed person as 'depressed' than classify a person who is not depressed as 'not depressed' as the former requires urgent medical attention and care. Therefore, out of all, XGBoost can best predict whether an individual has depression or not. For the sake of comparison, the improvement of XGBoost has been shown before and after hyperparameter optimization.

4.1 *Effect of Hyperparameter Tuning on Extreme Gradient Boosting Results*

From the confusion matrices in Tables 2 and 3, i.e. before and after hyperparameter tuning, metrics like accuracy, precision, recall, and F-measure at steps = 20 can be calculated. The differences between the two models—hyperparameter tuned XGBoost and regular XGBoost—are apparent. Starting with the confusion matrices, the former is unquestionably better as although it predicts only one person who is depressed, it is better than predicting no one as depressed, consequently having better precision. The number of False Negatives in the tuned XGBoost model is 0 compared to the regular XGBoost model with 1 False Negative. The former is better as no people who are depressed were wrongly predicted as not depressed.

Table 2 Confusion matrix of extreme Gradient Boosting before hyperparameter tuning

		Actual values	
		Positive (1)	Negative (0)
Predicted values	Positive (1)	281	4
	Negative (0)	1	0

Table 3 Confusion matrix of extreme Gradient Boosting after hyperparameter tuning

		Actual values	
		Positive (1)	Negative (0)
Predicted values	Positive (1)	270	15
	Negative (0)	0	1

The False Negatives are also better than False Positives for tuned XGBoost than regular XGBoost, which is good as it is better to wrongly predict a person who is not depressed as depressed than predicting that a depressed person is not depressed. The training accuracy being higher than testing accuracy is also seen in the tuned XGBoost model. All the other metrics—F1-score, recall, precision, and Gini and Kappa coefficients—show the closest to ideal results for the tuned XGBoost algorithm. Therefore, overall hyperparameter tuned XGBoost can be considered the best algorithm for MDD prediction.

5 Conclusion

Depression can be predicted using day-to-day lifestyle habits as well, apart from psychological methods. It is ingrained in the eccentric habits, nature, and way of living of the people. The research in this paper shows practicable methods to predict depression using an approach that only requires passive data like physical health, financial condition, family, and age. This novel approach is implemented using various machine learning algorithms, and the extreme Gradient Boosting model shows the best result. By using the same, we have been able to achieve an accuracy of 95.762% and an F-measure of 0.11765. However, the proposed method is not without its limitations. Firstly, the data set used is moderate in size and gives a generalized account of the rural demographic of India. A larger and more region-specific data set would be helpful further to investigate depression in different rural parts of India. Secondly, the likelihood of being depressed has been found based on financial and familial aspects. Information about an individual's emotional state, habits, and behaviour might give results closer to clinical diagnosis than our approach of an epidemiological one. One of the significant reasons for this research was to diagnose depression in patients in the initial stage to protect them from a higher risk of developing the disorder and create self-awareness of the condition. Knowledge of the same may divert their efforts to cure it in the right direction rather than resorting to religious hoaxes. This approach can be used as a rapid test by professionals to identify the risk of depression. The proposed study can be implemented in various rural areas with a lack of resources and medical help. It can be an appropriate step for early diagnosis and reduce the effects of depression for the inhabitants to look forward to hail and happy life.

References

1. Gao S, Calhoun VD, Sui J (2018) Machine learning in major depression: From classification to treatment outcome prediction. CNS Neurosci Therapeutics 24(11):1037–1052. ClerkMaxwell J (1892) A treatise on electricity and magnetism, 3rd edn, vol 2. Clarendon, Oxford, pp 68–73. Available: https://doi.org/10.1111/cns.13048
2. Khalsa SR, McCarthy KS, Sharpless BA, Barrett MS, Barber JP (2011) Beliefs about the causes of depression and treatment preferences. J Clinical Psychol 67(6):539–549. Available: https://doi.org/10.1002/jclp.20785
3. United Nations. UN health agency reports depression now "leading cause of disability worldwide" (Online). Available: https://news.un.org/
4. World Health Organization (2017) Depression and other common mental disorders: global health estimates. No. WHO/MSD/MER/2017.2. World Health Organization. Available: http://apps.who.int/
5. Panchal N, Kamal R, Orgera K, Cox C, Garfield R, Hamel L, Chidambaram P (2020) The implications of COVID-19 for mental health and substance use. Kaiser family foundation (Online). Available: https://www.kff.org/
6. United Nation Sustainable Development Group. Policy brief: COVID-19 and the need for action on mental health (Online). Available: https://unsdg.un.org/
7. Ettman CK, Abdalla SM, Cohen GH, Sampson L, Vivier PM, Galea S (2020) Prevalence of depression symptoms in US adults before and during the COVID-19 pandemic. JAMA Netw Open 3(9):e2019686–e2019686
8. Mwaura SN, Muluvi AS, Mathenge MK (2013) African leafy vegetables and household well-being in Kenya: a disaggregation by gender (No. 309-2016-5290). Available: https://doi.org/10.22004/ag.econ.161641
9. The World Bank Webpage (Online). Available: https://data.worldbank.org
10. Borthakur A, Singh P (2012) Agricultural research in India: an exploratory study. Int J Soc Sci Interdisc Res 1(9):59–74
11. Kapur R. Significance of agriculture sector in rural areas. Available: https://www.researchgate.net/publication/332013396_Significance_of_Agriculture_Sector_in_Rural_Areas
12. National Mental Health survey (Online). Available: http://indianmhs.nimhans.ac.in
13. Koshy J, Perapaddan BS. Addressing mental health in rural India. Available: https://www.thehindu.com/
14. Nemesure MD, Heinz MV, Huang R, Jacobson NC (2021) Predictive modeling of depression and anxiety using electronic health records and a novel machine learning approach with artificial intelligence. Sci Rep 11(1):1–9. Available: https://doi.org/10.1038/s41598-021-81368-4
15. AlSagri HS, Ykhlef M (2020) Machine Learning-based approach for depression detection in Twitter using content and activity features. IEICE Trans Inf Syst 103(8):1825–1832. Available: https://doi.org/10.1587/transinf.2020EDP7023
16. Bhakta I, Sau A (2016) Prediction of depression among senior citizens using machine learning classifiers. Int J Comput Appl 144(7):11–16. Available: https://doi.org/10.5120/ijca2016910429

17. Ryu S-E, Shin D-H, Chung K (2020) Prediction model of dementia risk based on XGBoost using derived variable extraction and hyper parameter optimization. IEEE Access 8:177708–177720. Available: https://doi.org/10.1109/ACCESS.2020.3025553
18. Kokane AM, Pakhare AP, Gururaj G, Varghese M, Benegal V, Rao GN, Anniappan Arvind B, Prasad M, Mitra A, Shukla M, Yadav K (2021) How healthy is state mental health system in Madhya Pradesh, India? An assessment of today to plan for a better tomorrow. Psychiatry J. Available: https://doi.org/10.1155/2021/6364321

Security Threats in Healthcare Systems—A Bibliometric Study

Saikat Gochhait and Amola Srivastava

Abstract Healthcare cybersecurity has surfaced as the industry's greatest serious extortions. The potential dangers of Web-based healthcare systems have rapidly increased, attracting the hacker community. When patient health records are digitized, there is a risk that sensitive health information will be compromised, and sensitive data may be leaked to hackers and used against them. Health data are a valuable asset in the cybercriminal world, making it a target for hacking. Because the well-being of sick people is jeopardized, healthcare entities are more in the offing to pay ransomware burdens.

Keywords Security · Health care · Threat · Risk

1 Introduction

The digitalization of well-being records is transforming the well-being care sector, ensuing in better medical consequences and healthcare practices. By delivering and exchanging information, digital healthcare information systems (ISs) benefit in various ways [1]. They lead to reduced social blunder, constant and independent patient intensive care, and better-quality presentation. However, as digital networks have grown in complexity, noteworthy security issues have arisen. In the last period, the severity and incidence of healthcare data breaches have increased, with prominent breaches affecting up to 80 million people [2]. In the last few years, 7.2 million data breaks have occurred in India's healthcare sector. According to the breach barometer survey, the number of data breaches discovered in 2019 increased by 48.6% over the

Amola Srivastava contributed equally to this work.

S. Gochhait (✉) · A. Srivastava
Symbiosis Institute of Digital and Telecom Management, Constituent of Symbiosis International University, Pune, Maharashtra, India
e-mail: saikat.gochhait@sidtm.edu.in

A. Srivastava
e-mail: amola.srivastava2022@sidtm.edu.in

V. K. Gunjan and J. M. Zurada (eds.), *Proceedings of 3rd International Conference on Recent Trends in Machine Learning, IoT, Smart Cities and Applications*, Lecture Notes in Networks and Systems 540, https://doi.org/10.1007/978-981-19-6088-8_2

previous year. When electronic health records (medical reports, pictures, and papers) are exchanged with an external entity in the new era of the Internet of Medical Things (IoMT), they become more vulnerable to data breaches [3].

2 Literature Review

See Fig. 1.

2.1 Cluster 1

2.1.1 Blockchain

The amalgamation of blockchain and smart health care will alleviate customary healthcare's pain points in information allocation, data security, and confidentiality maintenance. Also, it simplifies worker-centered smart healthcare structures and creates a cooperative medical association chain involving administration, businesses, and individuals to facilitate smart healthcare's industrial advancement. Blockchain has raised the interest of the entire industry at this time [2]. Recent research findings include fundamental blockchain bottom techniques, blockchain key management, a long-term authentication study of blockchain signatures, and so on. The research into the usage of blockchain in smart health care, on the other hand, is still in its

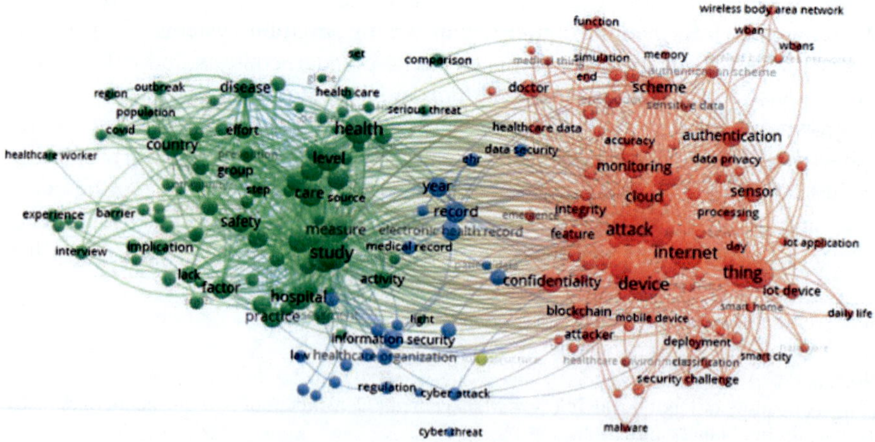

Fig. 1 Network visualization map of terms in title/abstract fields of documents in the health-related security threats literature. The map shows four clusters representing four research themes. Nodes with similar color represent a cluster of related terms. The map was created by VOSviewer

early stages. The present study's mainstream emphasizes coalescing blockchain with existing data expertise to create a new data application or framework, such as a blockchain-based electronic health system and blockchain-based data privacy and safety podium.

2.1.2 Authentication

The most common applications of IoT technologies in the healthcare industry are the checking of objects and individuals (staff and patients), acknowledgment and confirmation of people, and automated data collection and detection. IoT-based healthcare systems are highly vulnerable to attack due to various factors [2]. To begin with, device components are frequently left unattended, exposing them to a physical outbreak. Second, eavesdropping is more complex than wired because most communications are wireless. Finally, because most IoT modules have limited energy and computational power, they cannot enforce complex security measures. According to the Health Insurance Portability and Accountability Act (HIPAA), all sensitive medical data relating to a patient's health must be sheltered. However, the data aggregation scheme faces several security concerns that care must address. The most recent advancements in IoT show a lot of promise in terms of providing healthcare resolutions. However, preserving data concealment and discretion when aggregating data is a standard test in IoT-based healthcare systems.

2.2 Cluster 2

IoT Applications [1]: The growing adoption of Internet of Things (IoT) explanations in a variety of healthcare industries has expanded the IoT market in health care. Real-time health intensive care to manage chronic diseases is expected to drive demand for IoT in health care. Furthermore, advancements in communication technology, such as real-time data allocation, have given patients greater confidence in managing chronic disease and drug dosage. While IoT in healthcare benefits both patients and healthcare specialists, it also raises concerns about data security and privacy. The Internet of Things (IoT) has been used to implement healthcare systems that provide ubiquitous services and assistance to end-users/patients via their handheld devices/user apparatus.

2.3 Cluster 3

Cyberattacks in Healthcare Systems: The novel coronavirus (COVID-19) pandemic has had a substantial and long-term societal and fiscal impact on the world. It has carried out numerous cybersecurity trials and other potential encounters across various spheres, which must be spoken as soon as possible to defend sufferers and critical organizations. Cyberattacks/threats based on social engineering are among the most communal ways to cause havoc, primarily when they target crucial organizations such as hospitals and healthcare services. Cyberattacks focused on social engineering employ psychosomatic and systematic methods to achieve their goal. During the COVID-19 pandemic, the most collaborative social engineering-based strategies were phishing, scamming, spamming, and vishing, in conjunction with the most common socio-technical method: fake communications, websites, and mobile applications used as weapon platforms for directing active cyberattacks.

2.4 Cluster 4

Critical Infrastructure: While employees' shortages, technical apprehensions such as an absence of information principles and policy misperception around reporting requirements remain noticeable issues, the majority of encounters are due to the universal properties of public health agencies. Continued investment to promote technology implementation and the progression of the health informatics workforce and a revision to the interoperability scoring method would likely increase the sharing of electronic data between hospitals and civic health organizations. The (COVID-19) virus pandemic has exposed significant gaps in sharing dangerous material between hospitals and local public health agencies. As a result, it is critical to explain the specific issues hospitals may face when reporting to public health agencies to recommend besieged strategies to improve knowledge sharing for both the current pandemic and ongoing public health activities.

3 Research Methodology

The research methodology used in this study is bibliometric analysis. Cluster analysis is used to conduct a literature review. Four keyword clusters were extracted from relevant research papers. The final review included keywords provided by the authors of the paper that appeared more than five times in the VOS core database. Only, 105 of the 215 keywords passed the test. The most frequently used keywords were "attack" (total link strength 1514) and "authentication" (total link strength 627), which were closely related to "encryption" and "security analysis." Data privacy and confidentiality were two more keywords in authentication comparisons, each with a total link strength of over 769.

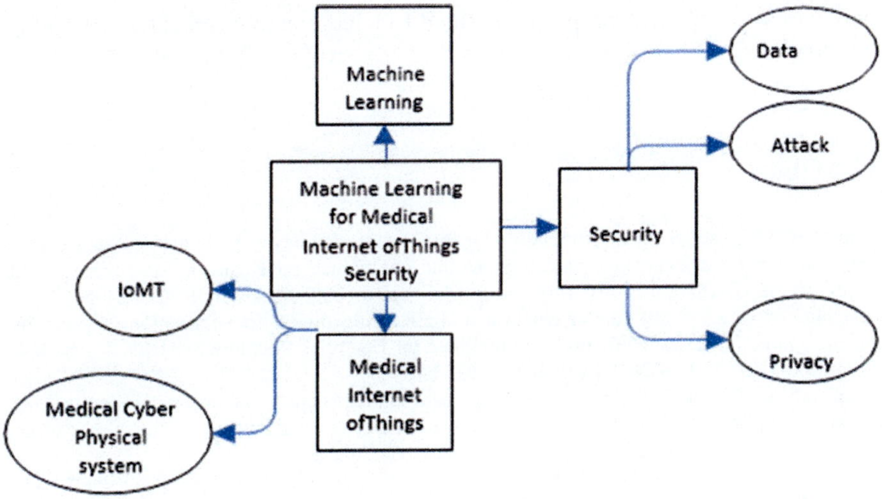

Fig. 2 Conceptual framework of security threats in healthcare systems

4 Conceptual Framework of Security Threats in Healthcare Systems

The primary objectives of this framework are to intricate on the safety situation, strive for constant upgrading, and eliminate communication risks. Firstly, the exchange of information between health organizations is a critical component of healthcare delivery, and all communications must be secure in a modern health system, whether analog or digital. E-mail and fax are two of the most common communication methods in the healthcare industry [4]. The number of cyberattacks on Web sites has recently increased due to the pandemic. This is because there have been increased requests for different levels of hackers to snip a patient's medical statistics to learn more about his treatment or the causes of death. Small-scale software sabotage is also supported by individuals posing as hackers (Fig. 2).

5 Conclusion

The bibliometric analysis introduces an innovative and stimulating theme for understanding security threats in healthcare systems. New findings in health care with the interface of technology have been observed using bibliometric analysis. The healthcare risk administration strategy should be a living manuscript that is regularly updated and better quality to reflect new risks, lessons learned, new information, and changes in the healthcare system and health practice. When these apprises and modifications are made, the proposal should include announcement and training requirements.

References

1. Rawat A, Gochhait S (2022) Iot enabled mental health diagnostic system leveraging cognitive behavioural science. In: 2022 international conference on Decision Aid Sciences and Applications (DASA), pp 1401–1405. https://doi.org/10.1109/DASA54658.2022.9765032
2. Gochhait S et al (2020) Implementation of EHR using digital transformation: a study on telemedicine. In: 2020 International Conference for Emerging Technology (INCET), pp 1–4. https://doi.org/10.1109/INCET49848.2020.9154146
3. Pathapati M, Gochhait S (2022) Intelligent data management to facilitate decision-making in healthcare. In: 2022 international conference on Decision Aid Sciences and Applications (DASA), pp 1–5. https://doi.org/10.1109/DASA54658.2022.9765260
4. Aich S, Tripathy S, Il Joo M, Kim HC (2021) Critical dimensions of blockchain technology implementation in the healthcare industry: an integrated systems management approach. Sustainability 13(9). https://doi.org/10.3390/su13095269

Impact of COVID-19 on Import and Export of Petroleum Products and Crude Oil in India

Mugdha Kulkarni, Juhi Tikyani, and Krishna Kumar Singh

Abstract Imports and export have a significant contribution to increasing the effect of economies of scale and industrialization and lead to an increase in foreign exchange earnings. The oil and gas industry is a prime factor of development in the Indian economy. The natural gas and petroleum sector of India contributes one-seventh percent of the South Asian countries. Petroleum products can be considered ranked among the top five contributors of export trade in India. With the spread of COVID-19 at an exponential rate leading to shut down of industries, manufacturing, transportation, and offices, it has widely impacted the global supply chain. This research aims at studying the effect of COVID-19 on the export and import volumes of petroleum products and crude oil in India. Along with that, the research also covers the changes observed in the consumption and production of petroleum products and domestic and international exchange rate of crude oil before and after the outbreak of COVID-19.

Keywords COVID-19 · Import · Export · Petroleum products · Crude oil

1 Introduction

No nation is independent to deliver every good and service, and this prompts exchange among countries. These exchanges have become an essential need for financial development, exports bring the availability of a wider market and earn foreign exchange, while imports provide goods and services which are insufficient inside the nation. After adopting the new economic policy introduced in 1991, India experienced

M. Kulkarni (✉) · J. Tikyani · K. K. Singh
Symbiosis Centre for Information Technology (SCIT), Symbiosis International (Deemed University) (SIU), Pune, India
e-mail: mugdha@scit.edu

J. Tikyani
e-mail: juhi.tikyani@associates.scit.edu

K. K. Singh
e-mail: krishnakumar@scit.edu

V. K. Gunjan and J. M. Zurada (eds.), *Proceedings of 3rd International Conference on Recent Trends in Machine Learning, IoT, Smart Cities and Applications*, Lecture Notes in Networks and Systems 540, https://doi.org/10.1007/978-981-19-6088-8_3

a higher rate of economic and export growth [1]. According to Macroeconomics and Elementary statistics, India exports more than 7500 commodities and imports more than 6000 commodities all over the world [2]. Mineral fuels including oil and petroleum products contribute about 48.3 billion US dollar having a value equivalent to 14.9 percent of total exports and 168.6 billion US dollar having a value equivalent to 33.2 percent of total imports of India in the year 2018. There are various factors that affect the imports and exports of a country. Foreign trade is an exchange of commodities among various countries. The rise in Domestic GDP and Foreign GDP often leads to a rise in domestic demand which encourages domestic firms to switch to foreign markets and the country will export more [3]. With the coronavirus outbreak, there has been a drastic change observed in the foreign trade across the world. Due to restrictions on travel and disruption in global supply chain, it is assumed to have a severe impact on foreign trade and the global economy.

1.1 Petroleum Products and Crude Oil

In 2017, petroleum products production amounted to 244 million metric tons which has impacted the private industry of petroleum immensely [4]. Petroleum products are essential parts of our imports and exports. The major petroleum products are Naphtha, FO Bitumen, LPG, MS, SKO, HSD, ATF, LDO, LSHS, Lubes, and others. These products are obtained by refining and fractioning of petroleum. Along with these petroleum products, crude oil and natural gas have a proportionate value in the total imports. Crude oil is a liquid source that is obtained naturally and often known as unrefined petroleum. Crude oil can be refined and used to produce petroleum products like gasoline, diesel, and other petrochemicals. India ranks third for importing crude oil for the year 2018 in the globe [5]. After coal, crude oil stands second and natural gas stands third for generation of energy sources in India [6]. Ministry of Petroleum and Natural Gas started working to minimize the import prices of petroleum products and crude oil by offering various incentives, building reserves for natural gas and crude oil storage, and increasing the refining of petroleum products [7].

1.2 Production and Consumption of Petroleum Products

Supply and demand are parts of price determination of the product. To keep the supply and demand of the product, it is necessary to keep producing the product as per the consumption. India ranks third in the consumption of petroleum products and crude oil across the globe in 2018 (Statista Research Department, 2020). With the growth in industrial sector, the production of petroleum products is insufficient, and so we import a major part of petroleum products from other countries. Oil reserves are being developed to increase the storage of petroleum products. Indian Strategic Petroleum Reserves Limited is responsible for maintaining the petroleum reserves.

The reason for the increase in consumption is due to the increase in petrol and diesel demand which in turn has been increasing due to more usage of private vehicles for transportation.

1.3 Prices of Crude Oil and Natural Gas

India is spending a critical expenditure of its trade abroad on importing unrefined fuel and natural gas, and this results in the steady issuance. India now has an excess of refining capacity which has led to transportation and fuel export to other countries like Europe and North America after meeting the domestic needs [8]. The exchange value of crude oil and natural gas affects the incentives spent on investment in energy using equipment. Various factors affect the pricing of crude oil and natural gas like seasonal fluctuations, weather changes, supply disruption, and inventory trends that are important for identifying the price relationship [9]. India imports oil from OPEC nations, so the raw petroleum costs are fluctuating as per Brent oil costs.

2 Objectives

The objective of the research is to understand the effect of a global pandemic on India's exports and imports of petroleum products and crude oil. Before and after the pandemic, the overall number of imports and exports of refined goods and crude oil will be analyzed. The identified factors are consumption of petroleum products, production of petroleum products, domestic price of crude oil, and international price of crude oil.

3 Literature Review

Crude oil is a major part of India's import and has a part in GDP. To recognize the relation among crude oil and GDP, studies formerly executed were referred. Paper by Gosh [10] estimates the interest for imported raw petroleum utilizing auto backward appropriated slack—ARDL testing approach. The demand of crude oil leads to the study of various factors of that determinants of exports to understand the demand and supply chain of petroleum products.

Shah [11] explores the determinants of India's export over the past thirty years and conducts the study for demand supply factors influencing the India's export behavior. The study suggested that government should consider export orient rate policy to take care of low and stable inflation. Two stage least squares (2SLS) substitution method was used for demand and supply equation model for export determination. Shah reasoned from the results that buoyant world demand plays a vital role on supply of Indian exports. The Indian exports price was observed to be more elastic as the worth of exports was higher on the supply side.

Asmundson Irena, Dorsey Thomas, Khachatryan Armine, Niculcea Ioana, and Saito Mika observed that the price of trade finance rose during the 2008 crisis [12]. Other factors which affected the trade decline were as follows: trade output relationship, financial crisis and recession, related party trade which had a decline in trade across countries, and among various sectors due to varying intra-firm trade.

Kotishwar studied that the India exports have grown a resilience during 2008 crisis despite of the financial crisis and global economy slowdown (Kotishwar 2010). Jain [13] studied about the concurrent reforms taken by the government for subsidies of LPG, diesel which leaves a small amount of taxes obtained for the government. Three exclusive coverage levers had been identified: retail price, tax rate, and subsidies. The drop by crude oil permits government authorities to adopt subsidy. The reforms had been released for the duration of years from 2014 to 2017 that decreased the under-recovery bill. Petroleum subsidy reforms had been observed to be a stability among the stakeholders. However, the exercising hike duty on petroleum and diesel has raised the overall level of governments' consumption from the petroleum sector. It is not always feasible for the government to offer subsidy in marketplace costs for all fuels if so, there must be minimal alteration of prices with minimum impact on the profits of the organization.

Aruna and Narasimhan [14] studied the global exchange of import and export rates for India. The study revealed that there may be a fluctuation in exchange volume for India's export charges suggesting a depreciating rupee price and bargaining skills of Indian exporters. While the exchange rates were higher than import charges. Initially, ordinary least square method (OLS) was used to estimate exchange rate with polynomial distribution, but since asset prices like exchange rate are constant, it violates the time series assumptions and creates false regression. So, means of co-integration and error correction method technique were used to solve the issue. Hence, the impact of change rate depreciation on trade deficit was unclear, and imported inflation was a major risk identified.

The study estimated the global economy impacts during and after the pandemic (Buheji et al. 2020). According to them, the COVID-19 outbreak will drop the economy by 2.8% in the 2020. This will result in global depression by −12.5% within the third quarter. The impact of this global economy will be on the prices, which are estimated to recover by mid-2022. The study raises an alert for new normal wave for poverty due to the pandemic.

Maryla et al. [15] calculated macroeconomic and trade impact using the Envisage model. Envisage model which is a global computational general equilibrium model (CGE) is used for estimating the global trade impact. The results from this model gave global chain values from linkages and assumptions. From the study, it revealed that the direct impact of pandemic results in (a) reduction in employment, (b) increase cost in international transactions, (c) drop in transportation and travel services, and (d) decline in demand for services that require proximity between people. It is examined the effect of social distancing coverage on financial activities and stock exchange

[16]. Additionally, they studied the effect of many other global crisis which occurred in the past.

4 Methodology

The data is collected from sources like the reports of Ministry of Petroleum and Natural Gas [17] and CMIE Economic Outlook [18]. After collecting the data, it is analyzed with the help of graphs and other statistical methods. The data is grouped into periods Before COVID-19 and After COVID-19. The data taken for the group Before COIVD-19 is from October 2018 to May 2019, and the data taken for the group After COVID-19 is from October 2019 to May 2020. Both the groups have sample size, i.e., 8 months. Independent Sample t test is used to find the correlation among each group and evaluate the results to discover the impact.

To measure the impact of COVID-19 on import and export of the petroleum products and crude oil, we have created groups of similar sample size to run Independent Sample t test. An Independent Sample t test aims at comparing the means of two sets of data. It measures the statistical evidence that the sets of data taken are significantly different [19].

For our research, the sample variables like import, export, consumption, production of petroleum products, and international prices and domestic prices of crude oil are taken. There are two groups created for each of these variables, i.e., Before COVID-19 and After COVID-19, and means of the groups are compared to find the impact. The survey period for both groups was from October to May (i.e., 8 months) for the financial year 2018–19 and 2019–2020, respectively. The null hypothesis states that there is no difference observed in the mean of groups, 'Before COVID-19' and 'After COVID-19'. Independent Sample t test values are stated in two sections which provide different information: (a) Levene's equal variances test and (b) t-means equality test.

5 Results and Data Analysis

To understand the results obtained from the Independent Sample t test, we look at each variable individually for both the groups: Before COVID-19 and After COVID-19.

Imports: As observed from Table 1, the group Before COVID-19 ($N = 8$) was associated with import volume $M = 22{,}105.50$ (SD = 1453.284). By comparison, the group After COVID-19 ($N = 8$) with numerically smaller import volume $M = 22{,}032.13$ (SD = 2069.164). As seen from the Table 2, the presumption of homogeneity of variances was checked and confirmed by Levene's F test $F\ (14) = 0.451$, $p = 0.513$. Statistically significant effect was correlated with Independent Sample t test with values $t(14) = 0.82$, $p = 0.936$. The Hedges' g value obtained $= -0.0388$

which suggests that the import After COVID-19 have decreased in value. Hence, there is negative but smaller effect of COVID-19 on the total import values.

Exports: As observed from Table 1, the group Before COVID-19 ($N = 8$) was associated with export volume $M = 5028.63$ (SD = 584.046). By comparison, the group After COVID-19 ($N = 8$) with numerically greater Export volume $M = 5745.00$ (SD = 602.704). As seen from Table 2, the presumption of homogeneity of variances was checked and confirmed by Levene's F test $F\ (14) = 0.089, p = 0.769$. Statistically significant effect was correlated with Independent Sample t test with values, $t\ (14) = -2.414, p = 0.30$. The Hedges' g value obtained = 1.1413 which suggests that the export values have increased by more than one standard deviation in After COVID-19. Hence, there is a large positive effect of COVID-19 on the total export values.

Consumption: As observed from Table 1, the group Before COVID-19 ($N = 8$) was associated with consumption values $M = 18{,}298$ (SD = 878.737). By comparison, the group After COVID-19 ($N = 8$) with numerically smaller consumption value $M = 16{,}509.6300$ (SD = 3022.786). As seen from Table 2, the assumption of homogeneity of variances was tested and satisfied via Levene's F test $F\ (14) = 4.906$, $p = 0.044$. Statistically significant effect was correlated with Independent Sample t test with values, $t\ (14) = 1.607, p = 0.146$. The Hedges' g value obtained = -0.7596 which suggests that the consumption values have decreased in After COVID-19. Hence, there is medium negative effect of COVID-19 on the consumption values.

Production: As observed from Table 1, the group Before COVID-19 ($N = 8$) was associated with production values $M = 21{,}824$ (SD = 877.791). By comparison, the group After COVID-19 ($N = 8$) with numerically smaller production value $M = 21{,}007.5$ (SD = 27,533.861). As seen from Table 2, the presumption of homogeneity of variances was checked and confirmed by Levene's F test $F\ (14) = 7.942$, $p = 0.014$. Statistically significant effect was correlated with Independent Sample t test with values, $t\ (14) = 0.799$, $p = 0.446$. The Hedges' g value obtained = -0.3777 which suggests that the production value has decreased in After COVID-19. Hence, there is smaller negative effect of COVID-19 on the production values.

International Price: As observed from Table 1, the group Before COVID-19 ($N = 8$) was associated with production values $M = 58.2375$ (SD = 6.969). By comparison, the group After COVID-19 ($N = 8$) with numerically smaller production value $M = 43.5625$ (SD = 16.143). As seen from Table 2, the presumption of homogeneity of variances was checked and confirmed by Levene's F test $F\ (14) = 10.724$, $p = 0.006$. Statistically significant effect was correlated with Independent Sample t test with values, $t\ (14) = 2.361$, $p = 0.41$. The Hedges' g value obtained = −1.1159 which suggests that the international prices have decreased by more than one standard deviation in After COVID-19. Hence, there is larger negative effect of COVID-19 on the international prices.

Domestic Price: As observed from Table 1, the group Before COVID-19 ($N = 8$) was associated with production values $M = 66.612$ (SD = 7.231). By comparison, the group After COVID-19 ($N = 8$) with numerically smaller production value $M = 48.775$ (SD = 17.976). As seen from Table 2, the presumption of homogeneity of variances was checked and confirmed by Levene's F test $F\ (14) = 14.105$, $p =$

Table 1 Group statistics of group before COVID-19 and after COVID-19

Group Statistics

Variable	Group	N	Mean	Std. Deviation	Std. Error Mean
Import	Before COVID-19	8	22105.50	1453.283	513.813
	After COVID-19	8	22032.13	2069.164	731.560
Export	Before COVID-19	8	5028.63	584.046	206.492
	After COVID-19	8	5745.00	602.704	213.088
Consumption	Before COVID-19	8	18298.00	878.737	310.680
	After COVID-19	8	16509.63	3022.786	1068.716
Production	Before COVID-19	8	21824.00	877.791	310.346
	After COVID-19	8	21007.50	2753.861	973.637
International_Price	Before COVID-19	8	58.2375	6.96951	2.46409
	After COVID-19	8	43.5625	16.14275	5.70732
Domestic_Price	Before COVID-19	8	66.6125	7.23078	2.55647
	After COVID-19	8	48.7750	17.97655	6.35567

Table 2 Independent Sample t-test values are stated in two sections which provide different information: (a) Levene's Equal Variances Test (b) t-Means Equality Test

Independent samples test								
Variables		Levene's test for equality of variances		t-test for equality of means				
		F	Sig.	*T*	*df*	Sig.(2-tailed)	Mean difference	Std. Error difference
Import	Equal variances assumed	0.451	0.513	0.082	14	0.936	73.375	893.971
	Equal variances not assumed			0.082	12.555	0.936	73.375	893.971
Export	Equal variances assumed	0.089	0.769	−2.414	14	0.030	−716.375	296.724
	Equal variances not assumed			−2.414	13.986	0.030	−716.375	296.724
Consumption	Equal variances assumed	4.906	0.044	1.607	14	0.130	1788.375	1112.959
	Equal variances not assumed			1.607	8.175	0.146	1788.375	1112.959
Production	Equal variances assumed	7.942	0.014	0.799	14	0.438	816.500	1021.902
	Equal variances not assumed			0.799	8.408	0.446	816.500	1021.902
International_Price	Equal variances assumed	10.724	0.006	2.361	14	0.033	14.67500	6.21654
	Equal variances not assumed			2.361	9.522	0.041	14.67500	6.21654
Domestic_Price	Equal variances assumed	14.105	0.002	2.604	14	0.021	17.83750	6.85055
	Equal variances not assumed			2.604	9.207	0.028	17.83750	6.85055

0.002. Statistically significant effect was correlated with Independent Sample t test with values, t (14) = 2.604, p = 0.28. The Hedges' g value obtained = -1.2309 which suggests that the domestic prices have decreased by more than one standard deviation in After COVID-19. Hence, there is larger negative effect of COVID-19 on the domestic prices.

Thus, from the results we can say there is a major drop in the prices of crude oil for international and domestic markets due to COVID-19. There is also a major rise in the export volume which is a good indication for growth for the country. The import value has decreased negligible due to COVID-19. Both production and consumption of petroleum products are moderately decreased due to coronavirus.

6 Discussion

Out of the total demand of oil and petroleum products worldwide, Indian oil and gas industry have a share of 5.2% in the market. With the outbreak of COVID-19, there is massive reduction observed in the demand and the prices of crude oil which is a growing concern for Indian Oil and Gas Industry. India's petroleum consumption dropped the lowest in 10 years, a direct sign of the country's decreasing demand for petrol, diesel, and other petroleum products due to coronavirus. Alone in March 2020, the consumption of petroleum products fell by 18% as compared to the identical month a year ago [20]. Both imports and exports of petroleum products and crude oil have fallen sharply along with the demand and prices from March 2020 itself [21]. However, with the collapse of global oil prices, the country has reduced its import bill with increasing volume at the same time. For the financial year 2019–2020, the oil imports have increased in volume by 4% when compared with the previous year. In the month of April'20 Indian refiners stuffed tanks with inexpensive oil, sold additional cargoes to the central authorities for strategic reserves and with this the crude oil import had been declared as force majeure. India's oil imports reached the lowest since October 2011 in May 2020 as refiners with storage purchases cut due to dropping demands for fuel. Indian oil imports are set to recover in June 2020 with the refiners raising crude processing, and with the resumption of transport and industrial activities, the demand is also significantly increasing. This is a favorable situation for India to stock up more oil and save for investment during the reverse economic times. The recovery of the Indian Economy for petroleum industry may be a U-shaped recovery or a W-shaped recovery depending on the second/third wave outbreak of the virus. Though COVID-19 has led to petroleum industry sit up, the leaders are moving toward a long-term view creating partner ecosystem for long-term resilience.

7 Conclusion

The current pandemic-driven global crisis at coronavirus has hit the Indian Economy hard. The entire world is in great uncertainty. Oil and gas industry is among the one sectors of India which have been majorly impacted due to the disruption in the global supply chain. From the research, we conclude that the import after COVID-19 has decreased in value. Hence, there is negative but smaller effect of COVID-19 on the total import values. However, the export values have seen to be increasing when compared with the previous year export volume, and there is a negligible fall in import volumes when compared with the previous year. There is a huge fall in the demand of the petroleum products leading to global drop in the prices. This raises a great concern for the Indian Ministry of Petroleum and Natural Gas.

References

1. Dhankar P, Chhikara KS (2018) India's foreign trade and impact of exports on foreign exchange reserves of India. Int J Res Humanities Arts Literature
2. Macroeconomics and Elementary Statistics (2008) V K Publications, New Delhi, India. Retrieved 25 Apr 2020
3. Ravikumar R, Arunhari P (2014) Compostion and direction of trade in goods and trade in servcies of India. EPRA Int J Econ Bus Rev
4. Petroleum Sector in India- Statistics and facts (2020) Retrieved 13 June 2020, from Statista.com: https://www.statista.com/topics/5621/petroleum-sector-in-india/
5. BP Statistical Review of World Energy (2019) BP p.l.c. Retrieved from https://www.bp.com/content/dam/bp/businesssites/en/global/corporate/pdfs/energy-economics/statistical-review/bp-stats-review-2020-full-report.pdf
6. Energy Statistics 2019. Central Statistics Office (2019) Ministry of Statistics and Programme Implementation
7. Energizing and Empowering India (2019) Ministry of Petroleum and Natural Gas. Retrieved from http://petroleum.nic.in/sites/default/files/AR_2018-19.pdf
8. Tuli D, Gupta R (2017) Production and consumption of petroleum and compressed natural gas in India, Oil and Gas Import, Development of Alternative Energy Sources, Price escalation pattern, and its effect on Indian economy. In: Sustainable biofuels development in India. Springer, Cham. https://doi.org/10.1007/978-3-319-50219-9_1
9. Hartley P, Medlock K, Rosthal J (2007) The relationship between crude oil and natural gas. James A. Baker III Institute for Publicly Policy of Rice University, vol 36
10. Gosh S (2008) Import demand of crude oil and economic growth: evidence from India. Int J Political Econ Plann Environ Soc Aspects Energy 37(2):699–702. https://doi.org/10.1016/j.enpol.2008.10.021Get
11. Shah A (2013) Deficit conundrums: the determinants of India's export behavior. Thesis, Haverford College, Economics
12. Asmundson I, Dorsey T, Khachatryan A, Niculcea I, Saito M (2011) Trade and trade finance in the 2008–09 financial crisis, International Monetary Fund
13. Jain A (2018) A fine balance: lessons from India's experience with petroleum subsidy reforms. Energy Policy 119:242–249
14. Aruna D, Narasimhan V (2011) Exchange rate pass-through: how much do exchange rate changes affect the prices of indian exports and imports. South Asia Econ J 12(1):1–23
15. Maliszewska M, Mattoo A, Mensbrugghe DV (2020) The potential impact of COVID-19 on GDP and trade. World Bank Group East Asia and Pacific region

16. Peterson O, Thankom A (2020) Spillover of COVID-19: impact on the global economy. SSRN Electronic J. https://doi.org/10.2139/ssrn.3562570
17. Production of Petroleum products by Refineries & Fractionators. Petroleum planning and analysis cell. Retrieved 05 2020
18. Prices of Crude Oil in Global Market (2020) CMIE economic outlook. Retrieved 5 2020, from 'https://economicoutlook.cmie.com
19. Kenton W (2020) T-Test. Retrieved 19 June 2020, from Investovpedia: https://www.investopedia.com/terms/t/t-test.asp
20. Abdi B (2020) Covid-19 impact: India's petroleum consumption falls to 10-year low
21. Keerlery S (2020) Export value of petroleum products from India. Retrieved from Statista.com: https://www.statista.com/statistics/624190/export-value-of-petroleum-products-india/
22. Wright C (2009) Testing homogeneity of variance. University of Newcastle, Australia, Statistics. https://doi.org/10.2139/ssrn.2953900

Stress Detection Using EEG Signal in Early Stage and Control Technique

Payal Ghutke, Sonali Joshi, Wani Patil, and Swati Sorte

Abstract This paper presents about the stress which is a significant issue in human world. Because of substantial remaining task at hand, cutoff time in office, nurses work in shifts, and so forth: Stress assumes a significant job in unevenness of human conduct and its tendency. Because of this, they feel furious, miserable, and loss of psyche and experiences high BP, increment in sugar level and now and again get heart diseases. Proposed configuration breaks down the feeling of anxiety by utilizing EEG signal and gives methods to decreasing the worry for improving their work. The main aim to reduce the stress by analyzing the EEG signal as it is high in human, and for that it needs to do the measurable investigation for checking the stress level reduction.

Keywords Electroencephalogram · Brain signal · Music model · Stress detection and reduction

1 Introduction

IDENTIFICATION of stress is very important to reduce it. Stress is the important factor in imbalance of human nature and behavior. Due to this, many people faced problems like not concentrating on work, headaches, heart burn, and hormonal changes, which leads to change in their behavior, they feel angry, sad, and loss of mind and get frustrated [1]

P. Ghutke (✉) · S. Joshi · W. Patil · S. Sorte
G H Raisoni College of Engineering, Nagpur, India
e-mail: payal.ghutke@raisoni.net

S. Joshi
e-mail: sonali.joshi@raisoni.net

W. Patil
e-mail: wani.patil@raisoni.net

S. Sorte
e-mail: swati.sorte@raisoni.net

V. K. Gunjan and J. M. Zurada (eds.), *Proceedings of 3rd International Conference on Recent Trends in Machine Learning, IoT, Smart Cities and Applications*, Lecture Notes in Networks and Systems 540, https://doi.org/10.1007/978-981-19-6088-8_4

To measure the stress level, there are different wearable sensors & bio-signal processing technologies are available like electroencephalography (EEG), electrocardiography (ECG), electromyography (EMG), blood pressure (BP), blood volume pulses (BVP), galvanic skin resistance (GSR), respiration, and skin temperature (ST) [2, 3]. Likewise, human physiological sign preparing innovation is utilized to gauge the feeling of anxiety from human physiological components. There is difference between individual's physiological features, its changes when individual response to troubling activities. Considering distinctive physiological features occur in human while individual is in stress/pressure. The assessment of stress is done by using bundle-based assessment methodology [1, 4].

The EEG signal is a non-stationary noninvasive technique which is used to measure and monitor the state of the brain signal. It is relatively inexpensive which will be helpful to do the research. EEG has number of medical uses ranging from normal to difficult situation for measurement and monitoring the brain waves which will contain unique information [3]. Different characteristics of an EEG signal which consists of many waves have been recorded with the electrodes placed in the scalp, distributed over the entire scalp [5]. Even a single electrode pair can record large amount of data [6]. The EEG recorded the electrical activity of the brain [7], which measured in microvolts (mV). Human EEG waves delta (δ), beta (β), alpha (α), and theta (θ) are the main parameters of frequency. Electroencephalogram (EEG) is one of the most widely recognized bio-signals which is used to study the conditions of the brain functions.

The main objective of this research is.

- To extract the data from the human scalp using EEG electrode.
- To analyze the EEG data and identify the stress.
- By providing the solution to reduce the stress level.

2 Proposed System

Proposed design analyzes the stress level of patient, and it gives solution for reducing the stress so that they can improve their performance at work or be relaxed. The main purpose of these work is to reduce the stress by using physiological signal processing technology into the system and to do the statistical analysis by identifying the stress level of human is reduced or not. In this approach as shown in the Fig. 1, the raw EEG signal is contaminated with noise from various structures and sources. To study the different brain functions and its conditions, EEG is one of the most common sources of information used to do that [8]. It may include (50/60 Hz) power lines noise. The work of classifier is to compare testing data with the baseline data. After preprocessing of EEG signals [7], feature extraction step needs to be followed. The advantage of feature extraction is to remove dissimilarities from the number of features in the datasets, and from existing features, it creates the new features. It improves the accuracy, data visualization, and speed up in training.

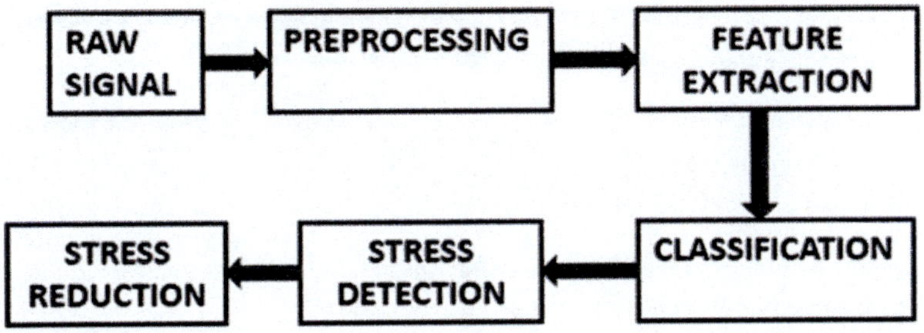

Fig. 1 Block diagram of stress detection

3 Stress Detection

The program is implemented using Matlab. Firstly, we collect the data of stressed person [9, 10]. We load the stress data, define its size, and set sampling frequency and period as 500 and 2, respectively. We plot the EEG data and EEG channels. Using fast Fourier transform (FFT), we plot the power spectrum of the EEG data. We plot the waveform band of each wave—delta, beta, alpha, and theta by the use of filters; we reduce the noise and generate the noise-free signals. Similarly, we did for the reference signal, and by observing threshold value of reference signal and the patient's EEG signal, we can conclude whether the patient is stressed or not.

Figure 2 shows the complete EEG signal and different channels like 1, 5, 10, 15, and 16.

Figure 3 shows waveform of delta band and single sided amplitude spectrum of delta. This wave has frequency less than equal to 4 Hz, and amplitude of delta is 20–200 uV. The power of delta wave increases during difficult condition.

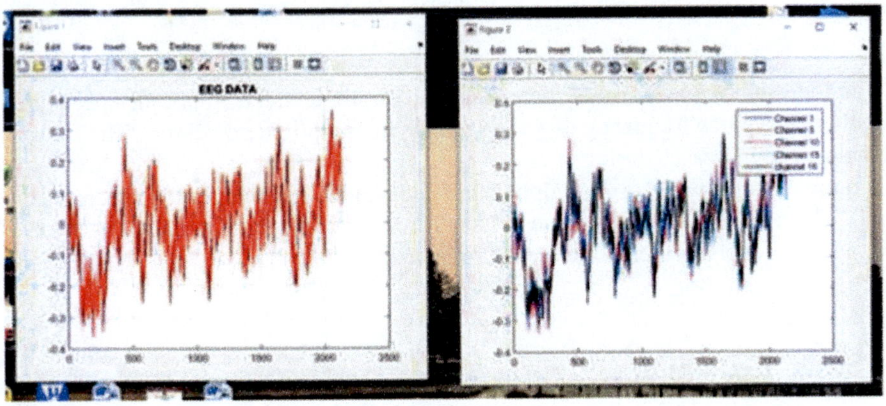

Fig. 2 EEG data

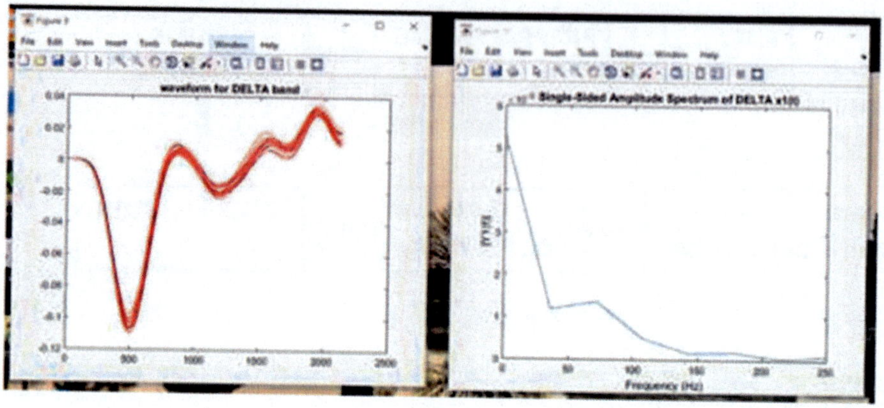

Fig. 3 Delta band

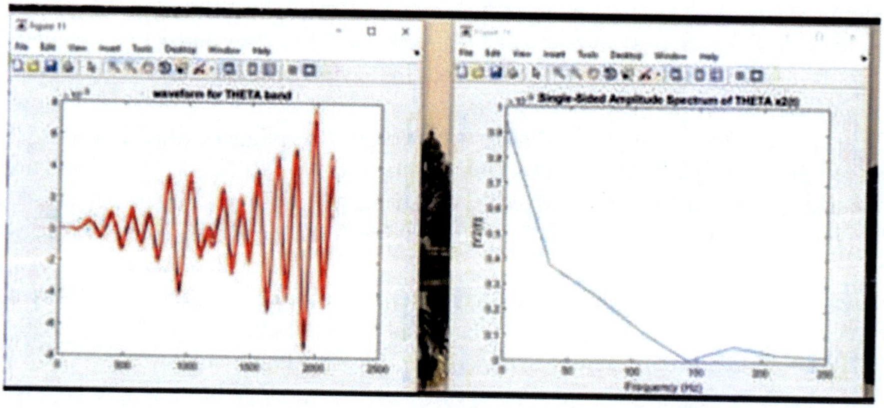

Fig. 4 Theta band

Figure 4 shows waveform of theta band and its single-sided amplitude spectrum. The theta wave has frequency 4–8 Hz with having amplitude of 20 uV. The power of theta wave increases during stress.

Figure 5 shows waveform of alpha band and single-sided amplitude spectrum of alpha. The frequency of alpha wave is 8–10 Hz, and amplitude of delta is 20–200uV. The power of alpha wave suppresses during stress (Table 1).

4 Stress Reduction

This application finds as a doorbell just as phone interloper music player. The fundamental work of the circuit IC1 APM993D is a customized 1023 note ROM memory, which creates 16 melodic tunes when it gets set off. The main component of this

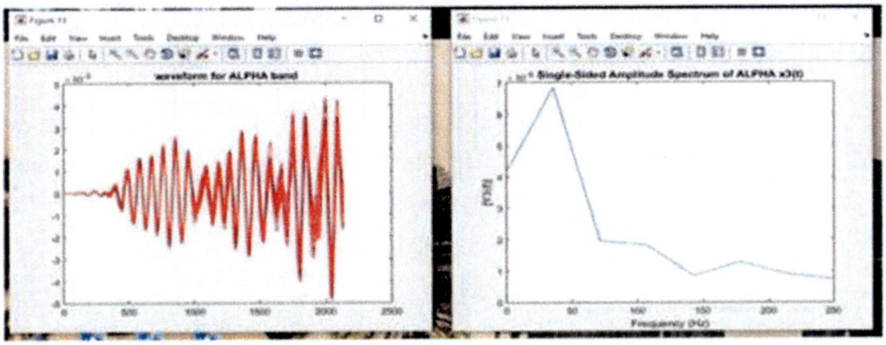

Fig. 5 Alpha band

Table 1 EEG wave comparison [11]

Parameter (signal)	*F* (frequency) (Hz)	*A* (amplitude) (μV)	Effect of power during stress
(δ)	<4	20–200	Power is increased
(θ)	4–8	Around 20	Power is increased
(α)	8–12	20–200	Power suppresses
(β)	13–31	5–10	According to task, power varies

IC is that it has inbuilt tone generator, musicality generator, modulator, oscillator, recurrence divider, and pre-amplifier [12]. This gives a benefit to interface not many number of parts remotely to construct an outer AF amplifier driver circuit [13] and to set up timing (Fig. 6).

5 Result

Interactive early stress detector is a device which detects the patients stress value. If he/she is suffering from stress, then it will give the measures to reduce the stress. In this circuit, we are using 2 electrodes to take the EEG waves and body register as an input to the A0 pin of Arduino. This input single through 10 M ground resisters to A1 pin of Arduino so that signal will go to the A1 without any disturbance, and A1 pin is used for LCD display from which we can note the stress values (Fig. 7). Whenever stress increases above the limit, the emitter, collector gate short circuit, and music start playing to reduce the stress.

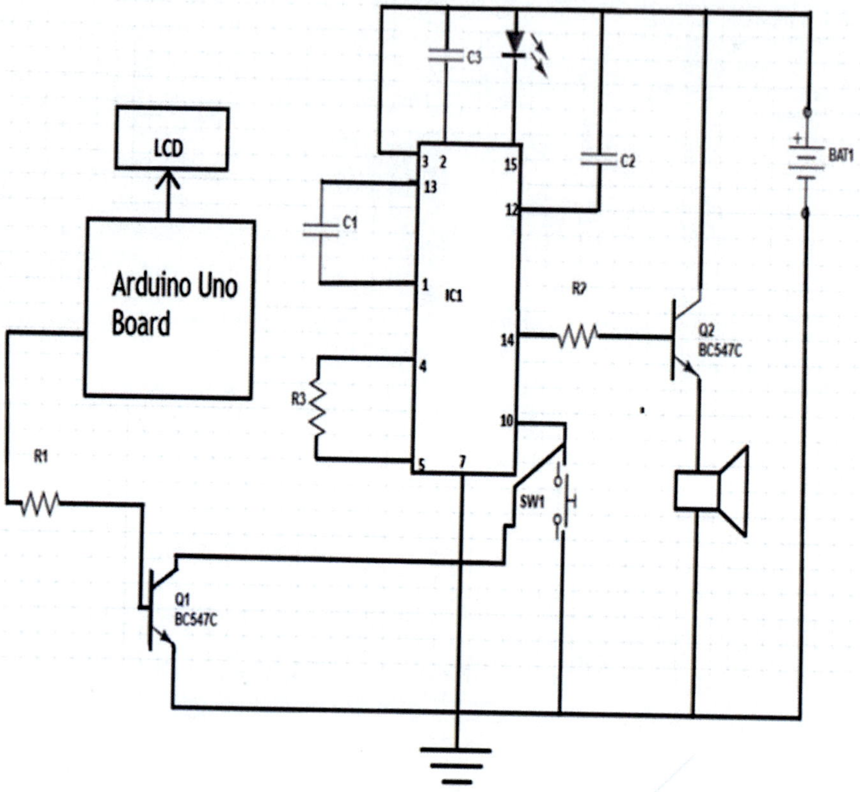

Fig. 6 Music model for reducing stress

Fig. 7 Hardware model for stress detection and reduction

6 Conclusion

This system consists of stress detection techniques by taking EEG signal from the human body suffering from stress. It will calculate the stress value which will compare with threshold value. As soon as stress value exceeds threshold value, the music system starts playing which will reduce the stress threshold value.

This system is useful for real-time stress monitoring for those who constant work under pressure in their daily life.

References

1. Patil NB, Mirajkar RP, Patil S, Patil P (2017) A method for detection and reduction of stress using EEG signal. (IRJET) Int Res J Eng Technol
2. Seo SH, Lee JT (2010) Stress and EEG. In: Convergence and hybrid information technologies, Marius Crisan, InTech, pp 413–426
3. Healey JA, Picard RW (2004) Detecting stress during real-world driving tasks using physiological sensors. IEEE Trans Intell Transp Syst 6(2):1–19
4. Ghutke P, Joshi S, Timande R (2021) Improving accuracy of classification of emotions using EEG signal and adaptive PSO. In: IOP conference series: materials science and engineering, vol 1170, International Conference on Advanced Materials Behaviour & Characterization (ICAMBC_2021), 24–26 Apr 2021, Tamil Nadu, India
5. Timande R, Ghutke P (2020) Improving accuracy of emotion detection using brain waves and adaptive swarm intelligence. Int J Innov Technol Exploring Eng (IJITEE) 9(6), ISSN: 2278-3075
6. Setz C, Arnrich B, Schumm J, La Marca R, Tr¨oster G, Ehlert U (2010) Discriminating stress from cognitive load using a wearable EDA device. IEEE Trans Inf Technol Biomed 14(2)
7. Agrawal R, Bajaj P (2020) EEG based brain state classification technique using support vector machine—a design approach. In: 3rd International Conference on Intelligent Sustainable Systems (ICISS 2020) organized by SCAD Institute of Technology, Palladam, India from 03–05 Dec 2020, ISBN: 978-1-7281-7089-3/20 2020 IEEE, pp 1523–1528
8. Krishnaveni V, Jayaraman S, Aravind S, Hariharasudhan V, Ramadoss K (2006) Automatic identification and removal of ocular artifacts from EEG using wavelet transform. Measurement Sci Rev 6(4) (Section 2)
9. Xu Q, New TL, Guan C (2015) Cluster-based analysis for personalized stress evaluation using physiological signals. IEEE J Biomed Health Inf 19(1)
10. Chen T, Yuen P, Richardson M, Liu G, She Z (2014) Detection of psychological stress using a hyper spectral imaging technique. IEEE Trans Affective Comput 5(4)
11. Thakor NV, Sherman DL (2013) EEG signal processing: theory and application, neural engineering. ISBN : 978-1-4614-5226-3
12. Patil W, Sorte S, Ghutke P, Joshi S (2021) Mapping of the Dobot M1(4axis) robotic arm for actual error free data. PENSEE J 51(6)
13. Sorte S, Patil W, Joshi S, Ghutke P (2021) Vision system check for authentication of quality of industry automation for detection of system parts using Raspberry Pi. In: International Conference on Research Frontiers in Sciences (ICRFS-2021), Feb 2021

Analysis and Rendering of Deauthentication Attack Using IoT Technology

Abhay Aggrawal, Isha Arora, and Animesh Giri

Abstract The Internet has developed exponentially over the years and has become an identity. Along with development, different kinds of networks such as LAN, WLAN, and WAN came into picture. Wireless local area network (WLAN) is the most used communication network in relatively smaller areas such as houses, schools, and offices. WLAN follows security such as WEP and encryption, but there are still vulnerabilities in the lower layer of the protocol stack. The vulnerabilities in the IEEE 802.11 have proved to be a medium for denial-of-service attack. Due to susceptibility of management frames, WLAN is vulnerable to medium access control-based (MAC) denial-of-service attacks. The results of these DOS attacks can range from man-in-the-middle attack to network congestion. By spoofing the MAC address of any legitimate user, deauthentication can be rendered to break the connection between the user and the Internet access point. There have been various amendments in the protocols to avoid the attacks on the MAC layer. But, much success hasn't been registered till date. Various algorithms such as medium access control spoof detection and prevention (MAC SDP DoS) perform spoof detection and prevention to render MAC layer attacks affectless. This paper talks about the technique for rendering the DoS deauthentication attack and also present the analysis of the packet that is sent to attack on MAC layer.

Keywords MAC · Denial-of-service · Deauthentication · Spoofing · IEEE 802.11

A. Aggrawal (✉) · I. Arora · A. Giri
Department of Computer Science, PES University, Electronic City, Bengaluru, Karnataka 560100, India
e-mail: abhayagarwal383@gmail.com

A. Giri
e-mail: animeshgiri@pes.edu

V. K. Gunjan and J. M. Zurada (eds.), *Proceedings of 3rd International Conference on Recent Trends in Machine Learning, IoT, Smart Cities and Applications*, Lecture Notes in Networks and Systems 540, https://doi.org/10.1007/978-981-19-6088-8_5

1 Introduction

The use of wireless network has drastically increase overtime because of the low cost, scalability, mobility, and low data error rates, etc., services offered by the technology irrespective of its wired counterparts [1]. Along with great feasibility and services, there are various vulnerabilities that are highly exploitable. Due to use of radio waves for communication, the data sent can be easily intercepted and modified by the attackers to perform various denials-of-service attacks, and also, information privacy can be breached. The data can be spoofed and masqueraded to provide ambiguous or false information.

Unlike the wired connection where there is physical connection to provide authenticity, wireless connection usually paired and connected through authentication mechanism such as shared key authentication or open system authentication. The authentication can only be confirmed by access points for the clients, but there is no way for a client to know if the network is genuine or not. This paves the way for fake and rogue access points to come into picture [2]. Only, after the successful authentication, data can be exchanged between the client and the access point. For the connection, various management and association frames are used between the client and access point which consist of both request and response.

The use of management and association frames opens up the overall network to various denial-of-service attacks.

A deauthentication attack is a type of denial-of-service attack that targets the communication between a user and an access point. The type of attack can paralyze the connection between the client and the access point, interrupting the communication and leads to the disconnection from the service offered by access point temporarily. The process of this attack occurs in the authentication. It is done by sending broadcast addresses and changing broadcast addresses on targets attacked [3]. The attack uses deauthentication management frame with subtype bits 1100, send from attackers machine (Fig. 1).

IEEE 802.11 specifies the set of media access control and physical layer protocols for implementing wireless local area network. IEEE had made amendments in the protocol over years, but still the protocol is considered to be vulnerable. The IEEE 802.11w standard provides protection against authentication/association requests in existing connection (Fig. 2) and against deauthentication/disassociation DoS attacks. It includes cryptographic MIC to protect against spoofed requests and spoofed deauthentication/disassociation frames [4].

Internet of Things (IoT) is the representation of physical hardware/objects that consist of sensors, micro-controller, software, and different other technology that connect and exchange data with homogenous or heterogeneous devices [5] The common way of communication for these devices is through wireless communication mostly using Wi-Fi. The selection of the communication medium used for transferring data between devices is also very important for any IoT device to work in real time.

Subtype bits	Subtype description
0000	Association request
0001	Association response
0010	Reassociation request
0011	Reassociation response
0100	Probe request
0101	Probe response
1000	Beacon
1001	Announcement traffic indication message (ATIM)
1010	Disassociation
1011	Authentication
1100	Deauthentication

Fig. 1 Management frames

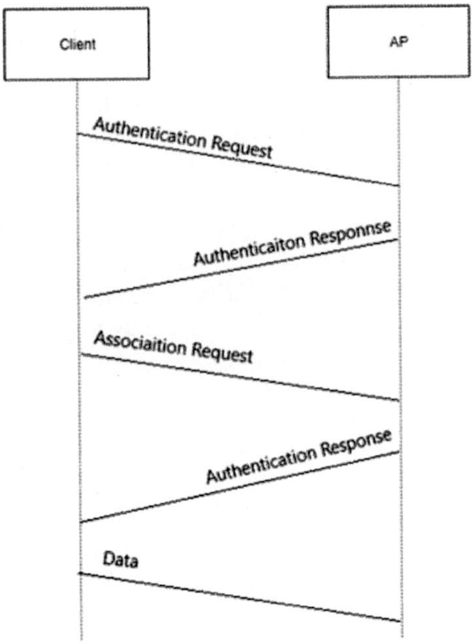

Fig. 2 Sequence diagram for authentication

2 Vulnerability in Security

Wireless local area networks consist of two types of vulnerabilities. Due to bad configuration or due to bad encryption, most of the vulnerabilities arrive in such networks. Due to ease of configuration, many flaws and configuration parameter are untouched; they result in a basic attraction for the attackers to exploit the network [6]. The non-encryption of management frames make it easier for attackers to get the value of parameters such as sequence number, frame id, subtype, and also the Internet protocol address and MAC address. These values are present in human readable format and are easy to exploit.

Multiple MAC layer attacks such as becon, deauthentication, and probe flood can rendered easily as the packet with the required fields can be easily constructed.

3 Deauthentication Attack

The attacks that take advantage of the way the 802.11 protocol handles management frames. The attack is one of the most popular attacks in IEEE 802.11 protocol. In protocol 802.11w, the management frames were encrypted, but the protocol then restricted speed and opened the door for various other attacks. IEEE 802.11w uses WPA pair-wise encryption key that is with no encryption. Most access points manufactured today also does not support this particular protocol and is the protocol is no longer considered [7].

When successful authentication consisting of two acknowledged authentication frames is completed, the client station requests for association (Fig. 3). Association response frames follow frame request association. Each frame is also recognized. The next step depends on the type of security used on Wi-Fi and determines how intrusive the deauthentication attack is. Then, all layer 2 management frames are broadcasted in plain text so that the closest device can find the network and request a connection. Many security problems arise from this lack of protection. If an attacker captures this plaintext management frame, she/he can fake a package that seems to come from the victim [8].

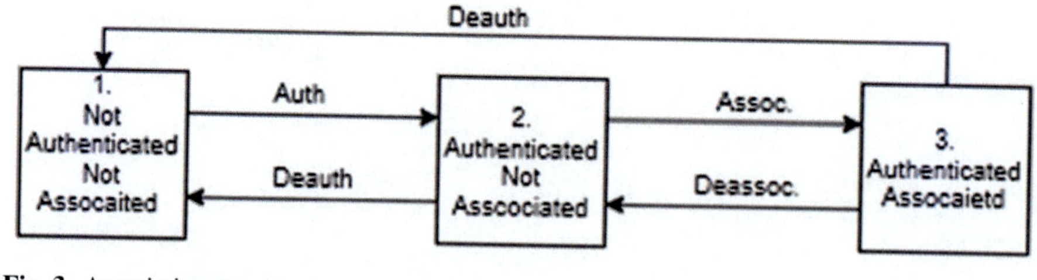

Fig. 3 Association sequence

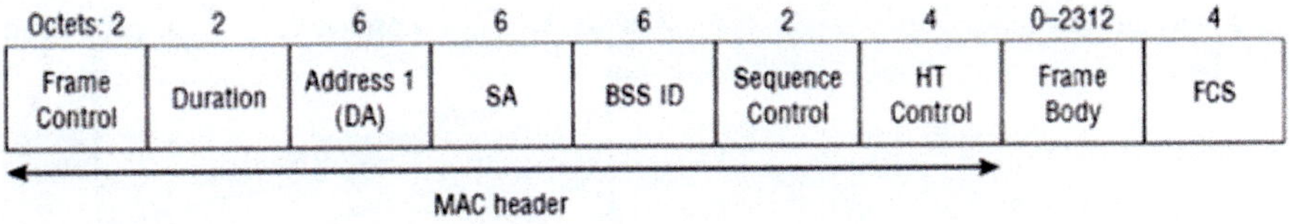

Fig. 4 IEEE 802.11 management frames

In the protocol provided by IEEE 802.11, all packets transferred/transmitted have the same type (homogenous) of header (Fig. 4). The similarities increase the chance of attack and make it easy for the attacker to guess the fields of the packet attacker want to create.

When a user/client or access point wants to disassociate, they send a deauthentication frame, which signifies that they no longer want to use the network at that instance. No authentication is taken care of about the sender of the frame, and the network gets disconnected as soon as they receive frames. The frame is identical, and subtype value use is 0x2c.

4 Research Method

The testing method used in this case when there is no knowledge of code is penetration testing. It is the technique used for assessing, evaluating, and analyzing systems based on security and information system. The attacks are stimulated on the basis of attacker's perspectives. The attackers use this method to gain unauthorized access and take mischievous actions [9]. The subtype of penetration testing used is external penetration testing. The external penetration testing is security assessment of any organization's perimeter systems. It consists of all the devices that are reachable directly through the organization's Internet. The aim of such testing is to find meaningful or ambiguous ways to compromise the existing systems and configurations. Penetration testing consists of planning, discovery, attack, and reporting.

With penetration testing and the IoT hardware, it aims to simulate a deauthentication attack on the home Wi-Fi access point. It deauthenticates all the devices connected to that particular access point. Access point is the target here, and IoT is the tool that will be used to stimulate the attack. The test data are recorded by using packet sniffer and analyzer.

Deauthentication frame construction is done by spoofing the MAC address of access point. After spoofing the MAC address, the frame is made by the microcontroller. The constructed frame is broadcasted to all nearby device to disconnect them from the particular network (Fig. 5).

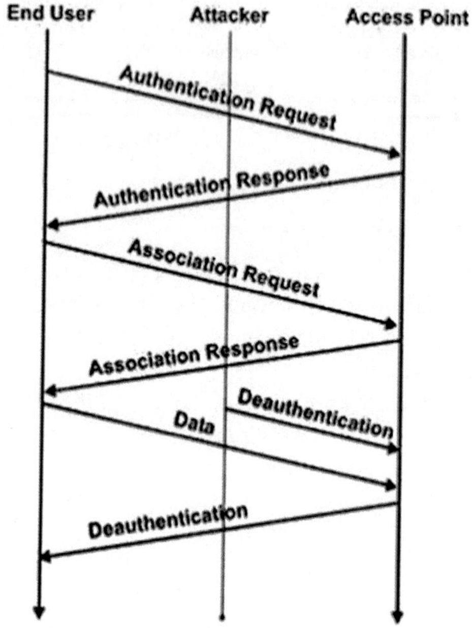

Fig. 5 Deauthentication sequence

4.1 *Test Tools*

1. Attacker Tool: It is an IoT-based simulation device for deauthentication attack using the Wi-Fi module (ESP8266), NodeMCU tools, C++ programming language, and Arduino for program compilation and upload.
2. Gateway: It is a network device that has a transmitter and a receiver. Device should be able to receive and send signals according to IEEE 802.11 protocol.
3. Target: The device used is a smartphone that is connected to IP access point. It is connected to Internet.
4. Packet sniffer and analyzer: It is software to capture transmitted and received package such as Wireshark, CommView.
5. Web Interface: To detect IP and MAC address of the target machine and spoof the mac address.

4.2 *External Penetration Test*

Scope and definition of the test should be properly defined to what the external penetration is capable of. The created scenario is to simulate the deauthentication attack on the device with IEEE 802.11, and the device should not have IEEE 802.11w protocol. The entire test aims to record the activities when the attack process occurs.

Table 1 depicts the actors involved in the test. The test starts with the target device having access to the Internet via access point. The attacker performs the attack and

Table 1 External penetration test scenario

Use case name	Deauthentication attack simulation
Description	Test, Target, Attacker
Purpose	Prove DoS deauthentication attack
Start State	Target device connected to access point
End State	Target device and access point disconnected

sends the deauthentication frame to all the devices in the range to get it disconnected from the access and is unable to use Internet. The initial state is connected, and the final state is disconnected.

To carefully understand the setup and transmission of packet, UML diagram is needed. Figure 6 depicts the network diagram of deauthentication attack simulation. The diagram represents all the components involved in the test scenario.

When the test target tries to use the Internet facility provided by access point, it cannot access because it has already been disconnected from the access point because of the deauthentication attack. The attack is rendered through ESP8266 using a NodeMCU with C++ and Arduino programming. The programming script will perform the deauthetication attack on the target. The device has an OLED screen and switches to select and spoof address and to attack. The device can also be connected to device supporting Web and Wi-Fi interface. It then scans for the SSIDs in the range of ESP8266. The access point is then selected from the search list and saved to copy its MAC address. In the background, the program creates the deauthentication frame using the saved SSIDs MAC address. When attack command is invoked, the created frames are then broadcasted to the nearby devices with the subtype as 0x2c.

Some basic requirement to render the attack includes basic service set identifier (BSSID) and the MAC address of the device from which target needs to be disconnected from. For analysis, estimated time of attack is also recorded. Throughput on the device is also recorded to analyze and make note that attack is successful.

Wireshark is used to collect all the network packets transfered between Wi-Fi router and the IoT device and CommView is used for capturing every packet on the air to display access point list, RSSI, etc.

5 Result and Discussion

The impacts of a deauthentication attack that occur on the test target are as follows:

1. The communication of the target is lost with the gateway or access point.
2. The MAC of the device is still registered with the device and is saved for future connection.
3. The target still tries to connect to the gateway/access point but couldn't connect because of the attack.

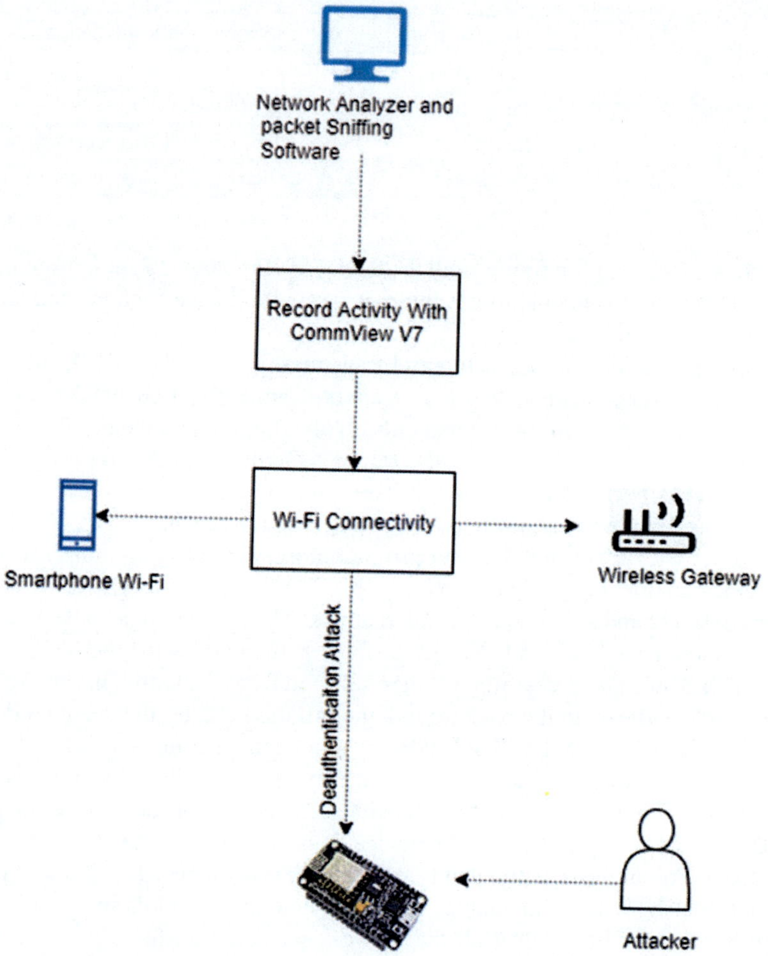

Fig. 6 Network diagram of deauthentication attack simulation

The proof of vulnerability of IEEE 802.11 to deauthentication attack is established by using penetration testing. The data revealed by the test is in form of data which can be used to draw out conclusion.

5.1 Attack and Results

Before the attack, the attacker needs to get the details of the Wi-Fi to disable target. By fingerprinting, we can perform an active and passive scans of Wi-Fi signals in

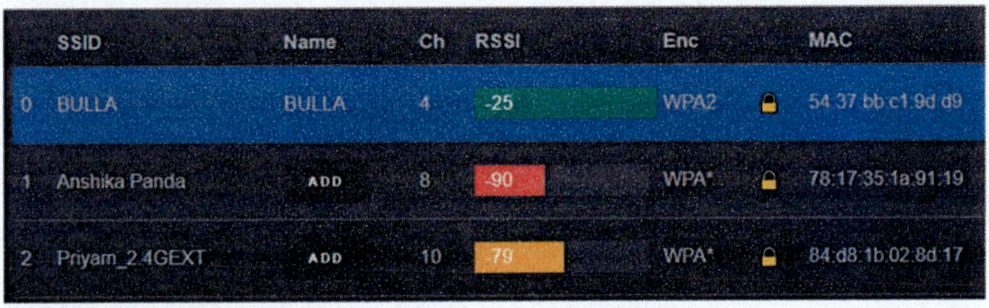

Fig. 7 Available SSIDs

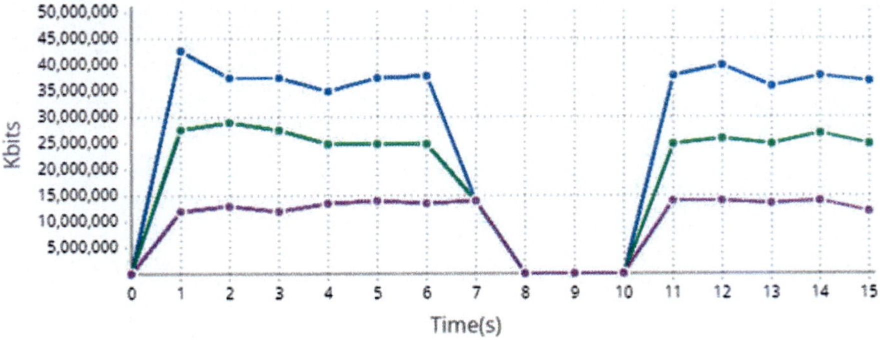

Fig. 8 Change in throughput

the range without authenticating with the network. After fingerprinting, the attack is performed by selecting the particular BSSID (Fig. 7).

Figure 8 demonstrates the change in throughput of the machine during the attack. The throughput completely drops down to 0 during the attack (Table 2).

5.2 *Mitigation*

To protect from the deauthentication attack, IEEE 802.11 should use protected management frames so that it will discard deauthentication frames sent from different access point [4].

Unique identifier can also be provided during association and that particular identifier should be checked when the deauthentication frame is sent.

Sequence number of the packet can be checked to determine the correct request for deauthentication.

Table 2 State conditions

State condition	Start	End
Data rate	3 Mbps	3 Mbps
Channel	5–2432 MHz	7–2443 MHz
To-OS bit	1	0
From-OS bit	0	0
Retry bits	0	1
Type	2- Data	0- 1- 1- Management
Sub-type	12- QoS NULL (No Data)	12-Deauthentication
Duration	414 ms	262 ms
BSSID	54:37:BB:C1:9D:D9	54:37:BB:C1:9D:D9
Source address	70:BB:E9:27:5B:E4	54:37:BB:C1:9D:D9
Destination address	88:66:39:A6:C4:5C	70:BB:E9:27:5B:E4
Sequence number	1672	0

6 Conclusion

The wireless networks using the protocol IEEE 802.11 are vulnerable to deauthentication and MAC layer attacks. The management frames are not encrypted, and sometimes, there true source is also not identified. The management frame with subtype 0x2c can be transmitted over the network using the same MAC address of the access point used by the target. The attack can disconnect all the devices in the range of the integrated IoT hardware. The attack caused communication paralysis and also allows devices to attempt reconnection and fail. The throughput also changes drastically at the time of attack. The sequence number of retry is enumerated from the start.

Future work should focus on the mitigation of management frames without comprising the speed and current security of the system. The research on the best configuration for the access point will also be very useful.

References

1. Matthew SG (2002) 802.11 wireless networks: the definitive guide. O'Reilly & Associates, 464 pages, ISBN: 0596001835
2. Wright W (2003) Detecting wireless LAN MAC Address spoofing. GCIH, CCNA, pp 1–5
3. Kristiyanto Y, Ernastuti (2020) Analysis of deauthentication attack on IEEE 802.11 connectivity based on IoT technology using external penetration test. CommIT (Communication & Information Technology) J 14(1):45–51
4. IEEE.org
5. Andrade RM, Carvalho RM, de Araujo LL, Oliveira KM, Maia ME (2017) What changes from ubiquitous computing to Internet of Things in interaction evaluation? In: International conference

on distributed, ambient, and pervasive interactions. Springer, Vancouver, BC, 9–14 July 2017, pp 3–21

6. Martinez A, Zurutuza U, Uribeetxeberria R, Fernandez M, lizarraga J, Serna A, Velez I (2008) Beacon frame spoofing attack detection in IEEE 802.11 networks. In: the proceedings of the third international conference on Availability, Reliability and Security (ARES08), Barcelona, pp 520–522
7. Gopinath KN (2009) A brief tutorial on IEEE 802.11w. Retrieved from airtight networks. http://blog.airtightnetworks.com/wp-content/uploads/2009/11/802.11w-Tutorial1.pdf
8. Milliken J, Selis V, Yap KM, Marshall A(2013) Impact of metric selection on wireless deauthentication DoS attack performance. IEEE Wireless Commun Lett 2(5):571–574
9. Course Technology Cengage learning (2011) Penetration testing procedures & methodologies. Nelson Education Ltd, USA

Development of Contextual Crop Ontology for Effective Information Retrieval

K. Ezhilarasi and G. Maria Kalavathy

Abstract Agriculture in India has a diversity of natural elements such as wider cultivable land, different agro-climatic zones, different soil types, and suitable seasons for cultivating a variety of crops. Indian agriculture is largely reliant on natural resources and favorable weather conditions. Therefore, under a vast agricultural diversity, strong information support and knowledge system are required for the farmer community to achieve a high yield. Accurate and timely information and instruction are vital to the agricultural processes that should be delivered to farmers properly. The available agricultural websites, mobile applications, and software provide only general practices to the farmers and do not address the solution for specific problems. So there is a need for a system that should give contextual information based on geographical area, climatic condition, soil nature, previous experience, and current state of the crop. The information provided by the system can be applied for their cultivation and marketing. The system is designed by using ontology for the agriculture domain and thereby providing information retrieval for this RDF repository. In this paper, farmer-based crop ontology is developed that will be useful for guiding the farmers by providing instructions and information related to crop cultivation and management of fertilizers based on their situation of the crop developing stage.

Keywords Ontology · Knowledge model · Agriculture domain · OWL

K. Ezhilarasi (✉)
KCG College of Technology, Chennai, India
e-mail: k.ezhilarasi@yahoo.com

G. Maria Kalavathy
CSE, St. Joseph's College of Engineering, Chennai, India

V. K. Gunjan and J. M. Zurada (eds.), *Proceedings of 3rd International Conference on Recent Trends in Machine Learning, IoT, Smart Cities and Applications*, Lecture Notes in Networks and Systems 540, https://doi.org/10.1007/978-981-19-6088-8_6

1 Introduction

The World Wide Web, scripts, and business reports generate the immeasurable source of unstructured information. It is necessary to structure the data in a computer-readable format for exploiting the knowledge from the information. A standard and systematic procedure is needed to use, categorize the information and to fetch hidden knowledge from this virtual repository. These activities can be carried out by combining technology like ontology, reasoning, and semantic web.

Agriculture in India is a multifaceted enterprise involves millions of different category farmers. Any small improvement in the agriculture sector has a strong impact on the entire economy. Although a farmer is a significant facet of a country, he is unable to exploit the financial benefits from agriculture to its full potential. To get the full financial benefits, farmers should learn knowledge about best practices for crop production and weather conditions. This information is present in multiple sources, but in different formats.

Farmers should be educated with seasonal climate, best agriculturist practices, and seed selection, information about pest and diseases and their control methods, harvesting and post-harvesting methods to make some prior decisions at different stages of the farming cycle. This type of information is available in government websites, brochures, and in video and audio files. This knowledge did not reach the farmer effectively due to its unstructured nature, use of different terms to explain the same concept, the same information in different formats, generic nature of information, and lack of suitable delivery methods. Numerous applications are available in India to help the farmers. Agricultural-related stakeholders are there to advertise information about their products and their activities. We have found out that farmers need relevant and specific information based on the location of land, budget, interests, and machinery available. This kind of information would be more relevant and appropriate to farmers' needs. It also impact on their decision-making process. So farmers need to access reliable, timely, and relevant information that becomes very important to get monetary benefits without harming the cultivating land.

1.1 Need of an Ontology for the Agriculture Domain

The paper [1] discussed the need of context-information for the farmers in India. They explained about the effectiveness of this knowledge since it improves the crop productivity and farmer's income. While developing new approaches for farmers, they recommended the subsistence of context-specific and relevant information. Ontology provides a systematic view of domain-specific knowledge. It acts as a repository of concepts. This structured view is needed to facilitate knowledge distribution, knowledge assemblage, information retrieval, and answering the question [2]. Ontology can be used to develop a better data model than other approaches, and as it is precise, it has well-defined semantic relationships, deduction mechanism, and interpretation support [3]. Therefore, ontology can be used to provide precise and relevant information to queries of a particular domain.

1.2 Related Work

In agriculture domain, there are several ontology repositories such as Thai Rice Ontology, ontology based on AGROVOC thesaurus, contextual ontology, and farmer centric ontology.

Ontology on Thai Rice explained the whole process from cultivation to harvesting of rice production at Thailand [4]. This ontology has been developed to facilitate the knowledge acquisition process and retrieval of information for research purposes. Currently, available research knowledge database for paddy production is unorganized; research policy administrators and researchers face many challenges in searching relevant previous studies for research and development. Thesaurus is the highest level ontology and it helps for building domain ontology. The AGROVOC Thesaurus of the Food and Agriculture Organization (FAO) [5] is well established and has the authoritative controlled vocabularies in the agriculture domain. However, there are several limitations with current vocabularies such as semantic ambiguity in definitions and usage of vocabularies; lack of high-level cross-domain concepts, and meaning of their relationships is not precisely defined [6]. In this approach [7], AGROVOC thesaurus was represented as linked data using OWL and SKOS.

Bansal and Malik described an agricultural ontology for the crop production cycle to provide relevant information for farmers based on AGROVOC vocabulary in semantic web [3]. This ontology helps the individual farmers to get relevant agriculture information. However, it didn't cover the context-based information retrieval.

The framework for information flow in agriculture domain by using stakeholders is explained in the paper [8]. In this paper [9], the author designed ontology to provide the needed information to the farmers at various stages of the farming life cycle. First Order Logic methodology is used to develop an agricultural ontology. The ontology covers for static information of paddy production and not simulated the dynamic information like what should be done when sudden rain and wind occurs?

This paper [10] explained how the ontology is developed for a vertical farm environment.

Currently, existing ontologies in the agriculture domain are too general and not specific. Even though context-based crop ontology was developed, dynamic aspects of crop production, interacting with agriculture experts, correlating production techniques and crop yields with land variability, specifying effective agricultural strategies and temporal crop development information are not considered. This research intends to develop a farmer's centric crop ontology by including dynamic details and suggestions of experts. This ontology gives the exact information and expert's instruction to farmers that will be useful for his decision making in crop production.

2 Analysis of Ontology Building Methodologies

There are abundant methodologies are there to build ontologies from scratch or by reusing other ontologies. The list of some important approaches are Cyc method, the Uschold and King's method, the Grüninger and Fox's methodology, the KACTUS approach, METHONTOLOGY, the SENSUS method, and the On-To-Knowledge methodology [11]. After going through the Ontology construction steps given in different approaches [12] and comparison of these approaches explained in [12, 13], it is evident that METHONTOLOGY approach provides the detailed steps for constructing domain ontology.

Reasons for choosing METHONTOLOGY approach [13]:

Level of detail: This approach explicitly states the methods and techniques which are used to perform different activities.

Associated Software Tool: Protégé Tool facilitates to perform all the ontology construction steps.

Conceptualization Phase: During this phase, this approach organizes and converts an informally perceived view of a domain into semiformal specification using set of intermediate representations based on tabular and graph notations. It is comfortable for ontology developers and researchers.

3 Ontology Construction—Methodology

Steps followed for building ontology by adapting METHONTOLOGY approach is represented in Fig. 1.

3.1 Knowledge Acquisition

All information related to crop production, protection, and harvesting process are gathered from different trustworthy knowledge sources like:

- Agriculture domain experts from Tamil Nadu agriculture universities, agriculture offices, and farmers in Tamil Nadu (by interviews, group discussions, and questionnaires);
- Websites [14, 15]
- Research articles

The collected knowledge resources were reviewed and categorized systematically.

All knowledge sources of crop ontology was studied, scrutinized, and represented in structural form. Final structural form verification was carried out by domain experts.

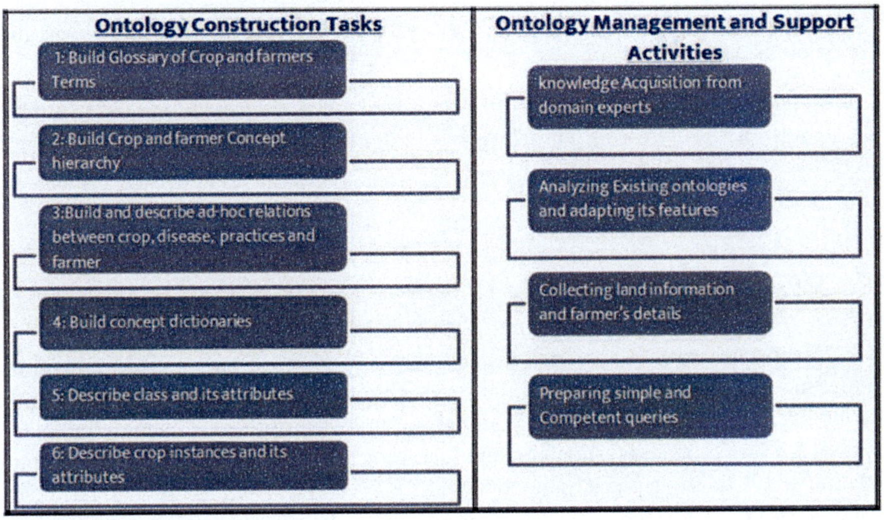

Fig. 1 CropOnt development steps

3.2 Ontology Design

The designed ontology should provide answers for "basic questions" [11] and "competency questions" [11]. These questions are useful to validate whether developed ontology achieves its purpose.

The purpose of basic questions is to explain the scope and purpose of the ontology. Some basic queries are:

- For which domain this ontology is developed?
- What is the purpose and need of ontology?
- Who are the users are going to use and maintain the ontology?
- What kind of information does it serve?

Competency questions provide answers by processing the knowledge base associated with "CropOnt" ontology. These competency questions need not be extensive. The answers to these questions may vary from farmer to farmer. These answers are used for evaluation to show how efficiently ontology-based semantic search provides better and needed answers to the farmer than keyword-based search.

Some sample competency queries are:

- During this period what type of crop variety should be cultivated?
- What is the market scope of that variety and yield prediction?
- Asking for any type disease how it should be treated?
- Which type of crop provides more profit based on farmer's soil and climatic condition?

- Which crop produces yield more the average yield during particular year and particular area?

The answers for these questions may be varied since it depends on land location, type of soil, instruments available, farmer's budget, climatic condition, and farmer's interest.

3.3 Ontology Development Process

Task 1. Build glossary of crop and farmers terms: The set of terms used in the crop production cycle is identified, and then a dictionary of terms with its description, synonyms, and acronyms is created. Sample of this description is given in Table 1.

Task 2. Build crop and farmer concept hierarchy: To find the hierarchical, disjoint, and equivalent relationship among concepts:

Subclass-Of: The concept C1 is derived from concept C2 only if every object of C1 is an object of C2. For example, the concept "Crop" is *Subclass-Of* concept

Table 1 Glossary of terms

Name	Synonyms	Acronyms	Explanation	Type
Plant			One type of living thing that grow in land	Concept
Practices			Procedure followed by the farmer to cultivate the crop	Functional concept
Crop			Crop is a type of plant that is cultivated in land by the farmer	Concept
Weed			Type of plant that grow along with crop	Concept
Land	Ground and farm land		Geographical location where agriculture is done	Concept
Farmer	Cultivator		Who cultivates the crop in their land	Concept
Disease			Plants are infected by viral, bacteria, and fungi disease that affects the growth and yield	Concept
NPK	Fertilizer components	Nitrogen and Potassium	Ingredients need for growth of crop	Instance

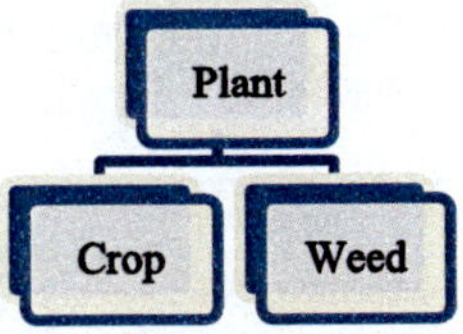

Fig. 2 Example for subclass-of relationship

"Plant" only if every instance of "Crop" must be an instance of "Plant" (shown in Fig. 2).

Equivalent-to: The concept C1 is equivalent-to concept C2 only when both concepts explain the same terms and reside in the same/different ontologies. For example, weed_name is a concept, which is in weed concept is equivalent to weed_name in Weed_Control_Method concept.

Disjoint classes: A concept C has a set of decomposed subclasses that have no common instance and need not to cover base concept C. Instances of the concept C can exist, even that are not instances of any of the decomposed sub concepts. For example, practices concept can be decomposed into Production_Practies, Protection_Pratcies, and Harvesting_Practices. All these subclasses are disjoint classes. One or more relationship hierarchy of concepts can be constructed based on class relationship.

Task 3. To build ad hoc relation: Functionality relationship other than hierarchical relationship among two concepts is explored and represented as a binary relationship between concepts of same or different ontologies. The binary relationship is shown in Fig. 3.

Every identified ad hoc binary relation should be described in detail. The outcome of this is mentioned in Table 2.

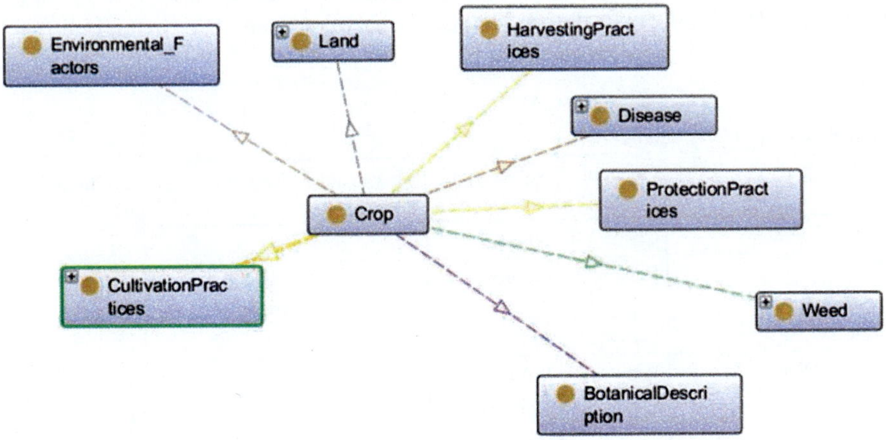

Fig. 3 Binary relationship for crop

Table 2 Object property details

Object property/relation	Domain	Range	Mathematical properties	Inverse relation
HasDisease	Crop	Disease	1:N	
hasPractices	Crop	CultivationPractices ProtectionPracties HarvestingPractices		isPracticeof
hasBotanicalDescription	Crop	BotanicalDescription	1:1	isBotanicalDescriptionOf

Table 3 Concept dictionary

Concept name	Instance	Object property (relations)
Crop	Paddy	hasCultivationPractices
		hasProtectionPractices
Weed	Weedname	hasWeed
		ControlledBy
Disease	Disease_Name	AffectedBy
	Symptoms	

Task 4: To build the concept dictionary: This is done by creating the instance for needed concept. Concept dictionary consists of concept, instances, attributes, and their ad hoc relations. Sample of this dictionary is given in Table 3.

Task 5: To describe class and instance attribute: All attributes mentioned in the concept dictionary must be described in detail. The sample outcome of this task is specified in Tables 4 and 5.

Constants are used to represent static information related to the domain. It takes the fixed value and used in formulas. The sample list of constants used in this ontology is given in Table 6.

Table 4 Data property details

Data property	Concept name	Value type	Measuring unit	Range	Cardinality
Actual_Yield	Crop	Number			1:1
ID	Person	Number		1000–10,000 (assumed)	1:1
Address	Person	String			1:N
hasWeedName	Weed	String			1:1
Area	Land	Number	Hectares		1:1
Deficiency_symptoms	Crop	Data range expression			1:N

Table 5 Class attributes

Class attributes	Defined for concept	Value type	Measuring unit	Values	Cardinality
BotanicalName	Paddy	String		Oryza sativa	1:1
Expected_Yield	Variety	Number	Tones		1:1
….,					

Table 6 Constants definition

Name	Value type	Value	Measuring unit
Maximum_Water_Needed	Integer	2000	Liters/kg of grain
Short_Duration	Integer	120	Days
Long_Duration	Integer	140	Days
Medium_Duration	Integer	170	Days
Max_Height	Float	1.5	M

Table 7 Instances definition

Instance name	Concept name	Attribute	Value
Sedge	Weed	Weed_Name	String
Broad-Leaf		Reproductive_organ	String
Grass		Growth_Environment	MultiValue String
Clap		Spread_Environment	Constraint String

Task 6: To define instances/individuals. For each instance, the ontologist should define: its name, the name of the concept it belongs to, and its attribute values, if known (given in Table 7).

3.4 Ontology Evaluation

The ontology evaluation plays evident role in the semantic web and ontology-enabled applications. There is no single best or preferred approach to do ontology evaluation. The choice of a suitable approach for ontology evaluation is depend on many factors like evaluation purpose, the application in which the ontology is to be used, and on what aspect of the ontology we are trying to evaluate. The performance of the developed "CropOnt—crop ontology" is evaluated by following methods:

- By using tools:

  The Ontology Metrics of "CropOnt" ontology is given in Table 8.

Table 8 Metrics

Axiom	570
Logical axiom count	334
Declaration axioms count	236
Class count	54
Object property count	50
Data property count	70
Individual count	131
DL expressivity	ALCHIQ(D)

ALCHIQ(D) : ALC (Attributive Concept Language with Complements) and add property hierarchy H, inverse properties I, and qualified cardinality restrictions Q, and data types (D).

- Validation by domain experts: The domain-specific experts validated the correctness of the developed ontology. This ontology is checked by going through the developed concepts and relationships related to rice production.
- Completeness verification by SPARQL (**SPARQL** protocol and RDF query language): The ontology is verified by executing the competent queries in Protégé SPARQL tab with sample triple dataset. These queries were framed during the development phase and mentioned in Sect. 2(Ontology Development Methodology). The completeness of this ontology is checked by executing all complex SPARQL queries in developed ontology.

Sample Query: To fetch details of crop and farmer who generate yield greater than average yield during particular year.

```
PREFIX owl: <http://www.w3.org/2002/07/owl#>
PREFIX rdfs: <http://www.w3.org/2000/01/rdf-schema#>
PREFIX xsd: <http://www.w3.org/2001/XMLSchema#>
PREFIX as: <http://www.semanticweb.org/ezhil/ontologies/2018/4/OntoCrop#>
SELECT *
        WHERE {
        ?ind rdf:type as:Farmer.
        ?land as:hasOwner ?ind.
        ?land as:hasCrop ?c.
        ?c as:hasCultivationProcess ?x.
        ?land as:hasCultivationProcess ?y.
        ?c as:hasAverageYield ?a.
        ?y as:yield ?b.
        ?y as:harvestingDate ?d.
```

Table 9 Experiment results

	Simple queries	Complex queries
Returned items	34	28
Relevant items retrieved	28	21
Relevant items	32	27
Precision	0.823529412	0.75
Recall	0.875	0.777777778

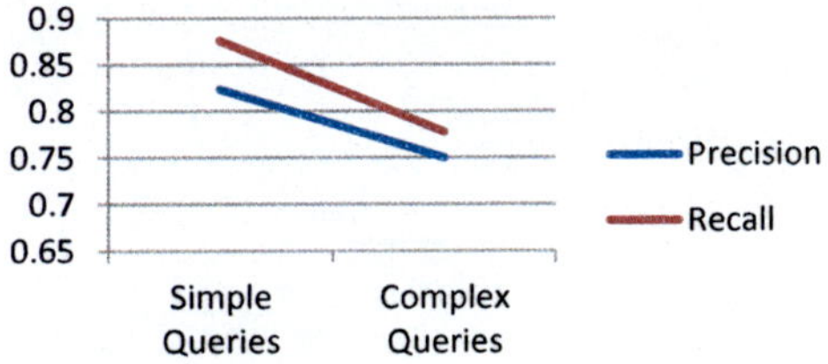

Fig. 4 Precision and recall for simple and complex queries

```
        filter ( (?x=?y) && (?a < ?b) && (?d>="2020-01-01"^^xsd:date
    && ?d<="2020-12-31"^^xsd:date)).
        }
Output: List the details of crop from the triples.
```

- Experimental Evaluation: By considering experiment bed has 100 Triples in crop ontology and executed 5 simple and complex queries each. The calculation of precision and recall is mentioned in Table 9 and the relationship is represented in graph Fig. 4.

CropOnt—crop ontology is developed using protege5.4.0 tool [16, 17] and its class's relationship is shown Fig. 5.

4 Conclusion

In the paper, we presented the development of crop ontology based on METHONTOLOGY methodology using SPARQL and description logic. The developed ontology is used for effective information retrieval. This ontology symbolizes context-based crop production knowledge. This structured knowledge can improve the decision-making process for farmers. This will increase crop yield, more profit to the farmer, and better use of agricultural methods and technologies. This development approach had the flexibility to add more concepts, relationships, and constraints

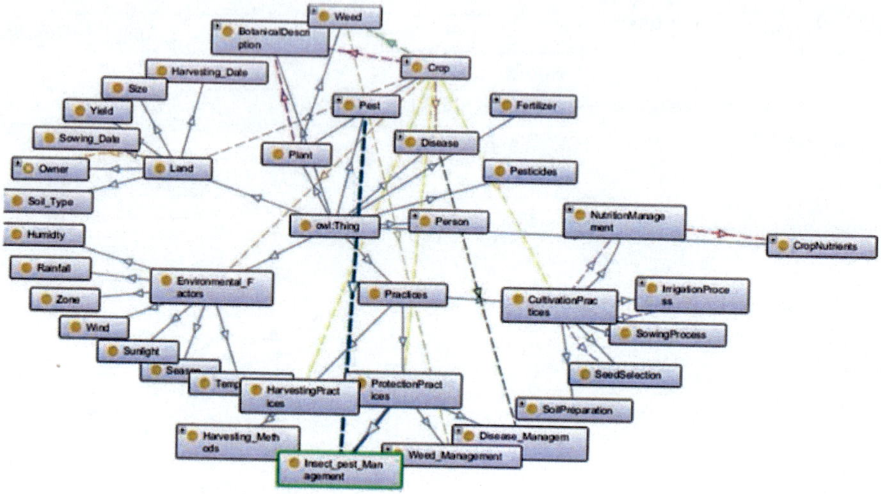

Fig. 5 CropOnt–crop ontology

at any time. The information retrieval effectiveness can be improved by applying machine learning algorithm on the crop ontology.

References

1. Glendenning CJ, Babu S, Asenso-Okyere K (2010) Review of agriculture extension in India are farmers' information needs being met? International Food Policy Research Institute
2. Gruber TR (1995) Toward principles for the design of ontologies used for knowledge sharing. Int J Hum Comput Stud 43:907–928
3. Bansal N, Malik SK (2011) A framework for agriculture ontology development in semantic web. In: International conference on communication systems and network technologies. IEEE Computer Society, pp 283–286
4. Thunkijjanukij A, Kawtrakul A, Panichsakpatana S, Veesommai U (2009) Rice production knowledge management: criteria for ontology development. Thai J Agric Sci 42(2):115–124
5. Food and Agricultural Organization of United Nations (online). http://www.fao.org/aims
6. Fisseha F (2002) Towards better Semantic Standards for Information Management AGROVOC and the Agricultural Ontology Service (AOS). UN FAO, Rome, Italy
7. Caracciolo C et al (2013) The AGROVOC linked dataset. Semantic Web 4(3):341–348
8. De Silva L (2014) Towards an agriculture information ecosystem, ACIS
9. Walisadeera, Indika A, Wikramanayake GN, Ginige A (2013) An ontological approach to meet information needs of farmers in Sri Lanka. In: International conference on computational science and its applications. Springer, Heidelberg
10. Kim T et al (2013) A study of an agricultural ontology model for an intelligent service in a vertical farm. Int J Smart Homes 7(4)
11. Gómez-Pérez A, Fernández-López M, Corcho O (2006) Ontological engineering: with examples from the areas of knowledge management, e-commerce and the semantic web. Springer Science & Business Media
12. Corcho O et al (2003) Methodologies, tools and languages for building ontologies. Where is their meeting point? Data Knowl Eng 46(1):41–64

13. Rosario M et al (2015) Methodologies to build ontologies for terminological purposes. Proc Soc Behav Sci 173:264–269
14. TNAU Agritech Portal: Home website (Online). Available http://www.agritech.tnau.ac.in/
15. Agropedia Website (online). Available http://agropedia.iitk.ac.in/
16. Protégé website (online). Available https://protege.stanford.edu/support.php
17. Patel-Schneider PF, Hayes P, Horrocks I. OWL web ontology language semantics and abstract syntax. W3C Recommendation, Webpage (online). Available http://www.w3.org/TR/owl-semantics

Detection of Hello Flooding Attacks on RPL in Internet of Things Networks Using Different Machine Learning Algorithms

Priyank Koul, Shevgoor Adithya Kamath, S. Akshatha, Nikita Ganvkar, and Animesh Giri

Abstract In Internet of Things (IoT), routing protocols by design are quite vulnerable to attacks. These attacks can be designed to reduce bandwidth, corrupt information, and/or threaten the integrity of the network. Hence, it is crucial to identify the attacks, mitigate them, and prevent further damage to the network. In this paper, we have proposed a methodology to identify Hello Flooding (HF) attacks using various machine learning classification techniques. The paper has a primary focus on dataset creation, followed by the implementation of various machine learning algorithms, using which we not only identify if a network is under attack, but also identify malicious nodes and the nodes affected by the malicious nodes.

Keywords IoT · Hello Flooding · Machine learning techniques · RPL

1 Introduction

IoT is quickly becoming an essential part of our lives. The idea that IoT would make one's life easier and smarter has been sold to the current society. The cost effectiveness of IoT devices available in the consumer market has enticed the curiosity of the consumers.

IoT is also being adopted by experienced companies to increase profits, by increasing data generated for analysis. Companies have always been competitive, which has led them in pursuit of better technology. Graduating from simple manual production to mass production technologies, to the adoption of computers to improve accounting (which led to faster and more accurate statistical analysis), proved to be beneficial for companies. The adoption of IoT is seen as another milestone toward their pursuit.

P. Koul (✉) · S. A. Kamath · S. Akshatha · N. Ganvkar · A. Giri
Department of Computer Science, PES University, Electronic City, Bengaluru, Karnataka 560100, India
e-mail: priyankkoul14@gmail.com

A. Giri
e-mail: animeshgiri@pes.edu

V. K. Gunjan and J. M. Zurada (eds.), *Proceedings of 3rd International Conference on Recent Trends in Machine Learning, IoT, Smart Cities and Applications*, Lecture Notes in Networks and Systems 540, https://doi.org/10.1007/978-981-19-6088-8_7

The IoT is a network of devices that connect and exchange data within the network and the Internet. Nodes in an IoT perform tasks with limited supply of energy. They are expected to predominantly communicate wirelessly. They need not have a preset path, that is, preset configuration in space.

Routing Protocol for Low-Power and Lossy Networks (RPL) is a routing protocol implemented in IoT. It is essential for designing a near optimal flow of data within the network.

Limited energy supply in IoT influences implementations of routing protocols. Energy consumption is controlled by limiting the up time of the node by keeping it in a low power mode the rest of the time. HF is one of the attacks on the IoT routing protocol which targets the limited energy resource of the IoT node. It floods DODAG Information Solicitation (DIS) packets to its neighbors, thereby increasing energy consumption of its immediate neighbors. This paper focuses on preventing exploitations from occurring, by catching the malicious nodes at an early stage.

2 Related Work

2.1 An Introduction to RPL

RPL is an IPv6 routing protocol (rules for flow of traffic) standardized by the Internet-Engineering Task Force (IETF) for IoT. In a 6LoWPAN (IPv6 over Low-Power Wireless Personal Area Network), RPL constructs a Destination Oriented Directed Acyclic Graph (DODAG) of the nodes in the Wireless Sensor Network (WSN), enabling bidirectional traffic between 6LoWPAN devices and the DODAG root. RPL has a tree-like architecture which is based on various optimization processes known as the objective functions.

RPL creates a DODAG with the help of control packets and the rankings of the nodes. The ranking is decided by the number of hops a node is from the root node, with the root node being first. There are four control packets used for the creation of a topology. The first packet is the DIS packet. It is sent by a new incoming node within a certain space to enquire if there is a network it can join to. The DODAG Information Object (DIO) packet is sent in response to the DIS packet; it is also sent at regular intervals of time to refresh the information in the topology. The DIO packet contains information about the network. The node then sends a Destination Advertisement Object (DAO) packet requesting the root for permission to join the network. The node sends a DAO-ACK (DAO Acknowledgment (DAO-ACK) in response to the DAO packet received.

The various stages of IoT that have marked the development of IoT in recent times was presented by Atzori et al. [1]. Various perspectives on the IoT paradigm as well as enabling technologies had been presented. They examined the role IoT can play in tackling the most pressing social issues, as well as the characteristics that should be expected from relevant solutions.

Contiki is a lightweight operating system for IoT devices with limited memory, bandwidth, power, and processing capability. Contiki supports low-power wireless standards such as RPL. Contiki's support for dynamic loading and unloading with limited resources and by keeping the underlying system compact was presented by Dunkels et al. [2].

2.2 Routing Attacks

Low power and Lossy Network (LLNs) are a class of networks with restricted power source, constrained network lifetime, and network dependability. These constraints give rise to vulnerabilities that can be leveraged for undesirable purposes.

Routing is a crucial function in all sorts of networks, guaranteeing secure routing is a must to ensure the operation's success. LLN lacks an infrastructure-based routing operation, encouraging multihop routing. Thus, making it a challenge to secure it.

Routing attacks take advantage of protocol flaws, router software incompatibilities, and weak authentication. Such attacks have an impact on network services and business operations while they occur.

Routing attacks focus on exploiting vulnerabilities in IoT networks and also the routing protocol logic itself, through the routing layer. Mayzaud et al. [3] established a taxonomy to classify attacks against the RPL protocol into three categories, namely resource, topology, and traffic. The impact on delivery ratio, overhead, inconsistencies in ranks, end-to-end delay, and loops was analyzed. It was observed that attacks on resources resulted in bogus control messages or the creation of loops, which limited the network lifetime.

HF is a resource attack. Blackhole (BH) attack is a topology attack where all the received packets are dropped by the malicious node, which results in isolation of its children from the network. Decreased-Rank (DR) attack is a traffic attack where the malicious node changes its rank to divert traffic.

Raoof et al. [4] investigated the effects and limitations of using RPL security methods in typical routing attacks. RPL performance (with and without security mechanism) under various types of attacks was compared using metrics such as average data packet delivery rate, average data packet delay, and average power consumption. They found that network performance in the pre-installed secure mode (without the replay protection mechanism) was equivalent to that in the insecure mode, but with the added benefits of attack mitigation. Enabling the replay mechanism in the pre-installed secure mode showed significant reduction in the effect of neighbor attacks on the Packet Delivery Ratio (PDR) and End-to-End (E2E) latency.

2.3 *Approaches Used to Detect RPL Attacks*

With the adoption of smart devices, there has been a significant rise in attacks targeting the IoT. As discussed previously, the RPL protocol is used by the nodes in a WSN for routing communications. This protocol is inherently susceptible and difficult to protect against various attacks due to the limited processing capabilities of the nodes.

Hence, many approaches have been proposed to not only prevent such attacks, but also identify and mitigate them. The primary focus is on the detection mechanisms of various RPL attacks.

Various intrusion detection systems have been implemented for RPL attacks. Rule-based techniques or protocols have been used for such systems. Raza et al. [5] used rule-based techniques to create a real-time novel intrusion detection named "SVELTE" on RPL attacks, such as selective forwarding and sinkhole, with the capacity to be extended to other attacks (as future work). The approach took advantage of RPL and the characteristics of the attacks.

On similar grounds, Wallgren et al. [6] utilized the heartbeat protocol to narrow down which nodes were being bogged down by heavy traffic, which aided in detecting the patterns of selective forwarding attacks in the network topology.

However, for complex systems, with various possibilities of unidentified attacks, a large number of rules and protocols are required. This makes it difficult to manage all the rules present in the system and is less of a feasible solution. A more general approach which can incorporate as many types of RPL attacks would be a more feasible solution.

With the advent of machine learning algorithms and its vast applications, several detection mechanisms make use of methods [7] such as traditional machine learning techniques: Support Vector Machine (SVM) [8, 9], decision trees, k-nearest neighbors (KNN) [10], and Kernel Density Estimation (KDE) [11]. For instance, Garafalo et al. [12] enhanced an existing intrusion detection system using threshold metrics, by creating a hierarchical architecture with a decision tree model embedded into the architecture itself. Various neural networks (Hidden Markov Models [13], Recurrent Neural Networks, and deep learning [14]) were also implemented to enhance existing approaches or to create novel methodologies.

Using the available parameters of the network topology, neural networks have been constructed to detect possible RPL attacks. Cakir et al. [15] focused on HF to build a Gated Recurrent Unit (GRU) deep learning model. Using parameters such as Radio Reception (RX), Radio Transmission (TX) and the Total Energy Consumption (TE), the neural network was implemented [16].

Yavuz et al. [17] used a deep learning model to deal with the IoT security by detecting the routing attacks successfully. A deep neural network was constructed and trained using their obtained datasets, which proved to be effective in detecting the attacks in the network topology.

3 Methodology

Previous approaches for identification of routing attacks emphasized on differentiating if a packet either belonged to a normal network, or a network which had been under an attack. They neither provide any information about the malicious nodes nor about the nodes which were affected by the malicious nodes.

The flowchart of the proposed methodology has been depicted in Fig. 1. For our dataset creation, Cooja simulator is used. To reproduce HF attack, changes are made to the "udp-client.c" and "rpl-timers.c" files to create the flooding nodes in the simulation. The simulation scenarios have a certain number of malicious and normal nodes, and all the simulations are allowed to run for a designated amount of time to capture sufficient packets for the creation of datasets. All the packets in the simulation are stored with the help of Cooja as a.pcap (Packet Capture) file. The parameters used for our simulations are compiled in Table 1.

The .pcap file is processed into a comma separated file, which contains rows with features such as source, destination, time, and type of packet for each individual packet. The data-points are labeled into three classes, and the datasets obtained from these simulation scenarios are merged and eventually fed into the machine learning models. Merging is done for the model to achieve robustness. With the available dataset, feature preprocessing is performed. The source, destination, and type of packet are encoded by mapping each value to a discrete integer. Next, the dataset is windowed into 1000 ms intervals. During this process, for each packet in the window, the features which are appended to each row are Transmission Rate (Transmitted Packet Count divided by window of time), Reception Rate (Received Packet Count divided by window of time), Transmission Average Time (Total Transmission Time divided by Transmitted Packet Count), Reception Average Time (Total Reception Time divided by Received Packet Count), Transmitted Packet Count (count of packets sent by the source node within the window of time), Received Packet Count (count of packets received by the destination node within the window of time in which the packet was received), Total Transmission Time (sum of time taken for

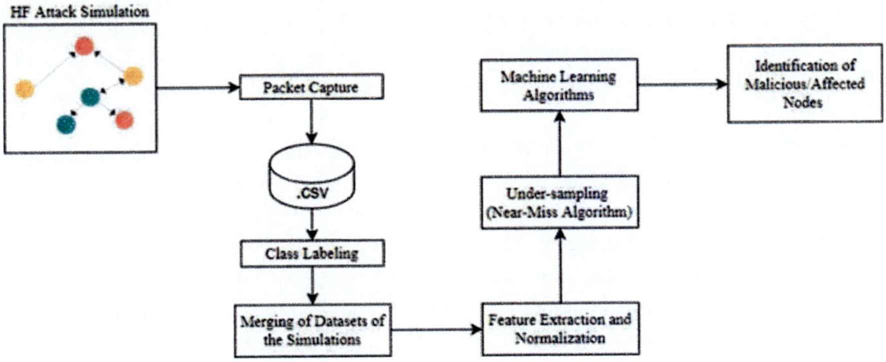

Fig. 1 Proposed methodology

Table 1 Simulation parameters

Parameter	Setting/Description
Attack	Hello Flooding
Simulation name	a/b/c
Tool	Cooja simulator (Contiki 3.0)
Type of mote	Sky Mote
Client/server source code	"contiki/examples/ipv6/rpl-udp/udp-server.c" "contiki/examples/ipv6/rpl-udp/udp-client.c"
Simulation time	10800 s
Number of sink nodes	1/1/1
Number of normal nodes	4/4/6
Number of attack nodes	1/2/2
Radio medium transmission range	Unit disk graph medium: Distance loss
Position topology	Random positioning

packets transmitted from the source node within the window of time), Total Reception Time (sum of time taken for packets received by the destination node within the window of time), DAO Packet Count (count of DAO control packet within the window of time), DIS Packet Count (count of DIS control packet within the window of time), DIO Packet Count (count of DIO control packet within the window of time), Normal/Malicious Label (labeled according to the requirements). Lastly, feature normalization is performed to provide better results when used for training.

The class labeling for the data-points takes place as follows:

- Data-points whose source nodes are benign/unaffected are labeled as 0.
- Data-points whose source nodes are malicious are labeled as 1.
- Data-points whose source nodes are in the full-transmission range of the malicious nodes are labeled as 2.

Figure 2 visualizes the class labeling of the data-points.

Seeing the distribution of these three classes, it was found out that the dataset had a huge number of data-points which were labeled as class 1 (packets sent by the malicious node), followed by data-points which were labeled as class 2, and lastly class 0. The way HF attack executes could possibly have accounted for such a distribution.

Creating a machine learning model with such a dataset which has an imbalance distribution would eventually end up having a bias toward the majority class. Undersampling is performed to solve this problem. It helps us balance the distribution of the classes of our skewed dataset. The main emphasis is to decrease the training examples which belong to the majority class, and equalizing the training examples of all the labels. After careful consideration, out of the different undersampling techniques, Near-Miss undersampling technique is performed because of its relevance to our approach. Those examples from the majority labels are selected which have

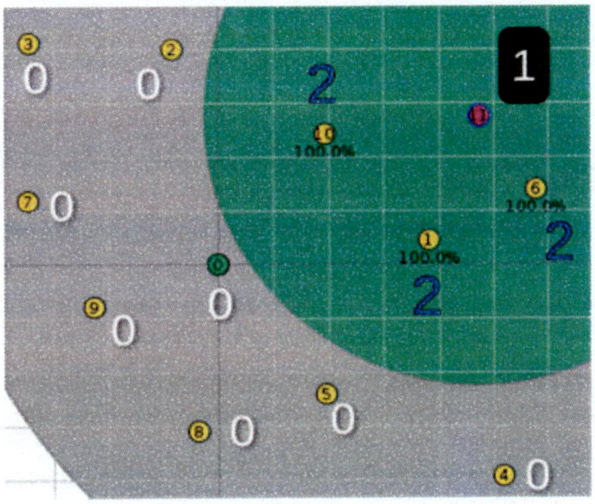

Fig. 2 Class labeling of the data-points

closer distances to the examples of the minority labels. Specifically, only those examples are chosen from the majority labels points which have the minimum average distance to three closest points of the minority labels.

A normalized dataset is achieved, with an equal number of points for each label in the dataset. For a train-test split of 0.80, different machine learning models are implemented.

4 Evaluations and Results

Different machine learning models were implemented for the created dataset. Their accuracies have been tabulated in Table 2.

One of the models trained was an ensemble hard voting classifier, combining logistic regression, Gaussian Naive Bayes, and Random Forest Classifier. The

Table 2 Model accuracies

Implemented algorithms	Kernel/description	Accuracy (%)
SVM	Linear	33.83
	Radial basis function	33.03
	Sigmoid	62.56
	Polynomial	66.47
Decision tree classifier		33.5
Gaussian Naive Bayes classifier		65.16
Ensemble learning model	Logistic regression Gaussian Naive Bayes Random forest classifier	79
k-nearest neighbors	$k = 5$	99.94

ensemble model gave better results than the individual classifiers, achieving a test accuracy of 79%.

K-Nearest Neighbors with $k = 5$ achieved the highest test accuracy of 99.94% among all the other machine learning algorithms. Robustness of the model was assessed using various other HF simulation scenarios.

5 Conclusion and Future Work

The proposed solution in practice is supposed to work on a device with a good amount of computational power. The data is collected at the sink and sent to this device which is either a server or any other resource with certain computational power. The further course of action can be to try for a solution which can partially be computed in the motes and later be sent to the sink. This would reduce the computational power required and increase the speed at which the malicious nodes can be found.

Another interesting take can be to try and implement a solution where nodes can compute a vector representing its individual self in the network. This vector can be sent at regular intervals of time to the sink. This concept is inspired by the idea of Graph Neural Networks (GNN), with the WSN representing the network, the nodes representing the vertices and the connections representing the edges. These vectors can be used to build models.

References

1. Atzori L, Iera A, Morabito G (2010) The Internet of Things: a survey. Comput Netw 54(15):2787–2805. https://doi.org/10.1016/j.comnet.2010.05.010
2. Dunkels A, Gronvall B, Voigt T (2004) Contiki—a lightweight and flexible operating system for tiny networked sensors. In: Proceedings of the 29th annual ieee international conference on Local Computer Networks (LCN '04). IEEE Computer Society, USA, pp 455–462. https://doi.org/10.1109/LCN.2004.38
3. Mayzaud A, Badonnel R, Chrisment I (2016) A taxonomy of attacks in RPL-based Internet of Things. Int J Netw Security ACEEE Division Eng Netw 18(3):459–473. https://doi.org/10.6633/IJNS.201605.18(3).07
4. Raoof A, Matrawy A, Lung C-H (2019) Secure routing in IoT: evaluation of RPL's secure mode under attacks. IEEE Global Commun Conf (GLOBECOM) 2019:1–6. https://doi.org/10.1109/GLOBECOM38437.2019.9013120
5. Raza S, Wallgren L, Voigt T (2013) Svelte: real-time intrusion detection in the Internet of Things. Ad Hoc Netw 11(8):2661–2674. https://doi.org/10.1016/j.adhoc.2013.04.014
6. Wallgren L, Raza S, Voigt T (2013) Routing attacks and countermeasures in the RPL-based Internet of Things. Int J Distrib Sens Netw. https://doi.org/10.1155/2013/794326
7. Buczak AL, Guven E (2016) A survey of data mining and machine learning methods for cyber security intrusion detection. IEEE Commun Surv Tutorials 18(2):1153–1176. https://doi.org/10.1109/COMST.2015.2494502
8. Kaplantzis S, Shilton A, Mani N, Sekercioglu YA (2007) Detecting selective forwarding attacks in wireless sensor networks using support vector machines. In: 3rd international conference on

intelligent sensors, sensor networks and information. ISSNIP 2007. IEEE, pp 335–340. https://doi.org/10.1109/ISSNIP.2007.4496866
9. Tang Y, Zhang Y, Chawla NV, Krasser S (2009) SVMs modeling for highly imbalanced classification. IEEE Trans Syst Man Cybernetics Part B (Cybernetics) 39(1):281–288. https://doi.org/10.1109/TSMCB.2008.200290
10. Rajasegarar S, Leckie C, Palaniswami M, Bezdek JC (2006) Distributed anomaly detection in wireless sensor networks. In: 2006 10th IEEE Singapore international conference on communication systems, pp 1–5. https://doi.org/10.1109/ICCS.2006.30150
11. Müller N, Debus P, Kowatsch D, Böttinger K (2019) Distributed anomaly detection of single mote attacks in RPL networks. In: Proceedings of the 16th international joint conference e-business telecommunication, vol 2, pp 378385. https://doi.org/10.5220/0007836003780385
12. Garofalo A, Di Sarno C, Formicola V (2013) Enhancing intrusion detection in wireless sensor networks through decision trees. In: Dependable computing. Springer, Heidelberg, pp 1–15. https://doi.org/10.1007/978-3-642-38789-0_1
13. Xu X, Sun Y, Huang Z (2007) Defending ddos attacks using hidden Markov models and cooperative reinforcement learning. In: Pacific-Asia workshop on intelligence and security informatics. Springer, Heidelberg, pp 196–207. https://doi.org/10.1007/978-3-540-71549-8_17
14. Diro AA, Chilamkurti N (2017) Distributed attack detection scheme using deep learning approach for Internet of Things. Futur Gener Comput Syst. https://doi.org/10.1016/j.future.2017.08.043
15. Cakir S, Toklu S, Yalcin N (2020) RPL attack detection and prevention in the Internet of Things networks using a GRU based deep learning. IEEE Access 8:183678–183689. https://doi.org/10.1109/ACCESS.2020.3029191
16. Yuan X, Li C, Li X (2017) DeepDefense: identifying DDoS attack via deep learning, 1–8. https://doi.org/10.1109/SMARTCOMP.2017.7946998
17. Yavuz FY, Ünal D, Gül E (2018) Deep learning for detection of routing attacks in the Internet of Things. https://doi.org/10.2991/ijcis.2018.25905181

Location Aided Secure Routing System in Ad-Hoc Networks

P. Sai Kishore Reddy, A. Hari Kishan Reddy, K. Vamshi Krishna Reddy, Lekha Suresh, and Animesh Giri

Abstract Nowadays Data protection is considered as the most important part in online communication. Every user looks for their private data to be safe in an online communication. Therefore, necessary measures and precautions should be taken to maintain confidentiality, Integrity, Availability (CIA) triad. Violation of any these would cause problems in an online communication in real time environment. The results and the conclusions have shown that using RSA cryptographic algorithm for securing the data while communicating between sender and receiver ensures the data is protected safely and using Location Aided routing (LAR) Protocol for sending the data packets from sender to the receiver in an Ad-hoc networks would reduce networks routing overhead and reduce traffic overhead in finding the path using fastest path using global positioning system. This ensures that CIA triad has not been violated and would help in achieving more security and a faster data transmission.

Keywords RSA · LAR · Ad-hoc networks

1 Introduction

Today's technology progresses at a fast pace, providing consumers with a wealth of enticing services to relieve the stress of large-scale data storage and upkeep. Nowadays, numerous online services, such as e-messaging, e-billing, e-transactions and e-mail, provide all services and data online. Many of these services allow users to electronically submit data to be processed. This data could include any sensitive information that the consumer wants to keep safe from malicious behavior, such as healthcare information, bank transactions, credit card information and so on. Because

P. Sai Kishore Reddy · A. Hari Kishan Reddy · K. Vamshi Krishna Reddy (✉) · L. Suresh · A. Giri
Department of Computer Science, PES University, Electronic City, Bengaluru, Karnataka 560100, India
e-mail: vamshik778@gmail.com

A. Giri
e-mail: animeshgiri@pes.edu

V. K. Gunjan and J. M. Zurada (eds.), *Proceedings of 3rd International Conference on Recent Trends in Machine Learning, IoT, Smart Cities and Applications*, Lecture Notes in Networks and Systems 540, https://doi.org/10.1007/978-981-19-6088-8_8

the disclosure of sensitive information can result in significant harm to the user, there is a high demand for data security and protection from unauthorized users. Because it prevents any unauthorized user from sniffing the data being communicated between two or more parties, authentication is a high requirement for preserving security criteria like data confidentiality and integrity.

One of the existing systems would include Bluetooth as Bluetooth can be used where there would be no network. Bluetooth allows users to send and receive messages over Ad-hoc networks but it provides very minimal data protection which is involved in communication. As Bluetooth provides very less security in transmitting messages, these data packets can be easily leaked and is vulnerable to many passive attacks. Bluetooth uses Ad-hoc On-demand Distance vector (AODV) routing protocol for transmitting messages from sender to receiver.

For security, Bluetooth uses symmetric key cryptographic algorithm where it uses Block cipher and more recently with AES algorithm. LAR routing protocol is the next modification of AODV protocol and this LAR uses GPS in order to locate nodes and this makes easier in decision making for a shorter route. Faster transmission of packets would also be considered as an important aspect in online communication as there might more traffic in forwarding the data packets. Using LAR would reduce networks routing overhead and reduce traffic overhead. In addition to this Bluetooth's data privacy can improved by using asymmetric key cryptographic algorithm as Symmetric key algorithm works on using the same key for encryption and decryption. Leakage of key from any end would lead to violation of CIA triad. Hence, using asymmetric algorithm for encryption and decryption would improve the above flaw as it uses two distinct keys for encryption and decryption. Using RSA cryptographic algorithm in Bluetooth would help in achieving more security and would help the user to achieve CIA triad (confidentiality, Integrity, Availability).

2 Related Work

The primary goal of the authors of one of the research works is to ensure secure data transmission between source and destination. With the help of 1000 mobile nodes, the simulation is run for various numbers of mobile nodes using the network simulator [1]. This model was compared to other models like DSR and AODV. In terms of packet delivery, packet drop and latency, this module as yielded the best results. Even if they are 5 rogue notes in the network the proposed model discarded 19% of packages.

Wireless network technology allows computers to communicate with one another without the use of a physical media. Wireless communication, as opposed to wired networks, enables cellular gadgets to connect to other local area networks or the Internet from any location and at any time [2]. Due to the benefits of flexible routing, global connectivity, and a highly adaptable potential, mobile Ad-hoc networks (Manet) are suitable for a wide variety of applications in both military and industrial ecosystems, including war zones, rescue operations, mobile information sharing, wireless smart phone networking and vehicle networks.

In Ad-hoc networks, a variety of security methods have been proposed to detect, prevent, or mitigate routing misbehavior attacks and their success-related assaults [3]. The bulk of these systems rely on encryption, rules, signatures, trust, or learning for detection and prevention. Other ways are proposed to counter assaults and the consequences of successful routing misbehavior attacks such as black hole, man in the middle, or denial of service attacks (DoS).

A Mobile Ad-hoc network is made up of nodes that can communicate with one another wirelessly. These nodes serve as an end system as well as a router, passing packets to other nodes without the requirement for any existing infrastructure or centralized administration [4]. These networks have a dynamic topology because any node can join or leave the network at any moment. These characteristics make MANET particularly valuable and practical in military and rescue applications, such as linking soldiers on the battlefield or constructing a new network at replace one that has failed due to a disaster such as an earthquake.

The Fundamental goal of such a basic Ad-hoc network routing protocol is to build a correct and efficient route between two nodes. So that packets can be delivered in a reasonable amount of time. The Bonding and continuous communication between its nodes is the idea behind this infrastructure less network, which means that instead of delivering data packets through a permanent base station, nodes are used to forward messages containing data from one node to another node [5]. The global positioning system is typically used by the location aided routing protocol to obtain location information.

3 Problem Formation and Design

The overwhelming percentage of today's online communications relies on infrastructure networks, under which all devices on a wireless network communicate among each other via an access point [6]. Packet transmission becomes slower as a result of the increased traffic in transmitting packets, and passive attacks on this combined network will steal information, resulting in data integrity and security violations.

To overcome this, we use Ad-hoc network where the wireless devices are not required to have an active internet connection to communicate. We use LAR routing protocol for communication between nodes. We use RSA Cryptographic algorithm in encrypting and decrypting the messages.

There are many advantages of using LAR Protocol and RSA Algorithm that include:

1. RSA is an asymmetric key cryptographic algorithm which uses two distinct keys.
2. The message is encrypted. The communication's confidentiality is ensured by the maltreatment receiver's public key, since no third party will be able to crack

Table 1 NS2_setup

Channel type	Channel/wireless channel
Radio-propagation model	Propagation/two ray ground
Network interface type	Wireless physical
MAC type	Mac/802 11
Interface queue type	Queue/drop trail/Priority queue
Layer type	Link layer
Antenna model	Antenna/omni antenna
Max packet in ifq	30
Number of mobile nodes	100 Nodes
Routing protocol	Location aided routing (LAR)
X dimension of topography (in m)	1400
Y dimension of topography (in m)	1000
Time of simulation end (in s)	230

it because the message will only be decrypted by the mistreatment receiver's personal key, which is only known to him.

3. LAR protocol reduces the control overhead.
4. The path search field in the LAR protocol is limited to the network segment where the route to the destination node is found.

The simulation works in the following way:

1. We have made NS2 Setup as shown in the Table 1.
2. Sender and receiver are selected.
3. Sender starts to find the route path to Receiver. Sender finds the path to the receiver by Using LAR protocol. LAR protocol uses GPS to locate the receiver's location. BY using LAR protocol, the traffic overhead is reduced.
4. Once the Sender discover the route to the destination. The data that need to be send to the destination is encrypted by RSA algorithm.
5. The encrypted data starts moving from sender node to the destination node.
6. While the data packet is traveling, there is a chance of passive attacks that can be done on the packet.
7. We have done the encryption, so that the original data is not leaked to the attackers.
8. The data packets send by the source node is now reached to the destination node.
9. At the destination node, the encrypted data is decrypted and the original data that has been sent by the sender has reached the destination node safely, without any data leakage to others.

These are the integral parts that are included:

Sender: The person who is going to start the communication with the receiver. Sender Class is a class created as a blueprint for the sender object. Each sender is considered as an object. This class represents a sender. Sender class is a class used to define the sender object. Sender has to select the destination node and the data that has to be sent the specified user.

Receiver: The person who is going to receive the data communication sent by the sender. Receiver Class is a class created as a blueprint for the Receiver object. Each Receiver is considered as an object. This class represents a Receiver. Each Receive has to perform that to collect and read the data that has send by the sender.

Send Message/File: It is the message or the file that is sent by the sender. Send Class is a class created as a blueprint for the sender object. Each sender is considered as an object. This class represents a send object. Each Sender has to select the destination node and the data that has to be sent the specified user. This class will send the encrypted data to the specified user.

Encryption: In the file that is send by the sender the data is encrypted to safeguard from the intruders while the data is transferred from the sender to receiver. Encryption Class is a class created as a blueprint for the Encryption object. Each Encryption is considered as an object. Each Sender has to perform that to send the data to the receiver in encrypted data form.

Decryption: The encrypted data is decrypted to human readable format. Decryption Class is a class created as a blueprint for the Decryption object. This class represents Decryption process. At the Receiver's end this has to be performed that to collect and read the decrypt the encrypted data that has send by the sender.

Receive Message/File: The message is received by the receiver which is send by the sender. The File is received by the receiver which is send by the sender. Receiver Class is a class created as a blueprint for the Receiver object each Receiver has to perform that to collect and read the data that has send by the sender. Then forward the data to the Receiver end.

4 Learning Methodology

RSA Algorithm

One of the most popular and safe public-key (asymmetric) cryptographic algorithms is the Rivest-Shamir Adleman (RSA) algorithm.

The technique takes advantage of the fact that there is no efficient way to factor very big (100–200 digit) values.

The algorithm utilizing (e, n) as an encryption key is as follows:

1. The message is represented by an integer ranging from 0 to $(n–1)$. Large messages are divided into blocks, each of which is represented by an integer in the same range.
2. Encrypt the message by multiplying it by the eth power modulo n, yielding cipher-text message.
3. Raise the message to a higher power d modulo n to decrypt it.

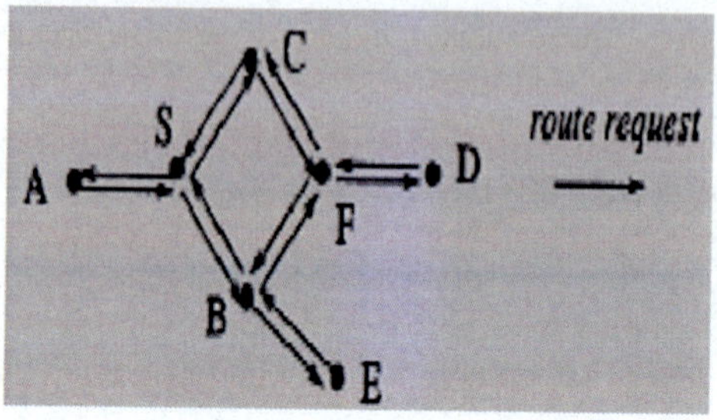

Fig. 1 Source and destination nodes

LAR Protocol

A Mobile Ad-hoc network is made up of wireless hosts that move around a lot. The migration of hosts creates a change in routes, necessitating the usage of a system to establish new routes. For Ad-hoc networks, several routing methods have already been suggested [7]. LAR protocol is a modification of the AODV protocol that uses GPS coordinates and node moving velocity to restrict flooding to a specific area that includes from start point to end point. Since only a portion of the network is scanned, this method decreases the amount of control traffic passing. This routing protocol helps in reducing control overhead message. LAR restricts the path search field to network segment where the route to the destination node is found.

The proposed Location Aided Routing (LAR) protocols use location information to restrict the search for a new route to the Ad-hoc network's "request zone." As a result, the quantity of routing messages has decreased dramatically. We offer two techniques for determining the request zone, as well as suggestions for how to improve them.

As shown in the Fig. 1, *S* and *D* are source and destination, respectively. The aim of this routing protocol is to send data packets from source to destination. LAR uses Global Positioning System (GPS) in locating nodes nearby *S* and *D*. LAR avoid the nodes which are present outside the request zone, this helps in reducing traffic overhead and helps in faster transmission of data packets from source to destination.

Location Aided Routing (LAR) protocol mainly works on (i) Route Discovery and (ii) Route Maintenance as shown in Fig. 2. In Route Discovery, the main function is finding the request zone and sending the route request to all nodes in that zone and then it retrieves the route path from the destination [8]. Finding the most efficient and shortest path between Source and Destination would play a crucial role in selecting routing protocols. Route Discovery helps in finding the path to send data packets from sender node to destination node. Request zone would be the area which includes source and destination node. All route requests are sent to the nodes which are present in the request zone. Then it retrieves the route path from destination and thus helps in sending and receiving the data [9].

Secondly, Route Maintenance first retrieves the route if the path is not found then it finds the request zone and send route request to all nodes in the zone present and then it retrieves the route path from the destination, lastly same as route discovery it sends and receives data using that path [10].

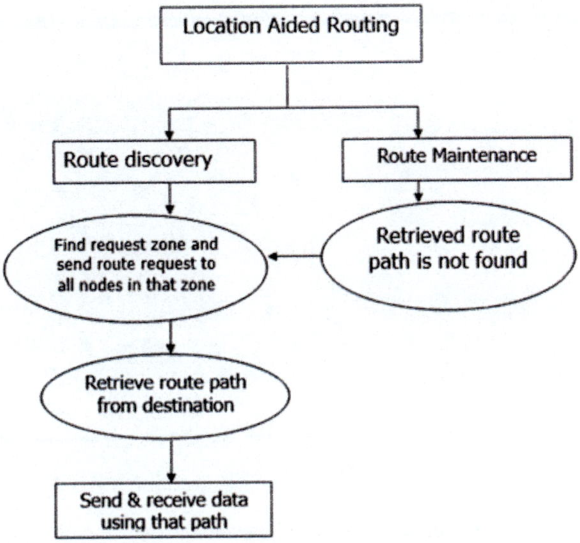

Fig. 2 LAR flow diagram

5 Results

Above shown figures from Figs. 3, 4, 5, 6 and 7 are the results obtained from NS2 simulation. Passive assaults seek to use information from the system but have no effect on system resources. Passive assaults are always in the form of eavesdropping or transmission monitoring. The opponent's primary purpose is to get information that is being delivered. Eavesdropping and traffic analysis attacks are both dependent on what the attacker hears in the network. In this type of assault, the attacker, on the other hand, does not have to compromise the real data. The attacker just listens in on network traffic in order to perform traffic analysis and determine the location of crucial nodes, the routing structure and even application activity patterns.

It does not actively engage in hostile behavior. By monitoring and listening in on interactions between MANET nodes, the malicious node in traffic analysis aims to extract crucial information from the system. If the malicious node notices that connections to a specific node are being requested more frequently than connections to other nodes, the passive attacker will learn that this node is critical for various MANET functions. Encryption, a critical layer of defense in cyber security architecture, makes it as difficult to exploit intercepted data as feasible. It may be used to protect information ranging from sensitive government secrets to personal credit card transactions. Data encryption software, sometimes referred to as an encryption algorithm or cipher, is used to construct an encryption system that can only be cracked with huge amounts of processing power.

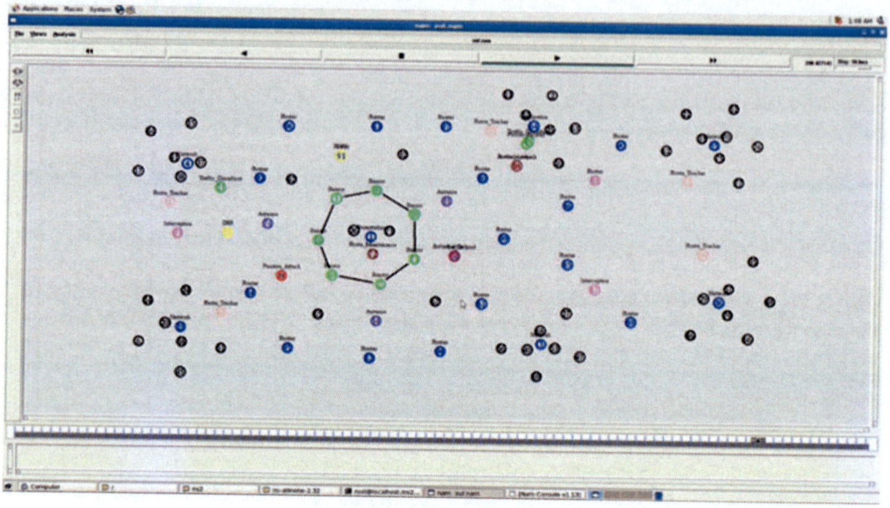

Fig. 3 Configured nodes with assigned work

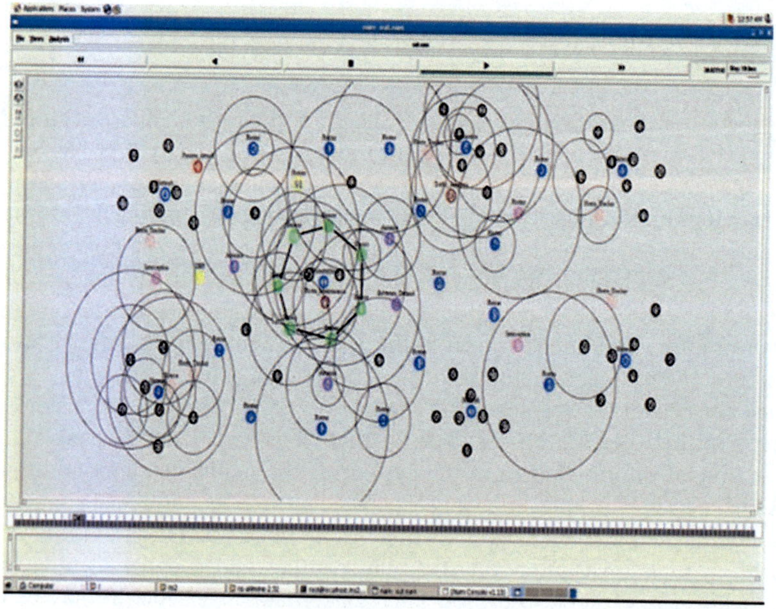

Fig. 4 Source node seeking a way to the target node

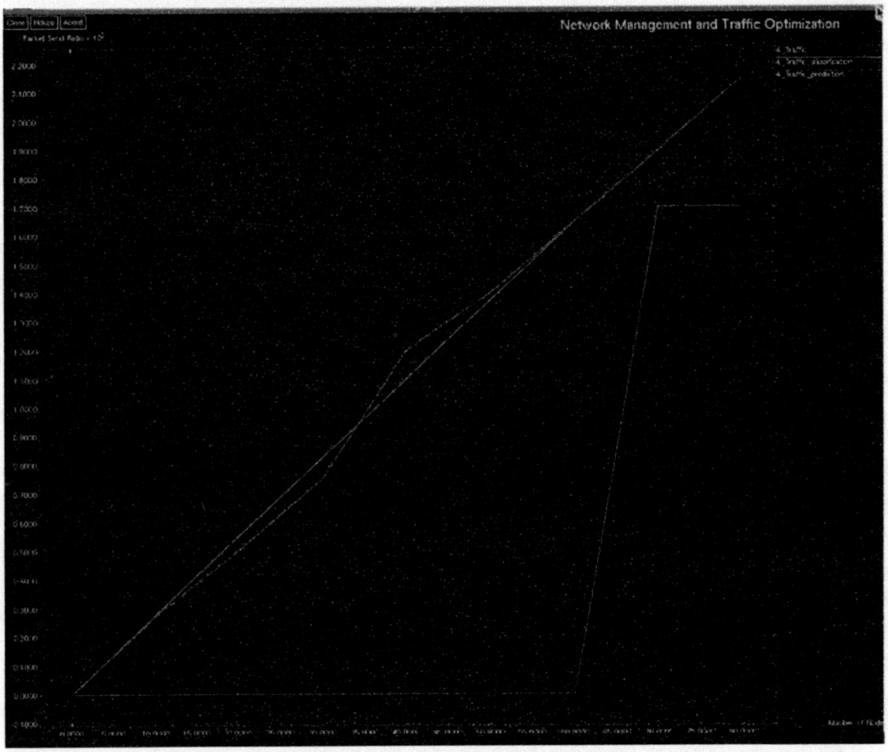

Fig. 5 Network management and traffic optimization. *X*-axis: Number of nodes, *Y*-axis: Packet loss ratio

6 Conclusion

Our main goal is to securely transmit the messages between sender and receiver while communicating. An extended literature survey on cryptography, types of cryptographic algorithms, types of routing protocols in Ad-hoc network and understanding LAR routing protocol was performed for better understanding of the architecture to gain a deeper insight for developing this proposed system.

Our new methodology aims to provide a better model for faster and secure transmission of messages. Network simulator (NS2) has been used in sending and receiving of data packets across different nodes between sender node and receiver node. So far, different types of attacks for stealing the data packets during communication are performed in order to show that there is no violation of confidentiality, integrity, availability (CIA) triad. Graphs which are generated using trace file depict the various characteristics of network traffic and many other parameters. This shows that there is minimal/ negligible stealing of data which would occur while communicating.

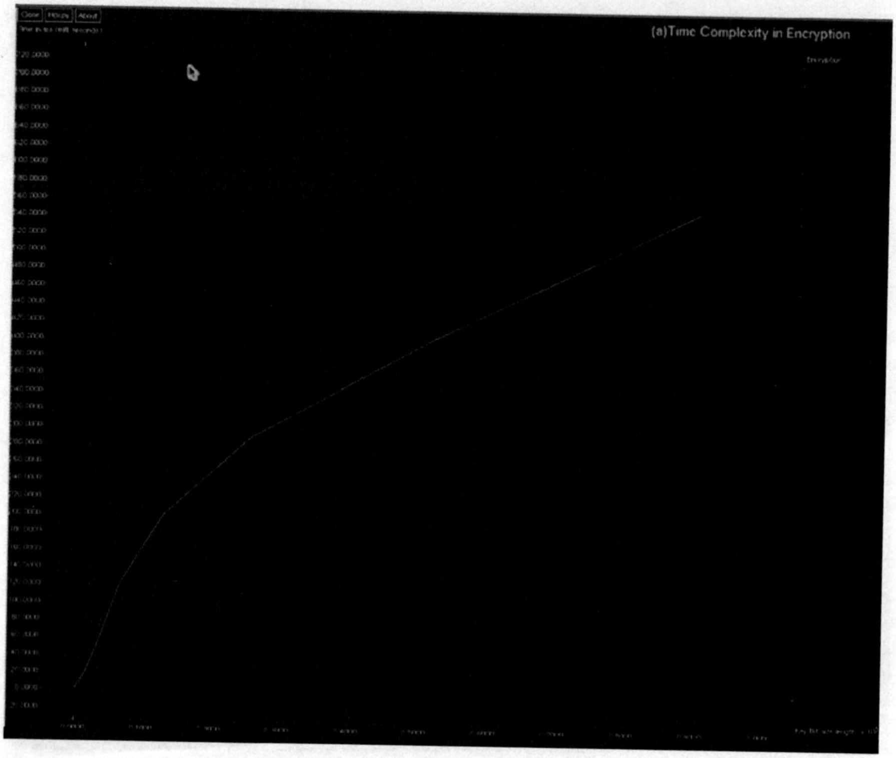

Fig. 6 Time complexity in Encryption. *X*-axis: Key size, *Y*-axis: Time in ms

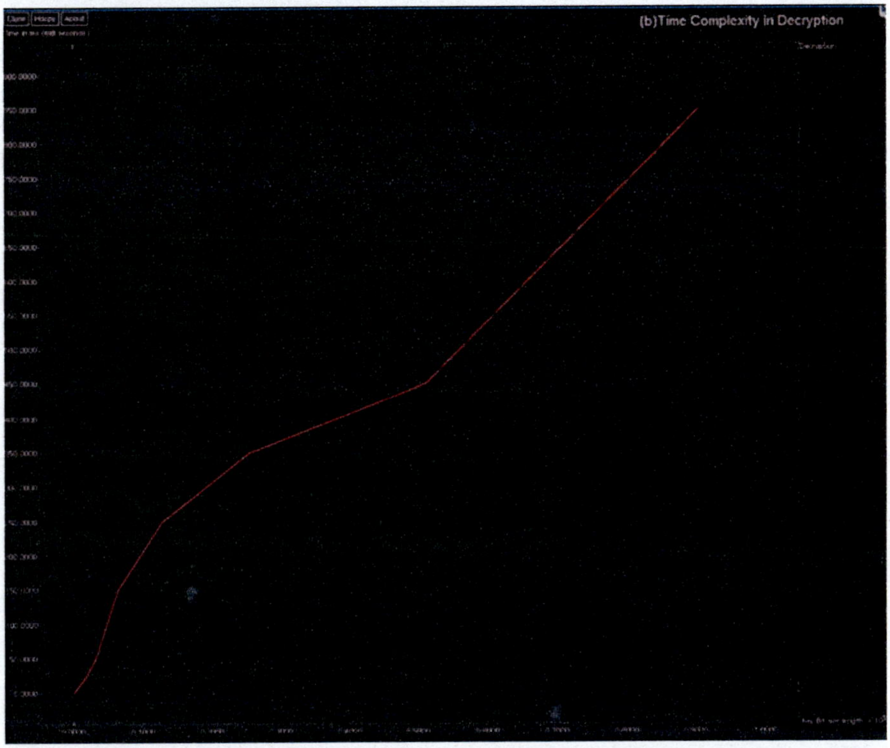

Fig. 7 Time complexity in Decryption. *X*-axis: Key size, *Y*-axis: Time in ms

References

1. Golla V, Rajaram M, Dhanalakshmi S (2018) New security algorithm for mobile Ad-hoc networks using zonal routing protocol. Ubiquitous Comput Commun 3
2. Yu PH, Pooch Texas UW (2011) Security and dynamic encryption system in mobile AD-hoc network. https://doi.org/10.5772/12843. S.No: 978-953-307-403-3
3. Basomingera R, Choi YJ (2020) Learning from routing information for detecting routing misbehavior in Ad Hoc Networks. Sensors (Basel, Switzerland). 20(21). https://doi.org/10.3390/s20216275.PMID:33158143;PMCID:PMC7663644
4. Rajeshwar J, Narsimha G (2017) Secure way routing protocol for mobile ad hoc network. Wireless Netw 23:345–354. https://doi.org/10.1007/s11276-015-1161-3
5. Prasad AK, Bhusan A, Gupta D (2012) Active source routing protocol for mobile network. IOSR J Comput Eng 7(3)
6. Papadimitratos P, Haas ZJ (2002) Secure routing for mobile Ad Hoc networks. In: Proceedings of the SCS communication networks and distributed systems modeling and simulation conference (CNDS 2002), San Antonio, TX, Jan. 27–31
7. Borsc M, Shinde H (2005) Wireless security and privacy. In: 2005 IEEE international conference on personal wireless communications, 2005. ICPWC 2005, pp 424–428
8. Bhatt D, Agarwal B (2017) Major challenges of mobile Ad-hoc networks. Oriental J Comput Sci Technol 10(2). https://doi.org/10.13005/ojcst/10.02.23

9. Yang H, Luo HY, Ye F, Lu SW, Zhang L (2004) Security in mobile Ad-hoc networks: challenges and solutions. IEEE Wirel Commun 11:38–47
10. Mukesh M, Rishi KR (2017) Security aspects in mobile ad hoc network (MANETs): technical review. Int J Comput Appl 12(2):37–43

A Review on Behavioural Biometric GAIT Recognition

B. Amogha and Rohini Deshpande

Abstract Vulnerability to individual authentication is a major concern in a number of real time contexts, for example: airport, hospital, workplace, vehicle registrations, metro stations, etc. This problem has had a huge effect on the need for surveillance monitoring systems. Behavioural biometric trait gait has established as a prominent means of surveillance monitoring method because of its inconspicuous and unperceivable nature. This is also helpful because can recognize anyone by their walking style. In this article we have formulated ten research papers which provide the brief information on the recent gait databases available, pre-processing methods, feature extraction and classification methods. We had formulated the factors influencing the recognition with research gap which helps for the new comers in the gait recognition system. In last the proposed framework is formulated in brief.

Keywords Biometric authenticator · GAIT · Machine learning · Deep learning

1 Introduction

In today's era the system need to be more secure and robust for authenticating the person in the public areas, institutions, state or national borders security systems, etc. The traditional or current authentication systems such as token based, password, card based has their own disadvantages and there is a chance of hacking or forgetting the passwords or losing the token. To address the above stated problems the biometrics authentication system, i.e. the authentication is going to happen by employing human parts such as voice, face, iris, signature, heart beats, gained the popularity as these can't be lost. Let us see about the types of the biometrics and these are divided into two types first is behavioural and physiological. The signature, Gait and Voice are

B. Amogha (✉) · R. Deshpande
Reva University, Bangalore, India
e-mail: amoghabandrikalli@gmail.com

R. Deshpande
e-mail: rohinideshpande@reva.edu.in

V. K. Gunjan and J. M. Zurada (eds.), *Proceedings of 3rd International Conference on Recent Trends in Machine Learning, IoT, Smart Cities and Applications*, Lecture Notes in Networks and Systems 540, https://doi.org/10.1007/978-981-19-6088-8_9

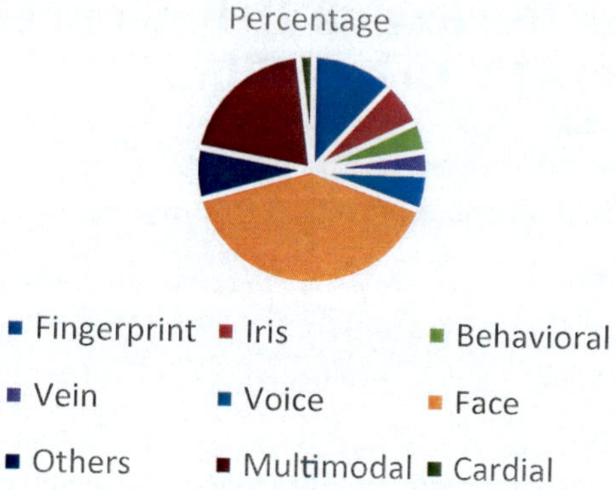

Fig. 1 Biometrics demands [5]

the examples for behavioural-based biometric system and iris, thumb, palm, face recognition is the example for physiological biometric system. These two types have their own advantages and disadvantages [1, 2].

Since many years there has been an increase in the automatic human recognition with the real time image or videos. In view of this, the Gait has gained the importance in the field of analysis of motion and computer vision for the following points:

- Works low resolution of the video and the authentication is possible.
- The video can be captured from the distance.
- No initial cost is essential as all building will be having CC TVs.
- Can detect whether person is wearing bags or not. (This can be employed in detecting the bags before and after leaving the monument places acts as bomb detection team as well.)
- The GAIT features are hard to mimic preferably its practically not possible to cover the GAIT features from input devices compared to facial recognition for example the book can be covered to face.

The authentication of the person is depending on the individual walking Style is titled as GAIT recognition [3, 4]. Figure 1 shows the demand for various biometric methods [4].

2 Methodology of Review Process

Table 1 summarizes the papers till the year 2021. In the survey the papers are referred from the IEEE transmission, Scopus, Springer, Web of science and last but not least Google Scholar.

The keywords employed in the survey are "GAIT motion" "GAIT recognition" "GAIT Machine learning" "GAIT Deep learning" "GAIT marker-based" "GAIT vision-based" "GAIT sensor-based" "GAIT occlusion" "Multiple GAIT" "Multi channel GAIT CNN", etc. By observing the keywords, a total of one hundred articles

Table 1 Overview of the carried review process

Publication	Year	References	Contribution
Journal	2016	[6–9]	Prepared a various database
Journal	2015–2017	[10–12]	Pre-processing methods
Journal	2015–2017	[12–14]	Feature extraction method
Journal	2016–2020	[15–24]	Classification method

are read and twenty papers are considered out of hundred articles focuses on vision-based GAIT recognition.

Outline of the paper

The rest of the survey paper is going to happen as follows Sect. 3 describes about the database preparation for the GAIT in less than five years involving sensor-based and vision-based data collection. Section 4 describes about the General Flow for GAIT recognition involving different feature extraction methods and various classification methods employed for GAIT recognition for last five years.

3 GAIT Database Preparation

The first step in GAIT detection is to collect the database by allowing the healthy volunteers to walk in indoor, outdoor, controlled and uncontrolled weather conditions such as lighting, cloudy, etc. [7–9]

Mihalis et al. [8] prepared the three databases with different conditions such as camera view angles, capturing videos indoors and outdoors. Database one (D-1) consists of 42 images captured with 30 Frames per second (FPS). D-2 prepared using two cameras with 20 and 70 feet apart from subject with 20 FPS. D-3 is captured using eight cameras with 135 and 60 degree view angles with 12 humans allowed to walk indoors on treadmill. In the same fashion Philippe et al. [7] employed the 36 persons to collect data indoors.

Gu et al. [9] prepared a database using 18 healthy volunteers out of them 16 are males. The input devices they employed are RGBD camera and motion capture system to record the video.

The Database collection or preparation is the primary step in any of the Machine leaning methods in the same way for the recognition of the person based on GAIT. In many of the cases the accuracy of the designed model directly relies on the collection methods. For the GAIT the data collection can be performed in two ways firstly vision-based and secondly sensor-based. The different forms of data acquisition are shown in Fig. 2.

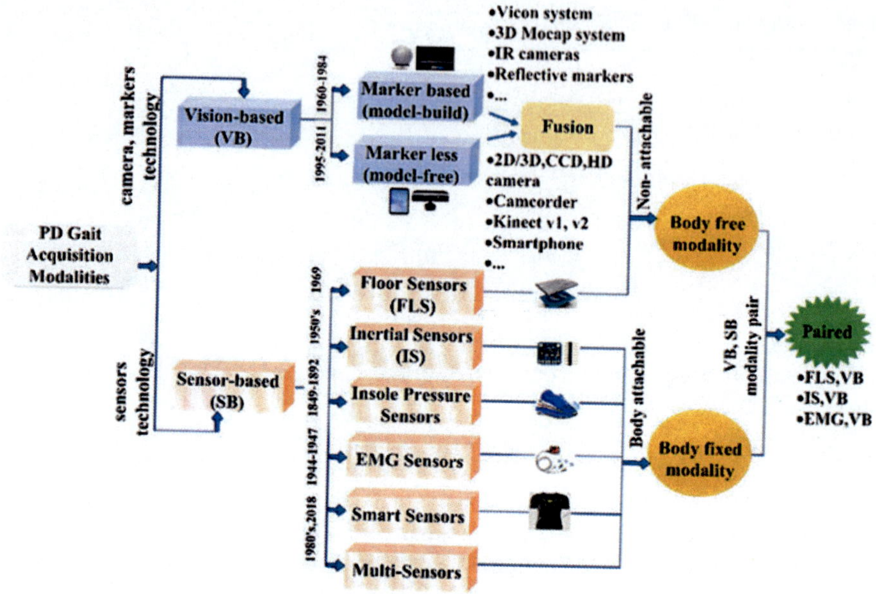

Fig. 2 Different database acquisition system for GAIT [25]

3.1 *Sensor-Based Data Collection*

In sensor-based the data can be collected from two forms of sensors such as floor and wearable. These sensors are directly attached to the body of human beings or kept on floor to acquire the signals. The floor sensors are the one which generates the pressure values when the person stepped on it [7, 11, 12]. The Philippe Terrier et al. [7] propose a methodology to collect the pressure values based on the centre of pressure trajectory by employing thirty six adults and they are allowed to walk on the treadmill for thirty minutes in order to collect the data.

Let us consider the wearable sensors where the sensors such as accelerometer, gyroscope and patch cards are attached to the body. The Siddhartha et al. [6] created the database by name MAREA using the accelerometer and considered about twenty persons in the database collection out of it twelve are males and eight are females. The sensors are placed at the waist, wrist and for ankles concentrating eleven subjects. The people are allowed to walk on the treadmill at different speed for different time duration (10 min with 4–8 km/h, six and three minutes for walk and run, respectively). Let us see the general flow for implementation for GAIT and in later section vision-based data collection is performed.

3.2 *Vision-Based*

In this method the input device is the camera, it is used to capture the video frames. The captured frames are processed to authenticate the person using GAIT.

4 General Flow for GAIT Recognition

The general flowchart for the GAIT recognition is shown in the Fig. 3.

4.1 *Pre-processing Methods Employed in GAIT*

After preparing the data-frames collected from the camera, removing the background and extracting the foreground, i.e. extracting the region of interest is the challenging task in the various detection such as speed detection, video surveillance. As in the rest of the applications in GAIT recognition the background removal is performed in order to obtain the silhouette of a person from video frames [10, 11].

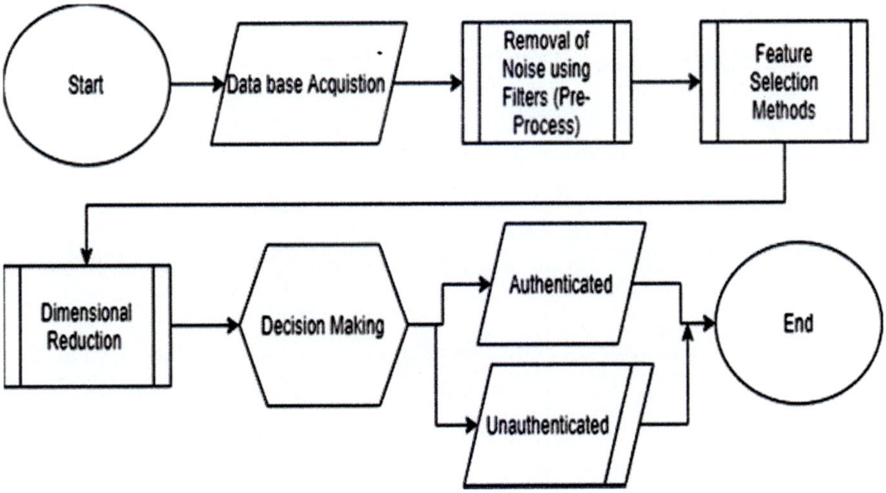

Fig. 3 Flowchart for GAIT recognition

4.2 Feature Selection for GAIT

Once the pre-processing of the frame is performed the next step is to select the dominant features. Based the feature selection the accuracy of the model is dependent. Instead of selecting the whole feature space the dimensionality reduction methods can be employed [10, 12, 26].

The most famous method titled as Principle component Analysis (PCA), which is used to reduce the number of features and selection the dominant feature. Abdul et al. [26] employed the Euclidian distance for the feature extract from the GAIT Energy images from the CASIA database later applied the PCA on the Silhouette images to reduce the dimensions of it.

Genetical Algorithm (GA) In addition to a genetic programming (GP), genetic algorithms (GAs) are used as a heuristic search technique for complex optimization problems. Feature modelling is an approach that is able to find the best possible solutions in a large-dimensional feature space it follows patterns derived from natural evolutionary design. Faezeh et al. [12] employed the Kernel Principal Component Analysis (KPCA) for feature reduction and employed GA for selecting the important Eigen values, classifiers were naïve Bayes and achieved an accuracy of 79.3% with GA and without employing GA the accuracy was 78% with the CASIA database. The summary of the feature extraction methods are tabulated in the Table 2.

Table 2 Summarization of Feature extraction methods available for GAIT recognition

Employed methodology	References	Remarks
Genetical algorithm	[12]	1. Decreases the processing data as dominant features are taken 2. Difficult to consider for correlated data
GAIT entropy image—GEI	[10, 12, 13, 15]	1. Accuracy depends on extraction of silhouette 2. Complex background, i.e. during the occlusion or in real time environment these features fail to provide good results
Principal component Analysis—PCA	[10, 12]	1. The accuracy of the system goes down for the correlated data 2. Data storage and processing time can be reduced if PCA is employed
Local depth texture pattern	[14]	Thermal cameras, sensors may include the initial costs
Spatial temporal features	[15, 16, 20]	Lesser numbers of humans are considered during the implementation (120 persons)

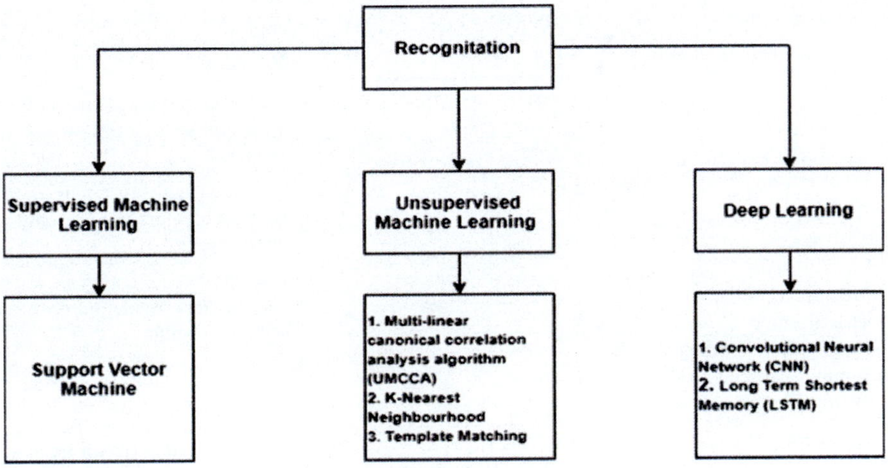

Fig. 4 Hierarchy of classifiers available for GAIT recognition

4.3 Classification Models for GAIT

The last step in GAIT recognition is to employ the classifiers of different machine learning techniques. The types in machine learning are supervised, unsupervised and Deep learning methods. Figure 4 shows the hierarchy of recognition and it is prepared by referring fifty to sixty papers which are summarized in Table 3.

5 Conclusion

GAIT-based biometric systems have readily attracting the more attention from few years. But by seeing towards the methodology applied on it or accuracy concern it is child compared to other existing biometric systems.

In these articles have formulated the papers based on the visual-based, sensor-based, features extraction methods and classification methods available.

The various databases are listed and these are available to carry out the research by starters. The databases are MAREA, CASIA—A and B, OU-ISIR, etc. Many of the researchers have used CASIA—A and B few have developed their own Database for the research. Lastly the article formulates the remarks of the literature review papers and given few opening paths in the field of GAIT.

Table 3 Summarization of classification methods for GAIT recognition

Employed methodology	References	Remarks
Gait entropy image	[15]	Gait energy image extraction varies for the real time scenario
Classifier—convolutional neural network	[17]	For video analysis long term shortest memory provides the good results
Classifier-multi-linear canonical correlation (UMCCA), support vector machine, K-Nearest Neighbourhood, template matching, artificial neural network	[8, 15, 18, 22]	Instead of HOG features, GLCM can provide more accurate results
Deep neural network	[8, 19–21, 24]	Model is ineffective for the changes in cloths and bag
Convolutional neural network	[7, 7, 13, 14, 17, 20–23, 23]	People allowed to walk in lab conditions Only 36 persons are taken for experiment For real time model with 36 members will provide the lesser accuracy

References

1. Jain AK, Nandakumar K, Ross A (2016) 50 years of biometric research: accomplishments, challenges, and opportunities. Pattern Recogn Lett 79:80–105. https://doi.org/10.1016/j.patrec.2015.12.013
2. Jain AK, Ross A, Prabhakar S (2004) An introduction to biometric recognition. IEEE Trans Circuits Syst Video Technol 14(1):4–20. https://doi.org/10.1109/TCSVT.2003.81834_9
3. Lee TKM, Belkhatir M, Sanei S (2014) A comprehensive review of past and present vision-based techniques for gait recognition. Multimed Tools Appl 72(3):2833–2869. https://doi.org/10.1007/s11042-013-1574-x
4. Bashir K, Xiang T, Gong S (2010) Gait recognition without subject cooperation. Pattern Recogn Lett 31(13):2052–2060. https://doi.org/10.1016/j.patre.2010.05.027
5. Find Biometrics Global Identity Management [online]. https://findbiometrics.com/yir-exciting-modalities-part-1-501110. Accessed 10 Feb 2019
6. Khandelwal S (2016) Evaluation of the performance of accelerometer-based gait event detection algorithms in different real-world scenarios using the MAREA gait database. Elsevier
7. Terrier P (2020) Gait recognition via deep learning of the center-of-pressure trajectory. MDPI Appl Sci 2020.
8. Mihalis A (2019) Editorial of special issue on human behaviour analysis "In-the-Wild". IEEE Trans Affect Comput 10(1)
9. Gu X (2021) Cross-subject and cross-modal transfer for generalized abnormal gait pattern recognition. IEEE Transactions Neural Networks Learning Syst 32(2)
10. Xing X (2015) Complete canonical correlation analysis with application to multi-view gait recognition. Elsevier

11. Yeoha TW, Daolioc F, Aguirreb HE, Tanakab K (2017) On the effectiveness of feature selection methods for gait classification under different covariate factors. Applied Soft Comput https://doi.org/10.1016/j.asoc.2017.07.041
12. Tafazzoli F (2015) Genetic feature selection for gait recognition. J Electron Imag 24(1):013036
13. Li C, Min X, Sun S, Lin W, Tang Z (2017) DeepGait: a learning deep convolutional representation for view-invariant gait recognition using joint Bayesian. Appl Sci 7:210. https://doi.org/10.3390/app7030210
14. Benedek C (2018) Lidar-based gait analysis and activity recognition in a 4D surveillance system. IEEE Trans Circ Syst Video Technol 28(1)
15. Nandy A (2016) A study on gait entropy image analysis for clothing invariant human identification. Springer Science Business Media New York
16. Chen X, Weng J, Lu W, Xu J (2017) Multi-gait Recognition based on Attribute Discovery. IEEE Transactions on Pattern Analysis and Machine Intelligence. https://doi.org/10.1109/TPAMI.2017.2726061
17. Castro FM, Marín-Jiménez MJ, Guil N, Pérez de la Blanca N (2020) Multimodal feature fusion for CNN-based gait recognition: an empirical comparison. ArXiv: 1806.07753v2 [cs.CV] 20 Feb 2020
18. Chen X, Xu J, Weng J (2016) Multi-gait recognition using hypergraph partition. Mach Vis Appl https://doi.org/10.1007/s00138-016-0810-6
19. An W (2020) Performance evaluation of model-based gait on multi-view very large population database with pose sequences. IEEE Trans Biomet Behav Ident Sci 2(4)
20. Ruben D-E (2018) An end-to-end multi-task and fusion CNN for Inertial-based gait recognition. IEEE Received November 26, 2018, accepted December 9, https://doi.org/10.1109/ACCESS.2018.28868992018, date of publication December 14, 2018, date of current version January 7, 2019.Digital Object Identifier
21. Tran L (2021) Multi-model long short-term memory network for gait recognition using window-based data segment. IEEE Access, Received December 21, 2020, accepted January 25, 2021, date of publication February 3, 2021, date of current version February 10, 2021. Digital Object Identifier.https://doi.org/10.1109/ACCESS.2021.3056880
22. Wang X (2019) Gait recognition using multichannel convolution neural networks. Neural Comput Appl
23. Das D (2016) Human gait recognition using deep neural networks. ICTCS '16, March 04-05, 2016, Udaipur, India c 2016 ACM. ISBN 978-1-4503-3962-9/16/03. . . $15.00. https://doi.org/10.1145/2905055.2905192
24. Yan C (2015) Multi-attributes gait identification by convolutional neural networks. 978-1-4673-9098-9/15/$31.00 ©2015 IEEE
25. Kour N, Arora S (2019) Computer-vision based diagnosis of Parkinson's Disease via gait: a survey. IEEE Oct 2019. https://doi.org/10.1109/ACCESS.2019.2949744
26. Matin A (2017) Segment based co-factor detection and elimination for effective gait recognition. 978-1-5090-6004-7/17/$31.00 ©2017 IEEE

Vital Role of 2D CNN in Brain Malignancy

Y. Vijayalata, Susmitha Valli Gogula, Vandana Yalla, M. Ananya Varma, Salunke Savitha, and Sudarsi Namrata Ravindra

Abstract Malignancy is abnormal cell proliferation in body tissues. In brain, if proliferation happens uncontrollably, then it is brain malignancy. It has to be detected at an early stage to improve the patients' scope for survival. Carcinomas tumors are to be treated and removed surgically whereas benign ones do not cause much damage to the brain tissues. A cancerous tumor grows aggressively or recursively damaging the entire structure. An magnetic resonance image (MRI) is the most prevalent way of detecting them. Manual tumor identification procedure is time-consuming and is more prone to human mistakes. Computer technologies assist more than humans in identifying micro-changes in brain tissues. Now-a-days, machine learning algorithms are playing an important role in evaluating medical images and data. Segmentation and classification of tumor region identified from MRIs assist professionals by extracting and detecting specific locations of infected regions in the brain. Convolution neural networks (CNNs) are the best source to identify tumors from MRIs which will be observed in this work. Our findings are based on rectified linear unit (ReLU), hyperbolic tangent (TanH), Sigmoid added with ELU (Exponential linear unit) functions resulting in mean and f-score accuracy of 99.44%.

Keywords Brain · Carcinoma · Convolution neural network · Malignancy · Proliferation · Tumor

1 Introduction

Image processing is the manipulation of the images using digital computers. It has point operations where each pixel value is replaced by some other value that is

S. V. Gogula · V. Yalla (✉) · M. A. Varma · S. Savitha · S. N. Ravindra
Department of CSE, GRIET, Hyderabad, Telangana, India
e-mail: vandanayalla321@gmail.com

Y. Vijayalata
Department of Computer Science and Engineering, KG Reddy College of Engineering & Technology, Hyderabad, India

V. K. Gunjan and J. M. Zurada (eds.), *Proceedings of 3rd International Conference on Recent Trends in Machine Learning, IoT, Smart Cities and Applications*, Lecture Notes in Networks and Systems 540, https://doi.org/10.1007/978-981-19-6088-8_10

obtained from the previous value. The procedures include preprocessing, segmentation, and feature extraction. The preprocessing technique elevates important things; through segmentation, the task becomes easy for feature extraction in an image. It is difficult for a novice to use typical machine learning approaches [1].

CNN has had exceptional success in image segmentation, despite the fact that fewer approaches have been used due to several intrinsic limitations. It helps to extract the pixels directly with nominal amount of preprocessing. CNN along with activation functions is used to extract patterns and their relationships from the data. It can also be used with image processing that yields effective results. The activation functions used are explained in full length in methodology. A wider range of datasets is available in which Kaggle being on the top end; this database is chosen for this problem [2].

Section 2 discusses about the related research carried on CNN approaches. Section 3 explains the process of the proposed machine learning approach, and Sect. 4 provides results and discussions on the proposed approach followed by the conclusion in Sect. 5.

2 Literature Survey

Image processing and CNN are a very popular combination for many visual recognition problems.

Kesav and Jibukumar [3] proposed an simple methodology for tumor detection in the brain using RCNN with two channel CNN models.

Özyurt et al. [4] proposed "An expert system for brain tumor detection: Fuzzy c-means with super resolution and convolutional neural network with extreme learning machine." The authors used fuzzy c-means to detect the resolution of the image along with CNN algorithms which incorporate additional machine learning approaches.

Choudhury et al. [2] proposed work involves in identifying the scanned image as either "TUMOR DETECTED" or "TUMOR NOT DETECTED." The authors used the techniques of deep neural networks and integrated it into a CNN-based model. With a f-score of 97.3, the model proved to be accurate with minimum score deviation.

Rammurthy and Mahesh [5] presented detection of brain tumors through the whale Harris hawks optimization which is a deep learning classifier.

Hashemzehi et al. [6] used a hybrid model of CNN with NADE to detect abnormal cell growth in the brain.

Goyal and Sharma [1] through their paper(2020) proposed that tumors may be distinguished using several picture division approaches, according to their study image processing techniques for brain tumor identification. The procedure for detecting brain tumors using MRI images is divided into four distinct areas: pre-planning, picture division, include extraction, and picture collection.

Zhou et al. [7] on CT images, a pre-trained InceptionV3 model was used to identify benign and malignant kidney tumors.

Dahab et al. [8] proposed a method to carry out an automated brain tumor classification utilizing MRI scans; a modified probabilistic neural network (PNN) model based on learning vector quantization (LVQ) with image and data analysis and manipulation techniques is presented. The parameters such as accuracy of classification, training performance, and computing time are all used to evaluate the updated PNN classifier's performance.

Swati et al. [9] proposed that categorization of brain tumors was done based on flattened pictures rather than three-dimensional volume because most clinical practices use two-dimensional slices with a high slice gape. Five-fold cross-validation was utilized to assess the performance. The proposed block-wise fine-tuning transfer learning procedure proposes a different way of using the pre-trained CNN as a feature extractor that trains a distinct classification algorithm. It shows that learning from natural images is also possible.

The work done by Deepak et al. [10] focuses on a three-class classification issue to distinguish between three forms of brain malignancies: glioma, meningioma, and pituitary tumors. To extract characteristics from brain MRI scans, the proposed classification approach uses deep transfer learning and a pre-trained Google Net. To classify the retrieved features, proven classifier models are used. On an MRI dataset from figshare, the experiment uses a patient-level five-fold cross-validation approach. The suggested system outperforms all current approaches with a mean classification accuracy of 98%. The input training data were created by randomly dividing the images in a ratio of 70:3.

Sheela and Suganthi [11] in this work presented an approach using the greedy snake method along with K-means to carry out detection of tumors in the brain.

Bernal et al. [12] proposed how several CNN designs have developed, outlined state-of-the-art techniques, and weighed the benefits and drawbacks of each.

Hemanth et al. [13] proposed a method that uses mean values in the standard objective function of the approach. The image processing approach in MATLAB was used to create and apply the technology. Unlike existing algorithms, the suggested CNN produces spontaneous results.

Hossam et al. [14] proposed a model where the tumor is classified into meningioma, glioma, and pituitary. The other one differentiates between three glioma. The dataset includes 233 and 73 patients with a total of 3064 and 516 images.

Suresh and Mohan [15] in this work, the proposed approach uses deep convolution neural network on lung nodules for locating tumors in the lungs.

Bal et al. [16] in this work, the proposed approach uses a combination of K-means and fuzzy logic to create a method that is not fully smooth.

Maruthamuthu and Gnanapandithan et al. [17] this work uses K-means with processing technique region of interest to locate tumors in specific locations.

Shaoguo et al. [18] presented a completely automatic segmentation method that contained vivo brain gliomas from the MRI data. This method can precisely segment the intratumor structure as well as pinpoint the entire tumor location. This research used a cascaded deep learning convolutional neural network with two sub-networks: a tumor localization network and an intratumor classification network. This method was tested using 54 low-grade glioma and 220 high-grade glioma patients from the multimodal brain tumor segmentation (BRATS 2015) datasets.

Sobhaninia et al. [19] used LinkNet network for tumor classifications. These classifications are of three forms; they are meningioma, glioma, and pituitary. For training, 2100 imagers were used out of which twenty percent were used for validation and the rest for testing purpose.

Sudhakar et al. [20] this paper outlines an effective strategy for detecting brain tumors automatically. Some Matlab techniques are crucial in tumor identification. To begin, three different edge detection algorithms (Canny, Robert, and Prewitt) are utilized to locate the tumor's points and edges in MRI images. For generating pictures from tiny components, certain preprocessing techniques such as morphological reconstruction and segmentation are used.

Joseph et al. [21] in this work, it is proposed brain MR image segmentation for detection of tumor using k-means clustering algorithm and that connected layer is based on the likelihood between 0 and 1.

3 Proposed Model

This model uses different types of activation functions in order to obtain high-level accuracy.

3.1 Convolution Neural Networks

Image scalability is impossible in a traditional neural network. Input, convolution, rectified linear unit (ReLU) layer, hyperbolic tangent layer (TanH), exponential linear unit (ELU), pooling layer, and fully connected layer make up the convolution neural network (CNN). The provided input picture is divided into several tiny sections in the convolution layer. The activation layers perform the element-by-element activation. There is an option of using or ignoring pooling since it used predominantly for downsizing. The class score or label score is computed in the last layer (i.e., fully linked layer) (Fig. 1).

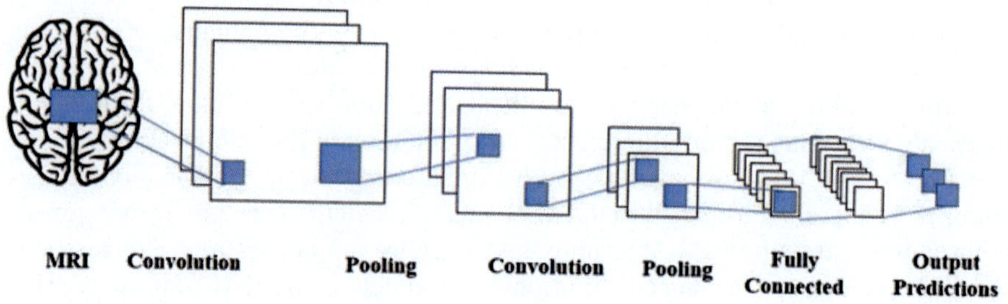

Fig. 1 Convolution architecture

3.2 *Model Description*

Some of the activation functions that are being used are rectified linear units (ReLUs), hyperbolic tangent function, ELU function, and sigmoid activation function.

Rectified Linear Unit (ReLU). It is a linear function which outputs the input directly if it is positive, otherwise outputs 0. It results in higher performance. The mathematical representation of ReLU is as follows:

$$f(x) = \begin{cases} 0 \text{ for } x < 0 \\ x \text{ for } x \geq 0 \end{cases} \tag{1}$$

Hyperbolic Tangent Function (TanH). The Tanh function is also known as hyperbolic tangent activation function. It resembles the sigmoid activation function and has an S-shape. This function accepts any real value as input, and the range lies between −1 and 1. The mathematical representation of this function is as follows:

$$f(x) = \tanh(x) = \frac{2}{1 + \mathrm{e}^{-2x}} - 1 \tag{2}$$

Exponential Unit Function (ELU). Exponential linear unit function considers negative values which assists it to lower mean unit activations toward zero. It is similar to what is done in batch normalization, but ELU is simpler with regard to computational complexity. This results in quick learning as the normal gradient shifts toward the unit natural gradient. The mathematical representation of this function is as follows:

$$f(x) = \begin{cases} \alpha(e^x - 1) \text{ for } x < 0 \\ x \text{ for } x \geq 0 \end{cases} \tag{3}$$

Sigmoid Function. The range of the activation function varies between 0 and1, represented by an S-shaped curve. The mathematical representation of this function is (Fig. 2; Table 1):

$$\varphi(z) = \frac{1}{1 + \mathrm{e}^{-z}} \tag{4}$$

4 Results and Discussion

The classification based on presence of malignancy in the image is evaluated based on parameters such as f-score, recall, and accuracy. This model achieved a training accuracy of 99.44% in 5 epochs with a training loss of 0.431% (Figs. 3 and 4).

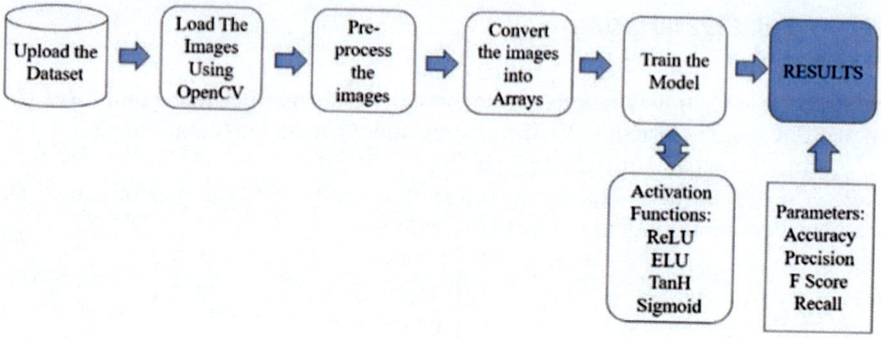

Fig. 2 Pictorial representation of the proposed process

Table 1 Flattened model summary

Layer (type)	Output shape	Params
batch_normalization	(≥1, 224, 224, 3)	12
conv2d	(≥1, 222, 222, 32)	896
MaxPooling2D	(≥1, 111, 111, 32)	0
conv2d_1	(≥1, 111, 111, 64)	18,496
MaxPooling2D_1	(≥1, 55, 55, 64)	0
conv2d_2	(≥1, 55, 55, 128)	73,856
MaxPooling2D_2	(≥1, 27, 27, 128)	0
conv2d_3	(≥1, 27, 27, 128)	147,584
MaxPooling2D_3	(≥1, 13, 13, 128)	0
conv2d_4	(≥1, 13, 13, 128)	147,584
MaxPooling2D_4	(≥1, 6, 6, 128)	0
Dropout	(≥1, 6, 6, 128)	0
Flatten	(≥1, 4608)	0
Dense	(≥1, 128)	589,952
dense_1	(≥1, 128)	16,512
dense_2	(≥1, 128)	16,512
dense_3	(≥1, 64)	8256
dense_4	(≥1, 64)	4160
dense_5	(≥1, 64)	4160
dense_6	(≥1, 32)	2080
dense_7	(≥1, 32)	1056
dense_8	(≥1, 32)	1056
dense_9	(≥1, 2)	66

Total Parameters: 1,032,238 Trainable Parameters: 1,032,232 Non-Trainable Parameters: 6

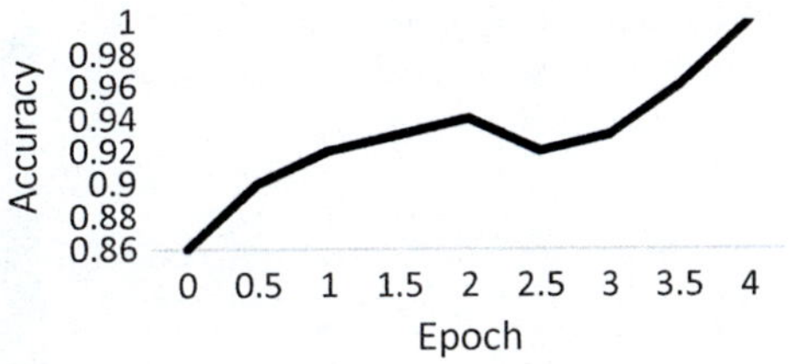

Fig. 3 Training accuracy of 2D model

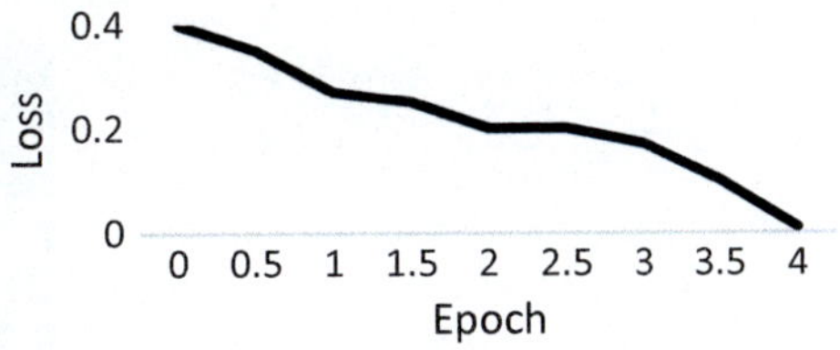

Fig. 4 Training loss of 2D model

4.1 Confusion Matrix

From the confusion matrix, the authors conclude that this approach gives an accuracy of 99.44% with an error of 2.54% (Figs. 5 and 6; Table 2).

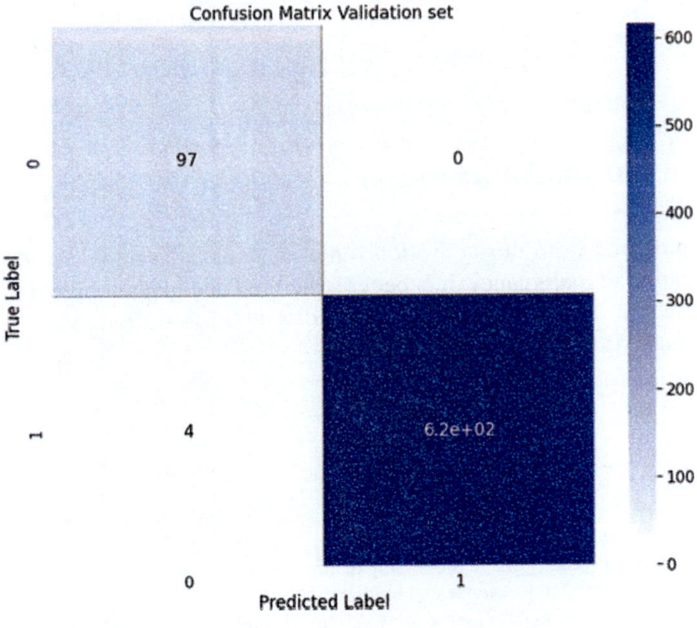

Fig. 5 Seaborn heatmap of the confusion matrix

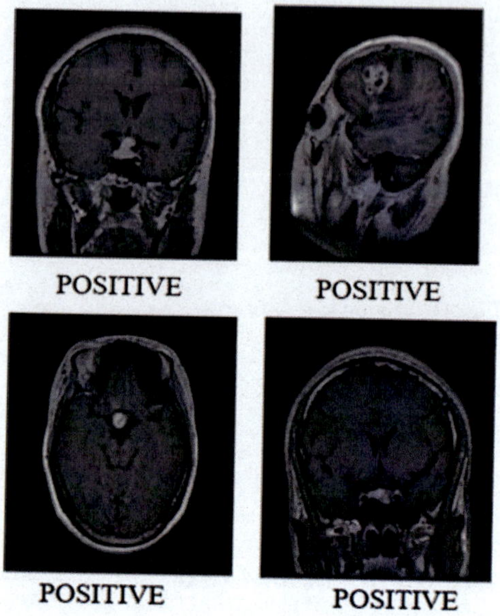

Fig. 6 Shows test images from the dataset along with the predicted labels

Table 2 Testing evaluations

Evaluation parameters	Resulted outcomes (%)
Precision	98
Recall	99
F-score	99
Error	2.98388

5 Conclusion and Future Scope

The result procured from this activation function gives an accuracy of 99.44% for the identification of malignancy. It is better structured and gives coherent results for brain malignancy. This work can be extended with a 3D CNN model for further concentric analysis of malignancy.

References

1. Goyal N, Sharma B (2021) Image processing techniques for brain tumor identification. In: IOP conference series: materials science and engineering
2. Özyurt F, Sert E, Avcı D (2019) An expert system for brain tumor detection: fuzzy C-means with super resolution and convolutional neural network with extreme learning machine. Med Hypotheses 134. https://doi.org/10.1016/j.mehy.2019.109433

3. Kesav N, Jibukumar MG (2021) Efficient and low complex architecture for detection and classification of brain tumor using RCNN with two channel CNN. J King Saud Univ Comput Inf Sci 2(5):99–110
4. Çinar A, Yildirim M (2020) Detection of tumors on brain MRI images using the hybrid convolutional neural network architecture. Med Hypotheses 139
5. Rammurthy D, Mahesh PK (2020) Whale Harris Hawks optimization based deep learning classifier for brain tumor detection using MRI images. J King Saud Univ Comput Inf Sci
6. Hashemzehi R, Mahdavi SJS, Kheirabadi M, Kamel SR (2020) Detection of brain tumors from MRI images base on deep learning using hybrid model CNN and NADE. Biocybern Biomed Eng
7. Zhou L, Zhang Z, Chen YC, Zhao ZY, Yin XD, Jiang HB (2019) A deep learning-based radiomics model for differentiating benign and malignant renal tumors. Transl Oncol. https://doi.org/10.1016/j.tranon.2018.10.012.Epub2018Dec17
8. Dahab DA, Ghoniem SSA, Selim GM (2019) Automated brain tumor detection and identification using image processing and probabilistic neural network techniques. Int J Image Proc Vis Commun
9. Swati ZN, Khan QZ, Kabir M, Ali F, Ali Z, Ahmed S, Jianfeng L (2019) Brain tumor classification for mr images using transfer learning and fine-tuning. Comput Med Imaging Graph 75:34–46
10. Deepak S, Ameer Brain PM (2019) Tumor classification using deep CNN features via transfer learning. Comput Biol Med 111
11. Sheela CJJ, Suganthi G (2019) Automatic brain tumor segmentation from MRI using greedy snake model and fuzzy C-means optimization. J King Saud Univ Comput Inf Sci https://doi.org/10.1016/j.jksuci.2019.04.006
12. Bernal J, Kushibar K, Asfaw DS, Valverde S, Oliver A, Martí R, Lladó X (2019) Deep convolutional neural networks for brain image analysis on magnetic resonance imaging: a review. Artif Intell Med 95:64–81
13. Hemanth G, Janardhan M, Suji Helen L (2019) Design and implementing brain tumor detection using machine learning approach. In: 3rd international conference on trends in electronics and informatics (ICOEI). IEEE, pp 1289–1294
14. Sultan HH, Salem NM, Al-Atabany W (2019) Multi-classification of brain tumor images using deep neural network. IEEE 7
15. Suresh S, Mohan S (2019) NROI based feature learning for automated tumor stage classification of pulmonary lung nodules using deep convolutional neural networks. J King Saud Univ Comput Inf Sci
16. Bal A, Banerjee M, Chakrabarti A, Sharma P (2019) MRI brain tumor segmentation and analysis using rough-fuzzy C-means and shape based properties. J King Saud Univ Comput Inf Sci
17. Maruthamuthu A, Priya GGL (2020) Brain tumour segmentation from MRI using superpixels based spectral clustering. J King Saud Univ Comput Inf Sci 32(10):1182-1193. ISSN 1319-1578
18. Shaoguo C, Mao L, Jiang J, Liu C, Xiong S (2018) Automatic semantic segmentation of brain gliomas from MRI images using a deep cascaded neural network. J Healthcare Eng
19. Sobhaninia Z, Rezaei S, Noroozi A, Ahmadi M, Zarrabi H, Karimi N, Emami A, Samavi S (2018) Brain tumor segmentation using deep learning by type specific sorting of images. ArXiv preprint arXiv: 1809.07786
20. Sudhakar T, Bethanney Janney J, Haritha D, Juliet Sahaya M, Parvathy V (2019) Automatic detection and classification of brain tumor using image processing techniques. Res J Pharm Technol. https://doi.org/10.5958/0974-360X.2017.00669.2
21. Joseph RP, Senthil Singh C, Manikandan M (2019) Brain tumor MRI image segmentation and detection in image processing. Int J Res Eng Technol

Early-Onset Identification of Stomach Cancer Using CNN

Susmitha Valli Gogula, Y. Vijayalata, H. Satya Akash, D. Thrinesh, M. Nagaraju, and Md. Mubasheer Uddin

Abstract Due to its ability to hide for a long time, Stomach cancer is considered to be difficult to find cancer among all other cancers that are present. It's a progressing case of cancer within our World. The most common method used across the world for the diagnosis of gastric cancer is Endoscopy. Endoscopy diagnosis could be a very specific and sensitive method. With high-resolution endoscopy, it's possible to detect mild discolorations, bulges, and structural irregularities over the surface of the Mucosa (a membrane in the stomach). However, due to the fact that procedures are performed in the presence of a doctor, it's possible that the cancerous areas are also missed and/or incompletely detected. Because the cancerous area can't be detected completely may result in the matter of cancer recurrence after a certain period of surgical intervention. So, in order to overcome this problem, a Computerized Decision Support System (CDS) is being implemented. Here, we are implementing a convolutional neural network (CNN) algorithm for us to spot the Stomach Cancer and classify it as either malignant or benign. This algorithm works as an assistant to gastroenterology doctors, helping them to spot the cancerous area within the endoscopic images of the scaffold, so as to require biopsies from these areas and to create a more vigorous diagnosis. We believe that the Gastric cancer identification plays a helpful role in determining the Cancerous area using the biopsy samples that are taken from the patient.

Keywords Benign · Biopsy · Convolutional neural network · Endoscopy · Gastric cancer · Malignant

S. V. Gogula (✉) · H. Satya Akash · D. Thrinesh · M. Nagaraju · Md. Mubasheer Uddin
Department of CSE, Gokaraju Rangaraju Institute of Engineering and Technology, Hyderabad, Telangana 500090, India
e-mail: susmitagv@gmail.com

Y. Vijayalata
Department of Computer Science and Engineering, KG Reddy College of Engineering & Technology, Hyderabad, India

V. K. Gunjan and J. M. Zurada (eds.), *Proceedings of 3rd International Conference on Recent Trends in Machine Learning, IoT, Smart Cities and Applications*, Lecture Notes in Networks and Systems 540, https://doi.org/10.1007/978-981-19-6088-8_11

1 Introduction

Cancer is considered to be the most problematic aspect for human life. Early identification and treatment of cancer is very crucial for the understanding of it because it has a high mortality rate compared to other diseases. Diagnosing the cancer and treating it is quite expensive with respect to the type of cancer [1]. According to the statistics from Ministry of Health, Stomach cancer is the second most common type of cancers in our country [2]. Gastric cancer is the fourth and seventh most common type of cancers to be found in men and women, respectively, across the world [3]. According to the data provide by the "GLOBOCAN", Gastric cancer is the third leading cause for cancer deaths throughout the world, followed by lungs and colorectal cancer in overall mortality rate. Therefore, Early detection, diagnosis and treatment of Gastric cancer is essential, but there may be a possibility that the Early-stage Gastric cancer is overlooked by the Physicians and Doctors. Certain cases also have reported that the false negative rate for Gastrointestinal cancer with endoscopy is ranging from 4.6 to 25.8% [4].

Certain genetic alterations and specific environmental conditions merge together resulting in Gastric cancer. Recent studies have also concluded that tobacco smoking increases the threat of Gastric Cancer. In order to identify this life threatening disease, Endoscopy reports utilized with Artificial Intelligence (here CNN) results in early identification of Gastric Cancer. MRI and CT hugely benefit and also enhance from their respective value to the diagnostic process, which in turn allows for increase in the precision and affirmation. Finally, the actual pathological diagnosis has shown guarantee that it can be majorly benefited from AI [5, 6, 7].

Since, Endoscopy is the strongest tool for detecting and diagnosing of Gastric Cancer, the accuracy in terms of identifying the cancer also plays an important role and it depends on the endoscopists and is made complicated by various factors of the Gastrointestinal (GI) tract [8, 9, 10].

Hirasawa et al. [11] developed an AI-based diagnostic system that was trained by >13,000 images of EGD to detect early and advanced Gastric Cancer. The developed CNN model had an overall sensitivity of 92.2% and a positive predictive value (PPV) of 30.6%. Seventy out of 71 lesions were correctly detected [11].

Recently, Artificial Intelligence (AI) based on deep learning through convolutional neural networks (CNNs) has made ground-breaking progress in various fields of study, including medicine [8, 12, 13]. CNN is widely known for its ability of image classification and recognition. Thus, we can implement a computer-aided decision support system using CNN for guiding doctors which in turn reduces the false negative rate in detecting Gastric cancer through endoscopy. Detecting cancers using AI has shown significant results in predicting cancers with high fidelity [14]. As an AI algorithm that automatically learns features from the data, the utilization of CNN is primarily for image recognition [15]. Recently, AI has also being applied in the detection, localization and segmentation of colonic polyps [16].

2 Literature Survey

Convolutional neural networks (CNN) is one of the majorly used deep learning techniques for image recognition process. It is a popular deep learning method for image recognition proposed by Szegedy et al. [8].

The accuracy provided by the CNN system is far greater than what is done and achieved manually. When a CNN-based diagnostic system is implemented [17], trained (over 14,000 images) and tested on approximately 2300 endoscopic images of gastric cancer, it correctly diagnosed 71 out of 77 gastric cancer lesions within the span of less than a minute, completing the process with an overall sensitivity of approx. 93%. By this, it was concluded that, CNN could evaluate and process a vast number of images within a short span of time, put forward by Hirasawa et al. [14]. Training a deep convolutional neural network (CNN) system based on a single-shot multi box detector, using 5360WCE images of lesions, resulted in correctly detecting the area under the curve of lesions with a confidence interval of 95% and accuracy of CNN system is 90%. Thereby, this system helped reduce oversights and burden on the physicians Tsuboi et al. [18].

This deep learning technique, CNN, has been active in the medical field over the past few years, considering the technological growth. Gastric cancer is considered to be one of the dangerous health hazards at present time, and finding the cancer through image identification has been less accurate lately since the number of images that need to be studied are large in number. So, when tested the endoscopy reports using CNN and by endoscopists (manually), the accuracy produced in terms of sensitivity, PPV and NPV, CNN performed more precisely compared to the manual work done by the endoscopists, put forward by Yasar et al. [11].

It's also been shown that Artificial Intelligence (AI) trained with endoscopic images could detect Gastric cancer accurately. CNN which is a deep learning method is used for learning optimum features from large amount of training datasets automatically.

Furthermore, the use of Graphical Processing Units (GPUs) has decreased the training time of deep learning methods. Three crucial characteristics make the convolutional neural networks different from other neural networks.

1. Sparse Connectivity
2. Shared Weights
3. Pooling (a form of down sampling)

Though there are other classified methods like SVM, ANN, etc., CNN produces more accurate results compared to these methodologies. The SVM classifier is also experimented with Linear, RBF and Polynomial Kernels. Even though the evaluation is done using three SVM types, CNN has higher overall accuracy, [19].

3 Methodology

3.1 *Flow*

The user interface is a website where the image in which cancer must be identified is uploaded and image uploaded is checked whether it is cancerous or not through CNN algorithm. Features are extracted from the image and lesions are identified and gastric cancer is detected. After the image is identified the output is shown in the webpage where the lesion is highlighted and cancerous area is identified, if any.

3.2 *Detailed Architecture*

3.2.1 Image Acquisition

The images are collected using endoscopic equipment at Vestre Viken Health trust (VV) in Norway. The VV consists of 4 hospitals and provides health care to 47,000 people. We have collected the images from Simula open datasets website [14]. The image data set is annotated and verified by medical doctors and specialized physicians. The images are with different resolutions ranging from 720 * 576 to 1920 * 1076 pixels. The image dataset consists of endoscopic reports both healthy persons and cancer affected patients [20, 21].

3.2.2 Image Preprocessing

As the images are of different resolution first the images are resized into a fixed resolution. As we need to extract the features such as lesion size, number of lesions from the images we need to have a fixed resolution. Hence the images are resized into 400 * 350 pixels.

The images are then converted into grayscale as it is easier to process the grayscale images as they have less complexity and are easier to process. The image is converted into grayscale by taking the average of the RGB values.

$$\text{Grayscale} = R/3 + G/3 + B/3 \tag{1}$$

We can take the combined average of RGB values but there is a chance that it may cause an overflow error, so we took the separate values and added them [11].

Then the median filter is applied on the images to reduce the noise. Median filter is effective because it reduces the noise without softening the image. Median filter is the most effective filter for reducing Salt and Pepper noise. We cannot use Gaussian filter because it blurs the image and we cannot identify the lesions as the image is blurred. Median filter is more effective because it reduces the noise while preserving

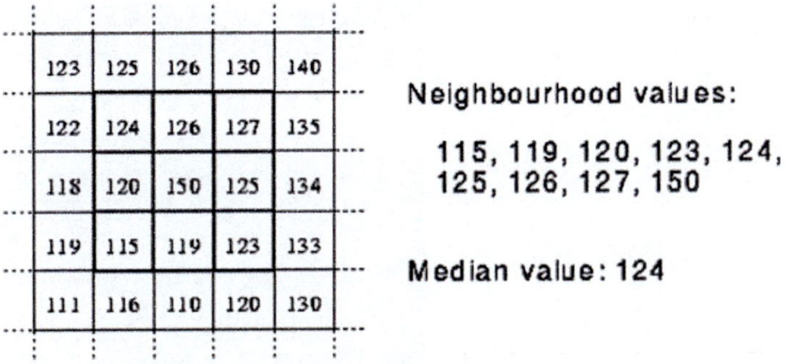

Fig. 1 2-Dimensional pixel matrix

Fig. 2 Original image

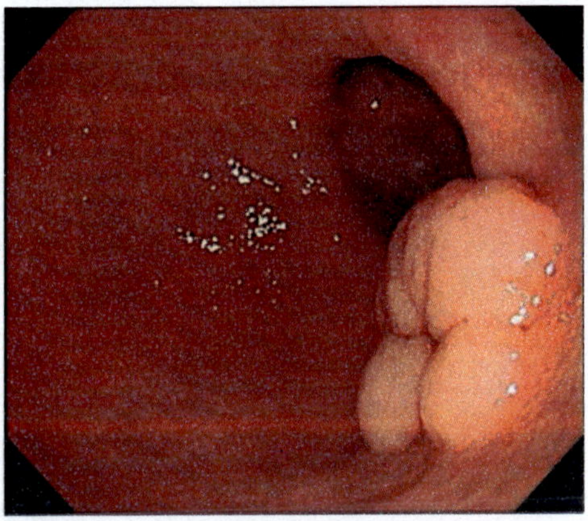

the edges which is very essential for lesion recognition. The median filter considers every pixel value and checks whether it is a representative of its neighbors or not and replaces the value by median of the surrounding pixels. The median is calculated by sorting all the neighboring pixels and finding the center or middle value of the all values (if there are even number of neighbors average of middle values is considered) [19].

Example: Consider a 2-Dimensional image with pixel values as follows shown in Fig. 1, consider 3 * 3 matrix as neighbors (Figs. 2, 3, 4 and 5).

3.2.3 Feature Extraction

We have extracted a total of 13 features from the image based on which the cancer is identified. The features extracted are:

Mean: Mean value is the average value of all the pixels in an image or can simply be said as the mean or average intensity of the image.

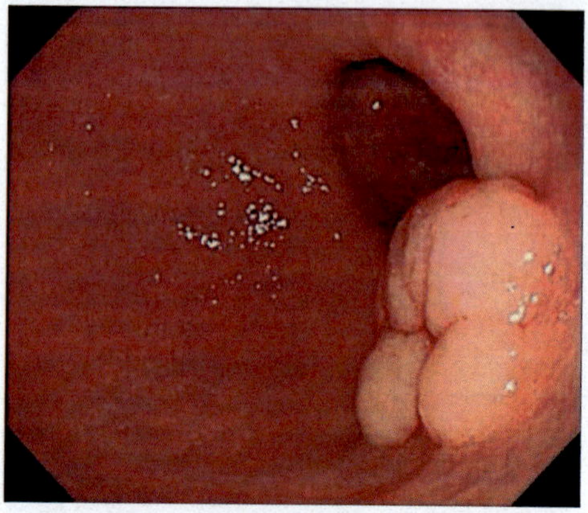

Fig. 3 Resized image (400 * 350 pixels)

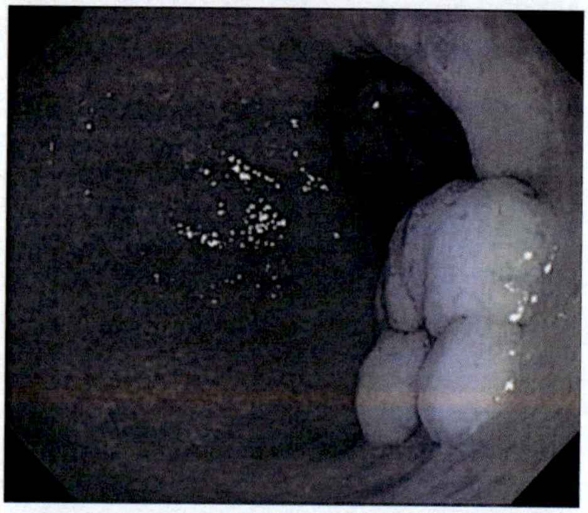

Fig. 4 Grayscale image

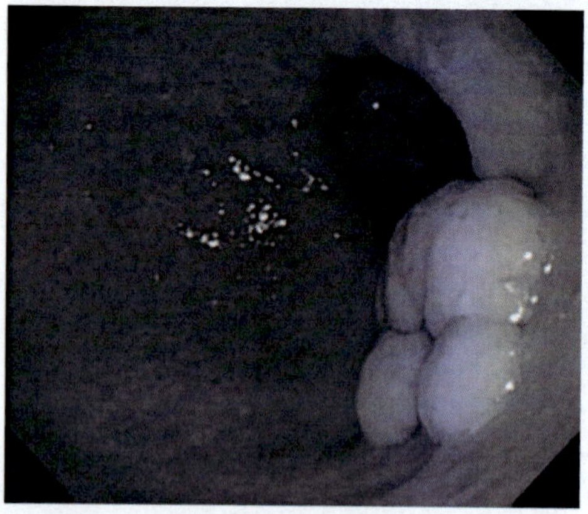

Fig. 5 Filtered image

$$\text{Mean} = \text{Sum of all elements/no of elements.} \tag{2}$$

Standard Deviation: Standard deviation is the measure of the amount of deviation of elements from the mean value of elements.

$$\sigma = \surd\left[\sum (x - \text{mean})^2 / N\right] \tag{3}$$

Entropy: Entropy is a measure of the degree of randomness of an image.

$$\text{Entropy} = \text{sum(p.} * \text{log2(p))}, \text{ where p is probability of each pixel.} \tag{4}$$

RMS: RMS stands for root mean square average. It is the arithmetic mean of the squares of all the values in the pixel matrix.

Variance: Variance is simply a measure of variability. It is calculated by taking the average of squared deviations from the mean. Variance is the square of standard deviation.

$$\text{Variance} = \sum (x - \text{mean})^2 / N \tag{5}$$

Smoothness: Smoothness is a measure of uniformity over an area in an image. It is high for areas with uniform pixel values.

$$\text{Smoothness } = 1 - (1/(1 + a)); \tag{6}$$

$$\text{a} = \text{sum(double(}G(:))), \text{ where } g \text{ is the pixel matrix} \tag{7}$$

Kurtosis: Kurtosis is a measure of whether a pixel or an area is peaked or flat when compared to the normal distribution.

Skewness: Skewness can be simply explained as a measure of symmetry. It is effectively the measure of lack of symmetry.

IDM: IDM stands for Inverse Difference Movement. It Measures the local Homogeneity of an image.

Contrast: Contrast is a measure of the intensity contrast between a pixel and its neighbor over the image. Contrast is also known as a variance.

Correlation: Correlation is a measure of how related a pixel is to its neighbor or how dependent a pixel is on its neighbor.

Energy: Energy returns the sum of squared elements in the GLCM. It is also known as uniformity.

Homogeneity: It is a measure of closeness of the distribution of elements in GLCM to the GLCM diagonal.

Fig. 6 Segmented image

3.2.4 Lesion Recognition

We are then identifying the lesion or tumor using segmentation. The image is segmented and features are then extracted. We have trained our model using this labeled data and then the model can detect the lesion and identify whether the cancer is malignant (harm causing) or benign (harmless like an ulcer). Also, the features such as mean, skewness, standard deviation, smoothness, are calculated (Fig. 6).

4 Results

We have developed our Stomach cancer identification model. The accuracy of the model is about 90% which is better than the identification using Statistical Region Merging method which is 80–85%. We have found the accuracy of the model while using different kernels to identify cancer. The kernels used are Linear kernel, Quadratic kernel, Polynomial kernel, RBF kernel. We have observed that the Radical Bias Function kernel gives more accuracy than other kernels because RBF is more flexible than others so it can easily incorporate outliers and identify cancer correctly. We have calculated the accuracy by iterating the same image for 100 times and based on the results accuracy is calculated as no of times correctly identified/total no of iterations (Fig. 7).

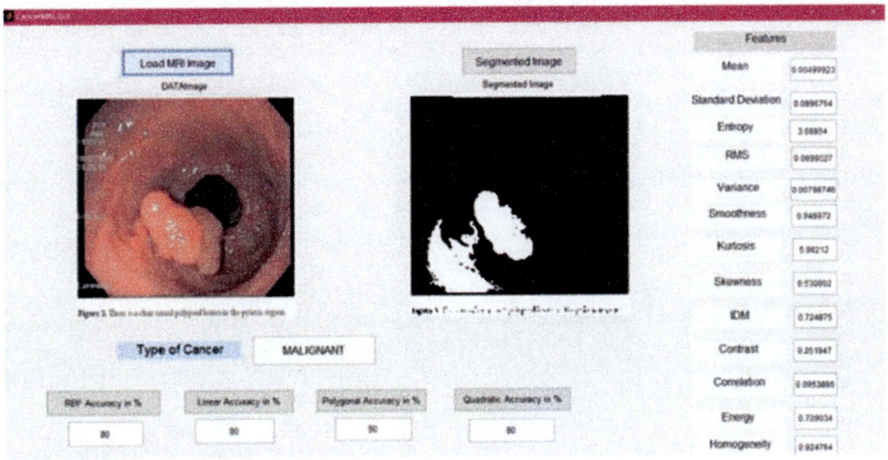

Fig. 7 Output screen

5 Conclusion

We have developed a Computerized Decision Support System (CDS) that can aid the physicians and doctors to easily and accurately detect the Gastric cancer at an early stage, which is proven to be very useful. This system is developed using CNN that can identify the lesions in the endoscopic images and can detect the Gastric cancer at an early stage. The accuracy of this model is improved with training and the final model is precise and reliable. We can say from the obtained results that this method can be very useful for endoscopic examination of cancers.

6 Future Scope

Currently our model is only limited to gastric cancer but in future we could develop this model such that all types of cancers can be detected using this Computerized Decision Support System (CDS) where the endoscopic reports are used to detect the cancers.

References

1. Alpert MA et al (1997) Cardiac morphology and left ventricular function in normotensive morbidly obese patients with and without congestive heart failure, and effect of weight loss. Am J Cardiol 80(6):736–740
2. Ural B et al (2016) Gastric cancer regional detection system. J Med Syst 40(1):1–9

3. Yu C, Helwig EJ (2021) Artificial intelligence in gastric cancer: a translational narrative review. Ann Translat Med 9(3)
4. Bray F et al (2018) Global cancer statistics 2018: GLOBOCAN estimates of incidence and mortality worldwide for 36 cancers in 185 countries. CA: A Cancer J Clinicians 68(6):394–424
5. Huang S et al (2020) Artificial intelligence in cancer diagnosis and prognosis: opportunities and challenges. Cancer Lett 471:61–71
6. Deo RC (2015) Machine learning in medicine. Circulation 132(20):1920–1930
7. Wong D, Yip S (2018) Machine learning classifies cancer 446–447
8. Voutilainen ME, Juhola MT (2005) Evaluation of the diagnostic accuracy of gastroscopy to detect gastric tumours: clinicopathological features and prognosis of patients with gastric cancer missed on endoscopy. Eur J Gastroenterol Hepatol 17(12):1345–1349
9. Menon S, Trudgill N (2014) How commonly is upper gastrointestinal cancer missed at endoscopy? A meta-analysis. Endosc Int open 2(02):E46–E50
10. Hosokawa O et al (2007) Difference in accuracy between gastroscopy and colonoscopy for detection of cancer. Hepato-Gastroenterology 54(74):442–444
11. Hirasawa T et al (2018) Application of artificial intelligence using a convolutional neural network for detecting gastric cancer in endoscopic images. Gastric Cancer 21(4):653–660
12. Ueyama H et al (2021) Application of artificial intelligence using a convolutional neural network for diagnosis of early gastric cancer based on magnifying endoscopy with narrow-band imaging. J Gastroenterology Hepatology 36(2):482–489
13. Ikenoyama Y et al (2021) Detecting early gastric cancer: comparison between the diagnostic ability of convolutional neural networks and endoscopists. Digestive Endosc 33(1):141–150
14. Szegedy C et al (2015) Going deeper with convolutions. In: Proceedings of The IEEE conference on computer vision and pattern recognition
15. Xia J et al (2021) Use of artificial intelligence for detection of gastric lesions by magnetically controlled capsule endoscopy. Gastrointest Endosc 93(1):133–139
16. Wang K-W, Dong M (2020) Potential applications of artificial intelligence in colorectal polyps and cancer: recent advances and prospects. World J Gastroenterol 26(34):509
17. Saito H et al (2020) Automatic detection and classification of protruding lesions in wireless capsule endoscopy images based on a deep convolutional neural network. Gastrointest Endosc 92(1):144–151
18. Tsuboi A et al (2020) Artificial intelligence using a convolutional neural network for automatic detection of small-bowel angioectasia in capsule endoscopy images. Digestive Endosc 32(3):382–390
19. Yasar A, Saritas I, Korkmaz H (2019) Computer-aided diagnosis system for detection of stomach cancer with image processing techniques. J Med Syst 43(4):1–11
20. Hasan M et al (2019) Comparative analysis of SVM, ANN and CNN for classifying vegetation species using hyperspectral thermal infrared data. In: International archives of the photogrammetry, remote sensing and spatial information sciences
21. https://datasets.simula.no/kvasir/
22. https://www.dynamsoft.com/blog/insights/image-processing/image-processing-101-color-spaceconversion/
23. Boyle R, Thomas R (1998) Computer vision: a first course

Multifold Secured Bank Application Authentication Service Using Random Visual Cryptography and Multimodal Steganography with Blockchain Technology

S. Jahnavi and C. Nandini

Abstract DATA protection is becoming increasingly critical as the demand for online services grows, which biometrics authentication can help with. To communicate biometric data authenticating services over the Internet in a secure channel is essential. To ensure the security and the existence of the biometric information over the communication channel Random Visual Cryptography and Multimodal Mask Steganography is used. The biometric features captured are embedded using multimodal mask steganography, considering mask image and cover image. The resultant stego image shares are created using Random visual cryptography technique. The resultant shares are placed in blockchain using blockchain technology to improve security aspects of the cloud application and reliable transmission of data. Combining proposed Random Visual cryptography and multimodal mask steganography methods with blockchain technology ensures high reliable authentication system of server room access or banking application.

Keywords Random visual cryptography · Multimodal · Steganography · Biometric features · Blockchain · Authentication

1 Introduction

In present situation most of the services are made online for the easy access to user. Users can access services online using an application. During this process user has to authenticate himself to access services available. The authentication process involves communication of biometric information, which is communicated over untrusted network. The biometric data can be hacked that leads to financial, business or personal threat. Hence there is a need of security mechanism to communicate biometric data

S. Jahnavi (✉) · C. Nandini
Computer Science and Engineering Department, Dayananda Sagar Academy of Technology and Management, Visvesvaraya Technological University, Bangalore, Karnataka 560082, India
e-mail: jahnavishankar.s@gmail.com

C. Nandini
e-mail: hodcse@dsatm.edu.in

V. K. Gunjan and J. M. Zurada (eds.), *Proceedings of 3rd International Conference on Recent Trends in Machine Learning, IoT, Smart Cities and Applications*, Lecture Notes in Networks and Systems 540, https://doi.org/10.1007/978-981-19-6088-8_12

over the Internet during authentication. The traditional authentication ways include pin, secret code or password to be given, which is replaced with biometric (face and fingerprint) as credentials in the authentication. Pertaining to this Random Visual cryptography and steganography technique is used. The biometric data captured during authentication are embedded in the cover image using steganography method [1]. To the resultant data visual cryptography method [2] to applied to create shares, used in communicating secret or sensitive data over the communication channel during authentication process. Blockchain was first launched in 2008; it is a peer-to-peer network that lies on top of the Internet. It's a public ledger made up of a series of blocks that keeps track of the network's transaction history [3, 4]. Each block's header contains the hash of the previous block, forming a chain or connected list. A timestamp, that contains details when the block was created, and a nonce, which is an arbitrary number that miners can modify periodically to get a certain hash value to solve a cryptographic puzzle [5]. Blockchain is making remarkable footprint in various application [6, 7] like Health care, Banking, Forensic and many more.

Smart contracts are self-executing code on the blockchain platform which are used to express business logic. It's an essentially programs that execute when such conditions are fulfilled and are stored on a blockchain. It's a program that automates business logic or consensus.

2 Related Work

This section focuses on the most relevant research, which are organized through usage cases like electronic medical reports, remote patient surveillance, pharmacy supply chains and health insurance claims.

A. *Securing Images with Fingerprint Data using Steganography and Blockchain*
   The author suggests a basic system based on blockchain and steganography to improve the transparency and non- repudiation of photographs posted on social media [5]. It enables the user to confirm the identity of the individual who took the shot. It comprises four phases: user user's fingerprint as well as image is captured in first phase. Fingerprint features are embedded into captured image in the second phase. Hash of the encoded image is created and stored in the body of the block with the header, containing hash of the previous block to track of image captured by authentic user in the third phase. In fourth phase the image is decoded to validate the fingerprint stored in the device. Upon authenticating, image is stored in gallery and uploaded to social media. If not image is discarded. Message-Digest 5 (MD5) is used to calculate hash value. Blockchain non-mutable indexing property ensures non-repudiation.
B. *Broadcasting Steganography in the Blockchain*
   The author presents an outline [8] to avoid modifications in the stego image communicated via insecure channel. During the block generation process, secret

data is embedded into the transactions within the block. To keep the data embedding mechanism within a block stable, we select a subset of transactions in a block based on a secret key and embed the secret data using a repeatable-address arrangement. Extracting the embedded data is impossible for an attacker. From the transaction pool, the miner picks n transactions. Then to the new block, transactions are placed in an arbitrary order. Then, using a key, he selects m transactions from the block and rearranges them so that the transaction list corresponds to the secret data to be embedded. After the stego block has been chained, a receiver could use the key to retrieve the secret knowledge from the stego transactions in the stego block.

C. *Joint Transaction-Image Steganography for high capacity covert communication*
The Author presents a framework [9] where the hint or manual (consist of partition number, image url, access time) to access the information hidden in a stegano image is placed in senders transaction block. If the manual pattern matches the transaction history, it is considered as stego block. Based on access time (part of hint) the stego image is swapped with normal image in cloud server. Since hacker can access the manual in transaction as long as till the network is alive. The receiver downloads the image to extract security part and content part of stego image. The Image is encrypted using RSA-AES method; the key with offset is stored in security part. The content part is trimmed using offset and using AES the real image is retrieved. Since the transaction pattern has to match the manual (hint) it consumes time to search and the swapping stego image with normal image in cloud based on access time could be used as a weak link to tamper data.

D. *Provably secure covert communication on blockchain*
In this paper [10] The sender sends one bit of data to blockchain by placing it in the least significant bit (lsb) of the transaction address till all the bits of data is sent, means one bit each transaction. So sender has to wait for the next transaction to send every bit. It is time consuming process, which is not an easy task to bring in real time application. The sender should be aware of receiver address to send a bit, hence shorter the address it is easy to track and maintain by the receiver.

3 Multimodal Mask Steganography and Random Visual Cryptography

The proposed multimodal mask steganography system [2] uses a mask image in the encryption process. Initially biometric data of a person are captured and the features are existing system. But it was more vulnerable to attack even with a minor changes made to a cover picture [2]. The proposed mask multimodal steganography overcomes this flaw and provides better security. Initially the features are extracted using Rotational bit operator method [2], the cover image is divided into 4 shares/sub bands. Then the extracted features are embedded in the cover image using mask multimodal steganography. Mask image can also be called as magic sheet.

In banking application, the captured face and fingerprint features are extracted. The cover image is divided into 4 sub bands. Considering a cover image, mask image generated, using mask steganography technique the features are embedded in four sub bands as shown in Fig. 1. The resulted stego image of sub bands are given as input to random VC technique to create colorful clueless shares. And thus those shares are transmitted over the Internet to reach the server for authentication.

In the server end the reverse process (decryption) is preformed, where the Random VC [1] is applied to all the shares to retrieve sub bands. The sub bands are then reunited to get a stego image as shown in Fig. 2. Then multimodal mask steganography [2] is applied in the decryption process considering stego image, mask image to extract the features of face and biometric yielding results as shown in Table 1. Thus fingerprint minutiae features, Facial biometric features are extracted using Viola–Jones and RBO algorithm. These extracted features are embedded using Mask Steganography using multimodal approach. And the resultant stego image shares are created, using Random Visual Cryptography technique.

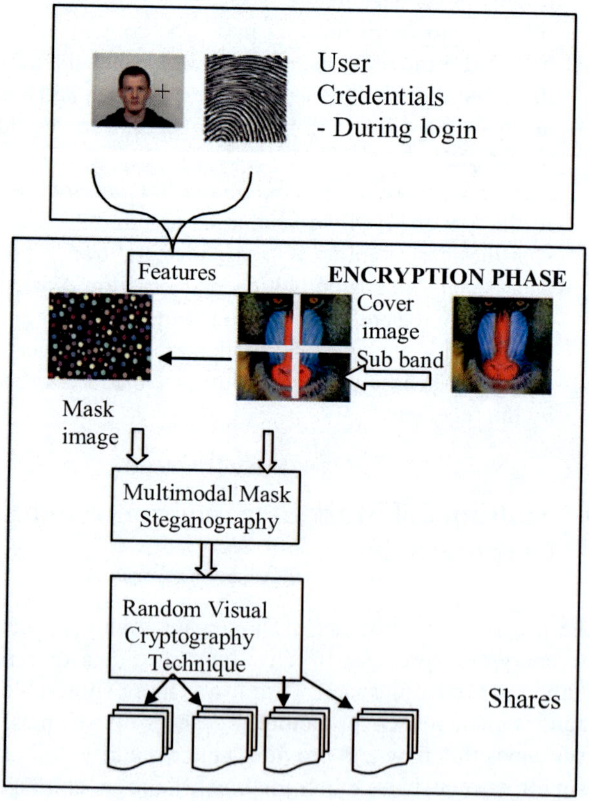

Fig. 1 Encryption phase of online banking authentication process

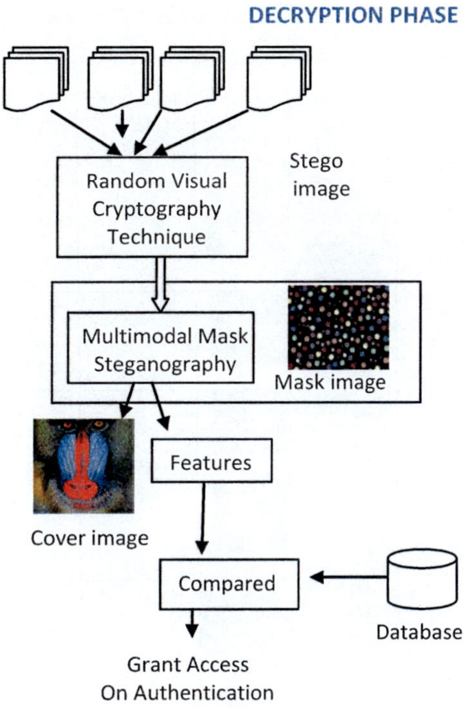

Fig. 2 Decryption at server side

The experiments are carried out with varying cover image, Fingerprint and face biometric data, recording Peak Signal to Noise Ratio (PSNR), Structural Similarity Index Measure (SSIM), Correlation coefficiènt (CC) and Mean Square Error (MSE) in the Table 1.

4 Multifold Security Using Blockchain Technology

Combining proposed authentication process with blockchain technology ensures high reliable authentication system of server room access or banking application. Blockchain creation process are shown in Fig. 5, where on receiving the information using Hashing Technique SHA256 root hash (RH) is generated. The block body is created by encrypting the Information using RSA public key, where the key if further encrypted using DES as shown in Fig. 3. The previous block root hash is fetched to create block header. Along with previous block root hash, Current block hash value, time stamp and 4bit random number called nonce is used to generate block header. The block body is compressed using compression technique and merged with block head to generate block as shown in Fig. 4. Then the block is stored in the blockchain.

Table 1 PSNR, SSIM, MSE and CC of multimodal mask steganography and random visual cryptography experiment

Cover image (KB)	Image to be hidden (KB)	Fingerprint data (KB)	PSNR	SSIM	MSE	CC
 Jeppe 26.8 800*600	 2.62	41.4	∞	1	0	1
	 2.92	43.3	∞	1	0	1
	 2.45	43.7	∞	1	0	1
	 2.66	43.1	∞	1	0	1
 Andreas 32.4 800*600	 2.62	37.4	∞	1	0	1
	 2.92	34.1	∞	1	0	1
	 2.45	35.7	∞	1	0	1
	 2.66	45.3	∞	1	0	1
 David 39.9 800*600	 2.62	44.2	∞	1	0	1
	 2.92	36.0	∞	1	0	1
	 2.45	33.6	∞	1	0	1
	 2.66	38.8	∞	1	0	1

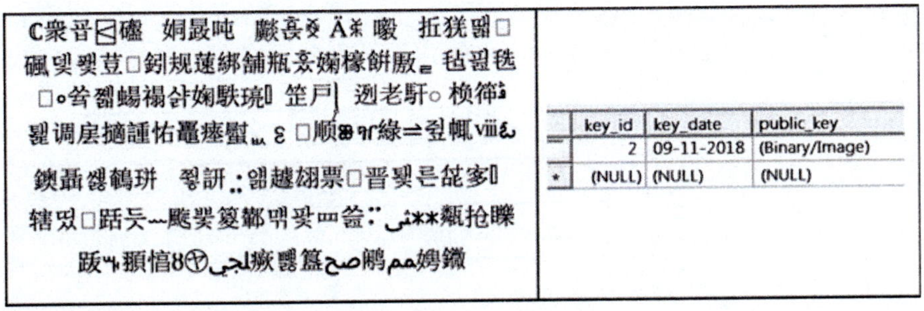

Fig. 3 Encrypted secret key and public key

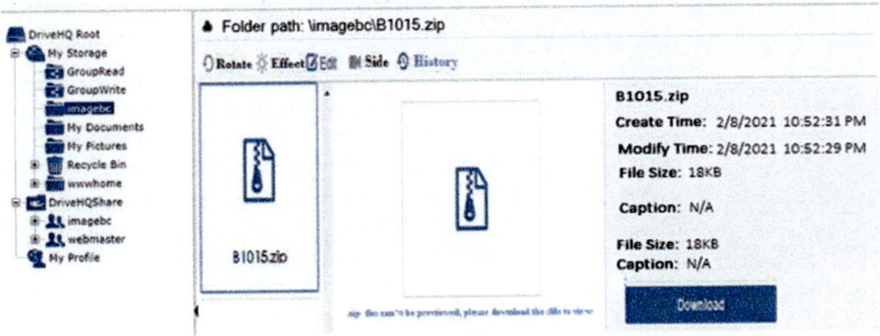

Fig. 4 Block body together with header compressed and stored in blockchain application

Consider a traditional bank application process, the person register him in the bank producing required document to open an account along with aadhar number and biometric data like face and fingerprint. The user biometric data along with fingerprint are stored in the database for further online authentication process. When user tries to access the bank service by providing biometric credentials the request will be directed to server to provide/reject access.

Block header and body is created using following steps:

Block Header

Step 1: Generate hash code for each and every file or information using Hashing technique.
Step 2: Generating random value (nonce).
Step 3: Taking the time stamp (Current data and time of the system).
Step 4: If the block is first block (Genesis) store all these values (CurrentHashValue + TimeStamp + Nonce) in a string.
Step 5: Write that value inside the confidential file (C_blockname.txt).

Fig. 5 Block creation process

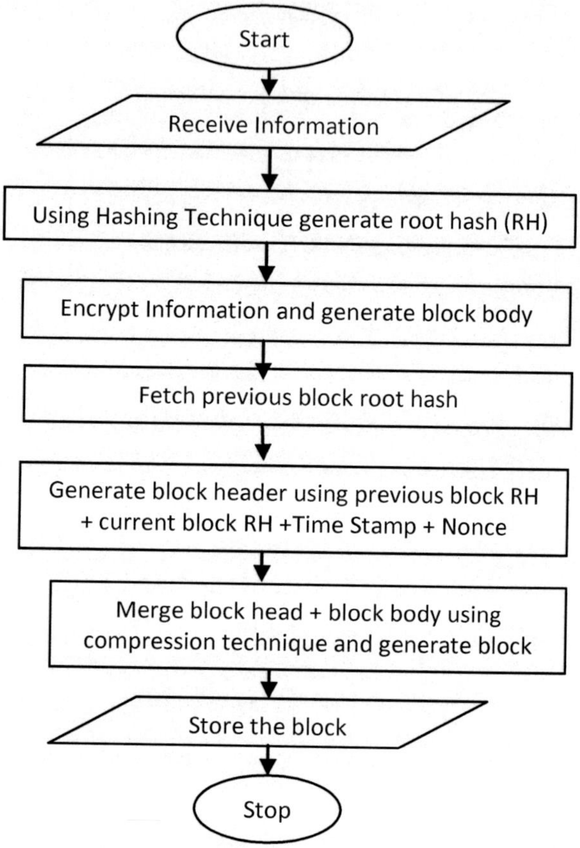

Step 6: If block is not the first block then take the previous block hash value and store all the contents (Previous Hash value + TimeStamp + Nonce + CurrentHashValue) inside String value.
Step 7: Write details into confidential file.

Block Body:

Step 1: Encrypt the Information.
Step 2: Add Encrypted Information to.zip file.
Step 3: To that.zip file add confidential file also.

Than just transferring the encrypted data over the Internet it will be stored as blocks in the blockchain with their aadhar number last digit as block numbers, shown in Fig. 5. So the authorized or intended server will access the encrypted data from the blockchain.

Then the server decrypts and extracts face, fingerprint features to authenticate user. On successful authentication the server grants permission to access the services in the application. So user can perform normal online bank activities as shown in Fig. 6.

The server details are enrolled in the blockchain application to access blocks as shown in Fig. 7. On registering, the server will receive the private key to the registered mail id. The key will be encrypted using DES as shown in Fig. 3. This is decrypted and applied while accessing the blocks. Admin can add, modify and delete server details. Admin balance the load by adding new server details with location as shown in Fig. 8.

User on requesting access to online bank application, user details are added as shown in Fig. 9 to the blockchain application. To the added user details encrypted shares are uploaded to the cloud as shown in Fig. 10. The Admin can add, delete user details who requested bank application service as shown in Fig. 9. While adding user details unique Aadhar number is recorded. The shares generated are uploaded by naming the file with last 4 digit of aadhar number as shown in Fig. 10. The shares are stored as blocks containing body and header part. Body contains Encrypted data using public key as shown in Fig. 11. Header contains previous hash value, Time stamp, Nonce value and Current Hash value as shown in Fig. 12. The hash value of file is calculated using SHA256. Even if shares are downloaded from the cloud it will be in an encrypted format. The server can access the blocks of intended user by giving aadhar number as shown in Figs. 13 and 14. The server downloads file by providing key information as shown in Fig. 15 to download the shares saved as blocks.

Server on accessing shares performs reverse operation by combining those shares using Random VC to get stego image. The server then extracts the features of face and fingerprint features embedded to compare with the bank database containing user details. Initially registered user face and fingerprint data is compared with the results to grant or deny online bank services. If the attacker breaks the servers credentials and make any changes in the block which is not an easy task. The hash value will be changed, since it differs from the hash value stored in next block, it gives alert to server while accessing the block as shown in Fig. 16, this process is called smart contract.

Hash value generated using SHA256 (hash tag) are stored in the block header as shown in Fig. 17. With varying size of block (MB), time (ms) taken to create block in cloud space is shown in Fig. 18. Time increases with increase in block size. On evaluating the proposed method and existing method the Table 2 shows that required transactions (MB or KB), highest capacity and capacity per transaction is depicted. The proposed system shows better performance than existing with good payload capacity.

Server response time is the amount of time that passes between a client requesting a page in a browser and a server responding to that request, which is calculated using Time to First Byte (TTFB). After making an HTTP request, the time it takes to obtain the first byte of a page is measured in milliseconds. The initial iteration of a first user often has some client-side overhead and can have some server-side overhead. This can

Fig. 6 Online banking application authentication service using random visual cryptography and multimodal steganography with blockchain technology

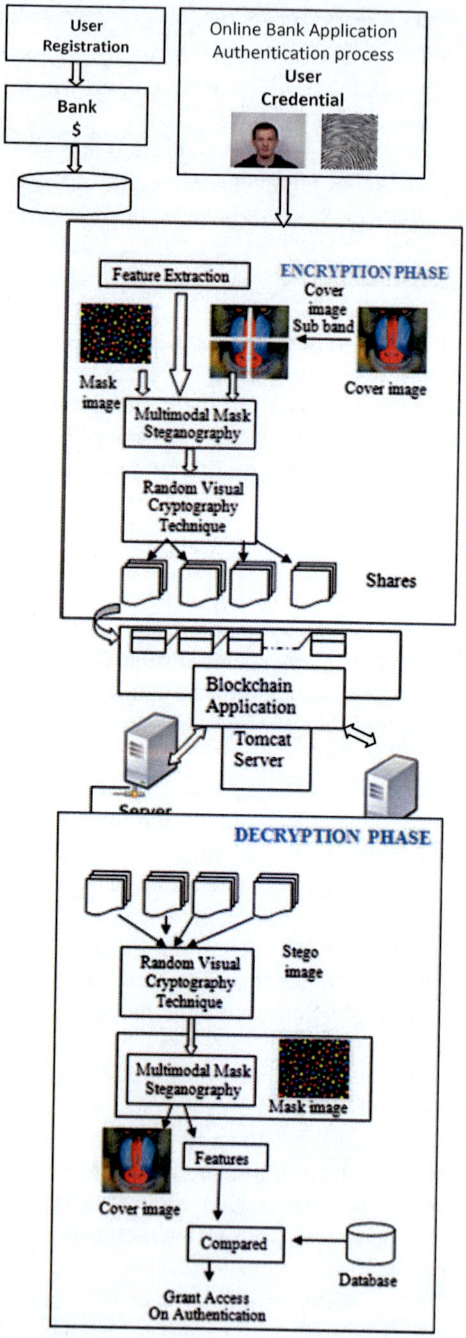

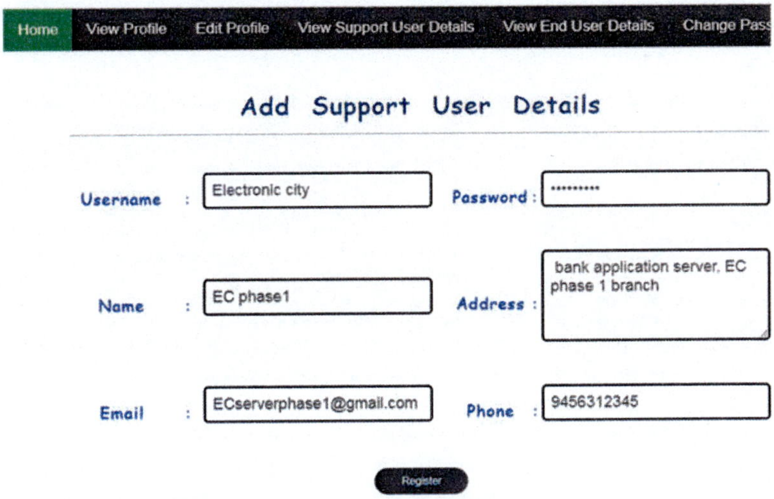

Fig. 7 Registering server details in application

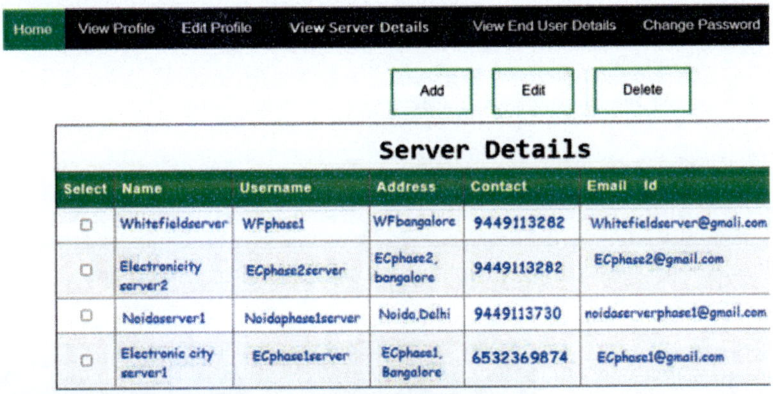

Fig. 8 Registered server details (to authenticate user)

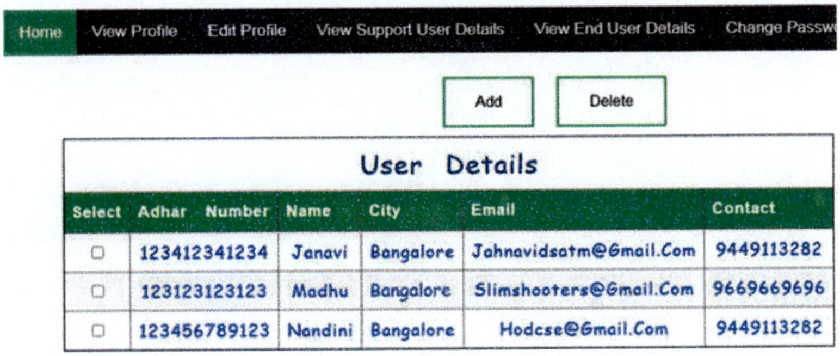

Fig. 9 List of user requested to access banking services

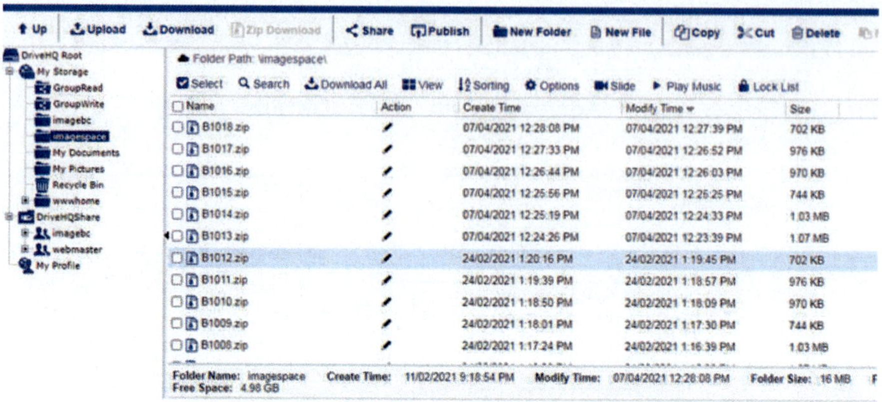

Fig. 10 Uploading Shares generated as blocks in the blockchain application

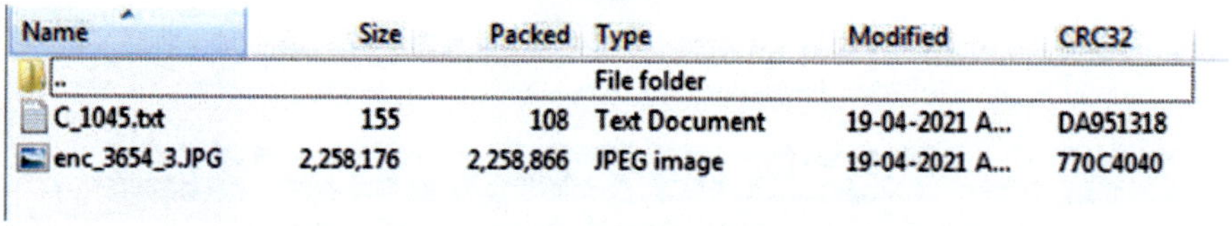

Fig. 11 Block header and encrypted user's share together stored as blocks

Fig. 12 Block Header consist of pervious block hash value, Time Stamp, Nonce and current block Hash value

Fig. 13 Server accessing user's biometric data stored as blocks by providing user's aadhar number

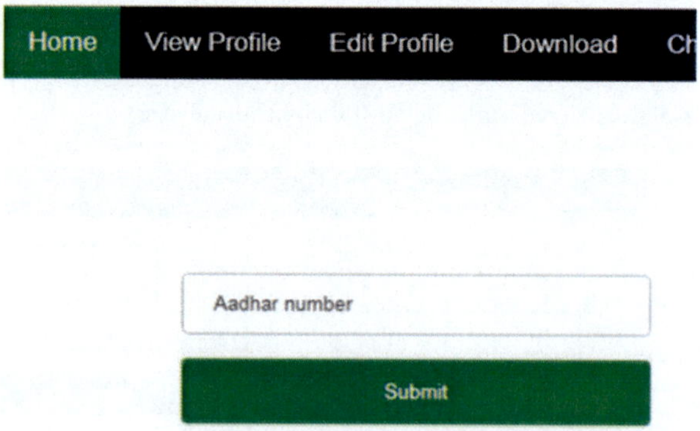

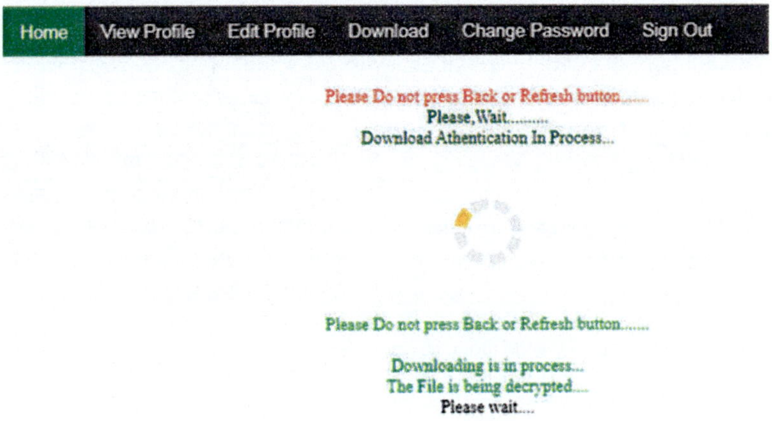

Fig. 14 Downloading user's data shares saved as blocks

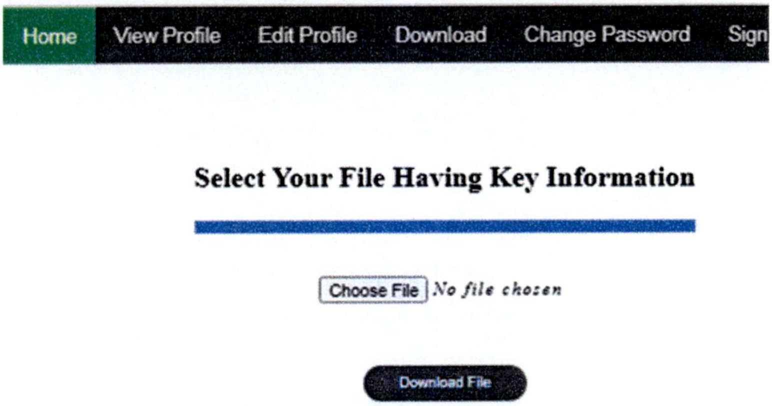

Fig. 15 Server providing private key to access block

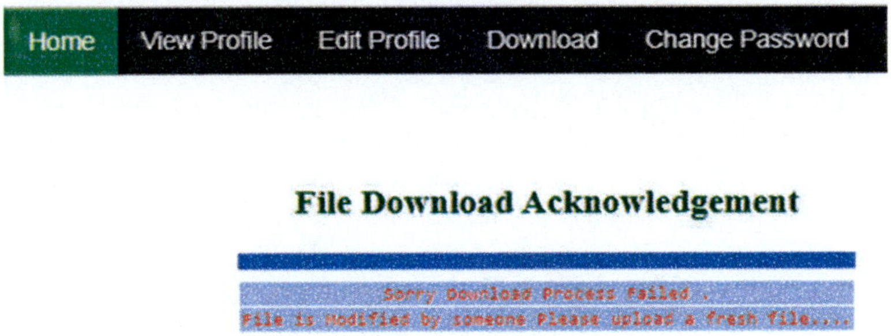

Fig. 16 Intimating server about change in block content: smart contract

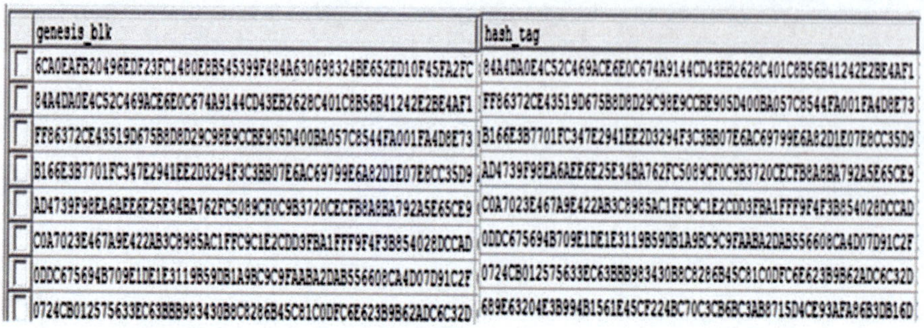

genesis_blk	hash_tag
6CA0EAFB20496EDF23FC1480E8B545399F484A630698324BE652ED10F45FA2FC	84A4DA0E4C52C469ACE6E0C674A9144CD43EB2628C401C8B56B41242E2BE4AF1
84A4DA0E4C52C469ACE6E0C674A9144CD43EB2628C401C8B56B41242E2BE4AF1	FF86372CE43519D675B8D8D29C98E9CCBE905D400BA057C8544FA001FA4D8E73
FF86372CE43519D675B8D8D29C98E9CCBE905D400BA057C8544FA001FA4D8E73	B166E3B7701FC347E2941EE2D3294F3C3BB07E6AC69799E6A82D1E07E8CC35D9
B166E3B7701FC347E2941EE2D3294F3C3BB07E6AC69799E6A82D1E07E8CC35D9	AD4739F98EA6AEE6E25E34BA762FC5089CF0C9B3720CECFB8A8BA792A5E65CE9
AD4739F98EA6AEE6E25E34BA762FC5089CF0C9B3720CECFB8A8BA792A5E65CE9	C0A7023E467A9E422AB3C8985AC1FFC9C1E2CDD3FBA1FFF9F4F3B854028DCCAD
C0A7023E467A9E422AB3C8985AC1FFC9C1E2CDD3FBA1FFF9F4F3B854028DCCAD	0DDC675694B709E1DE1E3119B59DB1A9BC9C9FAABA2DAB556608CA4D07D91C2F
0DDC675694B709E1DE1E3119B59DB1A9BC9C9FAABA2DAB556608CA4D07D91C2F	0724CB012575633EC63BBB983430B8C8286B45C81C0DFC6E623B9B62ADC6C32D
0724CB012575633EC63BBB983430B8C8286B45C81C0DFC6E623B9B62ADC6C32D	689E63204E3B994B1561E45CF224BC70C3CB6BC3AB8715D4CE93AFA86B3DB16D

Fig. 17 Hash value generated for the block information

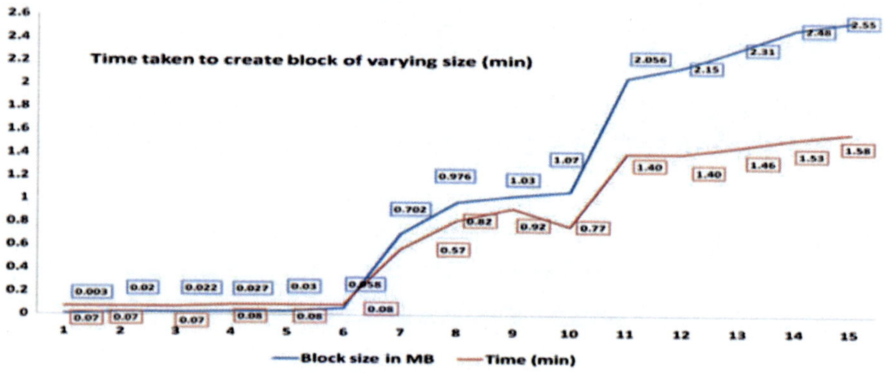

Fig. 18 Time (ms) taken to create block of varying size (MB)

Table 2 Comparison of required transactions (MB or KB), highest capacity and capacity per transaction of existing and proposed system

Method	Required transactions (1 KB)	Required transactions (1 MB)	Capacity/transaction	Highest capacity
Joint Transaction-Image Steganography for high capacity covert communication [9]	≤12	≤24	29 bits	759 KB
Proposed	1	1	KB–MB	Cover image

be overruled by subsequent iterations. As shown in Table 3 with increase in number of users the average response time initially increases and gets stabled as shown in Fig. 19. It is observed slight increase in transactions per second (TPS) with increase

Table 3 Performance Measure with load scenario

Virtual user (VU)	TPS (MB/s)	Average response time
20	18	763
30	22	231
50	23	120
100	24	119
750	18	123
200	25	108
250	27	91
300	42	85
350	41	83
400	43	81

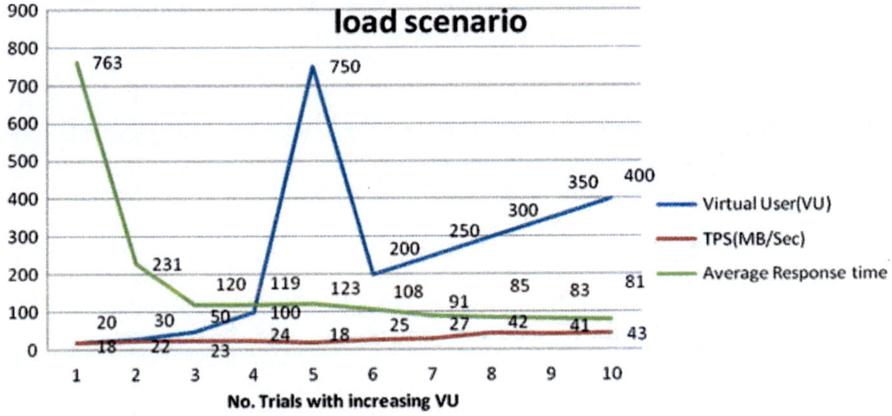

Fig. 19 Average response and TPS with increase in VU

in number of virtual users. The application is tested with system configuration of 4 GB RAM, 64 bit processor, I3 2.3 GHz.

Table 3 summarizes results obtained by conducting a performance test for a week. The servers are loaded with virtual users and analyzed the performance behavior of the application in handling multiple users. The application response time is affected by factors such as network bandwidth, number of users and number.

5 Conclusion

Communicating Biometric data securely over the Internet to authenticate services is very essential since the services are made online. Authenticating user to access application is an important aspect in banking sector. With regard to this the biometric features (face and finger print) are embedded in the cover image using multimodal mask steganography and mask image. The resultant stego image is passed to Random Visual cryptography technique to create colorful meaningless shares which are saved as blocks in the blockchain, than just passing information over the internet to reach server to authenticate user. The block contains shares, which are encrypted using RSA. The resultant is then combined with previous hash value, time stamp, nonce and previous block hash value are stored as single string in block.

The server access blocks by decrypting the key using DES and then applying the private key to get shares. The shares are then decrypted using Random Visual cryptography using to get stego image. The features of face and Fingerprint are then extracted using Multimodal Mask steganography. The extracted features are then compared with the bank database to authenticate user. Hence the biometric features using blockchain technology are tempered proof, since any changes in the block will be intimated to the server while accessing the block. On authenticating user, further he can access online bank application services. The application performance is tested with transactions per second (TPS), Average response time of an application with increasing the number of Virtual User (VU). And it observed the response time gets stabled with increase in VU and TPS increases with increase in VU. In future carrying out additional tests to evaluate its performance with even more scalable and realistic production environments can be exercised.

References

1. Jahnavi S, Nandini C (2019) A naive based share generation using random visual crypto technique. Int J Innovat Technol Explor Eng 8(12):5123–5129
2. Jahnavi S, Nandini C (2020) Novel multifold secured system by combining multimodal mask steganography and naive based random visual cryptography system for digital communication. J Comput Theoret Nanosci 17(12):5279–5295(17)
3. Nandini C, Jahnavi S (2020) Quantum cryptography and blockchain system: fast and secured Digital communication system. In: 4th international conference on intelligent computing and communication (ICICC 2020).
4. Knirsch F, Unterweger A, Engel D (2019) Implementing a blockchain from scratch: why, how, and what we learned. EURASIP J Info Secur 2019:2
5. S. Pramothini YVVS, Harini SPN (2019) Securing images with fingerprint data using steganography and blockchain. Int J Recent Technol Eng 7(4):82–85
6. Fekih RB, Lahami M (2020) Application of blockchain technology in healthcare: a comprehensive study. In: Jmaiel M (ed) Developing countries. ICOST 2020. Lecture Notes in Computer Science, vol 12157. Springer
7. Mokhtari M, Abdulrazak B, Aloulou H, Kallel S (eds) (2020) The impact of digital technologies on public health in developed and developing countries. In: 18th international conference, ICOST 2020, Hammamet, Tunisia
8. Xu M, Wu H, Feng G, Zhang X, Ding F (2019) Broadcasting steganography in the blockchain. In Wang H, Zhao X, Shi Y, Kim H, Piva A (eds) Digital forensics and watermarking. IWDW 2019. Springer

9. Basuki I, Rosiyadi D (2019) Joint transaction-image steganography for high capacity covert communication. In: International conference on computer, control, informatics and its applications (IC3INA), Tangerang, Indonesia
10. Partala J (2018) Provably secure covert communication on blockchain. Cryptography 2(3):18

Role of Blockchain in Health Care: A Comprehensive Study

Rashmi Pathak, Badal Soni, and Naresh Babu Muppalaneni

Abstract There has been a lot of interest in blockchain technology in a lot of different fields over the last decade, including finance, government, energy and health. This article gives an in-depth look at how blockchain can be used in the healthcare sector. In reality, research in this area is moving very quickly. People can now use blockchain-based technology for sharing remote patient monitoring, the supply chain for medicines, electronic medical records and so on. Under this paper, we want to show how blockchain can be used in the healthcare industry and how much research has been done in this field.

Keywords Blockchain · Health care · Blockchain use cases · Security · Blockchain applications

1 Introduction

Blockchain can be considered a distributed ledger in its most basic implementation. Distributed algorithms make transactions made on the blockchain nearly tamper-proof. Blockchain store ledgers in a peer-to-peer basis. The update would have to be applied to the majority of ledger holders, when it is implemented for a new transaction. In addition, this update must remove a proof-of-work notion that disables high-processing-power machines from rewriting the ledger history. An underlying principle for blockchain in its initial stages was that if records were distributed to a greater number of recipients, there would always be a ready-to-use source for ensuring the integrity and accuracy of a series of records. Until then, no actual blockchain-based utility had been built, and hence, the technology was also slower in developing. Bitcoin was introduced in 2009 via Satoshi Nakamoto's seminal white paper "Bitcoin: A Peer-to-Peer Electronic Cash System." According to Satoshi Nakamoto (Nakamoto, 2009), a currency existing only in a digital realm, based on a blockchain architecture, is what is meant by "existing strictly in a digital space." Every coin is totally

R. Pathak (✉) · B. Soni · N. B. Muppalaneni
National Institute of Technology Silchar, Silchar, Assam, India
e-mail: rashmi21_rs@cse.nits.ac.in

V. K. Gunjan and J. M. Zurada (eds.), *Proceedings of 3rd International Conference on Recent Trends in Machine Learning, IoT, Smart Cities and Applications*, Lecture Notes in Networks and Systems 540, https://doi.org/10.1007/978-981-19-6088-8_13

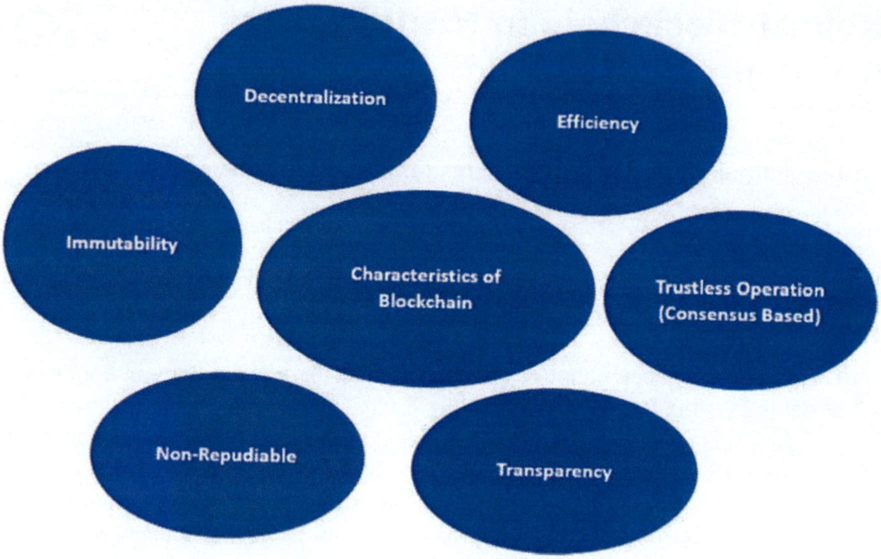

Fig. 1 Characteristics of blockchain

self-contained and can be confirmed by anyone who has previously kept a record of that currency [1]. Instead of being just another cryptocurrency, Bitcoin developed a reputation for itself, and this allowed for greater recognition of blockchain. Investigative agents began investigating the design, trying to find ways to implement it into other sectors. Blockchain technology was employed extensively throughout the early stages of Bitcoin's development. For speculators and researchers alike, it was an incredible find to discover a practical blockchain application available to practically everybody. By keeping an eye on the patterns of Bitcoin, which enables their upward scalability and stability, businesses can have a better understanding of blockchain-based applications. By developing an effective use case, researchers opened the door for academics to ask for funding to pursue and elaborate on broader and more complicated applications of blockchain (Fig. 1).

The decentralised, open, and permissionless nature of the blockchain makes it a promising technology for health care. The technology's increased applicability opens the door to new areas of health care, such as wearables and medical research. The blockchain's immutability is a critical feature for healthcare data. It is capable of safeguarding medical records, clinical trial results, and ensuring regulatory adherence. This shows how blockchain can support real-time patient monitoring and medical interventions through the use of smart contract technology. The supply chain in pharmaceuticals and the development of counter-counterfeit drug measures are other blockchain applications. Though there are significant costs associated with the development of new drugs, the use of smart contracts helps to streamline informed consent procedures while also improving data management and quality. Giving patients control over their own identity allows the informed consent procedure to be integrated

Table 1 Blockchain-based healthcare information system

Name of application	Type of blockchain	Platform type
Supply chain for pharmaceutical	Consortium/private	Ethereum
Electronic health records (EHRs)	Consortium	Health nexus
Data on finances and medicine	Consortium	Hyperledger
Prescription from a doctor	Public	Ethereum
Patient details	All	All
Managing medical data	Public	Ethereum

while still protecting the privacy of patients' health data. In this paper, we will look at a variety of blockchain-based healthcare applications and elaborate on a few of them one at a time. Table 1 shows some of the healthcare applications using different types of blockchain implemented on various platforms.

1.1 Key Blockchain Concepts

Blockchain technology provides a distributed ledger, data integrity and confidentiality, decentralised data management, anti-tampering and anti-forgery, highly efficient, relatively inexpensive and programmatic features that promote stability and agility. The numerous types of blockchains include public blockchains, private blockchains and consortium blockchains (hybrid blockchain). Because each shape has its own set of benefits and drawbacks, it can be utilised for a number of reasons. If you use a public blockchain, anyone on the network can transact in an anonymised and transparent manner. A public blockchain, also including bitcoins, has unrivalled decentralisation. Because the relies mainly on user consensus to function, there is no vulnerability. Public blockchain, on the other hand, is open to intrusions from the outside world. (b) Private Blockchain: the transactions are secret; data are not available for public view, but members are known. An attacker, for example, may reconstruct and properly chain all of the blocks that had been modified without the participants noticing. A participant cannot read or write the blockchain in a private blockchain network unless the participant has been granted permission or has been invited to join. Large corporations typically employ private blockchain, which has permissions set up amongst the many parties involved in the company's blockchain. It is possible for a bank, for example, to establish its own private blockchain network, with only 5 limited access to its many stakeholders, such as clients, workers, and suppliers. A private blockchain network is an option for businesses or organisations that want to share data amongst consortium members (such as banks, institutions and other enterprises or firms) (Fig. 2).

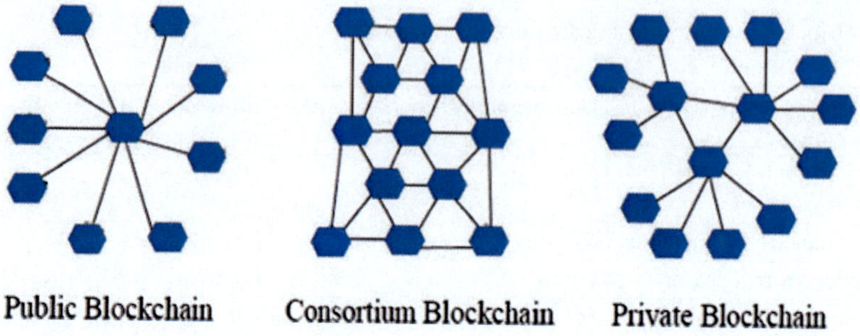

Fig. 2 Types of blockchain

2 Blockchain Use Cases in Health care

The healthcare industry can make extensive use of blockchain technology. The ledger technology makes it possible to securely transfer patient medical records, manage the medication supply chain and assist genetic code unlocking by the researchers.When it comes to health care, the use of blockchain has the potential to deflate the present expenditure bubble while also protecting patients' data and enhancing their overall experience. Using the technology, everything from safely encrypting patient data to managing disease outbreaks is now possible. Because blockchain technology is comprised of data blocks including hashes (digital fingerprints or unique identifiers), recent transactions and earlier blocks, it is referred to as distributed ledger technology (DLT) [2]. According to this design, each block is connected in order since each block contains the data from the preceding block and so on. A blockchain is a technology that consists of a series of interconnected blocks.

A modification to a block in the middle of a chain will not be possible since all blocks behind the converted block will have to be rotated at the same time, making it impossible. The data on the blockchain network do not change as a result of this method. It is possible that the data in one block hash have been mismanaged or altered, causing all subsequent block hashes to become invalid. This information is available to anyone who wishes to view it.

In the healthcare industry, many characteristics of blockchain technology, including the integrity of data stored in blockchain, have piqued the interest of researchers and practitioners.

Medical data and insurance claims will be managed more efficiently and precisely as a result of blockchain technology. It will also hasten the progress of clinical and biomedical studies. Due to the numerous advantages of blockchain technology, such as decentralised record management, immutable data audit trails, data provenance, robustness and better security and privacy, all of this is now achievable.

In order to benefit from blockchain technology in health care, medical information must be accessible to data subjects outside of the hospital setting. In contrast to traditional institution-driven contact, patient-centred interaction involves the participation of the patient in the decision-making process. Numerous obstacles arise from patient-centred interactions, including data standards, security and privacy concerns, in addition to technology-related concerns such as speed of delivery, financial incentives and governance. Blockchain technology has the potential to ease the move from institution-driven to patient-centred contact, thereby assisting in the resolution of these issues. Patients can grant access to their medical data through the use of blockchain technology, for example, by enabling particular researchers to access portions of their data for a short period of time. Patients can also connect to other hospitals, allowing them to automatically collect their medical data. Patients can offer their complete medical information to hospitals, which is then shared in a centralised database; when necessary, hospitals can directly retrieve information from that database, such as past diseases, allergies, past treatments and medical history, from that database [3]. It is impossible to modify or own such information because it is decentralised and immutable, meaning that no one has access to it. The information is available to anyone who has an identity and wishes to gain access.

This technology also provides a centralised database for pharmaceutical businesses and patients, allowing them to be assured that the drugs they purchase are genuine and not counterfeit. Due to the fact that the blockchain has all of the information about a product, and that as it moves from one location to another, the information about the product is updated on each block; hospitals are prevented from replacing genuine drugs with counterfeit ones. This precludes anyone from making changes to the expiration date, price or contents listed on the packaging.

Previously, hospitals may not have been prepared to communicate information about patients on blocks with anyone who possessed an identifier unless they were paid a charge for doing so. Patients, on the other hand, now have the ability to easily exchange their records because to this technology. The material is up to date and conclusive in regards to the medical records of the patients. Patient data sales for research purposes are also prevented by this technology, as patients have the ability to sell their data in exchange for a monetary return, in contrast to the typical practice of hospitals. One of the most valuable use case that can be thought of is that of electronic medical records.

Health data management needs to be prioritised in order to transform health care, with the goal of connecting heterogeneous systems and improving the accuracy of electronic health records (EHRs). Despite the fact that the terms electronic medical records (EMRs) and electronic health records (EHRs) are often used interchangeably, there is a distinction to be made. The term electronic medical records (EMRs) was first coined, and it refers to a digital version of the paper charts kept by clinicians. An electronic medical record (EMR) keeps track of all of a patient's medical history and treatment. As opposed to paper medical records, electronic health records (EHRs) are more concerned with the patient's overall health, including information gathered outside of the provider office and from other sources. According to the results of the mapping study, blockchain technology can help manage electronic health records (EHRs). The MedRec implementation proposed a decentralised approach to managing authorisation, permissions, and data sharing between healthcare stakeholders in this context [4]. In order to provide patients with knowledge and information about who has access to their healthcare information, MedRec utilises the Ethereum platform. FHIRChain (Fast Health Interoperability Records + Blockchain) is a second app that integrates EHR. It is a healthcare record management blockchain application built on Ethereum for the exchange of clinical data. Patients with FHIRChain's health information technology solutions that are in line with ONC's requirements.

Additionally, MedShare is an Ethereum app for systems that have trouble sharing data between cloud services because of the dangers associated with disclosing private data in the public domain (PIV). Medical data can be shared in cloud repositories with Medshare, which provides data provenance, auditing, and control between big data entities. MedBlock and Block-HIE are two other EMR blockchain applications. MedBlock has a record search feature. To keep track of a patient's records, the proposed system organises them by healthcare provider or department and keeps track of their addresses. Each patient record in the blockchain database has a unique identifier. Block-HIE is a blockchain-based healthcare platform.

MedBlock and Block-HIE are two other EMR blockchain applications. MedBlock has a record search feature. To keep track of a patient's records, the proposed system organises them by department or healthcare providers and keeps track of their addresses. Each patient record in the blockchain database has a unique identifier. Block-HIE uses both off-chain storage (data are stored in external hospitals' databases) and on-chain verification to maximise the use of existing databases. In the blockchain, records are stored as hashed values. FAIR FIRST and TP and FAIR, two fairness-based transaction packing algorithms, are proposed by the authors to improve fairness and throughput. Using Ethereum smart contracts to achieve data privacy, security, access control, and EMR interoperability, Ancile is another healthcare blockchain-based framework. In the distributed model of personal health records management, omniPHR maintains an interoperable single view (PHR). An elastic, interoperable, and scalable PHR data architecture underpins the proposed solution. OmniPHR evaluation may also ensure that PHR is divided into data blocks and distributed in a routing overlay network (Table 2).

3 Blockchain in Pharmaceutical Supply Chain

The pharmaceutical industry has also found use for blockchain. As a result, pharmaceutical manufacturers and distributors all over the world are looking into ways to enhance the security and traceability of their supply chains. In the last several years, blockchain technology has emerged as a major topic of discussion in this context. Although the medication distribution chain has expanded in extent and complexity, it has lagged behind in terms of digital technology and management systems needed to make it simpler and safer. These risks exist at every step when pharmaceuticals change hands before reaching patients, and pharmaceutical companies and other stakeholders have limited access into how items are being tracked to ensure their validity. With barcode-tagged medications scanned and placed into secure digital blocks everytime they change hands, blockchain could give major benefits in this case. Authorised parties, including patients at the end of the supply chain, might access this live record

Table 2 Application of blockchain technology in health care

Applications	Summary
Electronic health record	Blockchain provides integrity from generating data to the retrieval of data without any kind of human intervention when a digitalised EHR on distributed ledger is given access to the blockchain
Clinical research	With the blockchain decentralised feature any information can be collaborated securely and can be shared amongst the valid group of researchers
Medical fraud detection	One of the properties of blockchain is that it do not allow any duplication or modify any transaction, this property of being immutable provides the gateway for transaction that is secure and thus brings transparency
Neuroscience research	Blockchain brings several innovation in this area that includes thinking and augmentation of brain, and there are many more upcoming applications
Pharmaceutical industry and research	Counterfeiting of medicines and tracking of medicine at each stage is enabled using blockchain

at any moment. When patients receive substandard or counterfeit medications, it can have life-threatening consequences. This problem can be solved using blockchain technology, which has already been proven. QR codes can be used during the manufacturing process to combat the problem of counterfeit drugs and track them down. By scanning their QR codes, we will be able to see where they have been next. Even though a QR code can be copied and used on a counterfeit drug, this solution will not eliminate the problem of drugs being tracked or being faked completely. Our model is based on blockchain, using hyperledger fabric, and the manufacturer creates and uploads the details of a drug on this blockchain to make it decentralised. Following that, these medications will be approved by the government [5]. After that, hospitals and pharmacies can make requests for the approved drugs that are currently on the market, based on their specific needs. The blockchain network will allow patients to request medication in the future, and the request will be sent to a nearby hospital or pharmacy so that the patient can pick it up. Due to the drug's visibility on the blockchain network at every stage, using a network like this has several advantages. The blockchain network is also closed, so no one can steal drugs using it.

4 Blockchain in Remote Patient Monitoring

It is possible to collect medical data through handheld phones, body sensor and IoT devices in order to maintain track of a patient's condition using remote patient monitoring (RPM). Biomedical data collected from remote locations must be securely stored, shared and retrieved, and the usage of blockchains is important in this process. In this scenario, mobiles are being designed to send content to a blockchain application that is built on the hyperledger fabric platform. The use of Ethereum smart contracts to provide real-time patient monitoring applications, while also permitting automated interventions in a safe environment, has only been demonstrated in a few research. IoT has immense promise in a wide range of sectors, and it is already being intensively utilised and employed in the field of e-health care. IoBHealth, a model created that integrates the Iot technologies and blockchain, has indeed been suggested as a means of accessing, storing and managing e-health data in this approach [6].

5 Blockchain in Health Insurance Claims

The purpose of this paper is to examine how some of the blockchain-based use cases could provide extrinsic and intrinsic improvement the basic operations and business models of an insurer and to analyse their feasibility and ramifications. The use case scenario for an insurance company's job functions, as well as contacts with suppliers, intermediaries, and policyholders, all address how to provide better customer experience, boost product quality and provide the framework for greater consumer choice in the market [7]. The purpose of this paper is to examine how some of the blockchain-based use cases could provide extrinsic and intrinsic improvement the basic operations and business models of an insurer and to analyse their feasibility and ramifications. The use case scenario for an insurance company's job functions, as well as contacts with suppliers, intermediaries and policyholders, all address how to provide better customer experience, boost product quality and provide the framework for greater consumer choice in the market.citeg. Because of this, the goal is to minimise expenses while simultaneously enhancing operational efficiency and strengthening bonds with insured parties. Listed below are a few examples of frequent applications for blockchain technology.

5.1 Transforming Comprehensive Health Records into Interoperable Systems

It is possible that this technology, because of its greater security and ability to foster trust between entities, will be more effective in solving the interoperable challenge than currently existing technologies, for a variety of reasons.

5.2 Smart Contracts for Management and Strategy

Using blockchain technology, smart contracts might be used to collect records of agreement, transactions and other important information sets in real time, and then link the knowledge together and act on the data as needed.

5.3 *Increasing the Accuracy of Fraud Detection*

A life or healthcare insurance company may get fraudulent information through phoney claims, fraudulent applications or other routes. Smart contracts can help in assessing whether the information given is authentic or not.

5.4 *Improving the Accuracy of the Directory of Service Providers*

Specialised vendor listings could reap the benefits of blockchain consensus protocol in order to allow policyholders and providers to amend listings more rapidly and simply than they currently do.

5.5 *Making the Application Process Easier for Clients*

By making medical records on a blockchain easier to access and more thorough, we can bring comfort and peace of mind to a procedure that for many people has become intrusive and depressing.

5.6 *Improving the Insurer/client Relationship*

If patient data are safely stored on a smart contract, insurers and customers may be able to incorporate a huge spectrum of wellness-related actions into their interactions.

When it comes to blockchain technology, health and life insurance should take risks. Rather than simply making incremental changes to existing business models, the greatest potential may lie in leveraging blockchain's unique characteristics to develop totally new types of interactive rules and launch creative services that add value and grow the business.

6 IoT-Based Healthcare Monitoring Using Blockchain

During the previous decade, smart healthcare monitoring systems have undergone steady evolution in the industrial healthcare sector, and they have the ability to fundamentally alter present healthcare delivery methodologies. However, while smart health monitoring systems can automate patient monitoring activities while also improving patient procedure management, their effectiveness in clinical procedures and their efficiency in clinical procedures are still up in the air. Also, being carried out will be a thorough evaluation of the efficiency and clinical acceptability of the current healthcare monitoring systems, as well as strategies and proposals for enhancing them. Patients in faraway locations can benefit from the modern notion of new IoT-based healthcare monitoring, which can be carried out remotely for them. As a result, remote healthcare monitoring systems will be a significant development in the healthcare sector arena in the near future. The key advancements in the design level of IoT-based healthcare monitoring systems have been reviewed in light of the current difficulties that healthcare providers are experiencing [7]. As a result of this finding, we may conclude that there will be potential obstacles to the healthcare monitoring industry when compared to other similar systems. The Internet of Things (IoT)-based solutions for remote healthcare monitoring are composed of three primary components: the cloud, IoT gateways, and data gathering units. Data acquisition is the first aspect that largely relies on the smart devices and sensors of the end-system users to function properly. Patient data are collected from users' smart devices and sensors, and it is then processed at various levels in the IoT gateway before being forwarded to a cloud-based storage service. Diverse data analysis approaches are even being used in the cloud to gather and interpret important information in order to collect and understand valuable information. Medical practitioners then use this information to conduct more analysis and evaluation on the patient.

7 Healthcare Data Security Using Blockchain

Health care has become an irreducible aspect covering most of the part of our daily living, and as a result, data related to medicines, such as doctor's suggestion and past medical records, have become an increasingly important part of the patient's diagnosis and subsequent processes medical records. The traditional method of storing medical details as keeping the hard copy, it was susceptible to being harmed and transformed as time progressed. As a result, it was required to store the information in electronic form. The medical database, on the other hand, may be altered with or completely wiped. Then, there was the issue of information blocking, which caused some concern. Blocking of information occurs when a person or entity gains access to data that should only be seen with the express permission of the patient and/or the hospital concerned. This may occur with or without the individual's knowledge or consent. Even when it comes to improving the quality of care or resolving difficulties such as resource allocation and information blocking, technology has always played an important role. In the case of medical-care data exchange, technology has had to evolve through time. Patients may have access to a wide range of healthcare providers, including primary care physicians, experts and even practitioners. Health records must be sent securely and without tampering because a disease can be triggered by an earlier illness2. The patient does not need to be a professional or have a good memory to remember all of the information accurately if all of the data is safely stored and transferred. On a regular basis, patients must update their personal medical records. Article 21 of the Indian Constitution, as well as Articles 19(1)(a) and (b) of the Constitution of India, safeguard the fundamental rights to life and liberty and freedom of expression and movement of a patient, who may consult with or be transferred to another hospital for treatment. Again, the desire of the patient to divulge his information is sought after. A hospital must also obtain authorisation from the patient before sharing his or her information for research purposes. Again, if authorisation is obtained, the data transfer process will take a significant amount of time to complete. There are other concerns with time, speed, storage and security when data are transferred in paper form or even over an email message system. As previously mentioned, there are various limitations to storing data in a database, such as limited storage space and the possibility of cyberattacks. It is possible that an attacker will get access to the system and steal personal data from patients. It is also not possible to rely on a centralised database because it is difficult to implement practical access controls for different users, a search operation across an encrypted channel, a big memory for medical data storage and so on with a centralised database. In addition, Al Omar and colleagues explored a variety of issues that can arise when data is stored in an encrypted fashion. Similar to how bitcoin was regarded secure in transactions, a distributed ledger for such interchange of health information in a fair manner can provide data security. First and foremost, the patient will receive precedence because this is a medical device. A patient-centred approach strategy must be developed with a specific goal in mind. All patients' records must be kept private and adaptable, with the patient informed of which knowledge is being shared with insurance companies and with the blood bank as an example.

Medical experts have come up with an innovative way to keep medical data safe from manipulation and leaking using blockchain technology, which they want to implement in the near future. As a result, this system has the capability to safeguard data and guarantee reliability. Concerns about data storage can be resolved if this technology is used in combination with cloud computing. Cloud storage and management are credible sources of database management systems [8]. Additionally, the blockchain has the potential to address cloud security concerns. It is true that medical data exchange and archiving in a blockchain-based cloud can handle a wide range of challenges related to medical data. This technology can also be integrated into the patient's wearable gadget, which will allow him to keep track of his information. A scan of the patient's body will be performed in order to obtain the biometric signature. During the scanning of this wearable device, which is equipped with Internet of Things technology and stores data for later analysis, this signature will be required. Hence, blockchain has showcased amazing potential in the transformation of conventional healthcare industry.

8 Blockchain Technology in Indian Healthcare System

It is becoming increasingly common for the healthcare system to transition away from the conventional density-based medical system and towards a new style of therapy termed value-based care. Value-based health care, as opposed to volume-based health care, focuses on providing more customised services that are more directly related to the overall health of the patient. Affordable therapy and enhanced healthcare facilities data driven are two of the most significant revolutions in the healthcare industry. The Fourth Industrial Revolution (Industrie 4.0) and health care—preclinical studies and drug testing, as well as patient database administration—both rely on big data. The collection and management of such massive volumes of data, on the other hand, is proving to be prohibitively expensive. Because of privacy issues, health data have been mishandled, and data breaches have weakened public confidence in the current health regulatory system. As per the World Economic Forum, by 2025, blockchain will be responsible for 10% of global gross domestic product. The integration of blockchain with electronic health records (EHR) across all health providers would be able to provide stakeholders everywhere with quick access to patient medical records, hence reducing the time and costs associated with recurring medical procedures. Therefore, hospitals and other interested parties will save money because they won't have to spend as much on specialised software and databases.

Additionally, there are security problems that can be alleviated by utilising blockchain. Blockchain has three key characteristics: a decentralised network, transparency and immutability. By keeping track of prior transactions, it functions as a distributed ledger. Transaction records are spread throughout the network, acting as nodes on various participating systems. The transactions are stored in blocks of data that are encrypted and have a hash to identify them. As long as a hacker tries to change the contents, the hash value (i.e. signature) will change, and the transaction will be nullified. All participants will be informed if the records are tampered with [9].

For safe data access, users will have to choose between public, private and permissioned blockchains. Private blockchain allows data to be shared on a limited basis amongst a select group of players, whereas public blockchain is genuinely open and transparent, allowing any user to participate in network activities. It is a hybrid of the public and private blockchains, called a permissioned blockchain.

You can think of this as a big change for healthcare systems because patients will get a unique ID inside the blockchain-based network, and also, any novel diagnostic data they give will be linked to that ID in the EHR-integrated blockchain network. The person who owns the data will be the only one who can see it, giving them total ownership over their information. A similar thing to what we said before: The Health Utility Network (HUN) from IBM is a based on blockchain environment that acts as an ally between competitors and lets healthcare organisations come up with customer-focused payment processing solutions, share them and use them. This allows for the safe and easy interaction of healthcare information. There are a lot of companies that have worked together, like Aetna and Anthem. PNC Bank, Cigna, Sentara Health care and Healthcare Service Corporation are just some of them.

Security, scalability and legal requirements are just a few of the roadblocks to widespread adoption of blockchain. Despite this, Estonia was the first country to use blockchain technology on a national basis in 2016 in order to address these issues in detail. Their national healthcare database, e-Health records, integrates data from various healthcare service providers. Using a mechanism called e-Patient portal, clinicians can view a patient's records even if the service providers use separate EHR systems [10].

As a result, guardtime was able to deploy its Keyless Signature Infrastructure (KSI)-based blockchain technology on a wide scale across the country, establishing confidence with government institutions. Traditional blockchains had two key flaws, notably scalability and settlement time, which KSI helped eliminate. Instead of growing linearly in complexity with the number of transactions, KSI scales complexity in a time-dependent and independent manner. Due to the restricted number of players, the settlement time is decreased to a few seconds, making it possible to obtain data consensus synchronously. As a result of the implementation of the Personal Data Protection Act (PDPA), which was passed in 2008, the usage of blockchain is also in line with Estonians' rights and trust when it comes to personal data.

EU: The General Data Protection Regulation is a law in the European Union. Japan: The Act on Personal Information Protection (APPI) is a law in Japan, as is California's Consumer Privacy Act (CCPA). When it comes to data protection, the United States doesn't have a comprehensive law such as the Europe's General Data Protection Regulation (GDPR).

The Indian government issued a National Health Policy (NHP) in 2017 with the goal of improving the well-being of all Indian citizens and guaranteeing everybody has access to high-quality healthcare services, regardless of their financial ability to do so. One of the many goals of the National Health Programme is to establish a digitised health technology ecosystem. In order for the platform to grow, the collection of data, its storage and its distribution, and also the protection of user information and their right to privacy, will be key factors to consider [11]. It was proposed in 2017 that the Digital Information Security in Healthcare Act (DISHA) be enacted in order to protect patient privacy and the security of health information. In the future, DISHA will unify and regulate the complete data retrieval process, as well as protect users' rights, in order to ensure smooth electronic health record (EHR) system of people who have easy access to them. For example, when it comes to the exchange of health information (HIE), a federation-based permissioned network in blockchain may be able to overcome the knowledge gap that exists between policymakers and practitioners.

It is possible, using Estonia as an example, to construct a central repository for patient healthcare data that can be accessed from anywhere, at any time. However, India must undergo a significant transition in its technology infrastructure before introducing blockchain-based healthcare solutions. In comparison to other developing countries, India still has an energy deficit. At the end of 2017, the US used roughly 12 units of energy, while Japan used about 7.3 units, Estonia used about 6.5

units, China used about 4.5 units and India used about 1.2 units of energy (Units are defined as kWh per person per annual consumption). Multiple servers spread across various locations are required for blockchain technology to maintain the security of the transmitted information. Public–private partnerships between the government and business sector can assist to solve this shortcoming, which will speed up the adoption of blockchain in health care and other industries.

Preliminary attempts have been taken by the Indian government in this area. Using ThynkBlynk's patented blockchain technology, ChainTrail, Hyderabad-based start-up CallHealth, which provides healthcare platform, has teamed with ThynkBlynk to connect and securely share healthcare service providers' data on the network [12].

9 Research Challenges and Opportunities

The proposed prototypes and developed apps allow us to discover the many limits of healthcare blockchain-based applications, which we can then address. One of the major shortcomings of EMRs is that they do not consider semantic interoperability. As a result, medical and health data professionals will need to do manual assessment and mapping of specified ontologies. Another issue is that at this level, clinical misconduct is impossible to control [13]. Also, of note, concerns of scalability and interoperability are at the forefront of current and future research in this area. The lack of standards for designing healthcare apps based on blockchain technology is revealed by the interoperability difficulty. This may result in a lack of interoperability across the many applications that have been built. Additional issues with blockchain-based healthcare systems include scalability, which is particularly problematic when dealing with large amounts of medical data. Since healthcare data are extremely large in volume and cannot be stored on-chain, it is not feasible to do so on the blockchain since doing so would result in significant performance reduction. The speed with which transactions are processed, as well as the amount of data being loaded off-chain, all contribute to latency in a blockchain-based system. Finally, another flaw is related to the immutability of blockchains and the self-execution of code, as smart contracts may become exposed to hacking attempts in the future. Attacks on smart contracts, such as the decentralised autonomous organisation (DAO) assault, have resulted in the loss of millions of dollars in assets as a result of the attacks between 2016 and 2018 [14]. When it comes to using blockchain technology in the health sector, policymakers should consider four criteria (Fig. 3).

9.1 Blockchain Is Well-Suited for Its Intended Use

Blockchain is an enabling, general-purpose digital technology. Upon consideration of its benefits, it should be implemented in situations where it is the most appropriate answer for the problem at hand, after being compared to alternative alternatives [15].

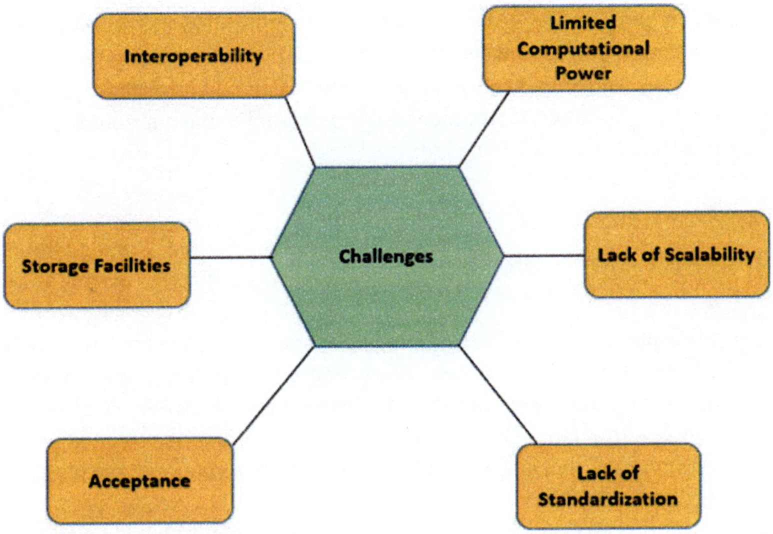

Fig. 3 Challenges of blockchain in healthcare domain

9.2 *Alignment with Governance and Regulatory Frameworks*

Blockchain-based solutions have unique characteristics that must be reviewed in terms of the solution's compliance with applicable laws, regulations and data governance frameworks.

9.3 *Integration in a Piecemeal Basis*

Incorporating blockchain-based solutions into existing systems and technologies should be done in little steps, rather than all at once. Blockchain technology should be used to complement and leverage current systems, and it should be evaluated incrementally in a controlled setting before being used on a broad scale.

9.4 *Education, Awareness and User-Centred Design*

Blockchain necessitates a fundamental shift in the way we think about data and information. User education is required for those who will be interacting with this technology, including patients and the general public, in order to understand the features of the technology and the implications of its use in terms of data ownership, access and privacy.

Blockchain technology has the potential to revolutionise health care, but it is neither fully mature nor a panacea that can be deployed instantly in the current state of affairs [16]. Before a healthcare blockchain can be embraced by companies across the country, a number of technical, organisational and behavioural economics difficulties must be overcome.

10 Future Research Directions

To mention a few uses, blockchain technology possesses the power to improve health transmission of information while also fundamentally enabling greater data openness, safer patient care, increased healthcare effectiveness and more rigorous clinical research, to name a few [17]. Despite the potential benefits, there are a number of fundamental difficulties that must be addressed before a safe and successful mass application can be achieved.

Healthcare organisations must do an accurate evaluation of blockchain technology in the area of their needs and teach practitioners on how to use the tools effectively if they are to effectively employ blockchain in the scope of their demands. It would be foolish to believe that the implementation of a blockchain conceptualised information management will instantly result in the accrual of the benefits outlined above. Implementations will need to be able to bridge the gap between rigid criteria that allow for global adoption and adaptability that allows for regional variation in practice in order to understand their full potential. The focus cannot be entirely on technical solutions; it also would take into consideration human traits that might otherwise make any digital service difficult to use [18].

Solutions must also be distributed and engage specialists from a variety of disciplines to ensure that information interchange is optimised while ensuring patient safety. This is similar to how blockchain strongly resists a segmented approach through the principles of decentralisation. National Coordinator for Health Information Technology (ONC HIT) produced a report on interoperability policy and technology requirements for the United States, such as:

1. Infrastructure for a ubiquitous and secure network
2. All participants' identities and authentication must be verifiable.
3. There is a need for consistent portrayal of authorisation to access EHI (Electronic Health Information) [19]

Current technologies, on the other hand, are unable to fully meet these needs due to constraints in terms of security, privacy and full ecosystem compatibility. Blockchain technology offers unprecedented opportunity to minimise complexity, enable trustless cooperation and produce secure and immutable information. Learn more about blockchain technology here. The Department of Health and Human Services is correct in keeping track on this quickly expanding subject in order to discover patterns and identify areas where additional government funding may be required in order for the technology to fulfil its full potential in health care. To help define the future of blockchain, the Department of Health and Human Services should consider mapping and convening the blockchain ecosystem, building a blockchain framework to manage early adopters and funding a consortium for debate and discovery [20].

References

1. Gatteschi V, Lamberti F, Claudio D, V'ıctor S (2018) Blockchain and smart contracts for insurance: is the technology mature enough?
2. Ichikawa D, Kashiyama M, Ueno T (2017) Tamper-resistant mobile health using blockchain technology. JMIR mHealth uHealth 5(7):e111
3. Zhang P, White J, Schmidt DC, Lenz G, Rosenbloom ST (2018) FHIRChain: applying blockchain to securely and scalably share clinical data. Comput Struct Biotech J 16:267–278
4. Jamil F, Hang L, Kim K, Kim D (2019) A novel medical blockchain model for drug supply chain integrity management in a smart hospital. Electronics 8(5):505
5. Blockchain and healthcare: the Estonian experience. https://nortal.com/blog/blockchain-healthcare-estonia/. Accessed 20 May 2020
6. Grishin D, Obbad K, Estep P, Quinn K, Zaranek SW, Zaranek AW, Vandewege W, Clegg T, César N, Cifric M, Church G (2018) Accelerating genomic data generation and facilitating genomic data access using decentralization, privacy-preserving technologies and equitable compensation, blockchain in healthcare today, vol 1
7. Carter G, White D, Nalla A, Shahriar H, Sneha S (2019) Toward application of blockchain for improved health records management and patient care, blockchain in healthcare today, vol 2
8. Jin H, Luo Y, Li P, Mathew J (2019) A review of secure and privacy preserving medical data sharing. IEEE Access 7(61656):61669
9. Mirza FF, Nguyen H, Naeem MA, Nguyen M (2017) A review on IoT healthcare monitoring applications and a vision for transforming sensor data into real-time clinical feedback. In: 2017 IEEE 21st international conference on computer supported cooperative work in design (CSCWD), pp 257–262. https://doi.org/10.1109/CSCWD.2017.8066704
10. Paranjape K, Parker M, Houlding D, Car J (2019) Implementation considerations for blockchain in healthcare institutions. Blockchain in Healthcare Today, vol 2
11. Gordon WJ, Catalini C (2018) Blockchain technology for healthcare: facilitating the transition to patient-driven interoperability. Comput Struct Biotechnol J 16:224–230
12. Yue X, Wang H, Jin D, Li M, Jiang W (2016) Healthcare data gateways: found healthcare intelligence on blockchain with novel privacy risk control. J Med Syst 40(10):218
13. LLiang X, Zhao J, Shetty S, Liu J, Li D (2017) Integrating blockchain for data sharing and collaboration in mobile healthcare applications. In: Proceedings of the 2017 IEEE 28th annual international symposium on personal, indoor, and mobile radio communications (PIMRC), Montreal, QC, Canada, pp 1–5
14. Engelhardt MA (2017) Hitching healthcare to the chain: an introduction to blockchain technology in the healthcare sector. Technol Innov Manage Rev 7(10):22–34
15. Alam MMA, Kaur H, Jameel H, Mourya AK, Chang V (2018) A proposed solution and future direction for blockchain-based heterogeneous medicare data in cloud environment. J Med Syst 42(8). Art. no. 156
16. Halamka JD, Alterovitz G, Buchanan WJ, Cenaj T, Clauson KA, Dhillon V, Hudson FD, Mokhtari M(M), Porto DA, Rutschman A, Ngo AL (2019) Top 10 blockchain predictions for the (Near) future of healthcare, blockchain in healthcare today, vol 2
17. Yue X, Wang H, Jin D, Li M, Jiang W (2016) Healthcare data gateways: found healthcare intelligence on blockchain with novel privacy risk control. J Med Syst 40:453
18. Dilawar NN, Rizwan M, Ahmad F, Akram S (2019) Blockchain: securing internet of medical things (IoMT). Int J Adv Comput Sci Appl 10(1):82–89

19. Zhao H, Zhang Y, Peng Y, Xu R (2017) Lightweight backup and efficient recovery scheme for health blockchain keys. In: Proceedings of the 2017 IEEE 13th international symposium on autonomous decentralized system (ISADS), Bangkok, Thailand, 22-24 March 2017; pp 229–234
20. Griggs KN, Ossipova O, Kohlios CP, Baccarini AN, Howson EA, Hayajneh T (2018) Healthcare blockchain system using smart contracts for secure automated remote patient monitoring. J Med Syst 42(7). Art. no. 130

A Novel Three Dense Layered Deep Fully Connected Neural Network for Hyperspectral Image Classification

K. Kanchanadevi, M. Shyamala Devi, J. Arun Pandian, R. Aruna, Suhas Palukuri, and Uday Kocherla

Abstract With the advancement of satellite observations, the use of hyperspectral images has become more common. The accurate classification of ground features using hyperspectral images is an important research area that has received a lot of attention. The classification of hyperspectral images is a fantastic mechanism for analyzing diverse land cover in remotely sensed hyperspectral images. The need for the dimensionality reduction and inadequate data during training in the hyperspectral image remains a major challenge toward image classification in the remote sensing field. With this overview, a novel model is proposed for hyperspectral image classification as three dense layered deep fully connected neural network. The classification is done with logistic regression, SVM and decision tree, and the accuracy is compared with our proposed deep fully connected neural network model. Implementation results show that proposed three dense layered deep fully connected neural network model shows the accuracy of 95% compared to the existing classification algorithms.

K. Kanchanadevi · M. Shyamala Devi · R. Aruna · S. Palukuri · U. Kocherla
Computer Science and Engineering, Vel Tech Rangarajan Dr, Sagunthala R&D Institute of Science and Technology, Chennai, Tamil Nadu, India
e-mail: kanchanadevi95@gmail.com

M. Shyamala Devi
e-mail: shyamaladevim@veltech.edu.in

R. Aruna
e-mail: drraruna@veltech.edu.in

S. Palukuri
e-mail: vtu12540@veltech.edu.in

U. Kocherla
e-mail: vtu17359@veltech.edu.in

J. Arun Pandian (✉)
School of Information Technology & Engineering, Vellore Institute of Technology, Vellore, Tamil Nadu, India
e-mail: aparunpandian@gmail.com

V. K. Gunjan and J. M. Zurada (eds.), *Proceedings of 3rd International Conference on Recent Trends in Machine Learning, IoT, Smart Cities and Applications*, Lecture Notes in Networks and Systems 540, https://doi.org/10.1007/978-981-19-6088-8_14

Keywords CNN · Accuracy · Classification · Dropout · Hyperspectral

1 Introduction

Most recent advancements in optics and photonics have made a hyperspectral imaging sensor with better ghostly and spatial goal. The spatial and otherworldly data is proficiently taken advantage of to recognize the materials and articles on the earth surface. The otherworldly marks are demonstrated so that they will separate the different articles and materials. It is feasible to see the distinguishing proof of various materials, items and surface ground cover classes dependent on their reflectance properties as a grouping task, for example, the grouping of picture pixels dependent on their unearthly qualities. Hyperspectral image is notable for the identification of the substances on the world's surface. The hyperspectral images are classified and utilized in an extensive scope of uses, like horticulture, natural science, cosmology, observation, stargazing and biomedical imaging. However, the classified hyperspectral images has its own specific issues, notwithstanding such as high dimensionality, the modest number of tests which have been named and huge spatial variety of ghostly marks.

2 Literature Review

A large part of the latest work on hyperspectral information characterization follows the customary example acknowledgment strategy, which involves two unmistakable advances: first, nitty gritty high quality highlights are acquired from the first information input, and also, classifiers like support vector machines (SVM), neural networks (NN) [1], greatest probability [2], parallelepiped arrangement, k closest neighbors [3], least distance [4] and calculated relapse [5] have been utilized to gain proficiency with the removed elements. The "scourge of dimensionality" influences most of the calculations referenced previously. Any dimensionality decrease-based order approaches [6] have been recommended to deal with the huge aspect intricacy and little preparing tests of hyperspectral information. Band choice and change are different strategies accessible to manage dimensionality. As a rule, measurable learning procedures have been utilized to settle the huge dimensionality and inconstancy of high-layered hyperspectral information and when barely any preparation tests are available [7]. Much of the recent work on hyperspectral data classification follows the traditional pattern recognition method, which comprises two distinct steps: first, detailed handcrafted features are obtained from the original data input, and secondly, classifiers like support vector machines (SVM), neural networks (NN) [1], maximum likelihood [2], parallelepiped classification, k nearest neighbors [3], minimum distance [4] and logistic regression [5] have been used to learn the extracted features. The "curse of dimensionality" affects the majority of the algorithms mentioned above. Any dimensionality reduction-based classification approaches [6] have been suggested to manage the large dimension complexity and small training samples of hyperspectral data. SVM is impervious to the Hughes peculiarities and has helpless aversion to high dimensionality. In specific circumstances, SVM-based classifiers perform better contrasted with other generally involved example acknowledgment systems as far as order precision. These classifiers were state-of-the-art mechanical instruments for quite a while. As of late, spatial data has become exceptionally applicable for hyperspectral information grouping. In [8], applying SVM and a directed picture channel, a procedure for arranging hyperspectral pictures has been introduced. The directed picture channel is utilized to fuse the spatial highlights into the SVM classifier. In [9], the edge preserving channels, for example, respective channel and directed picture channel are incorporated to join the spatial elements to the SVM classifier.

The different profound learning models [10, 11] have been produced for grouping purposes. These models are prepared from different degrees of highlights. The significant levels highlights needed for preparing the model are gotten from low-level elements. These models robotize the extraction of highlights for any issue contrasted and any convolutional design acknowledgment strategy [12]. In explicitly, we recommend the utilization of a modified convolutional neural network (CNN) that plays out the activity of developing enormous level elements and a multilayer perceptron (MLP) which is utilized for the grouping of the picture.

3 Three Dense Layered Deep Fully CNN

The architecture of the three dense layered deep fully connected neural network is shown in Fig. 1. The following contributions are carried out in this paper.

- The Pavia University dataset consisting of 42,776 hyperspectral images with several bands from Weebly database repository.
- The hyperspectral images are first normalized along with data preprocessing.
- The exploratory data analysis is done with the hyperspectral images for analyzing the image classes. Model fitting is done with fully connected layers by forming classes for each pixel.
- The deep fully connected neural network is formed with three layers having the first layer forming 3*Dense(128) followed by Dropout of least important features and is shown in Fig. 2.
- Then the second layer of deep fully connected neural network with 3*Dense(64) followed by Dropout of insignificant features.
- Then the third layer of deep fully connected neural network with 3*Dense(32) followed by Dropout of insignificant features.
- The classification is done with logistic regression, SVM and decision tree, and the accuracy is compared with our proposed deep fully connected neural network model.

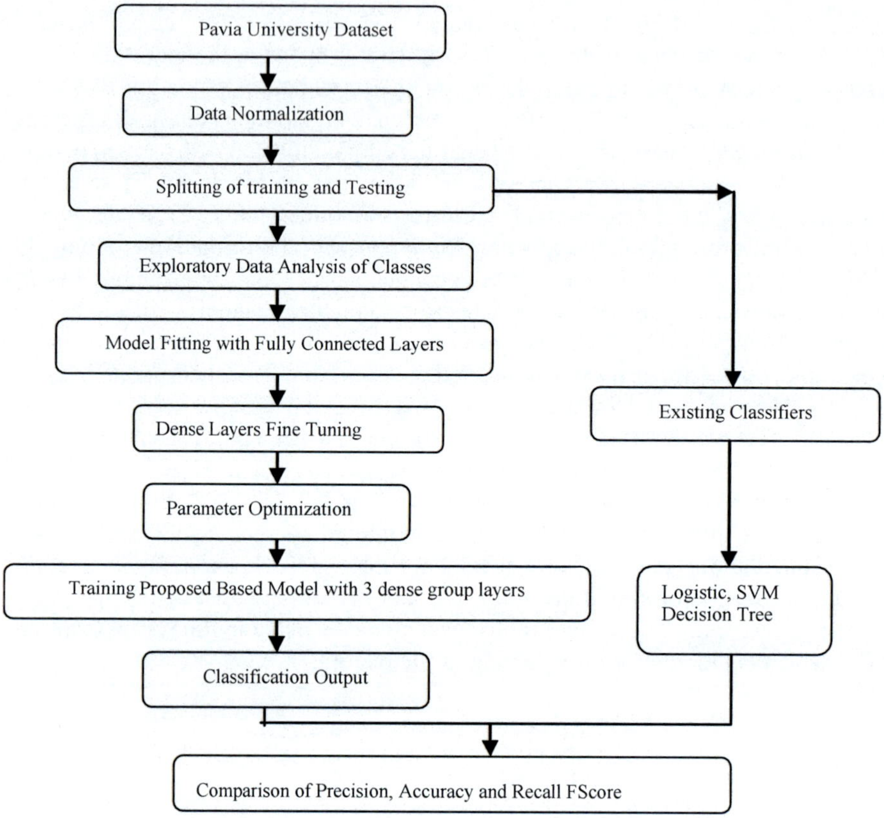

Fig. 1 Ten-layered deep CNN system workflow

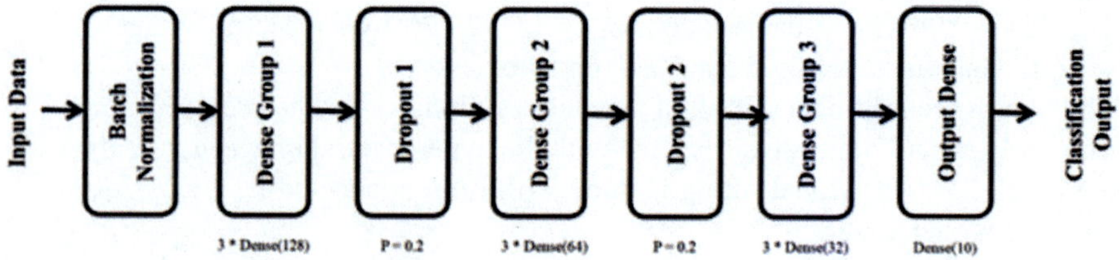

Fig. 2 Proposed three dense layered deep fully connected neural network

4 Implementation Setup

The Pavia University dataset consisting of 42,776 hyperspectral images with several bands from Weebly database repository. The hyperspectral images are normalized along with data preprocessing. The bands of the Pavia University dataset are shown in Fig. 3. The training dataset is formed with 38,498 images, with validation dataset of 4278 images and the system is tested with 4278 images. The training dataset is fitted with the proposed three dense layered deep fully connected neural network with the first layer having 3*Dense(128) followed by Dropout, and the second layer with 3*Dense(64) followed by Dropout and then the third layer with 3*Dense(32) followed by Dropout of insignificant features.

The classification is done with logistic regression, SVM and decision tree, and the accuracy is compared with our proposed deep fully connected neural network model. The training and validation loss, training and validation accuracy of three dense layered deep fully connected neural network is shown in Fig. 4. The ground truth and prediction image of three dense layered deep fully CNN are shown in Fig. 5.

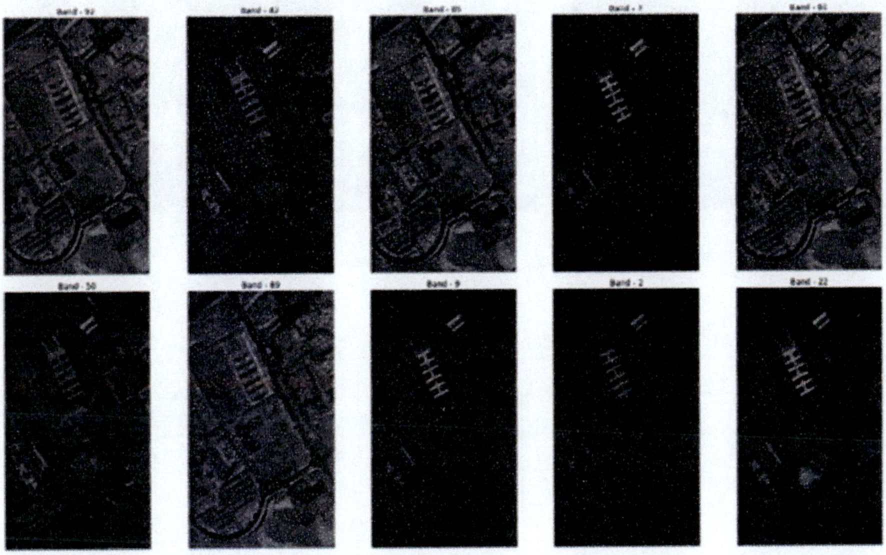

Fig. 3 Pavia University dataset of several bands

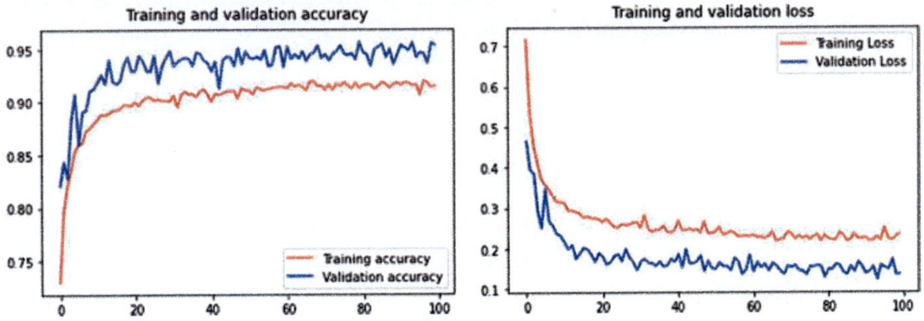

Fig. 4 Training, validation loss and accuracy of deep fully connected NN model

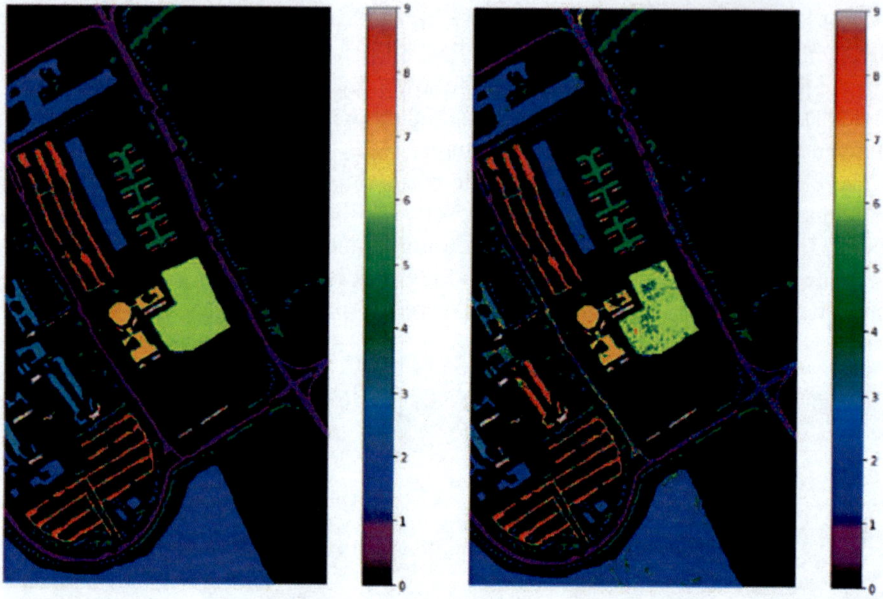

Fig. 5 Ground truth and prediction image

Table 1 Performance metrics of three dense layered deep fully CNN

Class type	Precision	Recall	F1-score	Accuracy
Leaf blight	0.81	0.90	0.83	0.82
Red leaf spot	0.87	0.91	0.88	0.87
Red scab	0.93	0.92	0.92	0.93
Healthy	0.95	0.94	0.94	0.95

The classification is done with logistic regression, SVM and decision tree, and the accuracy is compared with our proposed deep fully connected neural network model, and the obtained performance metrics are given in Table 1.

5 Conclusion

This paper attempts to provide the classification of hyperspectral image using the proposed novel three dense layered deep fully convolutional neural network. The project is implemented with Python under NVidia Tesla V100 GPU server with training epochs of 1000 and batch size of 32. This paper attempts to contribute in proving convolutional neural networks which performs more superior to hyperspectral image classification when compared to traditional machine learning techniques.

The classification is done with logistic regression, SVM and decision tree, and the accuracy is compared with our proposed deep fully connected neural network model. Implementation results show that proposed three dense layered deep fully connected neural network model shows accuracy of 95% compared to existing classification algorithms.

References

1. Li S, Song W, Fang L, Chen Y, Ghamisi P, Benediktsson JA (2019): Deep learning for hyperspectral image classification: an overview. IEEE Trans Geosci Remote Sens 57(9):6690–6709
2. Saqui D, Saito JH, Jorge LADC, Ferreira EJ, Lima DC, Herrera JP (2016) Methodology for band selection of hyperspectral images using genetic algorithms and gaussian maximum likelihood classifier. In: Proceedings of the international conference on computational science and computational intelligence, Las Vegas, NV, pp 733–738
3. Tu B, Wang J, Kang X, Zhang G, Ou X, Guo L (2018) KNN-Based representation of superpixels for hyperspectral image classification. IEEE J Selected Top Appl Earth Observat Remote Sens 11(11):4032–4047
4. Li W, Prasad S, Tramel EW, Fowler JE, Du Q (2014) Decision fusion for hyperspectral image classification based on minimum distance classifiers in the wavelet domain. In: The proceedings of the IEEE china summit and international conference on signal and information processing, Xi'an, pp 162–165
5. Bajpai S, Singh HV, Kidwai NR (2017) Feature extraction and classification of hyperspectral images using singular spectrum analysis and multinomial logistic regression classifiers. In: Proceedings of the international conference on multimedia, signal processing and communication technologies, Aligarh, pp 97–100
6. Sun Q, Liu X, Fu M (2017) Classification of hyperspectral image based on principal component analysis and deep learning. In: The proceedings of the international conference on electronics information and emergency communication, pp 356–359
7. Zhong S, Chang C, Zhang Y (2018) Iterative support vector machine for hyperspectral image classification. In: The proceedings of the international conference on image processing, pp 3309–3312
8. Shambulinga M, Sadashivappa G (2019) Hyperspectral image classification using support vector machine with guided image filter. Int J Adv Comput Sci Appl 10(10)
9. Shambulinga M, Sadashivappa G (2020) Supervised hyperspectral image classification using SVM and linear discriminant analysis. Int J Adv Comput Sci Appl 11(10)
10. Aydemir MS, Bilgin G (2019) Semisupervised hyperspectral image classification using deep features. IEEE J Selected Top Appl Earth Observat Remote Sens 12(9):3615–3622
11. Bidari I, Chickerur S, Ranmale H, Talawar S, Ramadurg H, Talikoti R (2020) Hyperspectral imagery classification using deep learning, in the proceedings of conference on smart trends in systems. Secur Sustainabil 672–676
12. Liu S, Shi Q, Zhang L (2020) Few-shot hyperspectral image classification with unknown classes using multitask deep learning. IEEE Trans Geosc Remote Sens

Research Trends in Artificial Intelligence and Nature Inspired Techniques

Shruti Aggarwal, Amit Verma, and Jaspreet Singh

Abstract Giving the machine the thinking capability is the most versatile of all human inventions. Artificial Intelligence is the most prominent research area in today's world. Nature inspired Algorithms are other most popular optimization techniques used in various domains. Researchers have implemented several nature inspired techniques using data mining functionalities for wireless sensors, Artificial Intelligence, etc. The latest research suggests the use of nature inspired techniques in Artificial Intelligence and for the Internet of things. In this paper, research trends for the application of nature inspired techniques in artificial intelligence are analyzed using data from Web of Science and Scopus databases. Data is analyzed globally. Cluster analysis of related keywords is computed along with link strength. Various other experiments are also conducted which help analyze research trends in Artificial Intelligence and nature inspired techniques.

Keywords Artificial intelligence · Nature inspired technique · Genetic algorithms · Data mining · Optimization

1 Introduction

Nature is always very stimulating for varied applications be it designing structures, manufacturing electronics or implementing algorithms to solve real world problems. The way the birds find food preys, the beautiful technology with which the water

S. Aggarwal (✉) · J. Singh
Department of Computer Science and Engineering, Chandigarh University, Gharuan, Punjab 140413, India
e-mail: drshruti.cse@gmail.com

J. Singh
e-mail: cec.jaspreet@gmail.com

A. Verma
University Centre for Research and Development, Chandigarh University, Gharuan, Punjab 140413, India
e-mail: amit.e9679@cumail.in

V. K. Gunjan and J. M. Zurada (eds.), *Proceedings of 3rd International Conference on Recent Trends in Machine Learning, IoT, Smart Cities and Applications*, Lecture Notes in Networks and Systems 540, https://doi.org/10.1007/978-981-19-6088-8_15

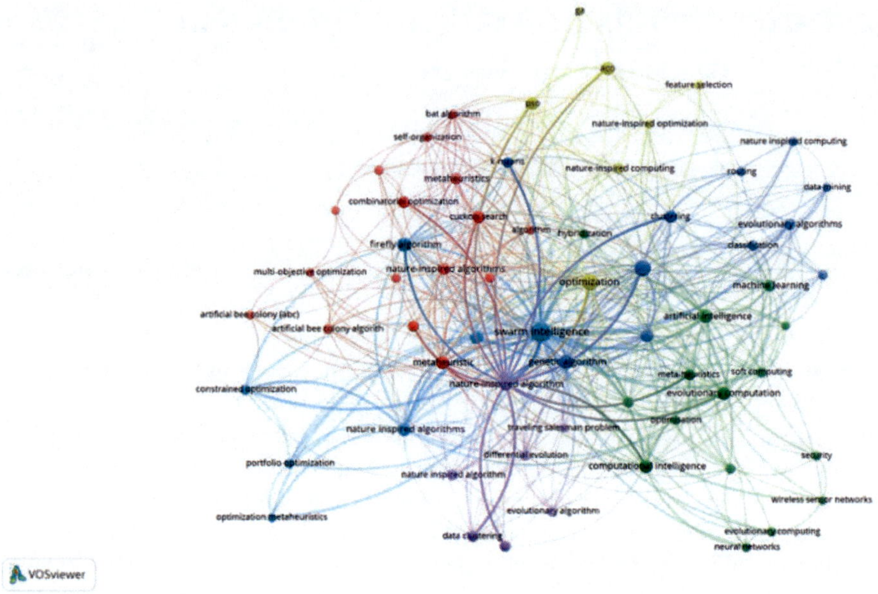

Fig. 1 Cloud density of artificial intelligence and nature inspired algorithms

droplets take smallest possible density, the analysis of king of the jungle, lion; all gives us various algorithms inspired by nature. Figure 1 gives the basic classification of Nature Inspired Techniques (NIT) [1], into ecology-based, population-based and evolution-based, depending upon the way the nature works. For example, Genetic Algorithms [2] are evolutionary in nature, bird mating, lion optimization [3], etc. are population-based techniques and behavior of ants, bees, etc. generating Ant Colony Optimization [4], Artificial Bee Colony [5], Particle Swarm Optimization Algorithms [6] are ecology-based NIT.

1.1 Artificial Intelligence-Based Nature Inspired Techniques

Artificial Intelligence [7] is the technique to make the machine think and learn based upon the training and testing. It has various applications such as pattern research, image processing, landmine detection. Artificial Intelligence also has wide application in optimization techniques [8]. Nature inspired techniques can be implemented using artificial intelligence [9] for creating smart grids, dew point temperature estimation, predicting covid cases in pandemic and many such applications.

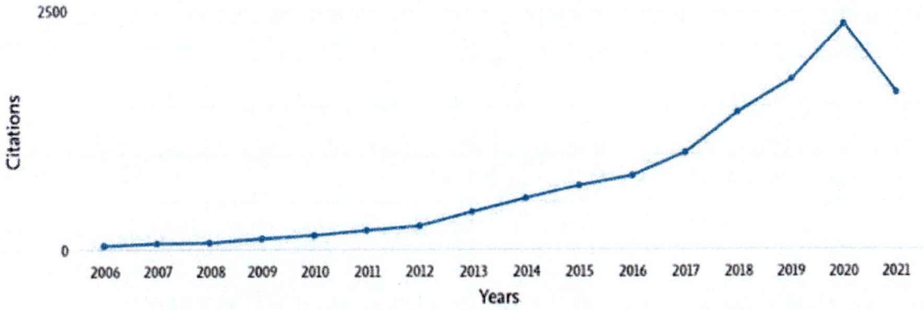

Fig. 2 Elsevier citation report using h-Index

2 Scientometric Analysis of Research Trends in Artificial Intelligence and Nature Inspired Techniques

Scientometric analysis of Artificial Intelligence and NIT with its allied terms is shown in this section. Year and country-wise analysis of the increasing research in this domain is also stated and described. Total link strength and occurrence analysis of popular terms is also computed and shown here.

2.1 Artificial Intelligence and Allied Techniques

A deep survey of research in the domain of Artificial Intelligence gives all the application areas [10] and related research of this functionality. Experiment is conducted using VOSviewer software tool using Scopus database for research in last two decades, as shown in Fig. 2.

Here high cloud density can be seen for nature inspired algorithms, with swarm intelligence, artificial intelligence, artificial bee colony algorithm, etc. Genetic algorithms, optimization techniques, meta-heuristic algorithms and other evolutionary techniques are highly correlated to Artificial Intelligence and thus define that plenty of research is conducted in this domain. Soft computing and machine learning are closely related to artificial intelligence. Various meta-heuristic techniques are also implemented using Nit and artificial intelligence [11].

2.2 Nature Inspired Techniques and Related Research in Artificial Intelligence

Figure 2 gives clear demonstration of the fact that nature inspired techniques [12] using artificial intelligence are closely used for optimization. This experiment is performed using VOSviewer for scientific landscaping of data using Scopus database for AI and Nature Inspired Algorithms [13] as epicenter of keywords.

The graph clearly depicts that the research trends in this domain are rising exponentially since past decade and last two years have been the highest research population for AI and NIT. The high bonding of Artificial Intelligence with nature inspired techniques depicts large volume of research in this area.

2.3 *Global Analysis of Artificial Intelligence and Nature Inspired Algorithms*

VOSviewer experimental data from Fig. 4 suggests that India has largely contributed in the global research in discussed area and researchers in Turkey, US and China have strong ties with Indians in this research. Jordan, Malaysia and Spain, etc. have also sparsely contributed and Malaysia, Brazil and Germany also have high research potential for implementing NIT [13] using Artificial Intelligence.

The experiment was conducted (Table 1) to check the global involvement of Artificial Intelligence with nature inspired techniques and it was concluded that 55 countries from the world have participated in research related to Artificial Intelligence and nature inspired techniques. Out of these 55 countries, 43 countries such as US, Iran, Spain, Turkey, have more than four publications in this domain and around 12 countries like Sweden, Qatar, etc. have two to three publications each in this domain. (Fig. 3).

2.4 *Year-Wise Analysis of Publication in Nature Inspired Algorithms and Artificial Intelligence Domain*

Another experiment was conducted using Scopus database, for last two and half decades from 1996 to 2021. The graph in Fig. 4 clearly indicated that the research prominently started in 2006 and since then there has been steady increase in research publications every year. The cumulative publications have risen sharply in last five years which clearly indicated that a lot of work is conducted in this area. More than 200 research articles can be found in the Scopus research database, with exponential increase every year.

Table 1 Global analysis of total link strength

S. No.	Country	Documents	Citations	Total link strength
1	India	184	2938	30
2	China	78	1369	18
3	United Kingdom	59	2506	32
4	United States	48	1083	21
5	Iran	34	324	6
6	Malaysia	34	500	17
7	Spain	26	184	2
8	Turkey	25	213	4
9	Serbia	22	891	18
10	Italy	20	369	6
11	Australia	16	224	1
12	Mexico	16	246	5
13	Czech Republic	15	72	0
14	Germany	15	140	2
15	Portugal	15	80	5

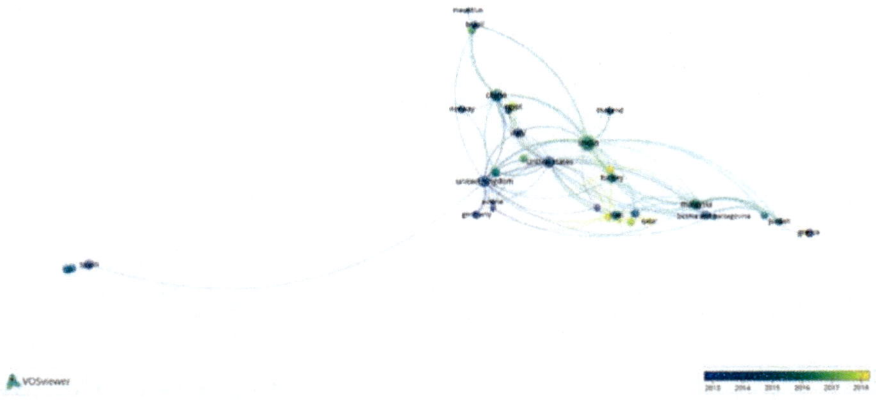

Fig. 3 Country-wise research

2.5 *Source-Wise Analysis of Published Work in Artificial Intelligence and Nature Inspired Techniques*

To map keywords with critical area of nature inspired techniques and clustering, source of citation is also taken into account as shown in Fig. 5. The Scopus database shares the various sources like Applied Soft Computing Journal, Advances in Intelligent Systems, Lecture Notes on Computer Science, ACM International Conference, Swarm and Evolutionary Computation, etc. Advances in Intelligent System, Soft

Fig. 4 Analysis of year-wise distribution of publications

computing, cluster computing are most prominent sources of research in the area of Artificial Intelligence and nature inspired technique.

An experiment was also conducted for research trends for each document based upon author. From various data mining techniques, Artificial Intelligence and nature inspired techniques [14] are used in abundance by researchers. Research trends analyze state that the optimization of nature inspired techniques is widely conducted using Artificial Intelligence algorithms. Numerous authors [15] in the past have worked not only in AI but also in collaboration in NIT. Figure 6 shares the document-wise citation details for the same.

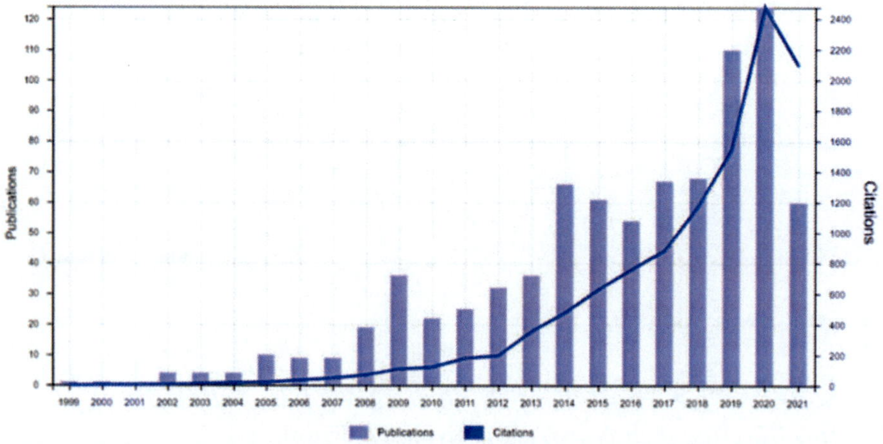

Fig. 5 Research trends of publications (*Source* Web of Science)

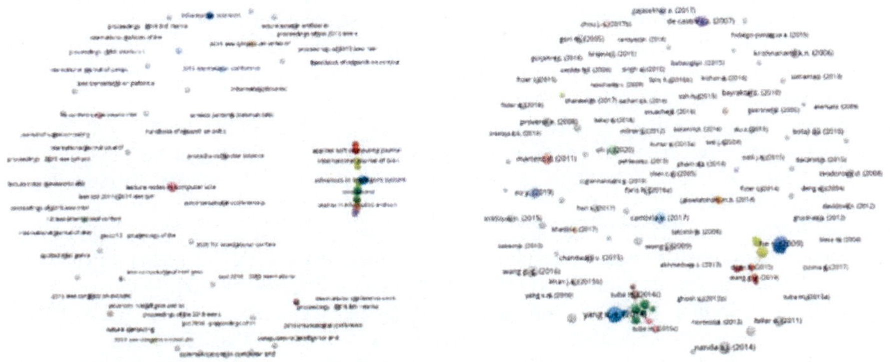

Fig. 6 Cloud density of source-wise citation and document-wise citation

3 Conclusion

Artificial Intelligence is the most popular technique to train machines to think and work like human beings. It is used for real time applications. Nature inspired techniques like genetic algorithms, swarm-based techniques, ecology and population-based methods are widely used for optimization of algorithms. Both Artificial Intelligence and nature inspired techniques are used hand in hand by researchers. In this paper, scientometric analysis of research trends related to Artificial Intelligence and nature inspired algorithms is conducted using Scopus and Web of Science databases. Various experiments are conducted in this paper to depict the research trends in this domain. The computed results are very supporting to researchers working in the field.

References

1. Zang H, Zhang S, Hapeshi K (2010) A review of nature-inspired algorithms. J Bionic Eng. https://doi.org/10.1016/S1672-6529(09)60240-7
2. Sivanandam SN, Deepa SN (2008) Genetic Algorithms. Introd Genet Lgorithms 15–37. https://doi.org/10.1007/978-3-540-73190-0_2
3. Yazdani M, Jolai F (2016) Lion optimization algorithm (LOA): a nature-inspired metaheuristic algorithm. J Comput Des Eng 3(1):24–36. https://doi.org/10.1016/j.jcde.2015.06.003
4. Rani B, Aggarwal S (2013) Optimization of association rule mining techniques using ant colony optimization 7109:340–344 [Online]. Available: http://inpressco.com/category/ijcet
5. Karaboga D, Akay B, Ozturk C (2007) Artificial bee colony (ABC) optimization algorithm for training feed-forward neural networks. In: Modeling decisions for artificial intelligence. Springer Berlin Heidelberg, Berlin, Heidelberg, pp 318–329
6. YS-I Connections and undefined 2004 (2018) Particle swarm optimization. marksmannet.com, Accessed: Sep. 05, 2018. [Online]. Available: http://marksmannet.com/RobertMarks/Classes/ENGR5358/Papers/pso_bySHI.pdf
7. Falk Anderson E, Pannu A, Tech Student M (2008) Artificial intelligence and its application in different areas related papers review on artificial intelligence: a boon or bane to humans

international journal of scientific research in science and technologists playing smart-artificial intel. Certif Int J Eng Innov Technol 9001L:10
8. Mccarthy J (2021) What is artificial intelligence? 2007, Accessed: Aug. 20, 2021. [Online]. Available: http://www-formal.stanford.edu/jmc/
9. Agbehadji IE, Awuzie BO, Ngowi AB, Millham RC (2020) Review of big data analytics, artificial intelligence and nature-inspired computing models towards accurate detection of COVID-19 pandemic cases and contact tracing. Int J Environ Res Public Heal 17(15):5330. https://doi.org/10.3390/IJERPH17155330
10. Naganna SR, Deka PC, Ghorbani MA, Biazar SM, Al-Ansari N, Yaseen ZM (2019) Dew point temperature estimation: application of artificial intelligence model integrated with nature-inspired optimization algorithms. 11(4):742. https://doi.org/10.3390/W11040742
11. Vassiliadis V, Dounias G (2011) Nature–inspired intelligence: a review of selected methods and applications 18(4):487–516. https://doi.org/10.1142/S021821300900024X
12. Aggarwal S, Singh P (2018) Cuckoo, Bat and Krill Herd based k-means++ clustering algorithms. Cluster Comput. https://doi.org/10.1007/s10586-018-2262-4
13. Singh P, Aggarwal S (2021) Software fault prediction using hybrid swarm intelligent cuckoo and bat based k-means++ clustering technique. Int J Adv Intell Paradig 20(1/2):1. https://doi.org/10.1504/IJAIP.2021.10016288
14. Agarwal P, Mehta S (2017) Empirical analysis of five nature-inspired algorithms on real parameter optimization problems. Artif Intell Rev 50(3):383–439. https://doi.org/10.1007/S10462-017-9547-5.
15. Zhang G et al (2020) Reinforced concrete deep beam shear strength capacity modelling using an integrative bio-inspired algorithm with an artificial intelligence model. Eng. with Comput 2020:1–14. https://doi.org/10.1007/S00366-020-01137-1

Comparative Analysis of Software Defect Prediction Using Dimensionality Reduction

Ruchika Malhotra, Sanyam Sharma, and Sanskar Aggarwal

Abstract Software defect prediction (SDP) is a dynamic research issue in the field of software development life cycle. It is very helpful in the testing phase of the life cycle of software development. It helps ensure the quality of the software being generated. In this particular paper, we have done a comparative analysis of various dimensionality reduction techniques such as principal component analysis (PCA), kernel PCA, incremental PCA, and sparse PCA, with random forests (RF) and artificial neural networks (ANN) as classifiers. We have collected the data from the android git repository and extracted metrics from 2 different versions of android. For comparing results, we have used three different metrics for a total of 10 experiments run on a dataset combined from 2 different versions of android. The result metrics we used are F1 score, area under receiver operating curve, and accuracy.

Keywords PCA · Kernel PCA · Sparse PCA · Incremental PCA · ANN · RF

1 Introduction

There has always been a constant attempt to develop more software-based systems with more functionalities and more reliable software hence better software quality. Because of such a sudden increase in demand for software, software quality has been the aspect that is most neglected and gets compromised. This is where the practice of software defect prediction comes in, which reduces the time and effort to debug and test the complete software to build high-quality software. Software defect prediction helps us identify the components of the software which are most likely to fail and helps prioritize their debugging. There is a range of parameters that can be considered while trying to predict the defect in the existing software. We will be focusing on the 12 initial parameters to help us identify the buggy module.

R. Malhotra · S. Sharma (✉) · S. Aggarwal
Department of Software Engineering, Delhi Technological University, Delhi, India
e-mail: sanyamsharma_2k18se115@dtu.ac.in

S. Aggarwal
e-mail: sanskaraggarwal_2k18se113@dtu.ac.in

V. K. Gunjan and J. M. Zurada (eds.), *Proceedings of 3rd International Conference on Recent Trends in Machine Learning, IoT, Smart Cities and Applications*, Lecture Notes in Networks and Systems 540, https://doi.org/10.1007/978-981-19-6088-8_16

Machine learning and deep learning techniques have helped solve complex and vast problems in identifying various patterns between various inputs and parameters [1]. Artificial neural networks are deep learning algorithms that work similarly to our biological neural networks. These networks constitute weighted edges connecting a number of nodes making up the neural network. The starting nodes are the inputs, and the weights on the edges define the amount of effect of the input coming through that edge. Neural networks here are used as binary classifiers and will classify the code module as defective or non-defective on the basis of input parameters. Another popular machine learning algorithm is the random forest which is a supervised learning technique that is being used as a classifier in our case. It is founded on ensemble learning, which basically means stacking various multiple classifiers to make a single classifier. So, in the case of random forest, it contains a number of decision trees that are based on subsets of the given dataset, and an average is taken to get the predictive accuracy of the dataset. So, the majority of decision trees make the decision in place of a single decision tree. Hence, the more the number of trees, higher the accuracy and the lesser the chances of overfitting on the given dataset.

Dimensional reduction techniques include principal component analysis (PCA), kernel PCA (KPCA) [2, 3], incremental PCA [4], and sparse PCA [5]. Having multiple input variables in a dataset makes a predictive modeling task more challenging to the model, which is the curse of dimensionality. This is where the dimensionality reduction techniques come in to reduce the number of input variables in a dataset. PCA is an unsupervised linear transformation technique, while KPCA or kernel PCA is an extension of PCA [6] using kernel methods and hence a nonlinear form of PCA. When in place of kernels, the extension of PCA is done using sparsity structures to the features, then it is sparse PCA (SPCA). Incremental PCA (IPCA) is a way of introducing the large dataset in the form of small batches in steps or batches and hence applying PCA individually on them.

The data from the dataset first goes through the dimensionality reduction technique and then is passed through the classifier to predict from the classes of being defective or being non-defective module. So, different combinations of dimensionality reduction technique and classifier are–

- PCA + RF
- PCA + ANN
- KPCA + RF
- KPCA + ANN
- SPCA + RF
- SPCA + ANN
- IPCA + RF
- IPCA + ANN

To compare all these combinations and find the most optimal combination for best software defect prediction on dataset metrics used are accuracy, F1 scores, and area under receiver operation characteristic (ROC) curve [7]. The weighted average of precision and recall considering both false positives and false negatives F1 score is calculated accompanied with accuracy, which can be misleading if considered alone. ROC curve analysis gives insights into the sensitivity and specificity.

2 Related Works

Dr. Ruchika Malhotra and Kishwar Khan analyzed software defect prediction (SDP) in the beginning phases of SDLC [8]. An examination is performed on nine open-source programming frameworks written in Java from PROMISE repository utilizing four, for the most part, utilized element extraction methods, for example, principal component analysis (PCA), linear discriminant analysis (LDA), kernel-based principal component analysis (KPCA), and autoencoders with support vector machine (SVM) as base AI classifier. The model approval is performed utilizing a ten times cross-approval technique, and the productivity of the model is assessed utilizing precision and ROCAUC.

Satya Srinivas Maddipati and Malladi Srinivas dealt with software defect prediction and diminished the aspects utilizing kernel PCA, and took care of class imbalance issues utilizing cost-sensitive class problem [9]. They performed experimentation on software defect datasets downloaded from the NASA dataset repository.

Rehan Ullah Khan, Saleh Albahli, Waleed Albattah, Mohammad Nazrul Islam Khan worked with state-of-the-art deep learning models and random forests [10]. They performed experiments with five different datasets and proved the effectiveness of deep learning in the field of software defect prediction leading to us utilizing ANNs in our experimentations.

Praman Deep Singh and Anuradha Chug worked with popular machine learning techniques to experiment with software defect prediction [11]. They used artificial neural network (ANN), particle swarm optimization (PSO), decision trees (DT), Naive Bayes (NB), and linear classifier (LC). They conducted their experiments on the NASA open-source repository for software defect prediction.

3 Dataset Preparation

There are plenty of datasets available online. We chose a dataset with varied attributes because a dataset with plenty of features will be a good example to test the working of dimensionality reduction techniques [12]. Also, we wanted to use the dataset of software with relevant usage in the real world to provide more credibility to the results of our experiments.

The dataset we chose consists of two different versions of android. It consists of Lollipop android version 5 and Marshmallow android version 6. The dataset consists of over 25 different apps consisting of a different number of classes. Overall, the dataset consists of 12 features.

We used a train test split ratio of 80:20. Our dataset consists of 9120 training data points and 2281 test data points (Table 1; Fig. 1).

Table 1 Feature description

Feature	Description
AvgCyclomatic	Average cyclomatic complexity for all nested functions or methods
CountClassBase	Average cyclomatic complexity for all nested functions or methods
CountClassCoupled	Number of other classes coupled to
CountClassDerived	Number of immediate subclasses
CountDeclInstanceMethod	Number of instance methods
CountDeclInstanceVariable	Number of instance variables
CountDeclMethodAll	Number of methods, including inherited ones
CountDeclMethod	Number of local methods
CountDeclMethodPublic	Number of local public methods
CountLineCode	Number of lines containing source code
MaxInheritanceTree	Maximum depth of class in inheritance tree
PercentLackOfCohesion	100% minus the average cohesion for package entities

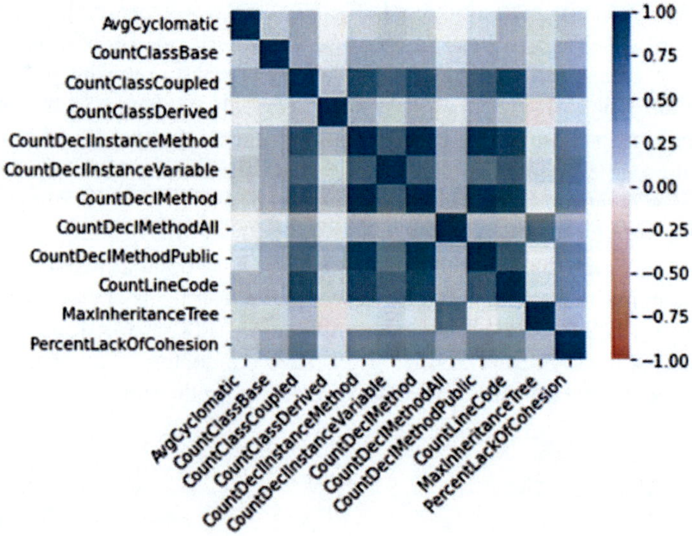

Fig. 1 Feature correlation heatmap

4 Research Methodology

In this section, we aim to provide textual and visual explanations of the techniques that we have used to carry out experiments for software defect prediction. PCA, KPCA, sparse PCA, and incremental PCA have been used for dimensionality reduction techniques and ANN and DT for classifiers.

4.1 Principal Component Analysis

PCA is a machine learning methodology based on dimensionality reduction. It is often used on large datasets and can reduce a large set of variables into a smaller set of variables that contains the most relevant information. Reducing the number of variables comes at the cost of reducing accuracy, but we are also trading in simplicity. Simplifying the dataset and problem helps us visualize better, makes the process faster, and helps us explore the dataset better.

4.2 Kernel Principal Component Analysis

Kernel principal component analysis is an extension of the principal component analysis technique. It differs from PCA such that it uses kernel methods. Using them, we can perform the originally linear PCA operations in reproducing kernel Hilbert space. Kernel PCA has been proven to be useful in the field of image denoising and novelty detection. Therefore, we chose to experiment with this particular dimensionality reduction technique as well.

4.3 Sparse Principal Component Analysis

Sparse principal component analysis (SPCA) is an extension of principal component analysis. It utilizes reduction of dimensionality in datasets by using structures known as sparsity structures to the features. It is used extensively in statistical analysis. It has many proven applications such as high-dimensional hypothesis testing, financial data analysis, and more. Its unique approach to reducing dimensions is the reason we included it in our experimentations.

4.4 Incremental Principal Component Analysis

Incremental principal component analysis (IPCA) is a substitution of the principal component analysis. It is used when the dataset to be reduced is too large to be fit in the memory. It deals with a limited batch size at a time and continues processing. Since the number of features and rows in datasets can be very large in the case of software defect prediction, we found it justified to use this particular technique in our experiments as well to see how it fares against other techniques and what will be the tradeoffs if we are dealing with memory constraints or large datasets.

4.5 Random Forests

Random forests combine the results of multiple decision trees to rule out final result. Decision trees are supervised learning algorithms that work on the principles of information gain and entropy. They are used as classifiers. The entropy of a dataset measures the contamination of the dataset, i.e., how scattered the informational collection is. The most basic part of decision tree calculation is the property determination strategy utilized at every hub of the tree since there are a few ascribes that split the information more simply than different properties. The calculation chips away at the rule of avarice, i.e., it searches for the arrangement that seems, by all accounts, to be best right now without taking a gander at the image at large. The decision tree calculation utilizes the information gain, which works out the decrease in entropy or gain in data, to divide the informational collection utilizing a specific trait. The calculation is beneficial as it does not need a lot of information cleaning and is not affected by anomalies and missing qualities to a reasonable degree.

4.6 Artificial Neural Networks

Artificial neural networks also work as classifiers. They are inspired by biological neural networks. They consist of various nodes that perform unit function and hold some constant value known as weight. During the training, we keep iterating through this network, and in case of wrong prediction, propagate backwards to correct the weights. After several iterations, the network is able to classify data points with good accuracy.

5 Performance Measures

5.1 Confusion Matrix

The confusion matrix is just a tabular representation of the performance of the trained model on the test dataset answers to which are known and are compared with the predicted ones. If there is a binary classification, then it can be represented as: True Positives (TP): The original answer is 1 and also predicted as 1 True Negatives (TN): The original answer is 0 and is predicted as 0 False Positives (FP): The original answer is 1 but is predicted as 0 False Negatives (FN): The original answer is 0 but is predicted as 1.

5.2 *Accuracy*

This is the metric which is the ratio of correctly predicted instances and the total number of predicted instances in the test dataset.

$$\text{Accuracy} = (\text{TP} + \text{TN})/(\text{TP} + \text{FN} + \text{FP} + \text{TN})$$

5.3 *F1 Score*

F1 score is the harmonic mean taken of precision and recall. Hence, the relative contribution of both is similar. The best F1 score is one, and the worst is 0. In the multi-class and multi-label case, this is the average of the F1 score of each class with weighting depending on the average parameter.

The formula for the F1 score is:

$$\text{F1} = 2 * (\text{precision} * \text{recall})/(\text{precision} + \text{recall})$$

5.4 *Area Under the Curve*

The performance of the predicted models was assessed by plotting the receiver operating characteristics (ROC) curve and evaluating the area under the curve. ROC curve, which is characterized as a plot of sensitivity on the y-coordinate versus its 1-particularity (it is characterized as the proportion of predicted non-flawed classes to the number of classes really non-flawed) on the x coordinate, is a successful strategy for assessing the quality of execution of predicted models [13].

6 Validation Measures

The cross-validation method is used, dividing the complete dataset into training and testing datasets in the ratio of 3:1. Hence, all the entries in the dataset are used, some which make up the training dataset are used to train the classifier, and the rest indices of the dataset that make up the test dataset are used for validation and finding the metrics for the trained classifier.

7 Results

Once the classifier is trained on, the training dataset is then run on the test dataset to get the required metrics for the necessary comparison between the various combinations of dimensionality reduction techniques and different classifiers considered. Hence, the following confusion matrices were produced (Table 2).

ROC curve plot for the respective combinations are (Table 3; Figs. 2, 3, 4, 5, 6, 7, 8 and 9).

Table 2 Confusion matrix for PCA + RF

	Actual NO	Actual NO	Actual YES	Actual YES
	Predicted NO	Predicted YES	Predicted NO	Predicted YES
RF	1165	42	262	812
ANN	1194	13	234	840
PCA + RF	1169	38	222	852
PCA + ANN	1171	36	137	937
KPCA + RF	1189	18	158	916
KPCA + ANN	1149	58	221	873
SPCA + RF	1178	29	229	845
SPCA + ANN	1175	32	135	939
IPCA + RF	1173	34	221	853
IPCA + ANN	1194	13	143	931

Table 3 Metrics

	Accuracy	F1 score	AUC score
RF	0.8667	0.8634	0.8606
ANN	0.8917	0.8890	0.8857
PCA + RF	0.8860	0.8837	0.8809
PCA + ANN	0.9272	0.9264	0.9277
KPCA + RF	0.8864	0.8847	0.8244
KPCA + ANN	0.9228	0.9211	0.9183
Sparse PCA + RF	0.8868	0.8844	0.8814
Sparse PCA + ANN	0.9281	0.9273	0.9273
Incremental PCA + RF	0.8882	0.8859	0.8830
Incremental PCA + ANN	0.9281	0.9271	0.9251

Fig. 2 Receiver operating characteristics PCA + RF

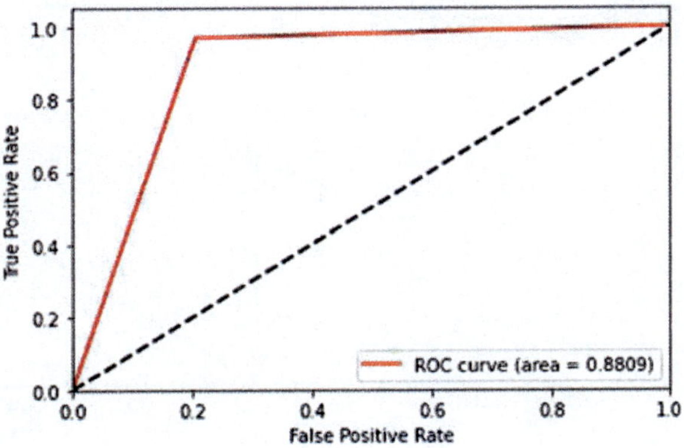

Fig. 3 Receiver operating characteristics PCA + ANN

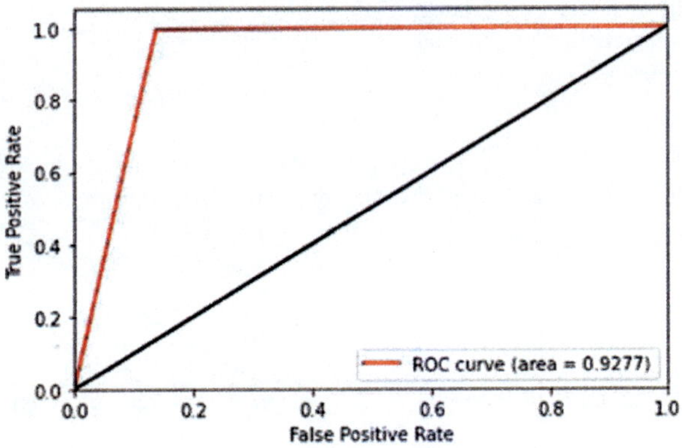

Fig. 4 Receiver operating characteristics KPCA + RF

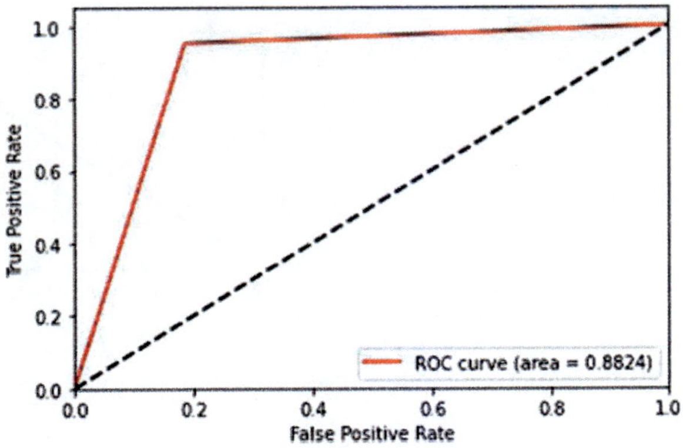

Fig. 5 Receiver operating characteristics KPCA + ANN

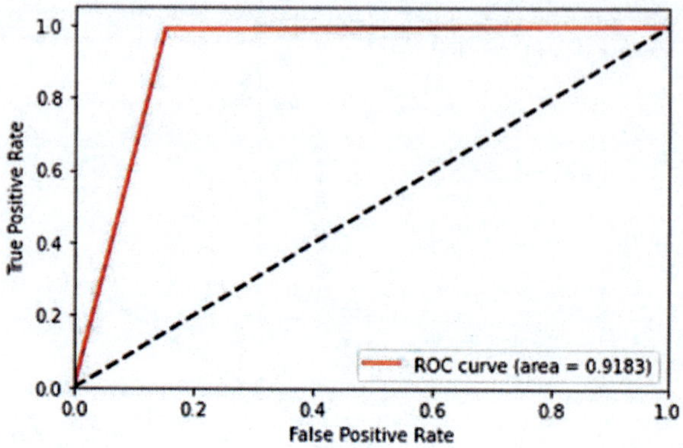

Fig. 6 Receiver operating characteristics incremental PCA + RF

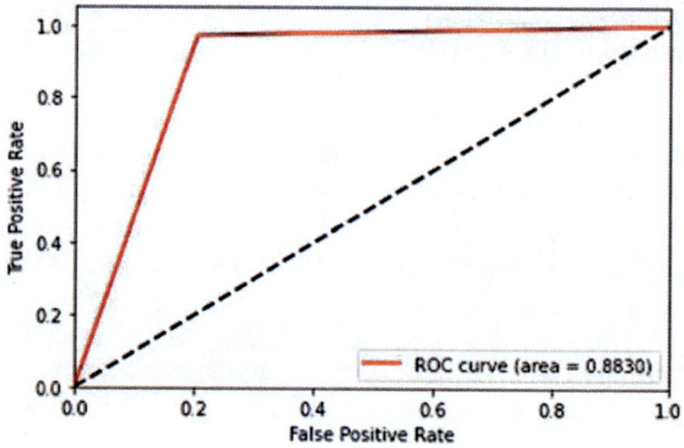

Fig. 7 Receiver operating characteristics incremental PCA + ANN

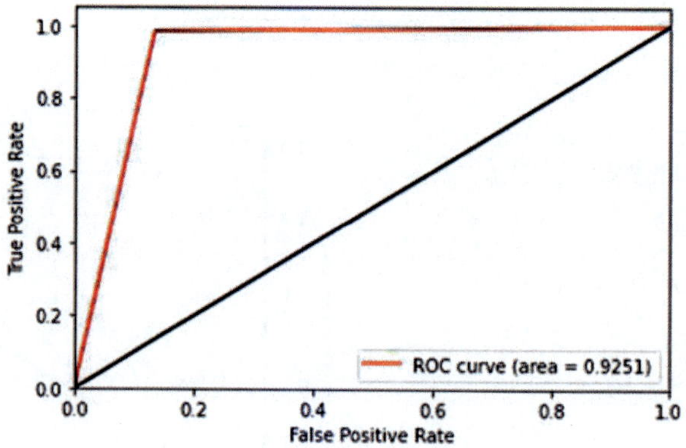

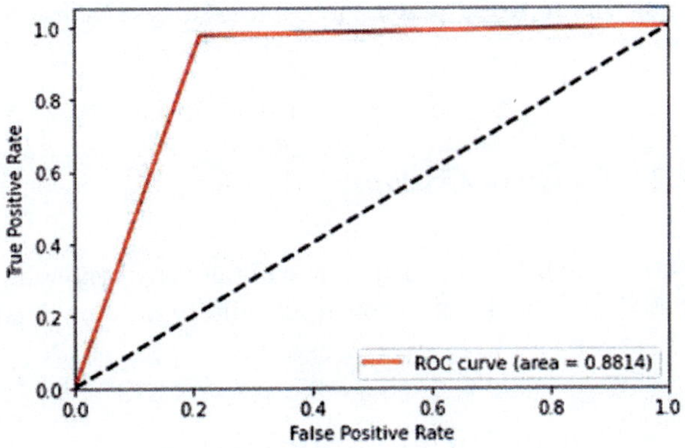

Fig. 8 Receiver operating characteristics sparse PCA + RF

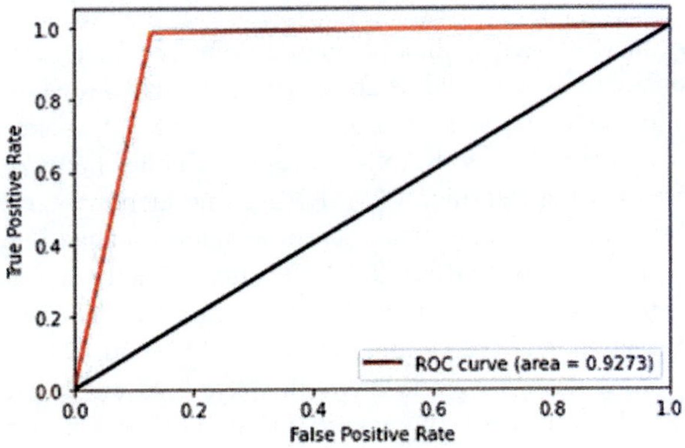

Fig. 9 Receiver operating characteristics sparse PCA + ANN

8 Discussion on Results

We conducted a total of 10 experiments and gathered results as listed in the results section. Solely utilizing classification techniques did not lead to good results as working on a lot of features can introduce unnecessary noise and requires dimensionality reduction or manual feature engineering. We experimented with four different dimensionality reduction techniques and received the best results from sparse principal component analysis and artificial neural network.

9 Threats to Validity

9.1 External Validity

Since the dataset used by us only contains open-source android repositories, it may not be able to generalize well over other software. However, this can be considered in the future work to generate a more diverse dataset to minimize external validity issues.

9.2 *Construct Validity*

We used well-known open-source libraries for conducting our experiments. However, it still carries a threat to construct validity to some degree.

10 Conclusion and Future Work

By the different compared metrics that are accuracy, F1 scores and AUC score, considering a cumulative comparison, we conclude that using ANN as the classifier followed by sparse PCA as the dimensionality reduction technique. Results from all the combinations with ANN as the classifier had better results as compared to random forests as the classifier. PCA, KPCA, and incremental PCA coupled with ANN also displayed promising results. The used neural net had 3 layers with 6, 6, and 3 neurons, respectively, using Relu, Relu, and sigmoid activation functions, respectively, in the layers. Thus, concluded while handling data with large input parameters like such of software defect prediction, dimensionality reduction techniques coupled with deep learning neural networks are highly effective and efficient.

In the future, we can explore further more techniques in dimensionality reduction like LDA, QDA, etc., as well as other supervised and unsupervised learning classifiers like Naive Bayes, kernel support vector machines, ensemble learning techniques coupling up multiple classifiers, and varied neural networks like recurrent neural networks with varied number of layers and parameters. Comparing their results with the existing combinations to get a better alternative.

Acknowledgements This research was supported by Delhi Technological University and Dr. Ruchika Malhotra of Software Engineering, Delhi Technological University.

References

1. Li J, He P, Zhu J, Lyu MR (2017) Software defect prediction via convolutional neural network. In: IEEE international conference on software quality, reliability and security (QRS). IEEE. https://doi.org/10.1109/qrs.2017.42
2. Kumar VA, Elavarasan N (2014) A survey on dimensionality reduction technique. Int J Emerg Trends Technol Comput Sci 3(6)
3. Mosci S, Rosasco L, Verri A (2007) Dimensionality reduction and generalization. In: Proceedings of the 24th international conference on machine learning—ICML '07. ACM Press. https://doi.org/10.1145/1273496.1273579

4. Lippi V (2019) Incremental principal component analysis: exact implementation and continuity corrections. In: 16th international conference on informatics in control, automation and robotics. https://doi.org/10.5220/0007743604730480
5. Zou H, Hastie T, Tibshirani R (2006) Sparse principal component analysis. J Comput Graph Stat 15(2):265–286. https://doi.org/10.1198/106186006X113430
6. Wang Q (2014) Kernel principal component analysis and its applications in face recognition and active shape models. arXiv:1207.3538
7. Hanley JA, McNeil BJ (1982) The meaning and use of the area under a receiver operating characteristic (ROC) curve. Radiological Soc North Am (RSNA) 143(1):29–36. https://doi.org/10.1148/radiology.143.1.7063747
8. Malhotra R, Khan K (2020) A study on software defect prediction using feature extraction techniques. In: International conference on reliability, infocom technologies and optimization (ICRITO) (trends and future directions). https://doi.org/10.1109/ICRITO48877.2020.9197999
9. Maddipati SS, Srinivas M (2021) Software defect prediction using KPCA & CSANFIS. Turkish J Comput Math Educ. https://doi.org/10.1016/j.matpr.2021.01.471
10. Ullah Khan R, Albahli S, Albattah W, Nazrul Islam Khan M (2020) Software defect prediction via deep learning. Int J Innovat Technol Explor Eng (IJITEE). https://doi.org/10.35940/ijitee.D1858.039520
11. Deep Singh P, Chug A (2017) Software defect prediction analysis using machine learning algorithms. In: 2017 7th international conference on cloud computing, data science and engineering—confluence, Noida, pp 775–781. https://doi.org/10.1109/CONFLUENCE.2017.7943255
12. Hinton GE, Salakhutdinov RR (2006) Reducing the dimensionality of data with neural networks. Science 313(5786):504–507
13. Hinton GE (2007) Learning multiple layers of representation. Trends Cognitive Sci 11(10):428–434. https://doi.org/10.1016/j.tics.2007.09.004

An Approach to Learn Structural Similarity Between Decision Trees Using Hungarian Algorithm

Ruchika Aggarwal and Nanhay Singh

Abstract The prevailing data mining techniques emphasize on discovering patterns and extract useful information from facts already recorded in various databases. Prior research substantiates the belief that the decision tree methodology is one of the most prevalent technique for classification systems, predictive modeling, and interpretation as well as data manipulations. In this paper, we intend to delve deeper to find out structural similarity measure among decision trees induced from dataset using a naive approach. The purpose of this approach is to quantify the similarity which is an important factor, especially for data clustering. Decision tree similarity can be broadly classified under two categories (1) Structural Similarity (2) Semantic Similarity. Thus, the main focus of this paper is to determine the homogeneity between decision trees as the similarity measures are the key contributors in solving pattern recognition problems.

Keywords Decision trees · Classification · Attribute operator value (AOV) · Structural similarity · Semantic similarity

1 Introduction

The decision tree model has been considered as one of the most powerful tools for classification systems, predictive modeling, interpretation as well as data manipulations, [1, 2] precisely because of their attractiveness which lies in the fact that they possess the property to work with large datasets, and in contrast to neural networks, decision trees have the ability to generate rules. These rules thus obtained can be easily manifested, and humans can understand and interpret them at their own discretion. Therefore, it can be used to derive a strategy to lead to a particular conclusion.

R. Aggarwal (✉)
GGSIPU, Delhi, India
e-mail: ruchika.aggarwal1989@gmail.com

N. Singh
NSUT, Delhi, India

V. K. Gunjan and J. M. Zurada (eds.), *Proceedings of 3rd International Conference on Recent Trends in Machine Learning, IoT, Smart Cities and Applications*, Lecture Notes in Networks and Systems 540, https://doi.org/10.1007/978-981-19-6088-8_17

Decision tree is a classifier in the form of a tree structure (Fig. 1), where each node is either:

- A terminal node/leaf node—indicates the value of the target attribute (class label), or
- An internal node—specifies some test to be carried out on a single attribute value, with one branch and a sub-tree for each possible outcome of the test.

A decision tree can be used to classify any example by starting at the root of the tree and moving down through it till a terminal node is encountered, which corresponds to the classification or labeling of the instance. Each internal node consists of an AOV triple (Attribute-operator-value) representing a test condition on that node, which leads to selection of a branch depending on the outcome of the test.

There are plenty of reasons to use similarity measure between decision trees. Decision trees can be generated from different datasets of varied sizes, and hence, different trees will be induced. It is useful to be able to compare the induced trees for similarity. Also, a very important use is to be able to compare the differences seen

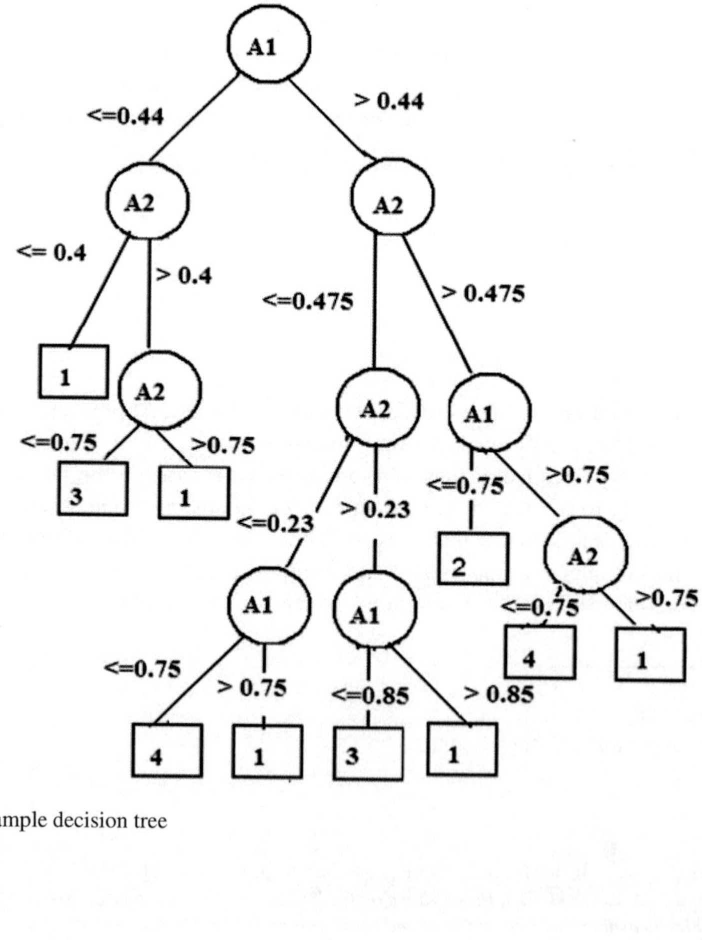

Fig. 1 Sample decision tree

in induction of decision trees from different training sets (coming from the same data generating distribution) structurally as well as semantically [3]. A decision tree (DT) can be decomposed into branches; each branch corresponds to a region. Hence, comparing decision regions is an important sub-problem when comparing DTs. A framework is proposed herein to compare two DTs for similarity, e.g., a sales manager might want to compare the sales patterns of different categories of products (high/medium/low value). Decision trees can be induced from data every week and change computed between weekly sales.

2 Background and Related Work

A lot of study has already been carried on semantic similarity between decision trees as well as datasets used for inducing the decision trees. Ntoutsi et al. [3] presented a general framework of similarity estimation based on partitions induced by decision trees on the attribute space of the datasets. It included as special cases, estimation of semantic similarity, as well as various forms of similarity estimation on classification datasets with respect to different probability distributions defined over the attribute-class space of the datasets. Touzet [4] gave a linear algorithm for comparing two similar ordered rooted trees with node labels. The method was based on tree edit distance, which is the number of edit operations required to convert one tree into the other tree.

In [5], Zager and Verghese applied an algorithm to find node similarity matrix to obtain matching between two different graphs. Perner [6] has introduced a similarity measure which is based on tolerance range and the position of cut point based on which the trees are compared. A domain ontology mapping method was proposed by Liu et al. [7]. Similarly, another approach for obtaining weather similarity analysis and feature engineering using a new model was proposed [8]. Yang et al. [8] transformed the tree-structured data into an approximate numerical multidimensional vector which encodes the original structure information to process similarity search. Another approach was presented to select the representative tree from an ensemble of decision trees induced from big data [9]. Ganti et al. [10] proposed a framework that measures the deviation using decision trees induced by two different datasets. The region probability estimation is computed between two datasets. Similar rationale is used by others. Turney et al. [11] proposed a measure for semantic similarity in case of classification models and named it: agreement and further did the estimation, empirically.

3 Methodology

In this paper, we intent to quantify similarity between two DTs by comparing corresponding regions in terms of their structural overlap. We may have two trees induced from the same training set S at different points in time, or we may have two trees coming from different training sets drawn from the same underlying population. Since in the first case, a region is being considered at different points in time, attribute test conditions, and their very presence may have changed within the time period, changing the associated space, distribution of the classes in the region. In the second case, region is considered coming from same data population, but training set is different; hence, different trees are induced.

3.1 Problem Definition

Let D be a d-dimensional dataspace spanned by the set of attributes $A = \{A_1, A_2, \ldots, A_d\}$, and let $C = \{C_1, C_2, \ldots, C_k\}$ be a set of class labels [3] which must be a discrete attribute. Also, let dom(C_i) where $i \in \{1, \ldots, k\}$ be the set of classes. Let the domain of attribute A_i be dom(A_i). $\forall i$, if dom(A_i) $\in R$, then $D \subseteq R_d$. S is a set of labeled training records, where $S \subseteq D$, from which decision tree T is induced using a learning algorithm. Thus, given two DTs, T_1 and T_2 induced from datasets S_1 and S_2, respectively, the aim is to quantify similarity between the DTs.

Definition 1 Attribute-Operator-Value (AOV)

$\langle$AOV$\rangle$ triple for a node x_k in a decision tree T represents the test condition at x_k. Here, A is the test attribute; O is the operator; $O \in \{<, \leq, >, \geq, =, !=\}$ and V can be any value based on which the selection of branch is made, $V \in$ dom(A).

E.g., $\langle$AOV$\rangle$ triple for the root node in Fig. 1 is $\langle A_1 \leq 0.44\rangle$.

Definition 2 Region

A region $R \in D$ is defined as a two-attributed entity: (set of $\langle$AOV$\rangle$, C_l), $C_l \in$ dom(C). Each leaf node in T corresponds to a region $R_i \in D \times C$.

Each $\langle$AOV$\rangle$ in a branch defines the range of that attribute defined by its domain and contributes a dimension to the space represented. A branch with d attributes will represent a d-dimensional space or region. A region is defined in terms of conjunction of all the AOVs of the corresponding branch, e.g., for the leftmost branch in Fig. 1, corresponding region is denoted by ($\{\langle A_1 \leq 0.44\rangle, \langle A_2 \leq 0.4\rangle\}$, 1) where 1 represents the class label.

3.2 Approach

To find the overall comparative analysis of two decision trees, the similarity measure can be performed in two phases:

Phase 1 revolves around structural similarity, and phase 2 depicts semantic similarity as shown in Fig. 2. Structural similarity is quantified by taking into account the commonality of comparable decision regions between the two trees. Two decision regions are comparable if they have the same class label. The semantic similarity for structurally similar regions can be quantified by comparing the probability distribution of the classes in the common decision region (in case of complete structural overlap).

Our final goal is to design a framework to quantify the similarity between two DTs. The metric should take into consideration both structural and semantic differences. In this paper, we have limited our research to structural similarity. The overall quantification framework must satisfy the following conditions:

1. The metric should have maximum value (which is 1) if the trees are structurally and semantically identical. Structural identicalness implies that each branch of a tree exists as a branch in the other tree, and this relationship is symmetrical. But as proposed, structural similarity is a precondition for semantic similarity, and semantic similarity is measured as the change in probability distribution of the classes at each of the leaves.
2. The metric should have the minimum value (which is 0) when there is no commonality between the structures of the two DTs, i.e., there is no branch common in the two DTs. Since the branches are not completely structurally similar (least structural similarity), there is no need to compute semantic similarity.

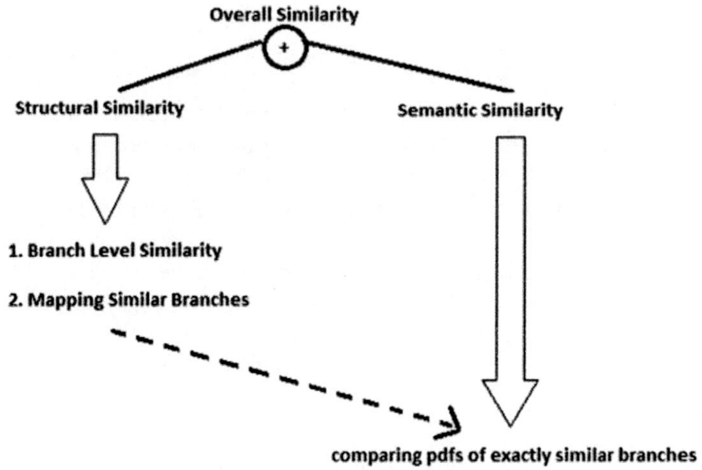

Fig. 2 Overall similarity computation

3. All intermediate scenarios must lead to the value of the metric ∈ [0,1]. The possible scenarios are as follows:
   (a) There is a strong structural similarity between the two trees, but difference in the semantic similarity.
   (b) Some branches of the two trees are both structurally and semantically similar, but some have differences of varying degree in similarity.
   (c) Partial matching between branches leads to varying degrees of structural similarity between the two trees.

Let T_1 and T_2 be two DTs induced from datasets S_1 and S_2, respectively. Let R_1 and R_2 be sets of regions (branches) in T_1, T_2, respectively, $|R_1| = m$, $|R_2| = n$. Without loss of generality, it can be assumed that $m \geq n$.

Define two functions θ and σ to compute structural and semantic similarity, respectively. Then, if number of comparable regions is n_c, for $i \in R_1$, $j \in R_2$, if class (R_{1i}) = class (R_{2j}),

$$S\,t_l = \theta\left(R_{1i}, R_{2j}\right)$$
$$S\,s_l = \sigma\left(R_{1i}, R_{2j}\right)\text{(if completely structurally similar)}$$
$$l \rightarrow 1 \ldots n_c$$

$$\text{Structural similarity } \theta(T_1, T_2) = \frac{2 * \sum_l S\,t_l - \sum_{j \in \text{unmapped-regions}} S\,t_l}{|T_1|+|T_2|}$$

$$\text{Semantic similarity } \sigma = (\Sigma_1 S\,s)/l$$
$$\text{Overall similarity } S = f(\theta, \sigma)$$

4 Structural Similarity

Structural similarity aims at computing the overlap between the two decision regions. First, the branch-wise overlapping region is found, and subsequently, comparable regions are mapped. Overall structural similarity is then obtained by aggregating mapped regions; unmapped regions contribute negatively to overall structural similarity.

Each leaf node in a DT corresponds to a decision region in the data space. The space represented by decision regions is inversely proportional to the number of attributes tested in the corresponding branch of the tree. The simplest way of quantifying the similarity between two trees is to look for common decision regions.

Example 1
Given the DT in Fig. 1, we can represent the regions in the DT as follows:

1. $(\{\langle A_1 \leq 0.44\rangle, \langle A_2 \leq 0.4\rangle\}, 1)$

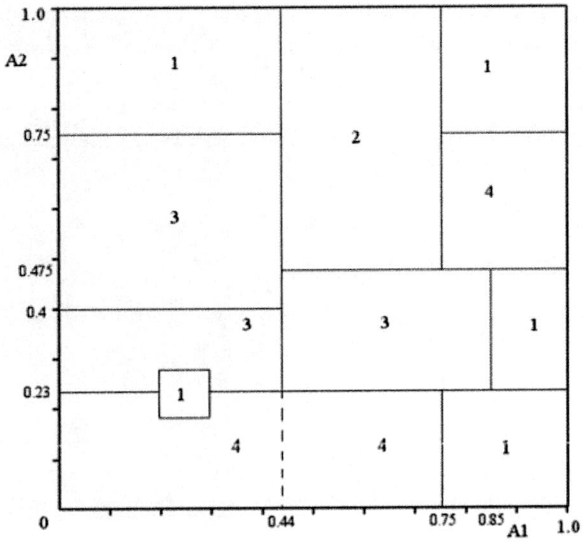

Fig. 3 Decision regions in two dimensions

2. $(\{\langle A_1 \leq 0.44\rangle, \langle A_2 > 0.4\rangle, \langle A_2 \leq 0.75\rangle\}, 3)$
3. $(\{\langle A_1 \leq 0.44\rangle, \langle A_2 > 0.4\rangle, \langle A_2 > 0.75\rangle\}, 1)$
4. $(\{\langle A_1 > 0.44\rangle, \langle A_2 \leq 0.475\rangle, \langle A_2 \leq 0.23\rangle, \langle A_1 \leq 0.75\rangle\}, 4)$
5. $(\{\langle A_1 > 0.44\rangle, \langle A_2 \leq 0.475\rangle, \langle A_2 \leq 0.23\rangle, \langle A_1 > 0.75\rangle\}, 1)$
6. $(\{\langle A_1 > 0.44\rangle, \langle A_2 \leq 0.475\rangle, \langle A_2 > 0.23\rangle, \langle A_1 \leq 0.85\rangle\}, 3)$
7. $(\{\langle A_1 > 0.44\rangle, \langle A_2 \leq 0.475\rangle, \langle A_2 > 0.23\rangle, \langle A_1 > 0.85\rangle\}, 1)$
8. $(\{\langle A_1 > 0.44\rangle, \langle A_2 \leq 0.475\rangle, \langle A_1 \leq 0.75\rangle\}, 2)$
9. $(\{\langle A_1 > 0.44\rangle, \langle A_2 \leq 0.475\rangle, \langle A_1 > 0.75\rangle, \langle A_2 \leq 0.75\rangle\}, 4)$
10. $(\{\langle A_1 > 0.44\rangle, \langle A_2 \leq 0.475\rangle, \langle A_1 > 0.75\rangle, \langle A_2 > 0.75\rangle\}, 1)$.

The decision regions belonging to the DT can be visualized in a two-dimensional graph (since there are two attributes) as shown in Fig. 3.

A decision tree is a recursive partitioning of the dataspace into regions, where each region is mapped one-to-one to a leaf node. The structural similarity between two decision trees is the aggregated structural similarity between the comparable regions of the two trees. Hence, the problem of finding structural similarity between two trees essentially requires:

- Finding structural similarities between all comparable regions.
- Subsequently, establishing the correspondence between the comparable regions of the two trees (Mapping).

Mapped regions are aggregated and contributed to structural similarity between two trees, while unmapped regions contribute negatively to similarity between two trees.

Similarity (or distance) between two regions is quantified in terms of the shared or common space between the two regions. We define a region as a set of ⟨AOVs,

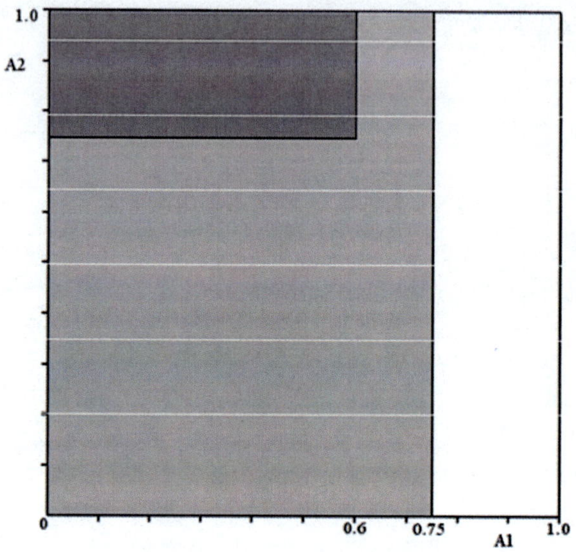

Fig. 4 Two regions of Example 2 mapped in 2 dimensions

Class⟩. Two regions are considered equal if they test the same attributes on same values, not necessarily in the same order and have the same class label. Apart from exactly alike regions, partially, overlapping regions are also considered.

Example 2
Given: A training set with two continuous attributes $\{A_1, A_2\}$ ($\mathrm{dom}(A_i) = [0,1]$) and class label $\in \{1,2\}$. Suppose we have the following two regions to be compared:

1. $(\{\langle A_1 \leq 0.75\rangle\}, 1)$
2. $(\{\langle A_1 \leq 0.6\rangle, \langle A_2 > 0.75\rangle\}, 1)$

We can visualize the data in two dimensions [12] as shown in Fig. 4. The darker shade of gray represents the overlap between the two regions and is the overlap.

As shown in Example 2, overlap between decision regions can be obtained using AOVs. In case of equal or completely alike regions, overlap will be 100%. Since a branch may test an attribute more than once, it is useful to convert AOVs into ALHs. An ALH represents the range of an attribute (⟨Attribute, Low, High⟩), helps in doing away with the operator. For each attribute, we have the range (min, max) readily available to compute overlap.

Example 3
For the AOVs $\langle A \leq 0.5\rangle$, $\langle A > 0.2\rangle$, the ALH triple is $\langle A\ 0.21\ 0.5\rangle$.

At this point, we redefine the region as an entity consisting of set of ALHs, class label. Notably, this redefinition is necessary for ease of computation. The similarity between two branches S_1, S_2 is a measure $\in [0,1]$.

- Similarity = 1, when $\forall$ ⟨ALH⟩ triples in branch S_1, there is an exact match in S_2, and $|S_1| = |S_2|$.

- Similarity = 0, when ∀ ⟨ALH⟩ triples in S_1, there is no matching triple in S_2, and the size of the branches is not important.

Intermediate Similarity: Compute overlapping fraction using ALHs using min (H_1, H_2) max (L_1, L_2).

Let $\langle A_1, L_1, H_1\rangle$, $\langle A_2, L_2, H_2\rangle$ be two ALH triples from branches S_1, S_2. Then, the overlapping region for attribute A is given by the equation

$$\text{Overlap}(A_k) = \min(H_1, H_2) - \max(L_1, L_2)$$

Overlap computed by the equation given above does not give an idea of extent of overlap with respect to the range of attribute A. We use the following equations to overcome the situation

$$\text{Range}(R_i) = \sum\nolimits_{\text{all-attributes } A_i} (\text{high}(\text{ALH}_i) - \text{low}(\text{ALH}_i))$$

$$(L_1, H_1) \in R_{1i}, (L_2, H_2) \in R_{2j}$$

$$\Theta(R_{1i}, R_{2j}) = \frac{2 * \sum \text{Overlap}(A_i)}{\sum \text{Range}(R_1) + \sum \text{Range}(R_2)}$$

4.1 *Mapping Comparable Regions Between Two Trees*

We need to obtain the best possible matching between the regions of the two DTs. This problem is essentially a matching problem, which can be generalized as an assignment linear programming problem. We use the Hungarian algorithm [4, 13, 14] given by H. W. Kuhn to solve the matching problem.

$$\text{Let } R_1 = \{r_{11}, r_{12}, \ldots, r_{1m}\}, \text{ and}$$
$$R_2 = \{r_{21}, r_{22}, \ldots, r_{2n}\}$$

be two sets of regions belonging to the DTs T_1 and T_2, respectively. Without loss of generality, it can be assumed that $m \geq n$.

R_1 and R_2 form a complete bipartite graph where the edge e_{ij} between regions $\{r_{1i}, r_{2j}\}$ has weight $w_{ij} = \text{sim}(r_{1i}, r_{2j})$ if regions are comparable, else -1. Note that $(m - n)$ regions do not have any mapping region in T_2 and represent distributions not present in T_2. Those regions cannot be compared and hence contribute negatively to the similarity between T_1 and T_2. Mapping is based on the principle of locality in the data space. If two regions r_1 and r_2 share locality, then they have overlapping non-zero hypervolume. After establishing the mapping regions, we compare the similarity in the distribution of the classes.

Hungarian algorithm is as follows:

Step 0: Initially, if required convert the problem from a maximum assignment to a minimum assignment. Thus, let X = maximum value in the assignment matrix. Replace each c_{ij} with $X - c_{ij}$.
Step 1: Now, each row is selected, and the row minimum is subtracted.
Step 2: Then, select each column, column minimum is subtracted.
Step 3: Try to cover all the zeros in the matrix using minimum number of lines using trial and error method. Suppose k lines are used.

- If $k < \min(|R_1|, |R_2|)$, let m be the minimum uncovered number. Subtract m from every unreserved number. Add m to every no. covered with two lines. Go back to the start of step 3.
- If $K = n$, go to step 4.

Step 4: Starting with the top row, work downward making assignments. An assignment can be (uniquely) made when there is exactly one zero in a row. Once an assignment is made, that row and column are deleted from the matrix.

If all $\min(|R_1|, |R_2|)$ assignments cannot be made and all the remaining rows contain more than one zero, we switch to columns. Starting with the left column, we work our way rightward making assignments.

We iterate between row assignments and column assignments until we have made as many unique assignments as possible. If we still have not made $\min(|R_1|, |R_2|)$ assignments and cannot make a unique assignment either with rows or columns, we make one arbitrarily by selecting a cell with a zero in it. We then try to make a unique row and/or column assignments.

Example 4
Let $R_1 = \{r_{11}, r_{12}, \ldots, r_{16}\}$, $R_2 = \{r_{21}, r_{22}, \ldots, r_{27}\}$ be two sets of regions in DTs T_1, T_2 shown in Figs. 5 and 6, respectively. The branches in the two DTs given in AOV notation are as follows:

T_1:

B_1—$\langle A_2 \le 0.5\rangle, \langle A_1 \le 0.7\rangle, \langle A_3 \le 0.4\rangle, C_1$
B_2—$\langle A_2 \le 0.5\rangle, \langle A_1 \le 0.7\rangle, \langle A_3 > 0.4\rangle, C_2$
B_3—$\langle A_2 \le 0.5\rangle, \langle A_1 > 0.7\rangle, \langle A_2 \le 0.3\rangle, C_2$
B_4—$\langle A_2 \le 0.5\rangle, \langle A_1 > 0.7\rangle, \langle A_2 > 0.3\rangle, C_1$
B_5—$\langle A_2 > 0.5\rangle, \langle A_1 \le 0.2\rangle, C_2$
B_6—$\langle A_2 > 0.5\rangle, \langle A_1 > 0.2\rangle, C_1$

T_2:

B_1—$\langle A_2 \le 0.5\rangle, \langle A_1 \le 0.7\rangle, \langle A_3 \le 0.4\rangle, C_1$
B_2—$\langle A_2 \le 0.5\rangle, \langle A_1 \le 0.7\rangle, \langle A_3 > 0.4\rangle, C_2$
B_3—$\langle A_2 \le 0.5\rangle, \langle A_1 > 0.7\rangle, C_2$
B_4—$\langle A_2 > 0.5\rangle, \langle A_2 \le 0.6\rangle, \langle A_3 \le 0.8\rangle, C_2$
B_5—$\langle A_2 > 0.5\rangle, \langle A_2 \le 0.6\rangle, \langle A_3 > 0.8\rangle, C_1$
B_6—$\langle A_2 > 0.5\rangle, \langle A_1 > 0.6\rangle, \langle A_2 \le 0.3\rangle, C_1$

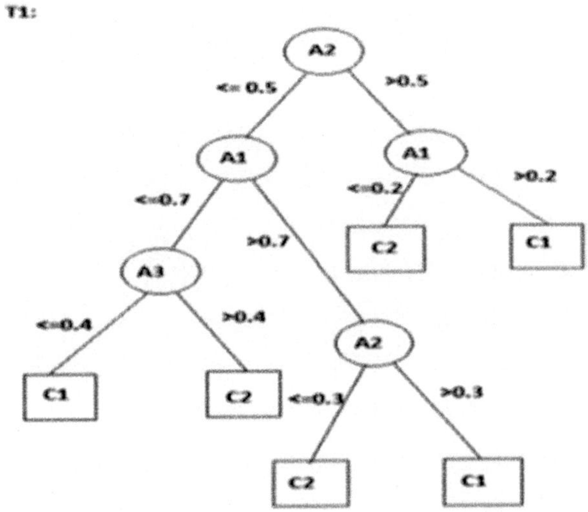

Fig. 5 Mapping example—first decision tree

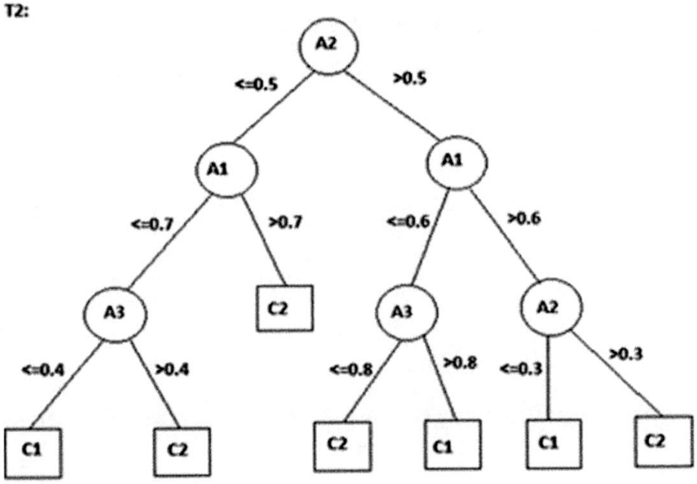

Fig. 6 Mapping example—second decision tree

B_7—$\langle A_2 > 0.5\rangle$, $\langle A_1 > 0.6\rangle$, $\langle A_2 > 0.3\rangle$, C_2

Converting the AOVs into ALHs, we get

T_1:

B_1—$\langle A_2\ 0\ 0.5\rangle$, $\langle A_1\ 0\ 0.7\rangle$, $\langle A_3\ 0\ 0.4\rangle$, C_1
B_2—$\langle A_2\ 0\ 0.5\rangle$, $\langle A_1\ 0\ 0.7\rangle$, $\langle A_3\ 0.41\ 1\rangle$, C_2
B_3—$\langle A_2\ 0\ 0.5\rangle$, $\langle A_1\ 0.71\ 1\rangle$, C_2
B_4—$\langle A_1\ 0.71\ 1\rangle$, $\langle A_2\ 0\ 1\rangle$, C_1

B_5—$\langle A_2\ 0.51\ 1\rangle$, $\langle A_1\ 0\ 0.2\rangle$, C_2
B_6—$\langle A_2\ 0.51\ 1\rangle$, $\langle A_1\ 0.21\ 1\rangle$, C_1

T_2:

B_1—$\langle A_2\ 0\ 0.5\rangle$, $\langle A_1\ 0\ 0.7\rangle$, $\langle A_3\ 0\ 0.4\rangle$, C_1
B_2—$\langle A_2\ 0\ 0.5\rangle$, $\langle A_1\ 0\ 0.7\rangle$, $\langle A_3\ 0.41\ 1\rangle$, C_2
B_3—$\langle A_2\ 0\ 0.5\rangle$, $\langle A_1\ 0.71\ 1\rangle$, C_2
B_4—$\langle A_2\ 0.51\ 1\rangle$, $\langle A_1\ 0\ 0.6\rangle$, $\langle A_3\ 0\ 0.8\rangle$, C_2
B_5—$\langle A_2\ 0.51\ 1\rangle$, $\langle A_1\ 0\ 0.6\rangle$, $\langle A_3\ 0.81\ 1\rangle$, C_1
B_6—$\langle A_2\ 0\ 1\rangle$, $\langle A_1\ 0.61\ 1\rangle$, C_1
B_7—$\langle A_2\ 0.31\ 1\rangle$, $\langle A_1\ 0.61\ 1\rangle$, C_2

- Computation of Structural Similarity:

Computing structural similarity between comparable regions of the two trees. For class label C_1,

$$\begin{aligned}
&\text{similarity}(T_1 : B_1, T_2 : B_1)\\
&\text{overlap}(A_1) = \min(0.7, 0.7) - \max(0, 0) = 0.7\\
&\text{overlap}(A_2) = \min(0.5, 0.5) - \max(0, 0) = 0.5\\
&\text{overlap}(A_3) = \min(0.4, 0.4) - \max(0, 0) = 0.4
\end{aligned}$$

$$\begin{aligned}
\text{similarity}(r_{11}, r_{21}) &= \frac{2 * \sum_i \text{overlap}(A_i)}{\sum_j \text{ALH}_j(T_1) + \sum_k \text{ALH}_k(T_2)}\\
&= \frac{2 * (0.7 + 0.5 + 0.4)}{(0.7 + 0.5 + 0.4) + (0.7 + 0.5 + 0.4)}\\
&= \frac{2 * 1.6}{3.2} = 1
\end{aligned}$$

Similarly, computing structural similarity between comparable regions of the decision trees T_1 and T_2, we get the matrix:

Matrix 1:

$R_1 - R_2$	r_{21}	r_{22}	r_{23}	r_{24}	r_{25}	r_{26}	r_{27}
r_{11}	1	− 1	− 1	− 1	0.43	0.39	− 1
r_{12}	− 1	1	0.4	0.54	− 1	− 1	0.19
r_{13}	− 1	0.4	1	− 1	− 1	− 1	0.51
r_{14}	0.35	− 1	− 1	− 1	− 1	0.78	− 1
r_{15}	− 1	0.16	− 1	0.53	− 1	− 1	0.58
r_{16}	0.39	− 1	− 1	− 1	0.57	0.33	− 1

Since our problem is a maximization problem, we convert it into a minimization problem using step 1. We get the matrix:

Matrix 2:

$R_1 - R_2$	r_{21}	r_{22}	r_{23}	r_{24}	r_{25}	r_{26}	r_{27}
r_{11}	0	− 1	− 1	− 1	0.57	0.61	− 1
r_{12}	− 1	0	0.6	0.46	− 1	− 1	0.81
r_{13}	− 1	0.6	0	− 1	− 1	− 1	0.49
r_{14}	0.65	− 1	− 1	− 1	− 1	0.22	− 1
r_{15}	− 1	0.84	− 1	0.47	− 1	− 1	0.42
r_{16}	0.61	− 1	− 1	− 1	0.43	0.67	− 1

Applying step 2 and subtracting row minimums, we get:

Matrix 3:

$R_1 - R_2$	r_{21}	r_{22}	r_{23}	r_{24}	r_{25}	r_{26}	r_{27}
r_{11}	0	− 1	− 1	− 1	0	0.4	− 1
r_{12}	− 1	0	0.14	0	− 1	− 1	0.35
r_{13}	− 1	0.11	0	− 1	− 1	− 1	0
r_{14}	0.43	− 1	− 1	− 1	− 1	0	− 1
r_{15}	− 1	0.42	− 1	0.05	− 1	− 1	0
r_{16}	0.18	− 1	− 1	− 1	0	0.24	− 1

Applying step 3 and subtracting column minimums, we get:

Matrix 4:

$R_1 - R_2$	r_{21}	r_{22}	r_{23}	r_{24}	r_{25}	r_{26}	r_{27}
r_{11}	0	− 1	− 1	− 1	0	0	− 1
r_{12}	− 1	0	0	0	− 1	− 1	0
r_{13}	− 1	0	0	− 1	− 1	− 1	0
r_{14}	0.25	− 1	− 1	− 1	− 1	0	− 1
r_{15}	− 1	0.31	− 1	0	− 1	− 1	0
r_{16}	0	− 1	− 1	− 1	0	0.20	− 1

Minimum number of lines required to cover all zeros = $6 \geq 6$. Hence, the solution is optimum. We obtain the following mappings:

T_1: $B_1 \rightarrow T_2$: B_1, similarity = 1
T_1: $B_2 \rightarrow T_2$: B_2, similarity = 1

T_1: $B_3 \rightarrow T_2$: B_3, similarity = 1
T_1: $B_4 \rightarrow T_2$: B_6, similarity = 0.78
T_1: $B_5 \rightarrow T_2$: B_4, similarity = 0.53
T_1: $B_6 \rightarrow T_2$: B_5, similarity = 0.57.

$$\sum_{\mathrm{CR}} \mathrm{sim} = 4.88$$

- Complete Structural Similarity

Branch-wise structural similarities are aggregated to obtain overall structural similarity according to the formula:

$$\Theta(T_1, T_2) = \frac{2 * \left(\sum_{i \in n_c} S_i - \sum_{j \in n_u} \max\left(S\left(b_j\right)\right)\right)}{|T_1| + |T_2|}$$

where n_u is the number of unmapped regions.

Hence, for the above example, overall structural similarity

$$\begin{aligned}\theta(T_1, T_2) &= \frac{2 * (4.88 - 0.58)}{6 + 7}\\ &= \frac{2 * 4.30}{13}\\ &= 0.66\end{aligned}$$

5 Conclusion

Hence, we defined similarity between two DTs T_1 and T_2 as a function of branch-level similarity (structural similarity and similarity in pdfs between branches).

$$\text{Overall Similarity } S = f(\theta, \sigma)$$

We intend to investigate the proper choice for the function *f* which is a concoction of both structural and semantic similarity in near future. Thus, the proposed methodology is an efficient approach that helps to compute similarity measures between decision trees which are induced either by a single dataset or different datasets. It further helps us to quantify the overall similarity taking into consideration both structural as well as semantic similarity. Therefore, our approach will help to improvise the measure by estimating similarity between various pairs of decision trees.

References

1. Safavian SR, Landgrebe D (1991) A survey of decision tree classifier methodology. IEEE Trans Syst Man Cybern 21(3):660–674
2. Cha SH (2007) Comprehensive survey on distance/similarity measures between probability density functions. Int J Math Models Methods Appl Sci 1(4):300–307
3. Ntoutsi I, Kalousis A, Theodoridis Y (2008) A general framework for estimating similarity of datasets and decision trees: exploring semantic similarity of decision trees. In: Proceedings of the 2008 SIAM international conference on data mining, Apr 2008. Society for Industrial and Applied Mathematics, pp 810–821
4. Touzet H (2005) A linear tree edit distance algorithm for similar ordered trees. In: Annual symposium on combinatorial pattern matching, June 2005. Springer, Berlin, pp 334–345
5. Zager LA, Verghese GC (2008) Graph similarity scoring and matching. Appl Math Lett 21(1):86–94
6. Perner P (2013) How to compare and interpret two learnt decision trees from the same domain? In: Proceedings—27th international conference on advanced information networking and applications workshops, WAINA 2013, pp 318–322. https://doi.org/10.1109/WAINA.2013.201
7. Liu J, Li Y, Tian X, Sangaiah AK, Wang J (2019) Towards semantic sensor data: an ontology approach. Sensors 19:1193
8. Yang R, Kalnis P, Tung AKH (2005) Similarity evaluation on tree-structured data. In: Proceedings of the 2005 ACM SIGMOD international conference on management of data (SIGMOD'05). Association for Computing Machinery, New York, NY, USA, pp 754–765. https://doi.org/10.1145/1066157.1066243
9. Zheng H, Wu Y (2019) A XGBoost model with weather similarity analysis and feature engineering for short-term wind power forecasting. Appl Sci 9:3019
10. Ganti V, Gehrke J, Ramakrishnan R, Loh WY (2002) A framework for measuring differences in data characteristics. J Comput Syst Sci 64(3):542–578. https://doi.org/10.1006/jcss.2001.1808
11. Turney P (1995) Technical note: bias and the quantification of stability. Mach Learn 20:23–33
12. Quinlan JR (2014) C4.5: programs for machine learning. Elsevier
13. Taha H. Operational research: an introduction, 8th edn
14. Castello B. Handout on Hungarian algorithm. John Hopkins University. www.ams.jhu.edu/castello

Design and Development of IoT-Based Intelligent Cattle Shed Management

Sanjay Mate, Prashant Dahiwale, and Vikas Somani

Abstract A village can be turned into a smart village by providing information exchange resources, i.e., technologies like the Internet of things (IoT) and the Internet of everything (IoE). To develop a smart village, many verticals need to be focused on health care, irrigation, climate monitoring, cattle surveillance, smart homes, smart lighting, wearable gadgets, etc. The main vein of the village-based economy is agriculture and the allied sector. IoT has a key role in improving crop yield agro-allied profits. It helps reduce the production costs in dairy, poultry, fishing, grains, fruits, and vegetables. IoT is helpful to reduce and manage time and workforce with the help of sensor and actuators-based mechanisms. A proposed approach in this paper is helpful to improve the socio-economic forum with maximum sustainability of natural resources and infrastructure while managing cattle sheds and outputs generated from cattle.

Keywords Internet of things (IoT) · Internet of everything (IoE) · Smart village · Smart cattle shed · Biofuel · Biofertilizers

1 Introduction

A vital task is to identify and evaluate village lifestyle, health, and economic infrastructure challenges. Table 1 shows that in 2020, 43.85% of the world's population and 69.07% of India's population residing in villages comparatively are less than all the previous years. The major reason behind the decrease in the overall percentage of the rural population is urbanization and declining income sources.

S. Mate (✉) · V. Somani
Department of Computer Science Engineering, Sangam University, Bhilwara, India
e-mail: matesanju@gmail.com

V. Somani
e-mail: vikas.somani@sangamuniversity.ac.in

P. Dahiwale
Department of Computer Engineering, Government Polytechnic Daman, Daman, India

V. K. Gunjan and J. M. Zurada (eds.), *Proceedings of 3rd International Conference on Recent Trends in Machine Learning, IoT, Smart Cities and Applications*, Lecture Notes in Networks and Systems 540, https://doi.org/10.1007/978-981-19-6088-8_18

Table 1 Rural population of the World and India [1]

Year	Rural population of the World (% of the total population)	Rural population of India (% of the total population of India)
1960	66.382	82.08
1970	63.446	80.24
1980	60.65	76.90
1990	56.971	74.45
2000	53.311	72.33
2010	48.354	69.07
2020	43.85	65.07

Table 2 Economy-based distribution of the rural population of the world in 2020 [1]

Economy (income)	World rural population w.r.t Income
Low	67
Low middle	58
Low and middle	49
Middle	47
Upper middle	32
High	18

Papua New Guinea, located at Melanesia in the south western Pacific Ocean, is the world's largest rural population country having almost 87% rural population of its total population. At the same time, many nations have no rural population at all, such as Singapore, Kuwait, Hong Kong SAR, and China. Table 2 shows that the rural population decreases exponentially as we move toward the low economy to the high economy group.

To turn a village into a smart village, reform is necessary that can be achieved with the help of IoT to raise agriculture and allied sector income. Figure 1 shows the smart village design cycle phases consisting of identification and mapping, decision-making, prototype development, evaluation, and scale-up. In each phase, the IoT components and systems for data processing are used as per the requirement.

1.1 A Smart Village Design Cycle

The first phase involves finding challenges and mapping them with solution generation processes. The final choice is made out of the possible options in the decision-making phase. Prototype development deals with designing a system processing the data received from sensors. A prototype helps an appropriate algorithm make precise decisions based on the data repository and received data. The purpose of evaluate and scale-up phase is to attain desired goals without disturbing natural resources

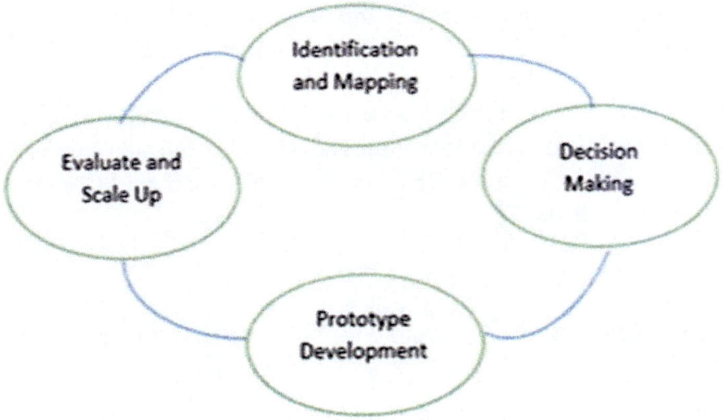

Fig. 1 Smart village design cycle

and infrastructure integrity and boost up the overall economy of the village sector. Figure 1 shows smart village design cycle.

This smart village system design continuously evolves on feed forward backpropagation neural network approach. After every iteration, it takes feedback from the previous output and works accordingly by minimizing the error component in the next iteration. Likewise, our system gets trained and generate optimum solution after several epochs.

Table 3 shows various verticals of the smart villages and their compatibility with the various technology platforms. Lower power wide area network (LPWAN) and cellular are the most widely used in recent years.

Some major factors needed to be considered in turning a village into smart village-like agriculture and allied sector, socio-economic framework, etc. Agriculture waste management can be improved with IoT by monitoring collection, transportation, and disposal treatment [2]. The real problem is the lack of education and hands-on knowledge about recent innovations in technology. Sectors related to agriculture or biotech, food processing, fishing, poultry, cattle-diary, honey harvesting, etc., can be improvised with technologies in the above domain. Finance and business models are key issues in smart village socio-economic reform [3].

Table 3 Smart village verticals and their IoT mapping

Smart village vertical	RFID	Zig-Bee	Bluetooth	Wi-Fi	LP- WAN	Cellular
Climate monitoring	–	–	–	–	Yes	–
Dairy	Yes	Yes	–	–	Yes	–
Smart lighting	–	–	–	–	Yes	–
Smart village home	–	Yes	Yes	Yes	–	Yes
Water and waste management	Yes	–	Yes	Yes	Yes	–

An effort is needed for better e-services and open and wide paradigms like multiple applications, agents, technology, and social innovations. It is time to redesign traditional unsustainable public infrastructure [4]. A good infrastructure is needed for clean, good pH value, oxygen, turbidity in water [5]. The Korean livestock, an Indus-based cattle sheds [6], fixed cycle algorithms and variable cycle algorithms were used in cattle shed management systems. BLE beacon-based smart cattle improves power consumption in cattle sheds [7].

2 Literature Review

In the era of high capacity milk parlor, automatic milking system various models were developed and studied based on the sensor for detecting clinical mastitis [8] and other health issues in the cattle. Mixed linear models or principal component analyzes were used to test lactation curves in dairy experiments [9]. Milking frequency requires specialized monitoring as per the lactation and milk production; several studies were made on udder health among conventional milking and automatic milking (AM) [10]. In the case of udder health management and herd somatic cell count (SCC), research has been done mostly in European countries. Udder health is improved with the help of some standard practices associated with lowering the rate of SCC. SSC rate can be lowered using various methods like blanket dry-cow therapy, sand bedding, parental selenium supplementation, frequent use of California Mastitis test, surveillance of dry-cow udder for mastitis, free stall system, udder hair management, and cleaning the Calvin pen after each calving [11]. The automatic milk system (AMS) is popular for having a keen focus on pros and cons regarding mastitis issues and milk quality [12]. AMS analysis is not only for milk but also for animal health issues and welfare [13]. Lameness influences the overall behavior of cattle. Changes occur in the cubical area when cattle is resting, interacting with the milking parlor, AMS [14]. AM farms were found on a large scale in European countries, and research on the relationship of technical, biological, economic issues is more [15]. No studies were found on pasture milk and milking interval (MI) and milking frequency (MF) on milk yield in pasture-based automatic milking system (AMS) cows. A few study analyzes found that compared to indoor AMS feeding cows, lower milking frequency (MF) is observed in pasture-based AMS cows [16]. Milk fatty acid and herbage were affected by botanical composition and grassland management, with less SFA and more polyunsaturated fatty acids (PUFA) found in grazing cows and biodiverse pasture. In contrast, house cows have more saturated fatty acid (SFA) and less PUFA [17].

Milk producers need to face loss due to cow diseases. Breeds with high productivity are more susceptible to udder inflammation than local breeds. Abnormalities in the construction of the udder were found with higher inflammation incidences. Some studies suggested that vitamin E, vitamin A, and selenium deficiency leads to an increase in the number of mastitis incidents [18]. Studies were made to connect technologies, processes, and practices within agriculture and allied industry [19].

Electrical conductivity (EC) was used to detect clinical mastitis, but it has variations in results [20]. AMS helps to increase wild and reduce human resources efforts [21]. Researchers are also trying to hassle-free milk parlors and AMS with advanced tools that will help detect diseases in early stages [22]. Some assessment of technology and strategic methods are helpful to detect diseases in the early stages in the AMS era [23]. Day by day, reducing labor effectiveness will help to reduce errors in milk parlors. On the other hand, AMS helps to improve the overall milking system and alerts about cow health, too [24].

It is observed that automation and allied innovation in milking a cow are becoming popular as it has several benefits compared to traditional approaches. Another important aspect is that it helps in the early detection of health issues like udder issues, mastitis, lameness, and it suggests remedies or first aid from its data repository. A sensor-based model is available for early detection of health issues which alerts stakeholders to take precautionary steps and generate reports on health issues. Overall, there are numerous benefits of automated milk systems and sensor-based health monitoring of cattle. Some novel thermal and non-thermal processes are available in pasteurizing milk, and advances in technology are available. In non-thermal ultrasonic, ultraviolet and irradiation were used while ohmic heating, microwave, and radiofrequency were used in thermal approach [25]. Non-thermal plasma as milk sterilization reduces the microbial cell numbers in the milk [26].

3 Proposed Model

This paper proposes smart cattle shed which monitors atmospheric values inside the shed, cow physical health, diet, and waste management. Smart food containers help monitor moisture and nutrition-based food supply to cattle at the shed and a fresh, pH-balanced, and clean drinking water supply in an acceptable manner.

In the case of nutritionally rich dairy animals, a well-balanced food supplement gives healthy life to cattle, leading to a good quantity of milk. Similarly, drinking water intake must be closely monitored as it must be clean, fresh, and of good pH value. Sometimes, few diseases in cattle spread due to sharing common water tanks as saliva mixed in water causes infection to other cattle, i.e., zoonotic diseases. Contaminated water is the source of a large outbreak of diseases like cryptosporidiosis, cholera, dysentery, etc. Cattle shed overall temperature and cattle temperature are essential aspects to monitor as even an eye temperature gets influenced by pain and heat stress in cattle [27, 28]. As ambient temperature exceeds its threshold value, heat stress is realized in cattle. Skin, eye, and rectal temperature also affect cattle [29]. The dairy farm [30] owners are considered the forefront of farming innovations [31]. The proposed system focuses on maximum efficiency and cattle shed monitoring and management. The primary vertices focused on the proposed work listed below and elaborated in Fig. 2. Monitoring involves overall supervision on all parameters like cattle shed's air ventilation, light, drinking water, food, i.e., nutritional diet, cattle behavior, and health. It can be achieved with the help of implanting wearable sensors.

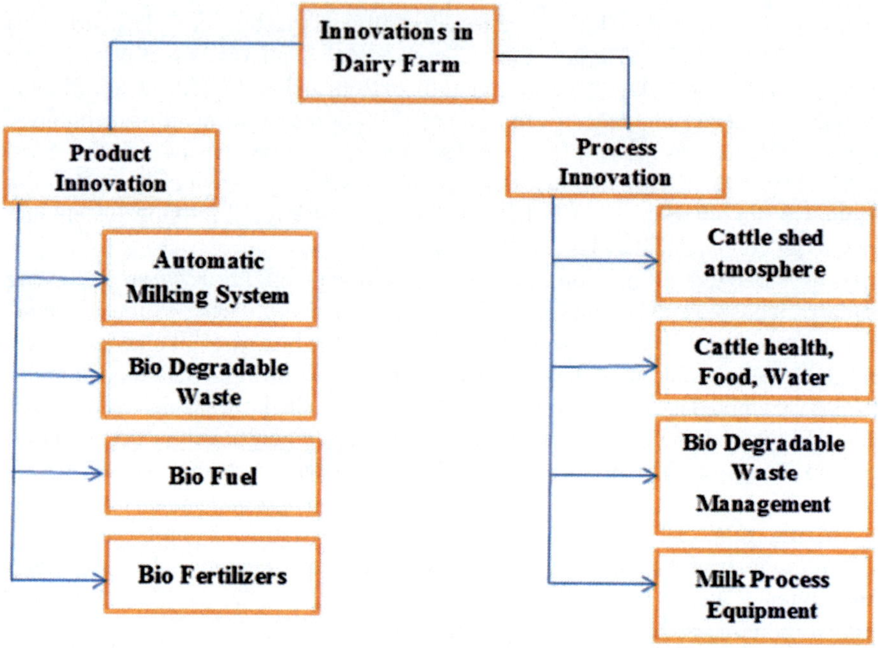

Fig. 2 Diagrammatic representation of the proposed model

3.1 *Cattle Shed Atmosphere Monitoring and Controlling*

Daylight and good airy atmosphere and suitable temperature are achieved via Arduino Uno and sensors-based model. A temperature sensor, humidity sensor, gas sensor (gases like O_2, CO, H_2O, methane), and the light sensor are used for atmosphere monitoring. Good ventilation and lights help to maintain cattle health and behavior too. Inputs from the sensor are processed, and if they reach a specific threshold value, the respective action plan gets executed. For example, as the particular gas concentration increase, the windows or shutters of the shed will get open automatically. The electrical light equipment will start in dim light, which may be due to a cloudy atmosphere. On the appearance of toxic fumes or gas detection, an alarm gets generated in short messages, email, ring-bell, etc.

3.2 *Cattle Health Monitoring*

Wearable sensors are used to get actual parameters of cattle health in real-time mode, and continuous monitoring is achieved. Various body parts of cattle were used to fix different sensors like temperature sensor and microphone at the neck, load sensor under feet, heartbeat sensor at vein on the neck, gas sensor near the nose, electrical conductivity sensor at the udder, accelerometer sensor at neck, feet, udder, and near the tail. These sensors are kept to get real-time data values from a cow, which help to monitor her health. Sensors are used to capture data that recognize diseases occurring in animals.

3.3 Cattle Food and Drinking Water Monitoring and Management

By placing a sensor-based feeding pot, moisture and overall quality of feed material like dry feed, wet feed, oil cake, and other nutritional supplements can be monitored. Weight sensor helps to know about feed consumption. Water containers sensors help to supply good pH-balanced drinking water to cattle. Logs can be maintained for water and feed consumed by cattle daily, and a balanced diet can be given accordingly. Cattle with dry and wet feed along with nutritional supplements remains healthy. A separate feeder water tank for each cow controls diseases usually spread by saliva, like ulceration and drooling.

3.4 Cattle Waste Management, Biofuel, Bio-Insecticides, Biofertilizers

Composting is a safe way to biodegrade organic wastage. Compost helps to remediate soil polluted by heavy metal. Several bio-insects have a vital role in compost to degenerate biomolecules. Some popular insects are black soldier fly (BSF), Milichiidae, Japanese beetles, etc. The few composting methods are Berkley rapid composting, Indian Indore composting, Indian Bangalore composting, sheet composting, status composting, vermicomposting, window composting, and vessel composting. The use of compost increases soil fertility, soil amendment, crop yield, and erosion control.

Biofuel in the form of biogas is prepared from cow dung with the help of the biogas plant, as shown in Fig. 3. Biogas plants can be advanced with the help of IoT technology, giving ease in monitoring, control, and overall management.

Biofertilizers were produced from the waste of biofuel. The popular biofertilizers are Rhizobium, Azospirillum, phosphate solubilizing microorganisms, and silicate solubilizing bacteria. There are six significant steps in making biofertilizers from biofuel such as.

A. Choosing active organisms.
B. Isolation and selection of target microbes.
C. Selection of method and carrier material.
D. Selection of propagation method.
E. Prototype testing.
F. Large-scale testing [32]

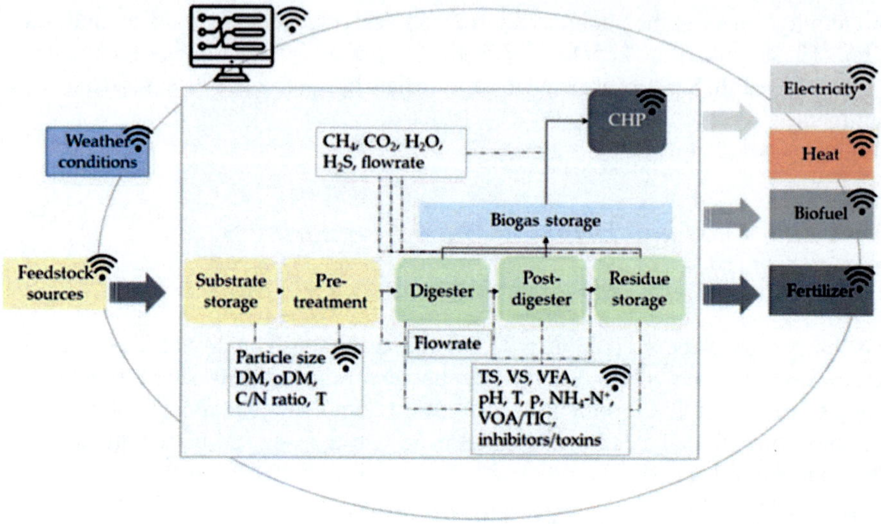

Fig. 3 By-products from cattle waste, i.e., feedstock sources [Cinar, Samet & Önen Cinar, Senem & Wieczorek, Nils & Ihsanullah, Sohoo & Kuchta, Kerstin. (2021). Integration of artificial intelligence into biogas plant operation. Processes. 9. 85. https://doi.org/10.3390/pr9010085.]

Sewage sludge ash is used to produce NPK fertilizers with the help of phosphoric acid as shown in Fig. 4.

3.5 *Milk Process Equipment*

Milk Process Equipment: In pasteurized milk ordinance (PMO), five things need to be accepted and managed in a tax-effective way. These five things are as follows.

(a) each particle of milk must be heated in harmony with the time and temperature criteria,
(b) the equipment or utensils or apparatus must be appropriately designed,
(c) the equipment or utensils or apparatus must be accurately operated,
(d) a temperature standard must be met, and
(e) a holding time or tenure to meet the standard must be achieved in complement to the temperature [34].

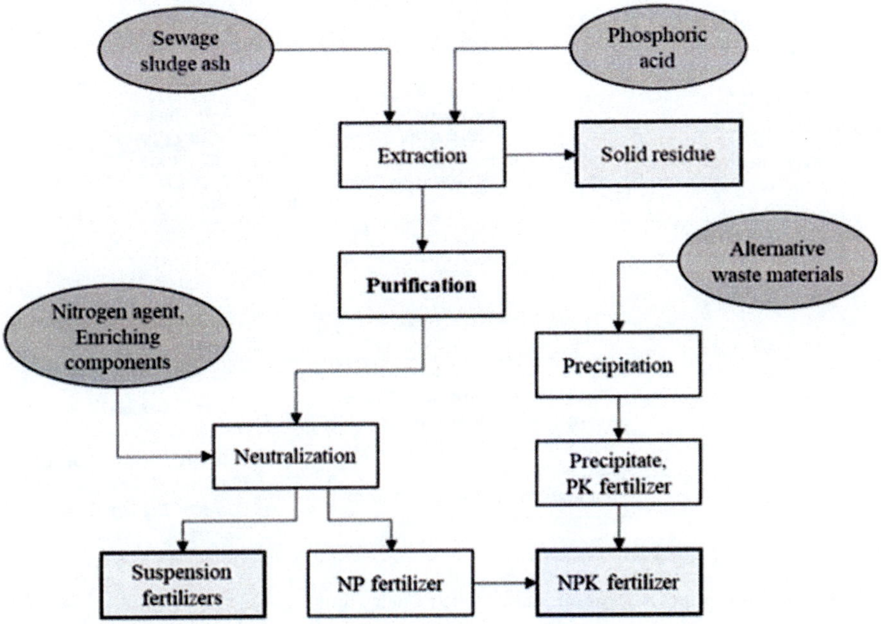

Fig. 4 Fertilizer (NPK) production using sewage sludge ash [33]

3.6 *Processing Milk and Robotic Milking*

For milk preservation, there are three basic approaches such as natural preservative, bio-preservative, and chemical preservative. IoT plays an essential role in monitoring, and controlling the storage unit milk is preserved by two approaches, mainly in a mechanical way, i.e., low-temperature treatment (i.e., refrigeration) and heat treatment method. The heat treatment method consists of five significant options: thermistation, pasteurization, sterilization or UHT, dehydration, and preservative agents.

In the last ten years in robotic milking or automatic milk processing systems, it improves the overall performance of milk parlors and cattle health care. Automatic milking system (AMS) has many advances over the labor-based approach. It helps to detect health issues in the early-stage like mastitis. To preserve milk, bacteria were killed with different methods and were stored in a cool refrigerator within the ideal temperature of 32–39 °F.

The model seen in Fig. 5 has mainly four-component as a dairy farm unit, data receiving unit, software system, and response administrative unit. System receives inputs from all different blocks like AMS or robotic milking, cattle waste management unit, cattle shed unit, and cattle health monitoring unit. AMS informs data from a milking parlor, an automatic milking system. Cattle waste management informs about sewage tank capacity and status, biogas operative status, biofertilizers status.

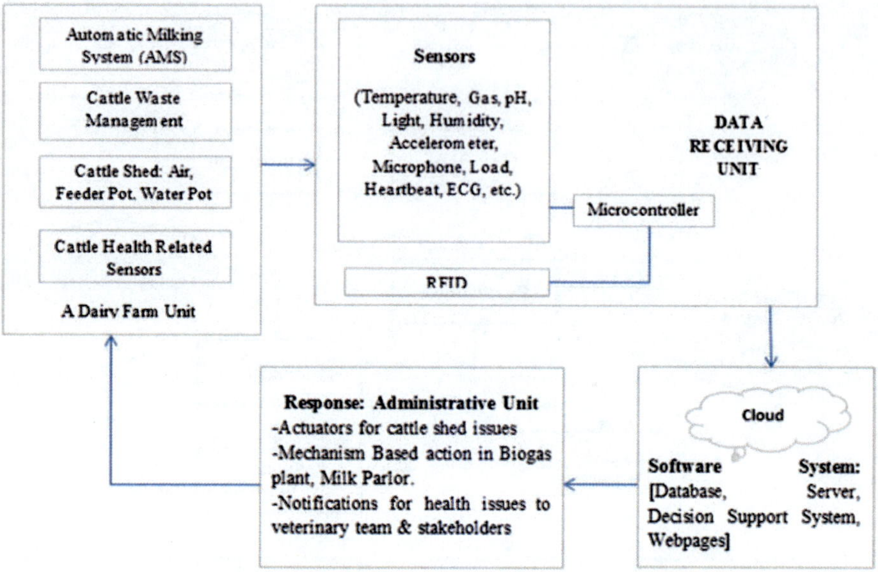

Fig. 5 A model for proposed stem for innovation in a dairy farm

Cattle shed atmospheric values and feeder data, i.e., water tank pH, level, intake by cow similarly feeder pot gives the status of total wet and dry feed consumed by cow. Cattle health-related sensors were giving continuous input from various sensors to the system. The receiving unit sends data in the desired format to a software system. A software system uses these inputs and processes according to the algorithm assigned to it further. It provides a wise solution or a decision. This decision cum action was performed with the help of actuators, for example, open water tap, feeder section, on or off lightening systems, air ventilation system, biogas refill tap on or off. On some specific actions, system automatically write an Email, SMS, dial a call-such different ways of notification to report or alert to veterinary team, doctors, or stakeholders on cattle health issues.

Arduino Uno microprocessor and sensors [35] are used for passing data to the system on a real-time basis. The system uses its AI-ML algorithms on received inputs, and actuators were used to respond with mechanism-based action or alerts via email, SMS like notifications.

4 Conclusion

A proposed model has a powerful strengthened solution for overall cattle health, milk parlor, and AMS effective utilization along with smart cattle shed leading to sustainable growth for the investors in cattle milk with eco-friendly approaches. As a solution-based mechanism for cow waste processing for biofuel, biofuel waste is processed to obtain biofertilizers. It gives environment-friendly solutions and by-products that improve soil nutritional values, fuel solutions, and cost-cutting via IoT-based monitoring and control. It is observed a growth in the cattle milk business and allied sector. The proposed model helps for sustainable and economic growth of cattle milk and, in some certain, agriculture and rural life.

References

1. https://data.worldbank.org/indicator/SP.RUR.TOTL.ZS?most_recent_value_desc=true
2. Gnanaraj AA, Jayanthi JG (2017) An application framework for IoTs enabled smart agriculture waste recycle management system. In: 2017 World Congress on computing and communication technologies (WCCCT), pp 1–5. https://doi.org/10.1109/WCCCT.2016.11
3. Anand PB, Nav´ıo-Marco J (2018) Governance and economics of smart cities: opportunities and challenges. Telecommun Policy 42(10):795–799
4. Saunders T, Baeck P (2015) Rethinking smart cities from the ground up. Nesta, London
5. Ramesh MV, Nibi KV, Kurup A, Mohan R, Aiswarya A, Arsha A, Sarang PR (2017) Water quality monitoring and waste management using IoT. In: 2017 IEEE global humanitarian technology conference (GHTC), pp 1–7
6. Choe J-G, Jang Y-H, Kwon Y-J (2016) Implementation of user's moving direction detecting system using BLE beacon and its application. In: Proceeding of Korean Society for internet information, pp 71–72
7. Yang S-S, Jang Y-H, Ju Y-W, Park S-C (2017) Design of smart Cattle shed system based on BLE Beacon to improve power consumption. Lecture Notes in Electrical Engineering, pp 77–82. https://doi.org/10.1007/978-981-10-7605-3_13
8. Hogeveen H, Kamphuis C, Steeneveld W, Mollenhorst H (2010) Sensors and clinical mastitis—the quest for the perfect alert. Sensors 10:7991–8009. https://doi.org/10.3390/s100907991
9. Macciotta NPP, Dimauro C, Rassu SPG, Steri R, Pulina G (2011) The mathematical description of lactation curves in dairy cattle. Ital J Anim Sci 10:213–223. https://doi.org/10.4081/ijas.2011.e51
10. Hovinen M, Pyörälä S (2011) Invited review: Udder health of dairy cows in automatic milking. J Dairy Sci 94:547–562. https://doi.org/10.3168/jds.2010-3556
11. Dufour S, Fréchette A, Barkema HW, Mussell A, Scholl DT (2011) Invited review: effect of udder health management practices on herd somatic cell count. J Dairy Sci 94:563–579. https://doi.org/10.3168/jds.2010-3715
12. Edmondson P (2012) Mastitis control in robotic milking systems. Practice 34:260–269. https://doi.org/10.1136/inp.e2660
13. Jacobs JA, Siegford JM (2012) Invited review: the impact of automatic milking systems on dairy cow management, behavior, health, and welfare. J Dairy Sci 95:2227–2247. https://doi.org/10.3168/jds.2011-4943
14. Varlyakov I, Penev T, Mitev J, Miteva T, Uzunova K, Gergovska Z (2012) Effect of lameness on the behavior of dairy cows under intensive production systems. Bulg. J. Agric. Sci. 18:125–132
15. Gaworski M, Leola A, Priekulis J (2013) Comparative analysis on effectiveness of AMS use on an example of three European countries. Agron Res 11:231–238
16. Lyons NA, Kerrisk KL, Garcia SC (2014) Milking frequency management in pasture-based automatic milking systems: a review. Livest Sci 159:102–116. https://doi.org/10.1016/j.livsci.2013.11.011
17. Elgersma A (2015) Grazing increases the unsaturated fatty acid concentration of milk from grass-fed cows: a review of the contributing factors, challenges, and future perspectives. Eur J Lipid Sci Technol 117:1345–1369. https://doi.org/10.1002/ejlt.201400469

18. Litwińczuk Z, Król J, Brodziak A (2015) Factors determining the susceptibility of cows to mastitis and losses incurred by producers due to the disease—a review. Ann. Anim. Sci. 15:819–831. https://doi.org/10.1515/aoas-2015-0035
19. Wigboldus S, Klerkx L, Leeuwis C, Schut M, Muilerman S, Jochemsen H (2016) Systemic perspectives on scaling agricultural innovations. A review. Agron Sustain Dev 36:46. https://doi.org/10.1007/s13593-016-0380-z
20. Khatun M, Clark CEF, Lyons NA, Thomson PC, Kerrisk KL, Garciá SC (2017) Early detection of clinical mastitis from electrical conductivity data in an automatic milking system. Anim. Prod. Sci. 57:1226–1232. https://doi.org/10.1071/AN16707
21. Jiang H, Wang W, Li C, Wang W (2017) Innovation, practical benefits and prospects for the future development of automatic milking systems. Front Agric Sci Eng 4:37–47. https://doi.org/10.15302/J-FASE-2016117
22. Hejel P, Jurkovich V, Kovács P, Bakony M, Könyves L (2018) Automatic milking systems—factors involved in growing popularity and conditions of effective operation literature review. Magy Allatorv Lapja 140:289–301
23. Penry JF (2018) Mastitis control in automatic milking systems. Vet Clin N Am Food Anim Pract 34:439–456. https://doi.org/10.1016/j.cvfa.2018.06.004
24. Pitkäranta J, Kurkela V, Huotari V, Posio M, Halbach CE (2019) Designing automated milking dairy facilities to maximize labor efficiency. Vet Clin N Am Food Anim Pract 35:175–193. https://doi.org/10.1016/j.cvfa.2018.10.010
25. Barbosa-Cánovas G, Bermudez-Aguirre D (2010) Other novel milk preservation technologies: ultrasound, irradiation, microwave, radiofrequency, ohmic heating, ultraviolet light, and bacteriocins. Improving Saf Qual Milk 1:420–450. https://doi.org/10.1533/9781845699420.4.420
26. Widyaningrum D, Sebastian C, Pirdo K (2020) Application of non-thermal plasma for milk sterilization: a review. In: Published under license by IOP Publishing Ltd, IOP conference series: earth and environmental science, vol 794, 4th International conference on eco engineering development 2020, 10–11Nov 2020, Banten, Indonesia, p 012146
27. Godyń D, Herbut P, Angrecka S (2019) Measurements of peripheral and deep body temperature in cattle—a review. J Therm Biol 79:42–49. https://doi.org/10.1016/j.jtherbio.2018.11.011
28. McManus C, Tanure CB, Peripolli V, Seixas L, Fischer V, Gabbi AM, Menegassi SRO, Stumpf MT, Kolling GJ, Dias E et al (2016) Infrared thermography in animal production: an overview. Comput Electron Agric 123:10–16. https://doi.org/10.1016/j.compag.2016.01.027
29. Yan G, Li H, Shi Z (2021) Evaluation of thermal indices as the indicators of heat stress in dairy cows in a temperate climate. Animals 11:2459. https://doi.org/10.3390/ani11082459
30. Akbar MO, Ali MJ, Hussain A, Qaiser G, Pasha M, Pasha U, Missen MS, Akhtar N (2020) IoT for development of smart dairy farming, Hindawi. J Food Quality 2020, Article ID 4242805, 8 pages, https://doi.org/10.1155/2020/4242805
31. Akbar O et al. Hindawi Journal of Food Quality Volume 2020, Article ID 4242805, p 8. https://doi.org/10.1155/2020/4242805
32. VS NK, Prabhuswamy MS, Mohanakumara KC, Harsha DN (2016) Performance study of production biofertilizer from biofuel waste. Int Res J Eng Technol (IRJET) 03(07):2145–2149, e-ISSN: 2395-0056
33. Gorazda K, Tarko B, Wzorek Z, Kominko H, Nowak AK, Kulczycka J, Henclik A, Smol M (2017) Fertilisers production from ashes after sewage sludge combustion–a strategy towards sustainable development. Environ Res 154:171–180
34. Rankin SA, Bradley RL, Miller G, Mildenhall KB (2017) A 100-year review: a century of dairy processing advancements—pasteurization, cleaning and sanitation, and sanitary equipment design. J Dairy Sci 100(12). https://doi.org/10.3168/jds.2017-13187
35. Awasthi A, Awasthi A, Riordan D, Walsh J (2016) Non-invasive sensor technology for the development of a dairy cattle health monitoring system. Computers 5:23. https://doi.org/10.3390/computers5040023

Review Paper on Technologies to Curb Noise Pollution in No Honking Zones

Nupur Kulkarni, Shubham Kadam, Apurva Jagtap, Omkar Walhekar, and Rohini Pise

Abstract The impact of noise pollution on the well-being of human beings is severe. It not only affects the health of human beings but also has several psychological, physical, and cognitive issues according to the statistics worldwide. This collectively results in negatively impacting the metropolitan cities, where noise pollution is a nuisance due to several reasons like continuous honking in no honking zones. Traditional methods have low efficiency to solve the problem. This problem can be solved efficiently using appropriate technologies like wireless communication, embedded systems, etc., which can make the complete process fully automated. This paper throws light on the severity of noise pollution, the limitations of existing measures to reduce it and the need of technology to curb the noise pollution. The paper further discusses the different technologies used to reduce noise pollution and presents a comparative analysis of the technologies based on several factors.

Keywords Noise pollution · No honking zones · Smart cities · Embedded systems · Internet of things

1 Introduction

Environmental noise, and especially road traffic noise, is a major environmental nuisance which affects the health and well-being of so many people across the globe. Drastic urban growth shows that there are no signs that the noise pollution in the urban areas will reduce. Instead, the majority of people residing in urban areas will be exposed to the noise pollution in upcoming years which is threatening to human life as well as wildlife.

Looking at the current data of the European Environmental Agency, 12,000 premature deaths and 48,000 cases of ischemic heart disease are caused due to environmental noise a year. Not only that but, around 6.5 million people suffer chronic high sleep disturbance, and about 22 million people suffer chronic high annoyance.

N. Kulkarni (✉) · S. Kadam · A. Jagtap · O. Walhekar · R. Pise
Department of Information Technology, PCCoE, Pune, Maharashtra, India
e-mail: nupursdkulkarni@gmail.com

V. K. Gunjan and J. M. Zurada (eds.), *Proceedings of 3rd International Conference on Recent Trends in Machine Learning, IoT, Smart Cities and Applications*, Lecture Notes in Networks and Systems 540, https://doi.org/10.1007/978-981-19-6088-8_19

Noise pollution has proved to be deleterious for school children as well. Due to aircraft noise, school children suffer reading impairment [1]. In a study conducted to know how much impact noise pollution can have on patients in hospital, majority of patients stated that they were affected by headache, palpitation, irritation, hypertension, nausea, temporary hearing loss, sleeplessness, nervousness, speech interference, mental stress, due to noisy condition which ultimately affected their overall health condition [2].

The policies to prevent noise pollution in these areas have not been strictly followed even in urban cities. Noise pollution standards are not even followed in the no honking zones like hospitals, schools, and libraries. Majority of noise pollution in metropolitan cities are caused because of the traffic noise. Statistics show that the area near ASHP Hospital in Mumbai is the quietest among the metros with a noise level of around 54 dB, but it still happens to be noisier than the limit that is 50 dB.

In a study conducted in 12 schools in Iraq, it was found that the noise levels outside the schools exceeded the upper limits suggested by the World Health Organization for a better learning environment. According to the World Health Organization, the noise level in and around schools should be in the range 40–50 dBA, but the actual noise levels do not reflect that. On an average, the actual noise levels were found to be between 51–73 dBA. As per one of the surveys done in libraries of colleges of arts, education, administration and economic, mathematics and computer science, and dentistry in a city, the noise levels turn out to be 61.7, 57.7, 52.3, 58.5, and 57.3 dBA, respectively [3]. This shows that the recommended noise level of 40 dBA is being surpassed even in libraries.

All these statistics suggest that in spite of the noise pollution standards and laws to reduce the noise pollution, the noise pollution has not been in control, not only in India but worldwide. Honking is one of the major reasons responsible for the same. If certain technologies to reduce the traffic noise in cities are implemented, it will bring noise pollution under control worldwide.

2 Present Measures to Curb Noise Pollution in No Honking Zones

Being highly concerned by the severe effects caused by noise pollution, especially in no honking zones, Governments all around the world are coming up with new initiatives.

2.1 Oren Horn Usage Meter

One of the honk activists in Mumbai, one of the noisiest cities, said that, until people pay for incessant honking, they will not stop doing it. To reduce the unnecessary honking in Mumbai, a group of engineers led by Jayraj Salgaonkar had developed "Oren Horn Usage Meter" in 2014, which did not prevent honking totally, but allowed a limited amount of honking. If the honking exceeded the limit, the vehicle's tail lights were flashed which alerted the police and then they issued a fine on the violators, thus ensuring that the noise pollution is reduced.

2.2 Project Bleep

Project Bleep, launched in 2014, includes a red button on the dashboard that beeps with a frowning face to alert the driver about his unnecessary honking. One of the creators of Project Bleep said that honkers do not realize that they are honking as it becomes habitual for them. This system was tested for around 30 drivers over the period of 6 months, and it resulted in about 61% of reduction in noise.

2.3 Horn Not Ok Campaign

One of the campaigns launched by the Government of Himachal Pradesh in 2018 is "Horn Not Ok" Campaign, in which fines were imposed on violators in case of unnecessary honking and pressure horns. "Shor Nahi" (No Noise) app was also launched to check the level of noise pollution and helped the citizens to register complaints related to noise pollution. Awareness regarding noise pollution was also spread in schools.

2.4 Use of Acoustic Cameras

According to a study in 2017 by the Central Pollution Control Board, it was found that honking was the biggest source of noise pollution in Delhi. In order to bring it under control, a Delhi-based NGO, purpose, had suggested a solution of using acoustic cameras to catch the honking drivers on busy streets in Delhi. Several microphones and sensors are used in these cameras to find the source of honking sound which is above 75 decibels. This was called the "Hands Off The Horn (HOTH)" Campaign. According to the World Hearing Index 2017, Delhi was the world's second city with the worst hearing degradation.

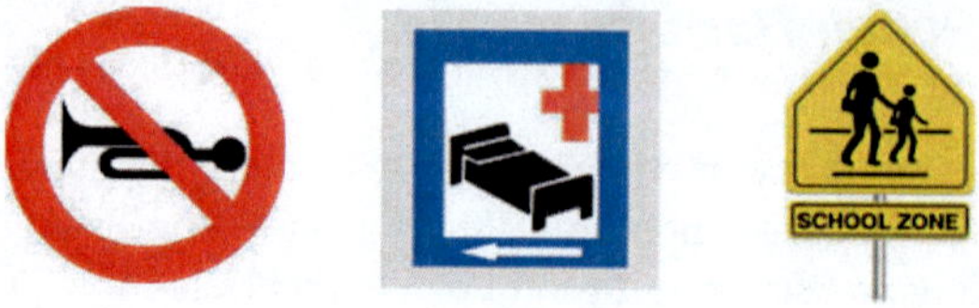

Fig. 1 Road signs for no honking, hospitals, and schools

2.5 No Honking Signs

According to the Central Pollution Control Board, honking is banned in a radius of 100 m around no honking zones like educational institutions, libraries, schools, religious places, and courts. For ensuring this, sign boards as shown in Fig. 1 are used in these places. However, all hospitals, schools, or libraries do not have sign boards, so it is difficult to enforce the rules. So, it is not an efficient way to reduce noise pollution.

2.6 IOT-Cloud-Based

Roads are very noisy due to congestion factors and traffic jams. Aggressive honking is the main factor for noise which is trending on the roads. Many residential places are adversely affected. To tackle the situation, an Arduino-cloud-based system is introduced [4]. Here with the help of Arduino UNO pin set, pulse count time is sent to the cloud-based database.

These data will be uploaded to the Internet application which will further generate bills as shown in Fig. 2. So this will prevent aggressive honking of the drivers.

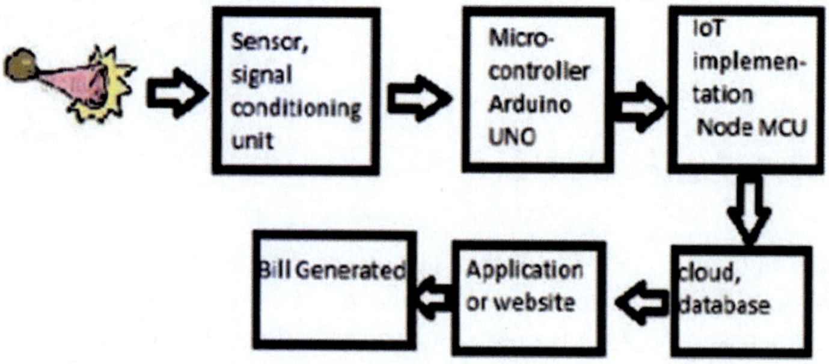

Fig. 2 IOT-cloud-based system to reduce noise pollution

3 Need for Technology to Curb Noise Pollution

As said, noise pollution can negatively affect our health in a variety of ways. All the methods and measures discussed above are truly innovative but need human intervention. The complete process is not automated; at some point, it requires humans to complete the process. Also the measures are costly. So there is a need for technologies to make the solution cost effective, accurate, efficient, and fully automated.

4 Technologies to Curb Noise Pollution

4.1 *In No Honking Zones*

Taking into consideration the concern of noise in no honking areas, several methodologies are proposed. This methodology is unique and also takes care of the safety of people [5]. This system has a simple working. RF transmitters and receivers are used in the system [6]. RF transmitters are placed near no honking areas. As soon as the vehicle enters into a no honking zone, the receiver will receive signals, and it will adjust the horn system in that way. Here, non-ionizing electromagnetic radiation is used.

4.2 *On Residential Roads*

As the number of vehicles increases on the road, there is an increase in noise pollution. To resolve the above problem, a novel approach [7] is introduced. In which infrared transmitter and receiver, global system for mobile (GSM) communications and human detector sensor are used. Proposed technology has a simple system. On a vehicle, the front side transmitter is embedded, and at the side, back receivers are embedded. To make vehicle-to-vehicle communication, IR beam is used. This system also takes care of emergency cases, if any human or animal arrives this sensor detects living things and automobile switches IR-to-HORN.

4.3 *Industrial and Commercial Areas*

Industries have shifted toward technologies that can help reduce noise pollution. There are various technologies such as quieter fans, low noise machines that help reduce noise pollution at the source itself (Fig. 3). There are also some machines that produce high levels of noise which are irreplaceable by other machines. In such cases, noise cancelation technologies are used to generate anti-noise to attenuate

Fig. 3 Silencer used for heavy machines

noise produced by such machines. Acoustic panel technology is also used against noise generated by heavy machines.

4.4 Highways

Residents near a busy highway suffer from heavy noise caused by vehicle honking and their commute. To reduce such noise, tires with low rolling resistance are used. Sound absorbent sheets (Fig. 4) are installed on both the sides of highway to prevent noise from disturbing people residing near to highway [8].

Fig. 4 Sound absorbent sheets

5 Performance Analysis

As noise pollution is a major concern worldwide, technologies can help to bring noise pollution under control. Not only that, but there are several measures implemented to reduce noise pollution, which require human intervention at some point and hence are less efficient and not recommended to use in day-to-day life. Comparative analysis of technologies used in no honking zones and residential roads (Table 1) as well as in industrial and commercial areas and highways (Table 2) have been shown in this review.

RF transmitters and receivers give high accuracy and adjust the horn system when vehicles enter into a no honking zone, but it does not eliminate honking completely. Also, emergency cases while driving a vehicle are not considered by this system, which is highly unreliable.

Table 1 Comparative analysis of technologies used in no honking zones and residential roads

	No honking zones	Residential roads
Technology used	RF transmitters and receivers	Infrared transmitters and receiver, GSM, and human detector sensor
Advantages of technology	Adaptable can be easily integrated into any vehicle	Good vehicle-to-vehicle communication and taking care about emergency cases
Disadvantages of technology	Not good for emergency cases	Lower sensitivity does not work effectively in corner regions
Installation	Integrated in vehicles	Integrated in vehicles
Need of human intervention	Negligible	Little
Cost of implementation	Less	Extremely high

Table 2 Comparative analysis of technologies used in industrial and commercial areas and highways

	Industrial and commercial areas	Highways
Technology used	Acoustic panel technology	Sound absorbent sheets
Advantages of technology	Good in absorbing sound	Customized for every place
Disadvantages of technology	If it is not designed properly, it can absorb large sound which can be disadvantageous and also blocks fresh air	Difficult to recycle
Installation	Installed in rooms	Installed along the either sides of the roads
Need of human intervention	Little	Moderate
Cost of implementation	Moderate	High

Infrared transmitters are embedded on the front, and receivers are embedded on the left and right side of the vehicles. As the count of vehicles is pretty high, it is hard to embed in each one of them.

Also, the cost for these transmitters and receivers is considerably high. And to use this system, it is necessary to embed it in each and every vehicle, which is not feasible. This system also takes emergency cases into consideration, for which a human detector sensor is embedded in every vehicle, but if this system fails to detect the human, then it will lead to immense problems.

Similarly, all the other solutions have their own set of disadvantages which makes it difficult to find a permanent solution to the problem of noise pollution, especially in no honking zones. We can still improve the existing systems to curb the noise pollution drastically.

6 Conclusion

After referring and analyzing various resources on noise pollution, it was observed that several technologies exist which can reduce noise pollution in no honking zones, residential roads, industrial, commercial areas, and highways. Various technologies like IoT [9], receiver-transmitters, acoustic panel technology, sensors, and sound absorbent sheets are used to control noise pollution to a great extent. But, certain factors like human intervention, cost of implementation, and ease of usage play a wide role in determining how effectively it can be used in day-to-day life to reduce the noise pollution in an impactful way. Hence, it can be concluded that the existing technologies can still be improved for better with respect to automation, installation, implementation, etc.

References

1. Qzar I, Azeez N, Alkinani S (2020) the-impact-of-noise-pollution-on-schools-students-of-Basra-city-Iraq-a-health-study-8210. Eurasian J Biosci 14:5197–5201
2. Rahman MM, Ali MA, Khatun R, Tama RA (2016) Effect of noise pollution on patients in hospitals and health clinics of Mymensingh Sadar Upazila 18:2028–9324
3. Alahmady K, Al-Annaz L, Al-Tammer M (2018) Analysis and assessment of noise pollution in libraries
4. Dey A, Majumdar A, Pratihar R, Majumder BD (2019) Design of a smart real-time excessive honking control system 14:8–12. https://doi.org/10.9790/1676-1406010812
5. Munaf S, Monika S, Monika G, Gopika V, Arthi S (2020) Auto honking control system for vehicles. Int J Eng Sci Res Technol 9. https://doi.org/10.29121/ijesrt.v9.i6.2020.13
6. Nerkar N, Nerkar M (2018) Green horn: novel design of honking to reduce the effect of noise pollution. Int J Res Appl Sci Eng Technol (IJRASET) 6(X), ISSN: 2321-9653, IC Value: 45.98, SJ Impact Factor: 6.887
7. Palankar AA, Gopi N, Hegde M, Naveen MR, Anguraja R (2017) A novel approach to reduce the noise pollution caused by honking of vehicles. Int J Eng Res Technol (IJERT) NConPCS—2017 5(18)

8. OECD Road Transport Research (2006) Road infrastructure rehabilitation and safety strategies in Central and Eastern European Countries. OECD Document (also available in Russian), Paris
9. Ahire B, Sakhare S (2021) Sound prohibited zone for smart cities using IoT. Tehnički glasnik 15:92–97. https://doi.org/10.31803/tg-20210205091148

News Channel Debate Analysis: A Detailed Insight

Anamika Chauhan, Sonal Bera, Ritik Garg, and Rishabh Sinha

Abstract The main concept behind this paper is to go beyond the surface and delve deep to perform a micro-analysis of the debates held by various channels. Although the debates are available in the public domain, an analysis such as this has not been carried out. It demonstrates the immense and vast use of NLP and ML techniques in various fields. The task at hand is to create a profile for a particular news channel based on the debates organized by it. Trends and various other parameters have been taken as features for profiling. This accords us a somewhat subjective result, open to interpretation up to a certain degree. Our interest and aim are to present an exhibit of what can be accomplished using these techniques.

Keywords Machine learning · NLP · SVM · NMF · TF-IDF · Topic modeling · NLTK · Classification · Vectorization

1 Introduction

In today's era of a technology-driven world, it is extremely important to keep in track and utilize the various state-of-the-art facilities to help organize our lives in a better and simple manner. Rarel do we ever stop and think about the accuracy of the news presented to us. Slowly but surely what we see and hear becomes a part of our personality and more essentially a part of the community we live in, and therefore by extension the community's ideology and creed. This is what makes digital media

A. Chauhan (✉) · S. Bera · R. Garg · R. Sinha
Delhi Technological University, New Delhi, India
e-mail: anamika@dtu.ac.in

S. Bera
e-mail: sonalbera_2k18it120@dtu.ac.in

R. Garg
e-mail: ritikgarg_2k18it099@dtu.ac.in

R. Sinha
e-mail: rishabhsinha_2k18it098@dtu.ac.in

V. K. Gunjan and J. M. Zurada (eds.), *Proceedings of 3rd International Conference on Recent Trends in Machine Learning, IoT, Smart Cities and Applications*, Lecture Notes in Networks and Systems 540, https://doi.org/10.1007/978-981-19-6088-8_20

a double-edged sword, that too a very delicately balanced one. Hence, it is very important for us to look in retrospect and ask ourselves when do we inspect the source of our beliefs and our principles/all this. One of the major driving forces in today's tech sphere is machine learning. Machine learning makes use of the huge amount of data that is produced by the people and for the people. One of the areas that work in close collaboration with ML is natural language text processing (NLP). NLP may loosely be defined as the working of ML techniques on text rather than data [1]. This study traverses a few functionalities and techniques of NLP to help understand how this cutting-edge technology can be used for getting insights down to every little detail. The central idea of this paper was to go beyond the surface and delve deep in order to perform a micro-analysis of the debates held by various channels. Although the debates are available in the public domain, an analysis such as this has not been carried out. It demonstrates the immense and vast use of NLP and ML techniques in various fields.

2 Techniques and Tools Used

This section covers the basic level details about the important techniques that were used in the implementation.

2.1 Non-negative Matrix Factorization

NMF is a matrix factorization method that requires the matrices to be non-negative [2]. Suppose, we factorize a matrix X into two matrices W and H so that $X \sim WH$. Now, suppose that X is composed of m rows, is composed of k rows, and is composed of m rows. Each row can be considered a data point. For instance, in the case of decomposing images, each row is a single image, and each column represents some feature. In this paper, we employ NMF to do topic modeling as an alternative to latent Dirichlet allocation (LDA), another topic modeling approach [3].

2.2 TF-IDF Vectorization

The term frequency inverse document frequency (TF-IDF) refers to the relevance of a word in a corpus or dataset. Term frequency (TF) and inverse document frequency (IDF) are two concepts in the TF-IDF. It is frequently used in information retrieval, text mining, and user modeling searches as a weighting factor. The TF-IDF value rises in proportion to the number of times a word appears in the document and is offset by the number of documents in the corpus that include the term, which helps to compensate for the fact that certain words appear more frequently than others in general. (TF-IDF, Wikipedia) [4].

2.3 Classification Algorithms—KNN and SVC

The K-nearest neighbors' algorithm (KNN) is a non-parametric classification and regression approach. The input is the k closest training instances in the feature space in both situations. Depending on whether KNN is used for classification or regression, the following is the result: The outcome of KNN classification is a class membership. We use KNN as a classification method in this paper. Support vector machines (SVMs) are supervised learning models using learning algorithms that evaluate data for classification and regression analysis in machine learning. It uses a statistical learning framework to give one of the most reliable prediction approaches. An SVM model is a representation of the instances as points in space, mapped such that the examples of the different categories are separated by a large distance. New instances are then mapped into the same area and assigned to one of the categories based on which side of the gap they land on [5]. We employ SVC as another classification purpose.

2.4 VTT to Transcript

The transcript was retrieved in video text track (VTT) format after data was obtained from a YouTube playlist page for a certain channel. These files required to be transformed into readable format before they could be used for any further text processing. WebVTT was popularized based on the parent paper—"Interface for smart audiovisual data archive" published in 2015. The purpose of such a system as the author states is to annotate inserted media (video or audio) and search in the previously annotated media files. The system is self-reliant and capable enough to perform all necessary steps to annotate media [6]. In the most recent advances, the paper "SPEECH RECOGNITION USING LONG SHORT-TERM MEMORY RNN" discusses the ways to improve the existing benchmarks in speech recognition through the use of LSTM and RNN for speech recognition [7].

2.5 Bokeh Library

For current Web browsers, Bokeh is an interactive visualization library. It provides simple, concise graphics architecture to enable high-performance interactivity across large or streaming datasets. Bokeh is a Python-based Web-based interactive visualization library. The procedural flow of how Bokeh assists in presenting data to a Web browser is depicted in the diagram below. This library was used to plot almost all of the graphs in the research, and they are all interactive to improve the user experience.

3 Literature Review

The paper "A Multimodal Approach for Early Detection of Cognitive Impairment from Tweets" brings forth a novel approach to attempt to detect and solve the problem of cognitive impairment. The paper stresses that an early identification would help prevent more serious diseases like Dementia and Alzheimer's. The author infers that the tweets shared by a person can be a source for identification of the condition. The paper utilizes NLP and ML concepts such as Levenshtein distance and fuzzy matching to score the tweets indicating the degree of CI and thus profile the person accordingly. Thereon, text mining approach was used to get the help of relevant local authorities to help facilitate the person [8]. The authors in "examination of fake news from a viral perspective: An interplay of emotions, resonance, and sentiments" discusses a new perspective on the problem of fake news detection. It introduces new parameters like sentimental resonance and emotions to help predict fake news. It also analyzes the impact of association of content and news article title attributes such as length on the probability of news being fake [9].

The paper "smart literature review: A practical topic modeling approach to exploratory literature review" aims to create a foundation for researchers to reduce the work of manually labeling a research paper into a category. It harnesses the LDA technique for topic modeling to perform an exploratory literature review in a faster, more transparent, and reliable way than manual exploration. It provides a good use case as the ease of use for modeling and the swiftness of the framework over the manual process seems very promising, though it has a few limitations [10].

"A Comparative Study of Sentiment Analysis Using NLP and Different Machine Learning Techniques on US Airline Twitter Data": In this 2021 paper, the author mainly describes the various ML classification methods available for the classification of text data which in this case is the US Airline Twitter data. The paper follows the basic outline of any NLP classification task. It first pre-processes the raw dataset by removing the punctuations and stopwords followed by tokenization and lemmatizing. Subsequently, it uses TF-IDF for vectorization. The best results for classification were achieved by SVC and logistic regression when using hold-out cross-validation with 25% data as test-set. This paper presents the most rudimentary summary of how to classify text data [11]. The 2016 paper "Statistical Topic Modeling for News Articles" is similar to [8] which also uses LDA for topic modeling. In this paper, the author uses Stanford Topic Modeling Toolbox for the topic allocation of 400 news articles. The paper then analyzes the results and possible conclusions [12].

4 Proposed Methodology

The entire study is carried out in three stages. The 1st stage mainly consists of preparing the data frame for topic modeling and classification. The 2nd stage consists of writing functions required for profiling a news channel on the basis of debates it has organized. The third and last stage consist of a thorough comparison of all news outlets in order to arrive at a valid conclusion.

4.1 Topic Modeling and Classification

4.2 Data Acquisition

The data to be acquired in this case is the transcript of every debate the channel has carried out to date. This paper was done with the help of the "YouTube Downloader" library. All those news channels which have a YouTube playlist especially dedicated to debates were taken into consideration; thereon, the debate's transcript was extracted using subtitles. It is very important to note that the dataset comprises only such debates whose subtitles were made available by the channel and uploaded on its YouTube page. To convert the ".vtt" files to readable format, WebVTT library was used. Another important note is that only selected debates from channels were taken as this will be a test dataset large enough so that when performing debate classification, we get a real-life situation. The dataset comprised 1600 debates approximately, which were a mix of debates across channels.

4.3 Stopwords List

These debates are real-life and live situations in which persons from different backgrounds have to present their views. As opposed to newspapers and online articles, this form of media is unfiltered and undergoes no pre-processing. Therefore, it is important that these words be filtered out before any further processing. A document-term-matrix (DTM) was created using count-vectorizer, and common stopwords of the English language were filtered out here itself by using the NLTK stopwords set. Thereon, the DTM was inspected, and very common words were added to the stopwords list. An interesting observation was that almost all words with less than 3 letters were insignificant and were added to the list as well. On the other hand, in this huge list (approx. 4500 words), if a word (ex. "BJP") was important in the context of topic modeling; it was removed from the list along with other meaningful words.

4.4 Topic Modeling

The dataset needed pre-processing and cleaning. TF-IDF vectorization was performed on the dataset to identify common words present in every debate. It was due to the reason that the stopwords we use in formal writing language are different from the stopwords we use in everyday conversations. Hence, a list of the most common and unnecessary words which held no significance in the context of the debate was filtered out using TF-IDF vectorization as well. A new DTM was produced using TF-IDF. TF-IDF was followed by topic modeling using the NMF technique. A simpler combination of using count vectorization followed by LDA was also tried but it yielded worse results. Thus, we used the former combination. The latter pair also takes probability as its genesis which is not the best-suited approach for this enormous text dataset. The top 40 words for each topic were used to identify the relevant topic. The number of topics was experimented with for better results, and it converged to 8 topics.

4.5 Classification

TO make a classification model that could predict a debate belonging to which topic, the model had to be trained first. Thus on an already labeled dataset, as inferred from above, different classification techniques were used for classification. A pipeline consisting of TF-IDF vectorization put together with the stopwords was followed by the classifier to be used. SVC classifier was seen to have the best accuracy and overall F1-score. Thus for further classification tasks, we use the SVC classifier as our model.

4.6 Profiling

A number of functions were created to give an OOP approach to the problem. The data (debates) would be encapsulated, and the entire process abstracted from the user. The user would simply have to use the functions mentioned in order to get a detailed profile of the channel. Each function served a unique and inherent purpose. Two of the major functions are discussed: • topic distribution and trend analysis—These functions combined create a profile for the distribution of topics for a given channel and its trend analysis for the topics. • Out-of-context—This function evaluates the percentage of out-of-context windows within a debate, calculated cumulatively for a particular channel. This function follows the principle of the sliding window method. Here is a given window (500 words) if after classification, the number of topic-related words present in the window is lower than a threshold (7 words), then the window is said to be out-of-context. The sliding length is 150 words. Through repeated sliding of this window, we traverse the entire debate and then consequently through all the debates of a channel, to create a new feature: out-of-context minutes percentage.

4.7 Comparison

For all comparisons between channels to give an insight, visual representation graphs have been used throughout. Bokeh and "matplotlib.pyplot" were the two libraries to be used for plotting. Bokeh provided the interactive graphs on a Web page. Since the number of debates in a given timeframe was variable for different channels, the absolute count was not taken as a metric for plotting. Instead, the percentage of debates during the period was selected as the Y-axis measure for comparison across channels as it would be channel-independent. On the X-axis for deciding a timeframe, 90 days or 3 months were taken as a unit. This was taken after considerable deliberation, as it would provide a balanced outlook for the reader's interpretability.

5 Results

See (Figs. 1, 2, 3, 4 and 5).

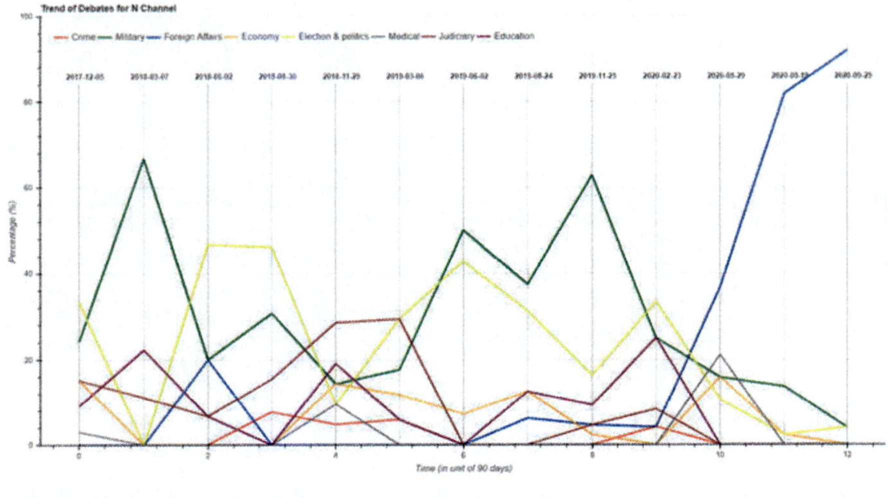

Fig. 1 Trend of debate for a leading news channel (N)

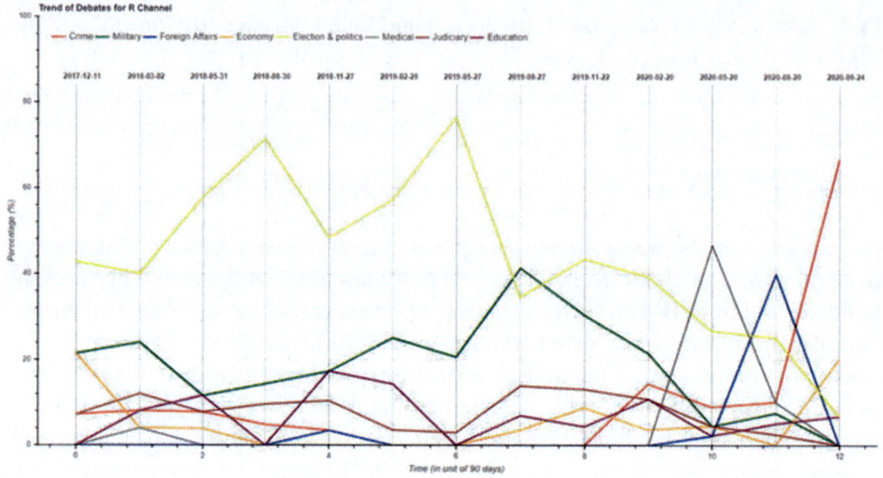

Fig. 2 Trend of debate for a leading news channel (*R*)

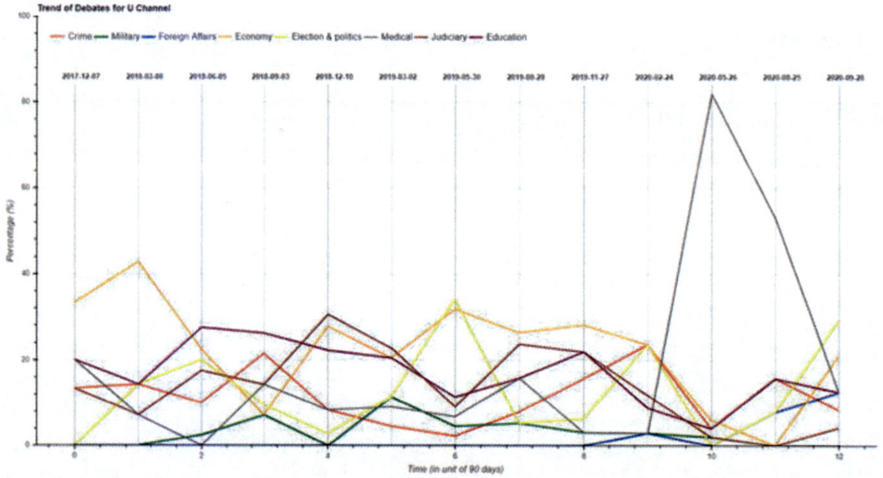

Fig. 3 Trend of debate for a leading news channel (*U*)

6 Analysis

The aforesaid profiling was done for 5 channels as shown above. 2 channels had a lesser number of debates than compared to the other 3 and did not have the same time period in which the trend analysis was carried out. For each channel, debates over a period of almost 3 years from Dec'17 to Sep'20 were taken for analysis. This time period was divided into 13 equally spaced units of roughly 90 days. For each

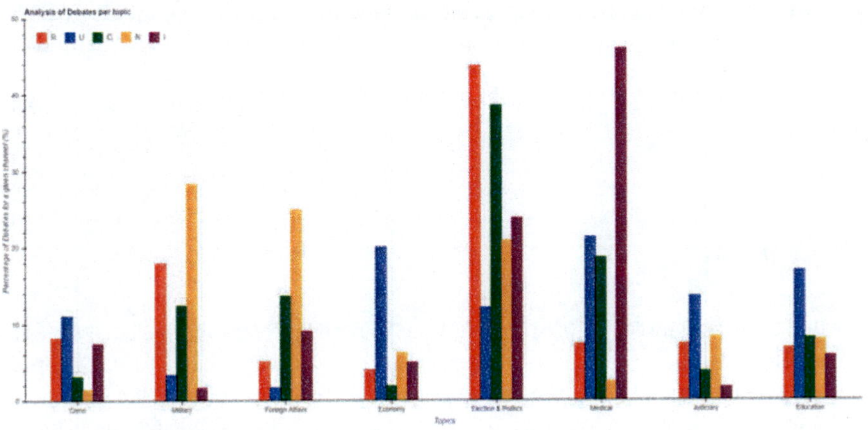

Fig. 4 Analysis of debates per topic

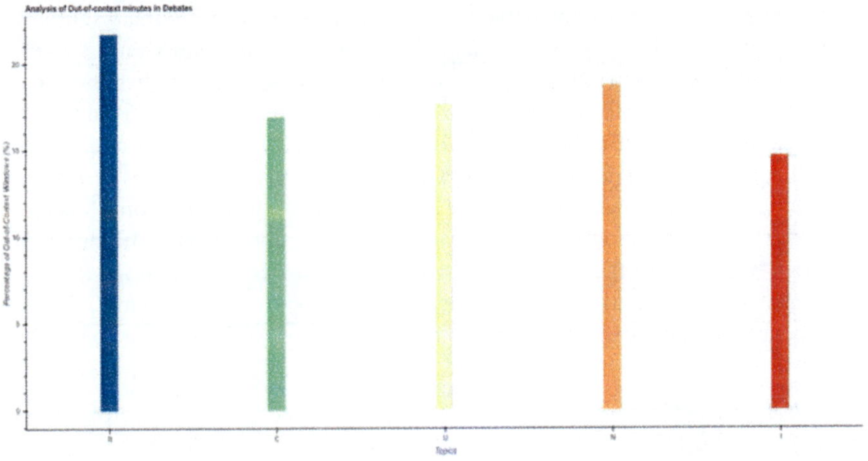

Fig. 5 Analysis of out-of-context minutes in the debates

time period, the debates conducted were divided into the respective topics and their respective percentage for that span was plotted to obtain a trend for 3 years.

The trend graph for Channel-R (Fig. 2) can be interpreted as follows:

From the beginning of the 2nd quarter of 2018 to the beginning of 2020, the number of debates related to election and politics increased, whereas the debates related to other topics like education and economy were not held by the channel. This stands consistent with the real-world fact as in that interval India had its elections, which may account for the sharp increase in debates related to Election and Politics. From 2020 onwards, we notice an increase in the number of debates for topics "Medical" and "Foreign Affairs". This also keeps in line with the reality as it was during that interval COVID-19 outbreak hit India, and India had to consider its relations with other countries. In between, we also observe some debates related to "Military" which can be related to the Indo-Chinese tensions that arose during that patch. A similar line of conclusion can be drawn for other channels using the trend graph (Figs. 1 and 3).

7 Conclusion

We see that one of the most uniform distributions of debates over time can be observed for Channel-U. While comparing the percentage debates attributed to each topic conducted by each channel over the entirety of 3 years (Fig. 4), we note that Channel-R conducts most of its debates related to "Election and Politics". While Channel-I had its focus on "medical"-related issues. The most balanced channel seems to be "M/U", which seems to focus on every topic except "Military" and "Foreign Affairs" a bit. An out-of-the-box feature "out-of-context minutes" (Fig. 5) shows the percentage of windows in the course of the debate that was not related to the topic being discussed. The figure plots this metric for each channel, and we see that while almost every channel has some windows (almost 18%) out-of-context, Channel-R has the most out-of-context windows with 22%, and Channel-I has the lowest with 14%. This study merely presents an example of how powerful and insightful ML, and NLP techniques can prove to be when used in different fields with proper considerations. We reiterate our view that any bias in news would impact a large chunk of the population. Hence, it is pertinent that we have neutral and unbiased and healthy discussions. This analysis is a tool to get an idea of the profile for a channel. With this, we can hope to have some form of rough idea as to what channels subject us to. In no manner, do we intend to conclude that a certain channel is better or worse.

8 Further Scope

This paper has been implemented in 3 stages with the scope of improvement in two of them. During the classification of debates, neural networks (ANN/DNN) can be used as the preferred algorithm for achieving high levels of accuracy. This would also incur an overhead of training the data on a large scale and a complex architecture to handle text. Secondly, to offer more insights, an n-gram profiling could also be done on the debates [13]. This could be incorporated in the graph as well, to enhance understanding regarding the sub-topics of a group of debates in the given timeframe. This type of analysis could also be cross-referenced with other countries' news channels, and the contrast or the similarity observed could help provide interesting insights. Nevertheless, this paper is capable enough to be extended across different fields and provides a framework for this type of analysis. A number of applications based on this framework could be designed. For example, similar profiling of newspapers could also be done. This paper essentially opens a portal for a bulk of data that is unlabeled (unsupervised) and waiting to be analyzed!

References

1. Sahin O (2020) A gentle introduction to ML and NLP. https://doi.org/10.1007/978-1-4842-6421-8_1
2. Wang Y-X, Zhang Y-J (2013) Nonnegative matrix factorization: a comprehensive. Review. https://doi.org/10.1109/TKDE.2012.51
3. Guven ZA, Diri B, Cakaloglu T (2019) N-stage Latent Dirichlet allocation: a novel approach for LDA. https://doi.org/10.1109/UBMK.2019.8907050
4. Trstenjak B, Mikac S, Donko D (2014) KNN with TF-IDF-based framework for text categorization. Procedia Eng 69:1356–1364. https://doi.org/10.1016/j.proeng.2014.03.129
5. Foody GM, Mathur A (2004) Toward intelligent training of supervised image classifications: directing training data acquisition for SVM classification 93(1–2):107–117. https://doi.org/10.1016/j.rse.2004.06.017
6. Koctúr T, Pleva M, Juhár (2015) Interface for smart audiovisual data archive. https://doi.org/10.1109/RADIOELEK.2015.7129036
7. Rizvi SA, Sundararajan M (2021) Speech recognition using long short term memory RNN. Int J Adv Res Eng Technol (IJARET) 10(6):315–320
8. Thakur N, Han CY (2021) A multimodal approach for early detection of cognitive impairment from tweets. https://doi.org/10.1007/978-3-030-85540-6_2
9. Nanath K, Kaitheri S, Malik S, Mustafa (2022) Examination of fake news from a viral perspective: an interplay of emotions, resonance, and sentiments. https://doi.org/10.1108/jsit-11-2020-0257
10. Asmussen CB, Møller C (2019) Smart literature review: a practical topic modeling approach to exploratory literature review. https://doi.org/10.1186/s40537-019-0255-7
11. Tusar MT, Islam MT (2021) A comparative study of sentiment analysis using NLP and different machine learning techniques on US airline twitter data
12. Suganya C, Vijaya MS (2016) Statistical topic modeling for news articles. https://doi.org/10.14445/22315381/IJETT-V31P242
13. Thakur N, Han CY (2021) Country-specific interests towards fall detection from 2004–2021: an open access dataset and research questions. https://doi.org/10.3390/data6080092

Disease Recognization of Plant Using Different Image Processing Algorithm

Gulbakshee Dharmale, Karan Kavatage, Pranav Kul, Parth Gite, and Sumit Ahire

Abstract Cultivation plays a vital role in development of the Indian economy. Almost 50% of the nation's workforce is employed in this industry. The Indian agricultural component is one of the important aspects of Indian economy. So, disease recognition of plants is very important so as to maintain sustainable agriculture. Different plants tend to show different symptoms when they are affected by variety of factors such as virus, bacteria, and climatic conditions. This paper presents survey on different image processing algorithm used for plant disease detection. The first step in plant disease detection is segmentation for which K-means clustering algorithm is used which retrieves textual features from input images. The edge detection of the leaves can also be done by canny edge detection which pays attention to the edges in a very robust way. Then, the feature extraction process is usually done using grey level co-occurrence matrix (GLCM). Followed by feature extraction, the classification comprises of K-nearest neighbour (KNN) which has accuracy about 90% for different inputs. Another classifier which can be used is support vector machine (SVM). Hence, machine learning plays an important role in the disease detection of the plants.

Keywords K-means segmentation · GLCM · KNN · Viral · Fungal · SVM · CNN · Machine learning · Deep learning

1 Introduction

India has a population of 1.39 billion approx., and agriculture is one of the most important aspects of Indian financial system. India has a population of about 95.8

G. Dharmale · K. Kavatage (✉) · P. Kul · P. Gite · S. Ahire
Department of Information Technology, Pimpri Chinchwad College of Engineering, Pune, Maharashtra, India
e-mail: karankavatage@gmail.com

G. Dharmale
e-mail: gulbakshi.dharmale@pccoepune.org

V. K. Gunjan and J. M. Zurada (eds.), *Proceedings of 3rd International Conference on Recent Trends in Machine Learning, IoT, Smart Cities and Applications*, Lecture Notes in Networks and Systems 540, https://doi.org/10.1007/978-981-19-6088-8_21

million of farmers. Agricultural sector produces 18% of India's GDP. So, be safe to suppose that if agronomy was advanced, it would profit nation significantly.

Plants are affected by various factors such as insects, bacteria, viruses, and different climatic factors leading them to various diseases. At the right time, farmers are not able to recognize the disease properly they may face a huge loss. As according to today's technology, we have various ML algorithms, AI, DIP techniques which would eventually be used practically to help farmers.

By using an external device, pictures of pretentious parts can be taken clearly, upload it to database, spot the specific disease, and suggest the precise dealing of the disease and pesticide. Utmost vegetation is disease-ridden by several factors such as climatic factors, fungous, and microbial diseases. DL practices are frequently used for form recognition as they have demonstrated to be operative. Our main motive is to recognize the plant disease using K-means clustering and classification of syndrome by means of KNN algorithm.

2 Literature Survey

Garima Shresta 2020 elaborated that disease on plant leaves can be detected with the help of convolution neural network and deep learning. She has also stated in her paper that convolution neural network is been separated into three layers input layer, unseen layer, and yield layer which helps us whilst finding [1].

The results show that after analyzing a database of 3000 images, final test accuracy was 88.80%, and train accuracy was 97.42%.

U. Mokhtar, M. A. S. Ali, A. E. Hassenian, and H. Hefny implemented the disease recognition of leaves of tomatoes. In this disease detection, the image segmentation was done using SVM classifier, and to clarify image, its background was removed by morphological operation. SVM is a direct model for sorting and regression complications. Thus, it can crack linear and non-linear difficulties and work fine for many real-world problems [2].

The outcome of this work has complete precision obtained by this method which is greater by Laplacian and Cauchy kernel functions that is 98 and 100% in comparison with Invmult Kernel that is 78%.

Pujari [3] used a variety of crop types for the fungal disease detection on the leaves of plant including distinct crops such as cereal, fruit, and vegetable crop. Different method has been implemented for each type of crop [3]. The segmentation in cereal crops was done using K-means clustering, and edge detection using canny was used for finding edges and textures.

For vegetable crops, the segmentation was done using Chan-vase method. For feature extraction, local binary patterns were used. SVM and KNN are used for texture feature extraction. The overall accuracy was about 87.2%. In fruit crops, K-means and ANN were used for segmentation and classification purpose, respectively. Overall accuracy was 90.72% in the case of fruit crops.

AI Hiary H, ALRahamneh Z, Bani Ahmad S, Reyalat M, Braik M proposed an automatic recognition and sorting of leaf disease. This method is based on K-means segmentation and ANNs as a classifier [4]. After comparing two to three models, the highest overall accuracy achieved was 94%.

Singh [5] used a hereditary algorithm as an image breakdown technique to automate and classification of the affected or diseased plants. He tested four sets of plant leaves which consisted of the leaves of rose, lemon, beans, and banana. Colour co-occurrence technique was used for feature extraction of plant leaves; this method extracted both colour and textual features. The image segmentation part was done using two algorithms that is SVM classifier and K-means clustering with an accuracy of 95.71% and 86.54%, respectively [5]. When the genetic algorithm and minimum distance criterion classifier both were combined the accuracy rises to 93.63%.

Kiani [8] used a fuzzy decision maker to identify virus-infected plants in a strawberry arena underneath outside conditions. The complete correctness whilst finding and separation of plant disease was 97%, with a dispensation time of 1.2 s for disease recognition [6].

Ali [], their research seeks to employ the ΔE colour alteration process to segregate the virus-affected zone in addition to diagnose diseases using colour histogram and textural cues, with an overall accuracy of 99.9% [7]. KNN, boosted tree, bagged tree, and cubic support vector machine were amongst the classifiers utilized. On RGB, HSV, and LBP features, the bagged tree was on top when compared to the remaining classifiers, reaching 99.5%, 100%, and 100% accuracy, respectively. Fine KNN, boosted tree, and cubic SVM classifiers all performed well, with accuracy of 88.9%, 90.1%, and 50.90%, respectively.

G. Saradhambal (2018) proposed a method for developing an automatic plant disease detection system. The K-means procedure and the classifier (Otsu's) used in study to guess the diseased part of leaves. The anticipated work extracted both figure and surface features. This work hauls out shape-oriented features such as area, colour axis length, peculiarity, firmness, and boundary, whilst surface-oriented features included distinction, association, and similarity. Finally, in this study, classification was performed using a neural network-based classifier [8].

Table 1 represents different techniques used for plant disease detection. A variety of different crops are taken into consideration to predict the disease with the help of different algorithms such as KNN, K-means, GLCM, ANN, and CNN. Whilst using these algorithms, it is found that different algorithmic techniques tend to show different result and accuracy.

3 Existing Methodology

The methodology in plant disease detection consists of four major parts as shown in the flow chart given in Fig. 1. The methodology for plant disease detection undergoes four main processes which we will discuss in detail. These four steps are as follows:

Table 1 Different techniques used for plant disease detection

Sr. No	Title of paper	Author and year	Methodology	Result and accuracy
1	CNN-based plant disease detection	Shrestha et al. [1]	CNN and deep learning were used for detection of diseases in the plant leaves. The input, unseen, and hidden were the three layers used in CNN	The final test accuracy was 88.80%, and train accuracy was 97.42%
2	Tomato leaves disease detection using SVM	Mokhtar et al. [2]	In this paper, disease of leaves of tomatoes was detected using SVM. To clarify the image, its background was removed by morphological operation	The outcome is greater than Laplacian and Cauchy Kernel functions that is 98% and 100% in comparison with Invmult Kernel that is 78%
3	Fungal diseases detection in plants using image processing	Pujari et al. [3]	In the case of vegetable crops, segmentation was done using Chan-vase method. In fruit crops, K-means and ANN were used for segmentation and classification purpose, respectively	The overall accuracy was about 87.2% and 90.7% in the case of vegetable crop and fruit crop, respectively
4	Accurate and fast disease detection of plants	ALRahamneh Z, Braik M, Al Hiary H, Reyalat M, Bani Ahmad S (2016)	This study comprises of involuntary recognition, organization of leaf illnesses has anticipated, and this technique is grounded on K-means as a segmentation method and ANN classifier algorithm	After comparing two to three models, the highest overall accuracy achieved was 94%
5	Plant leaf disease detection using image Segmentation and soft computing techniques	Singh and Misra [5]	Hereditary algorithm as an image breakdown technique was used in automation and classification for the affected or diseased plants. Four sets of plant leaves were tested	SVM and K-means clustering had an overall accuracy of 95.71% and 86.54%, respectively

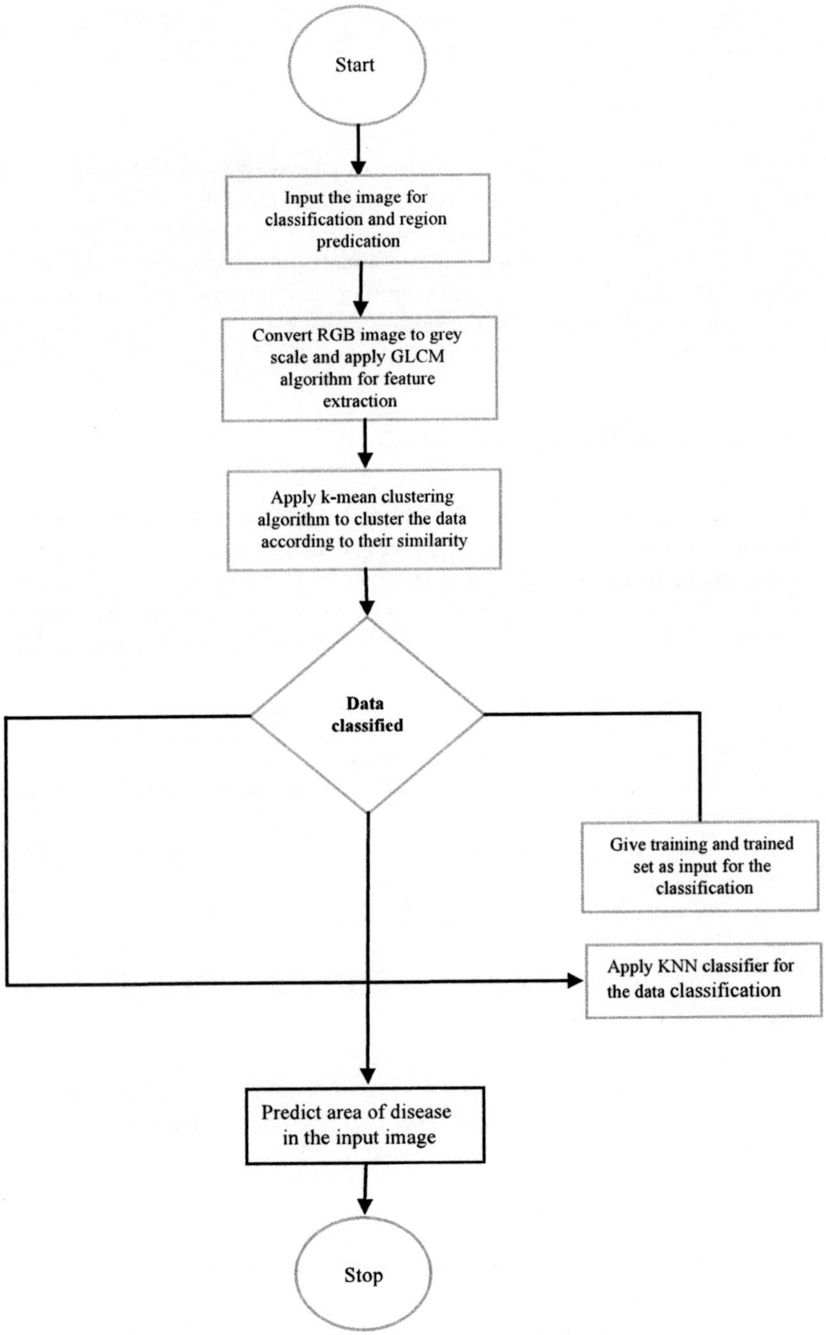

Fig. 1 Flow chart representing methodology in plant disease detection

(1) Image acquisition (fetching of image)
(2) Image segmentation (breaking of image)
(3) Feature extraction
(4) Classification

These four steps are very important in plant disease detection. The image acquisition requires a camera for fetching of the image; segmentation work is done us with the help of K-means. For feature extraction, GLCM algorithm is been used, and lastly, classification is done using different image processing algorithms such as ANN, CNN, and KNN. Different classification algorithms give different accuracy depending upon the disease, the plant is suffering from.

3.1 Image Acquisition

This process requires fetching of image via external media devices such as camera and mobile phone with appropriate size and clarity. The image database is important as it provides maximum accuracy of the used classifier in final part of disease detection. This process is the operation of acquiring an image from a source, usually hardware systems such as cameras, sensors, and so on, in image dispensation and machine visualization.

The following are the steps which undergo this algorithm:

Step 1: Choose 'k' primary cluster that is centre.
Step 2: Compute point to cluster centre space with other observation to all centres.
Step 3: The cluster should be assigned to each observation with closest centroid.
Step 4: Find and analyze average of the consideration with every cluster to obtain latest centre position.
Step 5: Repeat steps 2 to 4 until cluster allocation remain unchanged.

3.2 Segmentation of Image

Approach of segmentation process is presentation of image in such a way that it becomes easy to understand an analyse. This part is an important aspect of image processing. Usually, segmentation part is completed by the means of K-means segmentation.

K-means segmentation/clustering algorithm:

- The segmentation part is done by reducing the variation of distances between cluster centroids and intensities of image.
- This algorithm is also known as Llyod's algorithm, which is a repetitive procedure which divides the information & allocates 'n' annotations defined k clusters, also known as centroid.

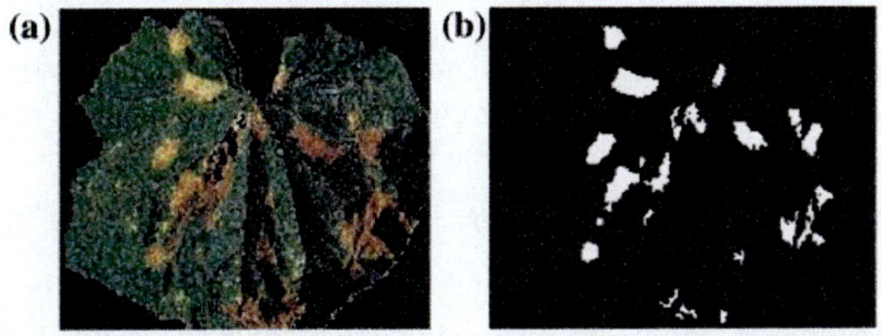

Fig. 2 **a** Original image. **b** Segmented image

- 'K' is the number of clusters you want to group your data points into, has to be predefined (Number of clusters)

3.3 Feature Extraction

IN this particular step, we have to extract characteristics from our area of study. This is done to acknowledge the sample image. GLCM algorithm can utilized for feature extraction. Colour, textures, and shape structures are inputs which are to be taken. Distinction, regularity, difference, and entropy can be involved to texture. For shape, chubbiness, region, strangeness, and concavity, features are recognized. GLCM can be used in number of application and is easy to use and implement when compared to other algorithms.

Feature extraction process works in following ways:

Step 1: Take image matrix and measure the number of pixels in it.
Step 2: Store the measured pixel value in the image matrix.
Step 3: Use histogram procedure to check resemblance between pixels in image matrix.
Step 4: Compute difference factor from the matrix.
Step 5: Normalize the elements by dividing the pixels (Fig. 2a and b).

3.4 Classification

This process is done after image segmentation. Classification is generally done using various algorithms. As it is one of the most important aspects in this report.

(a) Classification using KNN:

KNN is the classifier used for various classification difficulties. It does not need any information of SVM or any additional ML algorithm. It is not necessary to train new skills if new working form is bonded to existing set. Before new vector element classification, distance metrics should be used for comparison with training sample. It is 'K'-nearest neighbours which are further taken into consideration where most occurred class amongst neighbours are weighed via measurement of distance. Proper flow of system rests on appropriate collection of limits which is k that signifies number of neighbours to give a class to the newer elements.

(b) Rules and Phases of KNN Classifier:

It generally has two phases. The first one includes training phase where detailed labelling of leaf image with their classes is done, and other is evaluation in which images of leaves are unlabelled, and procedure produces the detail of K-nearest statistics opinion to label the unlabelled point and categorizes their modules. KNN instructions are given as follows:

- Kept value of testing and training statistics.
- Euclidian measurement distance is used as a distance framework to calculate the space between stored record and categorize

$$d\,(pi,\ pq) = \sum \sqrt{\left(p_{ir} - q_{jr}\right)^2} \tag{1}$$

- Obtain K near neighbours and utilize tags of closest neighbours to obtain the class tag of unspecified information by computing greatest poll.

3.5 Performance Measures

By means of recital of classifier, we calculate relating to training and testing. The performance of KNN classifier on the testing set is estimated by the four reserved constraints recall, F1-score, accuracy, and precision. ROC curve is utilized to compute the instructed classifier rendition. We use confusion matrix as to define the no. of true positive, true negative, false positive, and false negative which would eventually help us to evaluate the criteria.

$$\text{Accuracy} = \text{TP} + \text{TN}/\text{TP} + \text{TN} + \text{FP} + \text{FN} \tag{2}$$

$$\text{Precision} = \text{TP}/\text{TP} + \text{FP} \tag{3}$$

$$\text{Recall} = \text{TP}/\text{TP} + \text{FN} \tag{4}$$

$$\text{F1} - \text{Score} = 2(\text{Precision} * \text{Recall}/\text{Precision} * \text{Recall}) \tag{5}$$

Figure 3 represents different steps involved in plant disease detection. The first step is input image which undergo image segmentation in which the diseased part shows change in intensity of the colour. After colour segmentation, green channel

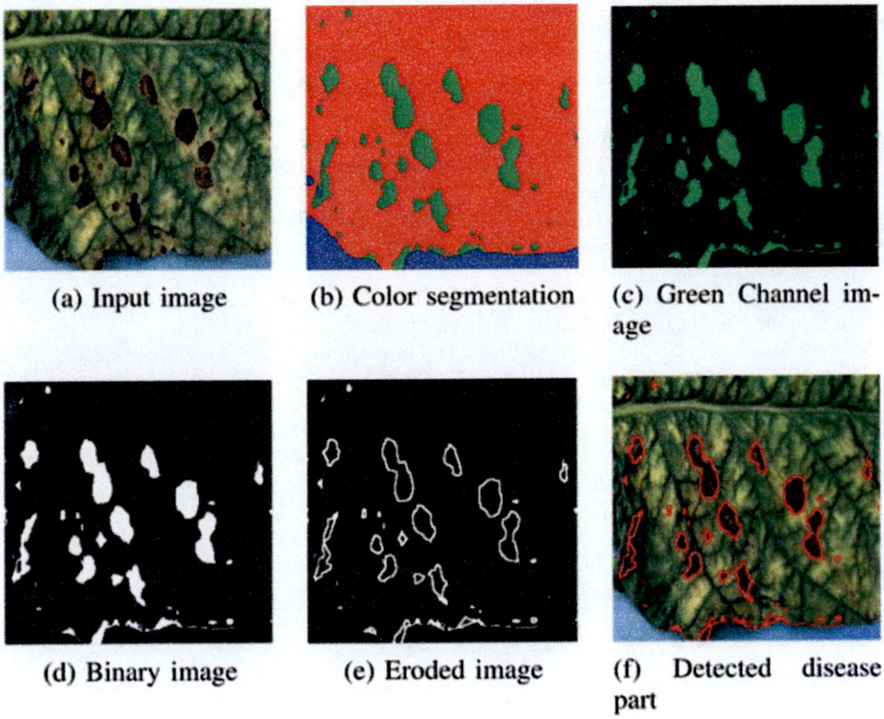

Fig. 3 Steps involved in plant disease detection

image is extracted from the original image which is then converted into binary image. Thus, after the encoding the binary image, the disease or affected part of the leaf is identified.

4 Conclusion

From the survey of different image processing algorithms for plant disease detection, it is stated that KNN method used to spot and categorize several diseases that is present in plant leaves tend to have maximum accuracy whilst compared with different image processing algorithms. The bifurcation of the diseased part is executed using K-nearest neighbour classifier, and GLCM texture features are being operated for the classification and feature extraction, respectively. KNN classifier grounded bifurcation outcome gives us optimal accurateness in plant disease recognition. Also, the classification criteria of the KNN on plant leaf virus likewise offer maximum precision and projected KNN, and GLCM grounded tactic has better results whilst comparing with other present approaches.

References

1. Shrestha G, Das M, Dey N (2020) Plant disease detection using CNN. In: 2020 IEEE applied signal processing conference (ASPCON), 07 Dec 2020, pp 109
2. Mokhtar U, Ali MAS, Hassenian AE, Hefny H (2015) Tomato leaves disease detection approach based on support vector machines. In: 11th International computer engineering conference (ICENCO), Cairo, Egypt, pp 29–30
3. Pujari JD, Yakkundimath R, Byadgi AS (2015) Image processing based detection of fungal diseases in plants. In: International conference on information and communication technologies, vol 46, pp 1802–1808
4. Al-Hiary H, Bani-Ahmad S, Reyalat M, Braik M, ALRahamneh Z (2011) Fast and accurate detection and classification of plant diseases. Int J Comput Appl 17(1):31–38
5. Singh V, Misra AK (2016) Detection of plant leaf diseases using image segmentation and soft computing techniques. Inform Process Agric 8:252–277
6. Bashish DA, Braik M, Ahmad SB (2010) A framework for detection and classification of plant leaf and stem diseases. In: International conference on signal and image processing, pp 113–118
7. Bhange M, Hingoliwala HA (2015) Smart farming: pomegranate disease detection using image processing. In: Second international Symposium on computer vision and the internet, vol 58, pp 280–288
8. Kiani E, Mamedov T (2017) Identification of plant disease infection using soft-computing: application to modern botany. In: 9th International conference on theory and application of soft computing, computing with words and perception, vol 120

A Detailed Survey on Sentimental Analysis on Social Media

Gulbakshee Dharmale, Shreenidhi Karjagi, Harshal Kotalwar, and Shakyadeep Khobragade

Abstract Public opinion or point of view matters a lot as it plays vital role in shaping country's image all around the world, may it be on certain products, decisions or their own feelings/expressions. At this time, people are more expressive on social media, more rather than in person, they express or share their daily tasks their feelings, opinions some good or bad incidents, and many more on social media, and we can collect this data and process into classifiable form. The data can be collected through API and then transformed into classifiable form. To classify data, we will be foreseeing three algorithms which are Naïve Bayes, maximum entropy, and lexicon-based. The main reason of gathering this data is to process and generate positive and negative notions of certain demography or specific person. This data can be useful in many ways such as to check if a person is in good mental health condition or suffering through depression through his/her positive negative posts on social media. Multi-national companies can collect insights on their products based on peoples review and opinion on social media. MNC and e-commerce stores can also product recommendation system based on those reviews. In short, sentiment analysis plays a very vital role in our economy as well as health care.

Keywords Opinions · Positive · Negative · Social media · Sentiments · Naïve Bayes · Maximum entropy · Lexicon-based · API's · Algorithms · Classifiers

1 Introduction

Social media comprehends of billions of users across the world; on this platform each and every topic such as current affairs economy, sports, reviews, opinions, and much more is discussed. In this discussion, the outcome is either positive or negative,

G. Dharmale · S. Karjagi (✉) · H. Kotalwar · S. Khobragade
Department of Information Technology, Pimpri Chinchwad College of Engineering, Pune, Maharashtra, India
e-mail: shreenidhikarjagi29@gmail.com

G. Dharmale
e-mail: gulbakshi.dharmale@pccoepune.org

V. K. Gunjan and J. M. Zurada (eds.), *Proceedings of 3rd International Conference on Recent Trends in Machine Learning, IoT, Smart Cities and Applications*, Lecture Notes in Networks and Systems 540, https://doi.org/10.1007/978-981-19-6088-8_22

for example, if Indian cricket team wins, Indians will share positive or happy posts, but in case of rival team, it could be otherwise. Trillions of data are generated each day, and some these are personal expressions such as if a person is feeling sad, he/she shares that he/she is feeling sad today and vice-versa. People are tending to be more expressive on social media the reason for such expressiveness towards social media was found out to be that netizens (abbreviation for people using Internet), and they would not judge them based on their ideology unlike in our surroundings if the ideology of a person is not plausible, then the person and his opinions are mocked or make them feel embarrassed. We can use this data and conclude positive or negative notions from them for many purposes as discussed prior. Data can be gathered through API which are free for developers so that they can work on such projects and make the best use out of it, for classification, we have three possible algorithms Naïve Bayes, maximum entropy, and lexicon-based.

2 Literature Survey

David Osimo and Francesco Mureddu research challenge on opinion mining and sentiment analysis using [5] Naïve Bayes classifier had an accuracy of 78.80%.

Adobe Social Analytics: This assesses the impact of social media on businesses by determining how online conversations and communities impacts on marketing performance. It compares the influence of those interactions with important business dimensions such as revenue and brand value after capturing and interpreting the ongoing dialogues. Aside from that, it tracks how businesses interact with their clients on social media, such as how Facebook posts influence site traffic and purchase decisions [6].

A Literature Survey on Sentiment Analysis Techniques Involving Social Media and Online Platforms: This survey depicts the possible techniques and importance of sentiment analysis. This survey consists of all information required for sentiment analysis such as need of sentiment analysis the importance theoretical knowledge of sentiment analysis [7].

Adaptive Co-training SVM for Sentiment Classification on Tweets: X. Cheng and H. Shen compare both SVM and maximum entropy and state that SVM [8] has 77% accuracy, and maximum entropy has 85%.

Table 1 represents the prior research on sentiment analysis and their outcomes to analyse which approach is better.

3 Approaches

There are many possible approaches for analysing and classifying these approaches for analysing natural language processing which extracts opinions off the statement and sends it for classification. For classifier, we will be looking at three possible

Table 1 Literature survey

S. No	Title of paper	Author and year	Methodology	Result and accuracy
1	Sentiment analysis for social media	Jayasanka et al. [1]	In this research, the data was classified using Naïve Bayes classifier	The results show that the data collected had an accuracy of 78.80%
2	Improved lexicon-based sentiment analysis for social media analytic	Jurek et al. [5]	The approach in this research was lexicon-based data for classification of the data	The results depict that the accuracy of the positive negative values is 78%
3	Sentiment analysis on Twitter using maximum entropy and SVM	Ermatita et al. [4]	The classifier used in this research was SVM and maximum entropy, and best amongst both is chosen	The result has two outcomes that are SVM of accuracy 77%, and maximum entropy has accuracy of 85%
4	Naive Bayes and sentiment classification	Stanford university (2021)	Method used was Naïve Bayes for spam detection	The result given out was precision of 86% and accuracy of 85.43%

approaches, there are more than three approaches, but the accuracy level precision is not up to the mark, that is, the reason we will be foreseeing only three approaches those are as follows:

I. Naïve Bayes
II. Maximum entropy
III. Lexicon-based

3.1 *Naïve Bayes*

It is a supervised learning method based on the Bayes theorem that is used to solve classification issues. It is commonly used in text classification with a large training dataset. The Naive Bayes classifier is one of the simplest and most effective classification algorithms available, and it assists in the development of fast machine learning models capable of giving accurate predictions. It is a probabilistic classifier, which means it makes predictions based on an object's probability. It is an algorithm used to find most probable value for classifying test data in most appropriate category. In this case, test data is document tweets. There are 2 document stage.

I. Training
II. Classification

$$\text{General formula}: P(H|X) = P(X|H) \times P(H)/P(X) \tag{1}$$

where

a. $(H|X)$ is the probable final probability (rearward probability) of hypothesis H occurs when given evidence E occurs.
b. $P(X|H)$ is the probability that when E occurs, it will affect hypothesis H.
c. $P(H)$ is the prior probability hypothesis H occurs regardless of any evidence.
d. $P(X)$ is prior probability evidence E regardless of hypothesis or other evidence.

In this theory, 2 variables are used as characteristics as hypothesis (H) and sentiment as evidence (E). The other 3 variables will be used as metadata as of sentiment. Sentence containing of many words which is very challenging in practice to decide which one might be called aspect or feature it is presumed each word is an aspect or feature. Application of Bayes theorem (Fig. 1):

$$P(F|K) \times P(K)/P(F)$$

F = Feature word
K = Category/sentiment value because the features or words that support the same category can be various, there are features F_1, F_2, F_3 can be transformed into

$$P(K|F_1, F_2, F_3) = P(F_1, F_2, F_3) \times P(K)/P(F_1, F_2, F_3) \tag{2}$$

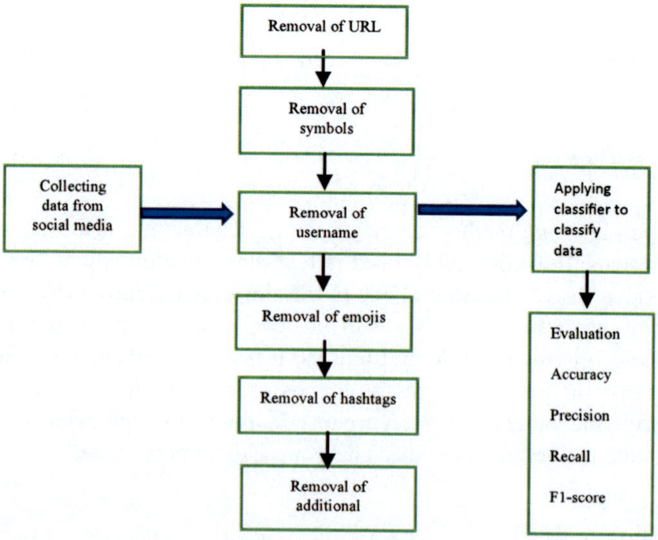

Fig. 1 Flowchart of classifier

It requires that the evidence (in this case is a feature or word) that exists is independent of each other, then formula can be changed to $P(K|F_1, F_2, F_3) = P(F_1|K) \times P(F_2|K) \times P(F_3|K) \times P(K)/P(F_1) \times P(F_2) \times P(F_3)$. If described in general, the formula is given as

$$P(K|F) = \prod P q i = (F_1/K)/P(F) \tag{3}$$

3.2 Maximum Entropy

It is a machine learning algorithm based on empirical data (information that comes from the research) and provides probabilities on which sentence belongs to specific class.

Unlike the Naive Bayes classifier, it does not assume that the data is independent; however, instead of using probabilities to define the model's parameters, it employs a search strategy to identify the set of parameters that will maximise the classifier's performance. The outputs of the procedure, like any learning technique, are dependent on the input dataset. There are no assumptions made about the features' relationships. The fundamental goal of the strategy is to optimise entropy in the system in order to predict label condition distribution in each class.

$$P_{\mathrm{me}}(c/b, \lambda)\frac{\exp\left[\sum \lambda_i f_i(c, b)\right]}{\sum_c \exp\left[\sum \lambda_i f_i(c, b)\right]} \tag{4}$$

where c is a class, b is social media like Twitter, λ is a weight of vector.

3.3 Lexicon-Based

This approach calculates sentiment function of whole document/data or set of sentences from semantic orientations of lexicons. The orientations can be positive negative or neutral. The dictionary of lexicons can be automatically or manually be generated, SentiWordNet dictionary is most commonly used. Firstly, lexicons are found from the dataset, and then, SentiWordNet can be used to discover the synonyms and antonyms to expand the dictionary. This technique uses adjectives and adverbs to discover the semantic inclination of text. For calculating inclination, the adjective and adverb union are extracted with their sentiment value. These can be transformed to single score for whole text.

So these are the three best approaches for classification in sentiment analysis; after classification, the process is not over we need to see how to import data how to make it classifiable (i.e. suitable for classification) then evaluate the data.

4 Procedure

The process of analysing basically consists of three phases.

1. Crawler/Web crawler
2. Sentiment analysis tool
3. Data mining (optional)
4. Evaluation

1. **Crawler**: The main motive of crawler is to collect data from Twitter or any social media platform to data source for easiness of classification and analysis. For importing, it has to use Twitter API to gain access to Twitter data. Twitter has many endpoints that have been customised for certain use cases like user stream, public stream, and site streams; for crawler, we will be using public stream.
2. **Sentiment analysis tool**: This is where the operations on data occur such as processing data into classifiable form, classifying data. The data here is same as the data that is crawled from the social media. It analyses sentiments and depicts it in accurate variance. It uses algorithms and techniques to derive most accurate output for given sentiments with the help of proper classifier. Further after retrieving, labelled text corpus is to extract features from it and to train the classifier. The entire system revolves around on how favourable this method of extraction is. Therefore, it uses several methods in sentiment analysis as follows.
   - Unigrams—Considers word individually in sentences as feature set of corresponding categories. In this case, it does not contemplate any relation between words.
   - Unigrams except stop words—It is similar to unigrams, besides it does not contemplate stop words that is a list of words which often appears in almost all sentences that has no context
   - Bigrams—Consider adjoining pair of words from sentences as a feature set of corresponding categories.
   - Bigrams except stop words—It is similar to bigrams feature set, besides from the words in the stop words list
   - Most informative unigrams and bigrams—Gets the feature set with unigrams and bigrams with most information and greatest frequency (occurrence of data).

Out of these methods, the most informative unigrams and bigrams are selected for data mining for forecasting future trends based on current analysed data.

3. **Data Mining**: During the product profiling, decision tree is used after correlating it with clustering technique, and for the trend analysis and forecasts it using Holt-Winters method, it is capable of analysing seasonal data and predict proper values for the future. (Holt-Winters method: it is a model of time series behaviour. Forecasting always requires a model.
4. **Evaluation**: After classification of data into positive negative or neutral sentiments, the most important task is evaluation. In evaluation process, we check for

accuracy precision and recall. The data which is more precise and accurate is preferable and suitable because the algorithm with more accuracy and precision is considered ideal. Operations performed in evaluation are

- Accuracy
- Precision
- Recall

$$\text{Accuracy} = \frac{\text{TP} + \text{TN}}{\text{TP} + \text{TN} + \text{FP} + \text{FN}} \tag{5}$$

$$\text{Precision} = \frac{\text{TP}}{\text{TP} + \text{FP}} \tag{6}$$

$$\text{Recall} = \frac{\text{TP}}{\text{TP} + \text{FN}} \tag{7}$$

- Accuracy: It is used to determine which model is best at identifying relationship and patterns between variables.
- Precision: It defines number of positive class that actually belong to positive class.
- Recall: It states how many true states were found.

The basic structure of the model would be as shown in Fig. 2.

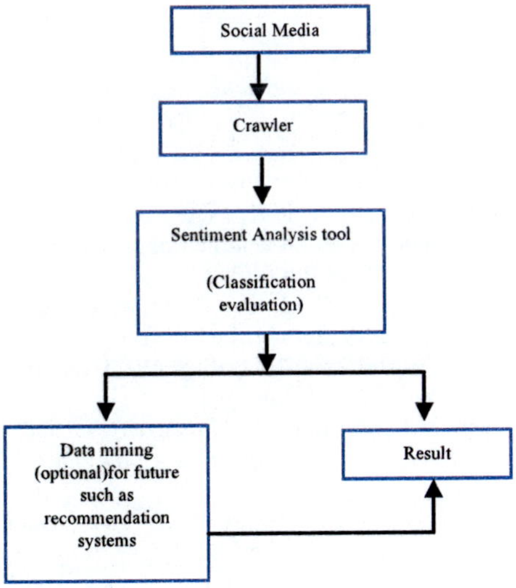

Fig. 2 Flowchart of model

5 Conclusion

Through this we can conclude that we have foreseen every successful feasible and best suited approaches for sentimental analysis. The main question is which is the best approach in case of Naïve Bayes approach the data or sentiments is assumed independent of each other which is quite difficult every word is related to each other, but still, the resultant has good accuracy and precision as it collects positive and negative words. In case of maximum entropy unlike Naïve Bayes, data is not assumed independent and consider a whole which would be good for accuracy, but it requires more time to train and also overfitting may occur. Lexicon-based approach is complete different than above two it searches for positive negative words extracts its sentiment scores through the help of SentiWordNet and evaluates the sentiment of the statement it calculates orientation of whole document which is time consuming, and the accuracy is low compared to two other approaches. Many researches have stated that both maximum entropy and Naïve Bayes give out same accuracy and precision, but maximum entropy has an issue of overfitting that may affect accuracy in further cases or datasets, so with this we can conclude that Naïve Bayes is the most preferable and suitable approach for sentimental analysis, and also the most important part is accuracy which depends on dataset, each data may have different accuracy; in some cases, maximum entropy may give more accuracy than Naïve Bayes, but in this case, Naïve Bayes performs better than Maximum Entropy.

References

1. Jayasanka SC, Madhushani MDT, Marcus ER, Aberathne, IAAU, Premaratne SC (2013) Sentiment analysis for social media. In: Conference: information technology research symposium, vol 4, Project: Market potential of thumba karawila, November 2013
2. Adobe social analytics US social analytics best practice guide (2016)
3. A literature survey on sentiment analysis techniques involving social media and online platforms. Int J Sci Technol Res 9(6):1–8
4. Ermatita E, Cindo M, dian Palupi Rini (2020) Sentiment analysis on Twitter using maximum entropy and SVM. Published 17 April 2020 Computer Science
5. Jurek A, Mulvenna MD, Bi Y (2015) Improved lexicon-based sentiment analysis for social media analytic. Sec Inform 4
6. Jurafsky D, Martin JH (2021) Naive Bayes and sentiment classification. In: Speech and language processing. Stanford University. Copyright © 2021. All rights reserved. Draft of December 29, 2021
7. Osimo D, Mureddu F (2012) Research challenge on opinion mining and sentiment analysis. Technical report, September 2012
8. Adaptive co-training SVM for sentiment classification on tweets. CIKM'13: Proceedings of the 22nd ACM international conference on information & knowledge management, October 2013, pp 2079–2088

Perceiving Correlation Among Spatiotemporal Gait Parameters and Verifying Its Relation Using Machine Learning Classification Technique Pilot Study for Indian Population

Neha Sathe, Anil Hiwale, and Archana Ranade

Abstract World is surrounded with key buzz emerging from technological use by common man. Technology lends a hand to achieve the fitness by providing assistance. Walking is predominantly and usually preferred exercise by mass. Walking pattern analysis or gait analysis expands the probable zone for researcher to contribute in gaining and maintaining fitness for common people. There are number of parameters influencing walking pattern of the person. The motive of this study is to verify the relation among various spatiotemporal parameters used in gait analysis. Paper tests and comments on correlation among selected spatiotemporal parameters on the basis of analysis performed on 50 healthy participants from Pune, India. Contributing participants are in the age group of 20–75 years combining male and female members. GAITRite Walkway is used to perform recording of the gait pattern. Pearson's correlation coefficient is used for the testing the correlation within two selected parameters. The correlation coefficient varies within range of −1 to +1, indicating positive, negative, or no correlation. Correlation results are verified through application of them in classification of subjects in two age groups. Support vector machine, logistic regression, and K-nearest neighbor machine learning approaches are tested.

Keywords Spatiotemporal parameters · Correlation coefficient · Machine learning · Gait analysis

N. Sathe (✉) · A. Hiwale
MIT World Peace University, Pune, India
e-mail: Neha.sathe@mitwpu.edu.in

A. Hiwale
e-mail: anil.hiwale@mitwpu.edu.in

A. Ranade
Deenanath Mangeshkar Hospital and Research Centre, Pune, India
e-mail: Archana801@gmail.com

V. K. Gunjan and J. M. Zurada (eds.), *Proceedings of 3rd International Conference on Recent Trends in Machine Learning, IoT, Smart Cities and Applications*, Lecture Notes in Networks and Systems 540, https://doi.org/10.1007/978-981-19-6088-8_23

1 Introduction

Understanding the importance to maintain or regain health in due time is now a wide spread thought, as the latest technology helps the layman in utilizing the technical features available for the understanding of wellbeing. For example, use of fitness band, support of various smart phone-based applications which are used to monitor day-to-day activity of the person, etc., increase awareness of people toward health wellness. Human locomotion or movement from one place to other, its relation with exercise, and benefits of performing routine exercise are dominant areas being explored by all age groups to contribute in maintaining healthy lifestyle. Understanding the normal walking pattern, i.e., gait pattern of each person and having quantitative analysis at preliminary stage provides the plenty assistance. If needed, auxiliary qualitative clinical analysis would append the details about requisite level of treatment. Eventually, we can use gait analysis as one of the tools to understand the appropriate way of walking as well as getting the guideline regarding few health conditions or in some cases intimations regarding particular diseases and fall risk prediction [1, 2]. "Walking pattern" or "gait pattern" of the human follows the defined stages specified as gait cycle. Gait cycle is defined as the sequence of events in the specified time during locomotion in which one foot contacts the ground to when the same foot again contacts the ground and involves propulsion of the center of gravity in the motion. The general phases of gait cycle are shown in Fig. 1 [3].

Broadly, gait cycle (GC) is divided into stance and swing phases as shown in Fig. 1. Stance phase covers approximately 60% of GC and swing contributes in approximate 40% of it. Stance phase provides the detail for the duration when foot is in contact with ground, and swing specifies the duration when foot is not in contact with the ground. To understand the basic components to be used for analysis, contribution of mentioned spatiotemporal parameters becomes important.

Fig. 1 Phases of gait cycle

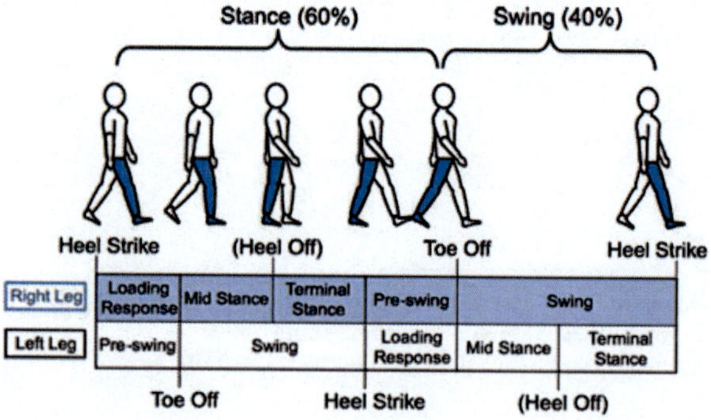

2 Literature Review

To understand the significance of gait analysis, several authors had contributed on different aspects influencing the necessity of gait analysis, Shirin Modarresi et al. had reviewed the relation of gait characteristic and dementia and conclude that increase in the double support time and variation in step length are linked with recurrent fall risk [4]. Morrison et al. had studied the type 2 diabetic patients and concluded that the advancing age in older, and type 2 diabetic cases show progressive increase in fall risk, reduction in leg strength and slow gait speed [5]. Sant'Anna et al. put forward the new symbol-based symmetry index and a symmetry measure to represent the events of gait cycle and compare the symbol method with traditional symmetry measure method and data obtained from cross correlation operation for concluding about better result and dynamic extraction of signal [6]. Anwary et al. with intent to identify the asymmetry in gait pattern, authors suggest use of automated gait feature extraction method. Selected features include distance covered in each iteration, velocity, cadence, step ration, etc., spatiotemporal parameters [7]. Kyung-Ryoul Mun et al. select foot length, foot width, height, and the foot arch curve along with the spatial–temporal parameters. Measurements are done using video and 4 uniaxial pressure sensor. Number of subjects considered for analysis is seventeen. Correlation was tested among medical longitudinal arch (MLA) and lateral longitudinal arch (LLA) along with stride length, step length, stride time, step time, and stance time [8]. Andrew et al. use a single 3D camera with movement capturing device. Author provides the comparative analysis between three types of technologies Virtuesense, GAITRite, and manual Pedograph. All the systems are able to find out good correlation factor among step length, stride length. Significant result with value around 0.9 is obtained in left and right category of all listed parameters [9]. Shogo Misu et al. use acceleration–angular velocity method; A-V method and angular velocity–angular velocity method; V-V method. Young adults (age 21+) are compared with older adults (age 80+) using body worn sensors and able to achieve positive correlation between stride time, step time, stance time, and swing time, but the range of all the result remains around 0.6 [10]. Orlowski, Katja, et al. perform comparative analysis parameter received through inertial sensor and Vicon System. Parameters show a good agreement between the systems in terms of a strong correlation [11]. Ramnani, Raj, et al. stated that the physical parameters such as height and weight of the person so also there are differences in areas/locations, culture (region specific) around the world. Each society has a slightly different gait pattern and so does our country. Even in India Gait, pattern changes with location [12]. Based on the literature reviewed, it had been noted that the gait analysis is accepted as a clinical tool. Features as step length and double support time are utilized in dementia analysis; step length, toe off, and heel strikes are used in identifying walking asymmetry, and influence of demographic parameters becomes key feature in gait analysis [14].

3 Methodology

The analysis process to build a classification model is divided into 4 stages as follows: selections of the features, training the classifier, predicting the target, and evaluate the outcomes. Feature is an individual measurable property of phenomenon being observed. In the process of data analysis, selection of various features and understanding the correlation between all of them is very important. To understand and establish the correlation coefficient among selected features, statistical measure was performed. To understand correlation among various selected parameters, different techniques are available like scatter diagram method, Karl Pearson's coefficient of correlation, Sparman's rank correlation coefficient, and method of least squares. The work put forwarded uses Pearson's correlation coefficient equation used in linear relation establishment in the process of calculation; the first step is to determine covariance between two variables used in the equation. In second step, calculate each variable's standard deviation. The covariance is reflected in such a way that it will show similar or opposite behavior of selected variables. Standard deviation is a quantity which member of group differs from mean value for the group. It is important to discover and quantify the degree to which variables in data set are dependent upon each other, as the performance of some algorithm can deteriorate if two or more variables are tightly related [14]. The Pearson's equation is as follows:

$$r = n\left(\sum xy\right) - \left(\sum x\right) - \left(\sum y\right) / \sqrt{\left[n\left(\sum x^2\right) - \left(\sum x\right)^2\right]\left[n\left(\sum y^2\right) - \left(\sum y\right)^2\right]} \quad (1)$$

A value 1 means there is a perfect positive relationship between two variables. For a positive increase in one variable, there is also a positive increase in second variables. Value -1 means there is a perfect negative relationship between two variables. This shows that the variable moves in opposite direction. For a positive increase in one variable, there is decrease in the 2nd variable. If the value is 0 means, there is no relation between two variables.

4 Findings and Discussion

As specified, the various groups are tested, and results are evaluated for understanding the correlation among different spatial parameters. The Pearson's linear relation is tested. While performing analysis, first iteration is performed by considering all available subjects irrespective of gender. All the subjects are divided among different age group span to understand the influence of aging on spatial–temporal parameters. Second testing was performed by dividing the subjects according to gender, and furthermore, they are rearranged in age group span.

4.1 Age and Walking Velocity

All the subjects are considered without any gender wise sub-classification. Figure 2 shows the relation between age and the walking velocity. Selected subjects are within age group of 20–75 years with uneven distribution of count/age. Considered walking style is normal walk of each person. From the observation, it can be seen that, as the age of the person increases, the walking velocity decreases. Observed decrease in velocity is linear in nature.

To understand the influence of aging, subjects are divided in various age groups, and its correlation factors are calculated and shown in Figs. 3 and 4. The relation established within age and walking velocity is verified, with the aging the walking velocity reduces. The negative correlation coefficient satisfies the result obtained considering age of the person and walking velocity and shows the negative value of correlation coefficient. Age group of 60–69 shows the dominance in the result with value −0.52. Further to verify the effect of aging on gender, the subjects are classified into male and female category. Instead of using four age spans, all the subjects are divided into two age groups as 25–50 and 51–75. The negative correlation coefficient satisfies the result obtained considering age of the person and walking velocity and shows the negative value of correlation coefficient. Age group of 60–69 shows the dominance in the result with value −0.52. Further to verify the effect of aging on gender, the subjects are classified into male and female category. Instead of using four age spans, all the subjects are divided into two age groups as 25–50 and 51–75. Influence of aging on decreasing walking velocity in male and female participants became clear. With considered subjects, comparatively, female participants show higher value of correlation coefficient than male participants in an age span of 25–50 years. While considering age span of 51–75, the correlation coefficient values do not show variation in it. Leading to the observational remark as, female participants show decrease in walking velocity from early stage of age in comparison with the male participants observed in same age group.

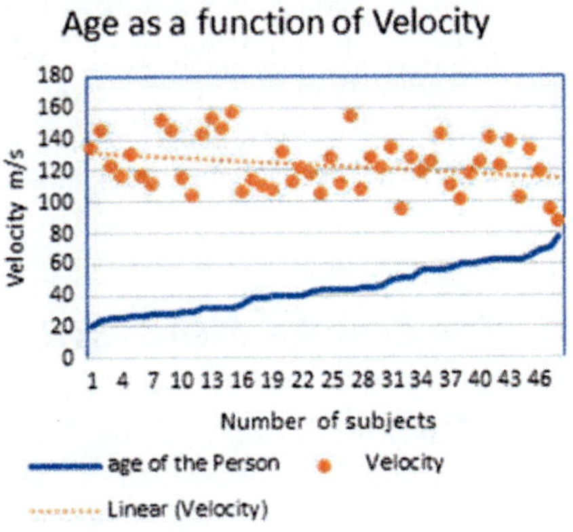

Fig. 2 Relation of human age and walking velocity

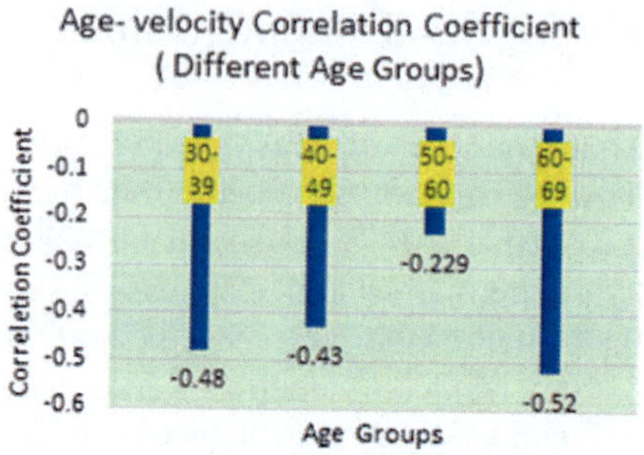

Fig. 3 Age-velocity correlation within age

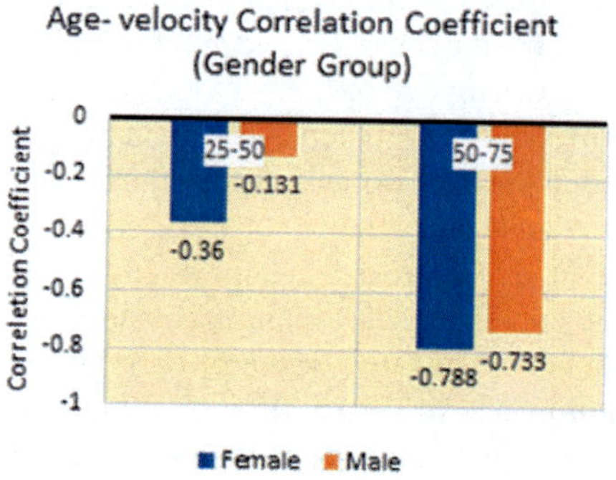

Fig. 4 Age-velocity correlation within gender

4.2 Age and Cadence

Figure 5 shows the correlation between age and the cadence. The cadence is measured in terms of number of cycles completed within minute. The cadence value is compared among different age groups, covering the number of completed cycles or rate at which a person walks, expressed in steps per minute. The correlation coefficient calculates over the complete range which shows the negative value, proving that with increased age, the number of cycles completed per minute will get reduced. The use of variation in velocity and cadence observed is used as feature for classification of subjects in two broad age groups—group 1 of 20–45 years and group 2 of 46–75 years. SVM, LR, and KNN are tested for classification accuracy and able to achieve 62% accuracy. The Fig. 6 shows the SVM decision surface.

4.3 Leg Length—Step Length Correlation Coefficient

Figures 7 and 8 are depicting the relation between the leg/limb lengths and step length. The step length considered for right leg. Parameters show the positive correlation coefficient between leg length and the step length. Step length is the distance covered between right and left leg placed in consecutive manner. The person having high

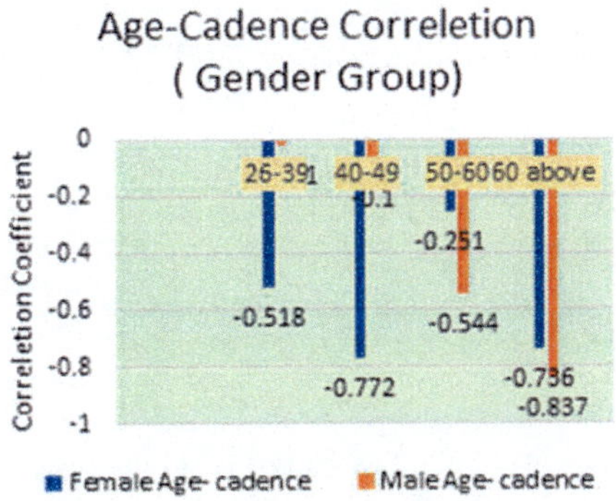

Fig. 5 Age-cadence correlation

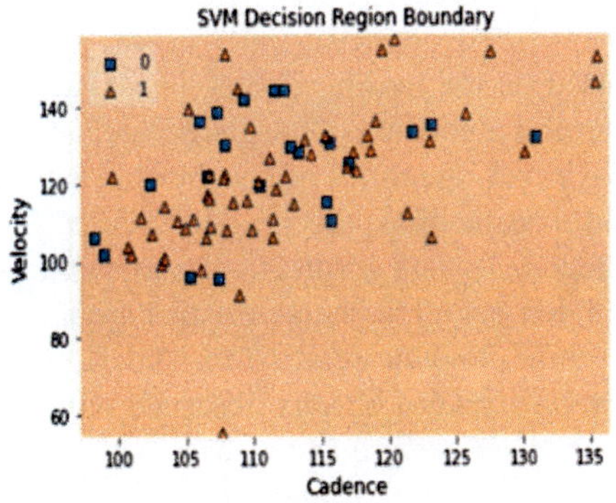

Fig. 6 Age classification using SVM

leg/limb length irrespective of gender covers high step length as shown in Fig. 7. Incrementing values satisfy the stated fact, but after the age of 60, the correlation coefficient shows reduction in the value due to influence of aging as observed in relation between step count and age. As seen in the Fig. 8, the effect of increment in step length with leg length is dominant in male compared to female subjects. But the relation satisfies the positive correlation in both cases. To verify its effect in female category, a separate study with higher count of female subjects needs to be performed.

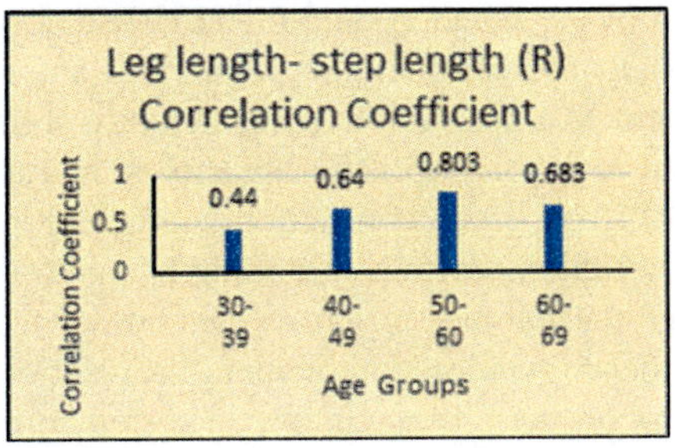

Fig. 7 Leg length–step length correlation based on age

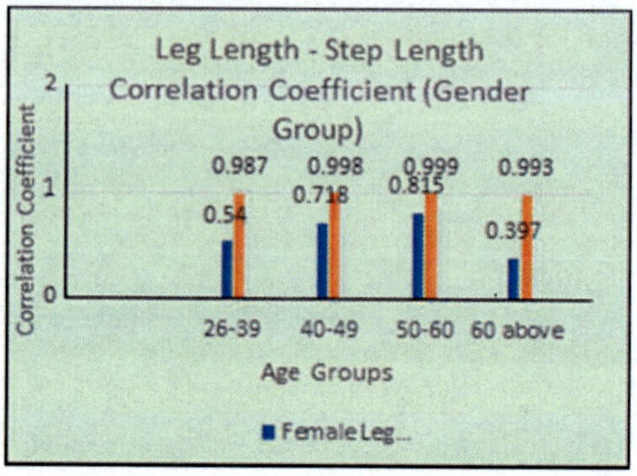

Fig. 8 Leg length–step length correlation based on gender

5 Test Cases for Correlation Verification

Based on the findings, some test cases are implemented and tested using machine learning algorithm. Considered scenario for verification is classifying the person in two broad categories of age group. Using derived features, few features are selected, and classification percentage is verified. To verify the result considered, categories are category 1—age group 20–45 years and category 2—age group 46–75 years. Gender and demographic characteristics are not considered. Performance of support vector machine, logistic regression, and K-nearest neighbor is tested for set of features. Flat response of 62% is observed with SVM and LR irrespective of set of features but KNN's performance varies with range of 62–75% classification accuracy. Step length, stride length, and single support time in association with velocity and cadence able to achieve 75% accuracy, while step length and distance provide 62%. Cases are tested for supplementary confirmation. Currently, broader ranges are considered due to a smaller number of participants in each age group. With increased number of contributors in small age span will definitely increase the classification accuracy as well as inclusion of demographic parameters and gender wise sub-classification will contribute in possible precision and accuracy.

6 Conclusion

From the selected data of the 50 subjects recorded through GAITRite Electronic Walkway, different gait parameters are considered to understand the relation amid them. An age and walking velocity show the linear relation. Decrement factor is linear with an age. Female shows the range of −0.36 to −0.78, while male shows −0.13 to −0.73 correlation coefficient values. Age and cadence show the negative correlation coefficient ranging from −0.2 to −0.5 indicating that with the increased age the number of gait cycles completed within minute starts reducing. Age and cadence relation complement the relation derived for age and walking velocity. To support these two relations, the step count is tested as a function of age and shows the expected result that, as the age increases, irrespective of gender the number of steps required to cover the same distance increases gradually, showing linear relation between step count and age on incremental scale. For few parameters, dominance is observed in female subjects than male. Leg length and step length are of such category, and parameters show positive correlation coefficient in both genders across all age groups. But female subjects range from 0.39 to 85, while male subjects maintain value around 0.99. To understand this behavior of female group, large of similar physic and age need to be studied. To emphasis on leg length, it is verified with stride length which shows the positive correlation in male but turn down to negative for female above age 60. After analyzing these parameters, the initial consideration of relating age and gender to analyze walking pattern becomes trustworthy. Walking velocity, cadence, and step count state the influence of aging clearly on male and female subjects.

Acknowledgements Authors would like to express sincere thanks to all the participants for their contribution in this study. Also special thanks to the team of Gait Laboratory of Deenanath Mangeshkar Hospital and Research Center, Pune, India for their valuable support by allowing us to use the GAITRite setup for the study purpose.

References

1. Walford GA (2019) Development of a gait assessment protocol for elderly Veterans with cognitive decline, using an instrumented walkway. J Mil Veteran Family Health 5:49–57
2. Kadoya S (2014) A precise gait phase detection based on high-frequency vibration on lower limbs. In: IEEE international conference on robotics and automation (ICRA)
3. Modarresi S, Divine A, Grahn JA, Overend TJ, Hunter SW (2019) Gait parameters and characteristics associated with increased risk of falls in people with dementia: a systematic review. Int Psychogeriatr 31(9):1287–1303
4. Morrison S, Newell KM (2019) Intraindividual variability of neuromotor function predicts falls risk in older adults and those with type 2 diabetes. J Mot Behav 51(2):151–160
5. Sant'Anna, Wickström N (2010) A symbol-based approach to gait analysis from acceleration signals: identification and detection of gait events and a new measure of gait symmetry. IEEE Trans Inform Technol Biomed 14(5):1180–1187
6. Anwary, Yu H, Vassallo M (2018) An automatic gait feature extraction method for identifying gait asymmetry using wearable sensors. Sensors 18(3):676–676
7. Mun K-R (2017) A study on the correlation between the foot features and gait characteristics during over-ground walking. ISBS Proc Arch 35:113–113
8. Strubhar AJ (2018) Concurrent validity of the VirtuSense® gait analysis system for the quantification of spatial and temporal parameters of gait. Int J Exerc Sci 11:934–940
9. Misu S, Asai T, Ono R, Sawa R, Tsutsumimoto K, Ando H, Doi T (2017) Development and validity of methods for the estimation of temporal gait parameters from heel-attached inertial sensors in younger and older adults. Gait Posture 57:295–298
10. Orlowski K (2016) Evaluation of gait parameters determined by InvestiGAIT against a reference system. Biosignals
11. Buchner DM (1996) Evidence for a non-linear relationship between leg strength and gait speed. Age Ageing 25:386–391

12. Ilias S, Tahir NM, Jailani R (2016) Feature extraction of autism gait data using principal component analysis and linear discriminant analysis. In: IEEE industrial electronics and applications conference (IEACon)
13. Sathe NP, Hiwale AS (2020) Verifying the effect of barefoot walk and comfortable shoe walk using gait velocity as a tool for analysis. In: 2020 2nd international conference on innovative mechanisms for industry applications (ICIMIA). IEEE
14. Caldas R et al (2017) A systematic review of gait analysis methods based on inertial sensors and adaptive algorithms. Gait Posture 57:204–210

Radial Basis Neural Network Trained Minimum Snap Trajectory for Quadrotor

M. A. Abitha and Abdul Saleem

Abstract This paper proposes radial basis neural network trained minimum snap trajectory for quadrotor unmanned aerial vehicle. The quadrotor is intended to fly over a set of waypoints with the least amount of jounce as possible. To do this, a minimum snap trajectory is created using velocity, acceleration, and jerk as boundary constraints. For the successful establishment of such a trajectory on real-world scenario, an accurate approximation approach is required. A compact design and a simple continuously differentiable data mapping are necessary for such an approximator. Radial basis neural networks are well-known for their qualities as universal approximators, making them ideal candidates for this task. As a result, the generated trajectory is trained using a radial basis neural network. Simulations are carried out using the MATLAB/SIMULINK software platform.

1 Introduction

Quadrotor UAVs have recently gained popularity due to their high agility and simple design. For increasing quadrotor flying performance, optimal trajectories can be constructed by taking into account the required minimization criteria such as velocity, acceleration, time, and so on. The optimal trajectories for UAVs are generated in [1, 2], which can complete the mission in the shortest amount of time. Reference [3] formulates the time-optimal trajectory with input and boundary-state constraints. The constrained trajectory is validated using a quadrotor model with three degrees of freedom (DOF). The quadrotor dynamics with six degrees of freedom is explained in [4]. Here, time optimal trajectory based on nonuniform rational B-spline curve is taken for multiple waypoint navigation. Reference [5] explains how to arrange time-optimal motions using a sequential quadratic technique with limitations on state, control, and jerk. The external environmental features includ-

M. A. Abitha (✉) · A. Saleem
Department of Electrical and Electronics Engineering, Government Engineering College Thrissur, Affiliated to APJ Abdul Kalam Technological University, Thrissur, Kerala, India
e-mail: mailabitha@gmail.com

V. K. Gunjan and J. M. Zurada (eds.), *Proceedings of 3rd International Conference on Recent Trends in Machine Learning, IoT, Smart Cities and Applications*, Lecture Notes in Networks and Systems 540, https://doi.org/10.1007/978-981-19-6088-8_24

ing windy conditions are taken into account for generating time and jerk optimal trajectory in [6]. Minimum jerk trajectory is generated in [7] based on differential flatness while transfering dynamic constraints of quadrotor to control point coinstraints. References [8, 9] propose minimum snap trajectory for quadrotor UAV. The optimal trajectories are trained using neural network in [10] for faciliating real-time control of quadrotor. The fourth derivative of position is intimately related to the aircraft's orientation, according to quadrotor dynamics. As a result, a minimal snap trajectory is used to produce grace without the use of unnecessary control inputs. In the real-world scenario, these trajectories are implemented using sampled data. As a result, precise approximation techniques with continuously differentiable data mapping are required. Neural networks, particularly radial basis neural networks, are well-known for function approximation. This paper focuses on the minimum snap trajectory for quadrotor UAVs and uses a radial basis neural network to train this trajectory.

The following is the outline for the paper: Sect. 2 is dedicated to problem formulation while Sect. 3 discusses minimum snap trajectory generation. Section 4 contains the simulation results and discussion. Section 5 contains closing remarks.

2 Problem Formulation

Consider a quadrotor that is meant to fly through the waypoints $w_0, w_1, w_2, \ldots, w_n$ at time instants $t_0, t_1, t_2, \ldots, t_f$ in a complex environment, as shown in Fig. 1. The quadrotor must create enough force and torque to fly through a trajectory with these way points. The force F that is generated is proportional to the acceleration. This acceleration is a function of orientation (ϕ, θ and ψ) in quadrotor dynamics. Consider the acceleration of a quadrotor with mass m in the direction of unit vector $\hat{i}$ as follows:

$$\ddot{p}_n = (-\cos\phi\sin\theta\cos\psi - \sin\phi\sin\psi)\frac{F}{m} \tag{1}$$

The accelerations in the direction of unit vectors $\hat{j}$ and $\hat{k}$ are calculated using similar equations. The torque needed to maintain the quadrotor's orientation in the intended trajectory is proportional to the attitude acceleration ($\ddot{\phi}$, $\ddot{\theta}$ and $\ddot{\psi}$). As a result, the torque is proportional to the snap, which is the position's fourth derivative. Thus, the use of a minimal snap trajectory to produce gracefulness without unnecessary

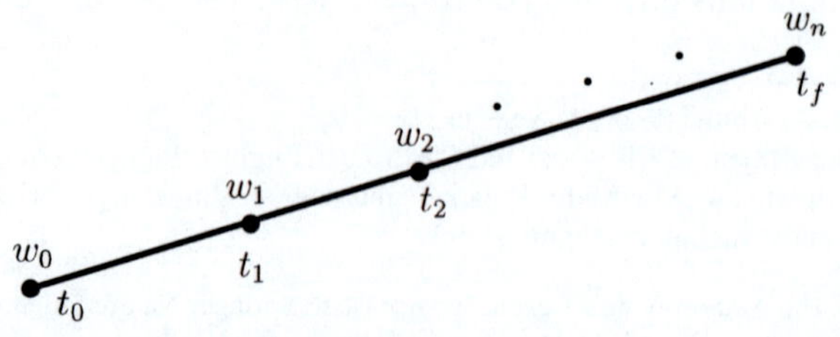

Fig. 1 Problem objective

control inputs [11]. Neural networks can learn any complex function and are highly fault tolerant. Among different neural networks, radial basis neural network (RBNN) outperforms the others in function approximation [12]. The goal of this paper is to create a RBNN trained minimum snap trajectory through the desired waypoints.

3 Minimum Snap Trajectory Generation

While a quadrotor fly through a series of waypoints, it is highly desirous that the transitions are smooth. This can be acheived by designing optimal trajectory. Minimum jerk optimal trajectory is obtained in [13] by minimizing the performance index

$$\mathcal{L}(t, r, \dot{r}, \ddot{r}, r^{(3)}) = ||r^{(3)}(t))||^2 \tag{2}$$

Here, r represents the position; $r^{(3)}$ is the third derivative of r. A snap is the fourth derivative of r. Thus, by extending Eq. (2) to fourth derivative of r, the performance index can be written as

$$\mathcal{L}(t, r, \dot{r}, \ddot{r}, r^{(3)}, r^{(4)}) = ||r^{(4)}(t))||^2 \tag{3}$$

Hence, the cost function for minimum snap trajectory for the interval $[t_0\, t_f]$ is,

$$C_s = \arg\min_r \int_{t_0}^{t_f} ||r^{(4)}(t)||^2 \mathrm{d}t \tag{4}$$

and the trajectory can be found from Euler-Poisson equation,

$$\frac{\partial \mathcal{L}}{\partial r} - \frac{\mathrm{d}}{\mathrm{d}t}\frac{\partial \mathcal{L}}{\partial \dot{r}} + \frac{\mathrm{d}^2}{\mathrm{d}t^2}\frac{\partial \mathcal{L}}{\partial \ddot{r}} - \frac{\mathrm{d}^3}{\mathrm{d}t^3}\frac{\partial \mathcal{L}}{\partial r^{(3)}} + \frac{\mathrm{d}^4}{\mathrm{d}t^4}\frac{\partial \mathcal{L}}{\partial r^{(4)}} = 0 \tag{5}$$

which is equivalent to

$$r^{(8)} = 0 \tag{6}$$

Thus, a minimal snap waypoint trajectory which satisfies Eq. (6) can be represented by a 7th order polynomial function $p(t)$

$$p(t) = q_7 t^7 + q_6 t^6 + q_5 t^5 + q_4 t^4 + q_3 t^3 + q_2 t^2 + q_1 t + q_0 \tag{7}$$

The coefficients $q_0, \ldots, q_7$ can be found by putting $t = t_0$ and $t = t_f$ in Eq. (7), and it is first three derivatives.

$$\begin{bmatrix} p(t_0) \\ p(t_f) \\ \dot{p}(t_0) \\ \dot{p}(t_f) \\ \ddot{p}(t_0) \\ \ddot{p}(t_f) \\ p^{(3)}(t_0) \\ p^{(3)}(t_f) \end{bmatrix} = \begin{bmatrix} 1 & t_0 & t_0^2 & t_0^3 & t_0^4 & t_0^5 & t_0^6 & t_0^7 \\ 1 & t_f & t_f^2 & t_f^3 & t_f^4 & t_f^5 & t_f^6 & t_f^7 \\ 0 & 1 & 2t_0 & 3t_0^2 & 4t_0^3 & 5t_0^4 & 6t_0^5 & 7t_0^6 \\ 0 & 1 & 2t_f & 3t_f^2 & 4t_f^3 & 5t_f^4 & 6t_f^5 & 7t_f^6 \\ 0 & 0 & 2 & 6t_0 & 12t_0^2 & 20t_0^3 & 30t_0^4 & 42t_0^5 \\ 0 & 0 & 2 & 6t_f & 12t_f^2 & 20t_f^3 & 30t_f^4 & 42t_f^5 \\ 0 & 0 & 0 & 6 & 24t_0 & 60t_0^2 & 120t_0^3 & 210t_0^4 \\ 0 & 0 & 0 & 6 & 24t_f & 60t_f^2 & 120t_f^3 & 210t_f^4 \end{bmatrix} \begin{bmatrix} q_0 \\ q_1 \\ q_2 \\ q_3 \\ q_4 \\ q_5 \\ q_6 \\ q_7 \end{bmatrix} \tag{8}$$

Equation (8) can be written in a compact form as,

$$P = AQ \tag{9}$$

The determinant of matrix A is equal to $144t_f^{16}$ which implies that A is a nonsingular matrix for all values of t except at $t_f = 0$. Since t_f indicates the final time of the quadrotor traversal, it holds only positive values. Thus, the inverse of matrix A always exists. The values of $[p(t_0),\ p(t_f),\ \dot{p}(t_0),\ \dot{p}(t_f),\ \ddot{p}(t_0),\ \ddot{p}(t_f),\ p^{(3)}(t_0),\ p^{(3)}(t_f)]^T$ are the initial and final boundary conditions of the position, velocity, acceleration and jerk in the respective directions. The generated minimum snap polynomial trajectory is basically a mapping between time and a function of time. In the real-world applications, this mapping can be viewed as input–output data pair $(t,\ p(t))$ instead of an explicit formula. This is the main objective of function approximation. Reference [14] states that radial basis neural networks are capable of universal function approximation.

3.1 Proposed RBNN Trained Trajectory

Radial basis neural network is a feedforward network which consists of two layers shown in Fig. 2. The first layer is the radial basis layer, and the second layer is the output layer. Here, I represents the input vector, and W^1 is the weight vector of the first layer. The net input to the neurons in the first layer is obtained by multiplying the Euclidean distance between input and weight vectors with bias b^1 as follows:

$$net^1 = ||I - W^1||b^1 \tag{10}$$

The activation function f for the radial basis layer be any nonlinear functions like Gaussian functions and multi-quadric functions. This paper considers Guassian activation function for the radial basis layer. Thus, the output of first layer is,

$$O^1 = e^{-(\text{net})^2} \tag{11}$$

Consider W^2 and b^2 represents the weights and bias of the second layer. As the second layer makes use of linear activation function, the output O^2 is given by,

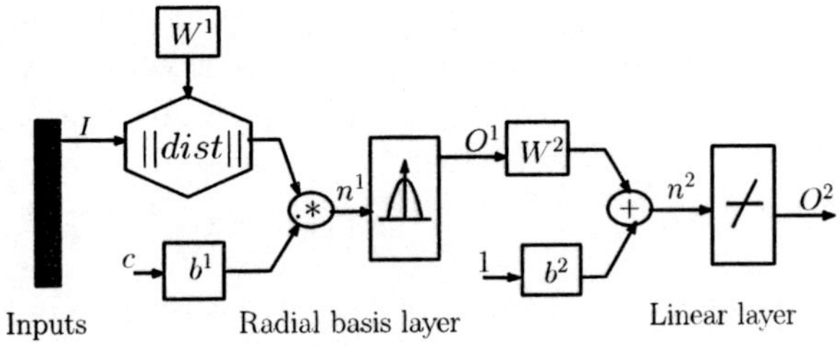

Fig. 2 Radial basis neural network

$$O^2 = W^2 O^1 + b^2 \tag{12}$$

3.1.1 Proposed Trajectory Training Methodology

For a give time interval, $[t_0, t_f]$, radial basis neural network learns the equation of minimimum snap trajectory $r(t)$ and their derivatives. Here trajectory is composed of a series of waypoints. Thus, the trained trajectory $r_T(t)$ be expressed in terms of peicewise polynomial function of degree 7 over m intervals.

$$r_T(t) = \begin{cases} \sum_{i=0}^{7} q_{i1} t^i, & t_0 \le t \le t_1 \\ \sum_{i=0}^{7} q_{i2} t^i, & t_1 \le t \le t_2 \\ . \\ . \\ . \\ \sum_{i=0}^{7} q_{im} t^i, & t_{(m-1)} \le t \le t_m \end{cases} \tag{13}$$

where $r_T(t) = (x_r(t), y_r(t), z_r(t))$ is the desired three-dimensional trajectory and q_{ij}, $j = 1, 2, 3, \ldots, m$ is the coefficient of the polymonial equation representing the jth segment of the trajectory. RBNN takes time interval as the input vector, $I = [t_i \; t_j]$ as the input vector for every j^{th} segment, and target is the trajectory computed by Eq. (7). Thus, the training of radial basis network is done for all the three-dimensional trajectory segments. In this work, we considered 402 number of training data.

4 Simulation Results and Discussion

The minimum snap trajectory can be generated by commencing with the quadrotor at the origin. Then, when $t = 1$ second, proceeded to the next waypoint in the same manner as before (5, 5). It finally arrives at waypoint (10, 10) at $t = 2$ s. As a result, a three-point polynomial trajectory with the least snap condition is generated. The boundary conditions are set so that the quadrotor's velocity, acceleration, and jerk values all equal zero at each waypoint, and the quadrotor comes to a complete stop at each waypoint before continuing on its journey. The northeast-down (NED) coordinate system is employed in this scenario. As a result, the x, y, and z directions are pn, pe, and pd, respectively. The 7th-order polynomial equation representing p_n and p_e positions corresponding to unit vector direction $\hat{i}$ and $\hat{j}$ can be solved as, for $t = [0,\ 1]$

$$r_T(t) = -100t^7 + 350t^6 - 420t^5 + 175t^4 \tag{14}$$

and for $t = [1,\ 2]$

$$r_T(t) = -100t^7 + 1050t^6 - 4620t^5 + 11025t^4 - 15400t^3 + 12600t^2 - 5600t + 1050 \tag{15}$$

The resultant trajectory is trained using a radial basis neural network. The network employs 402 neurons in the radial b-asis layer to accurately learn the minimum snap trajectory. The radial basis neural network is shown in Fig. 3. The performance of RBNN in approximating the desired function is shown in Fig. 4. The training is complete with a mean square error of 4.50928e − 12. A total of 400 epochs are required.

Figure 5 depicts the RBNN learnt trajectory's x and y locations as a function of time. The neural network analysis for velocity, acceleration, and jerk is shown in

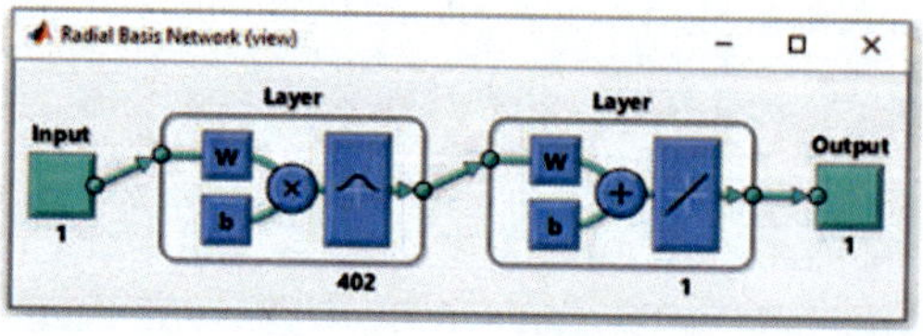

Fig. 3 Radial basis neural network

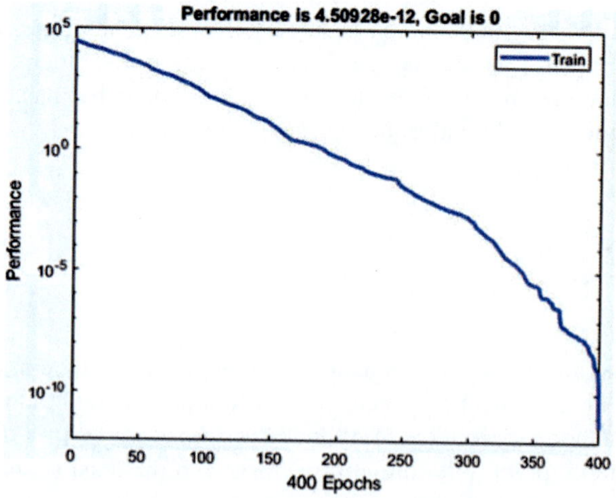

Fig. 4 Radial basis neural network performance

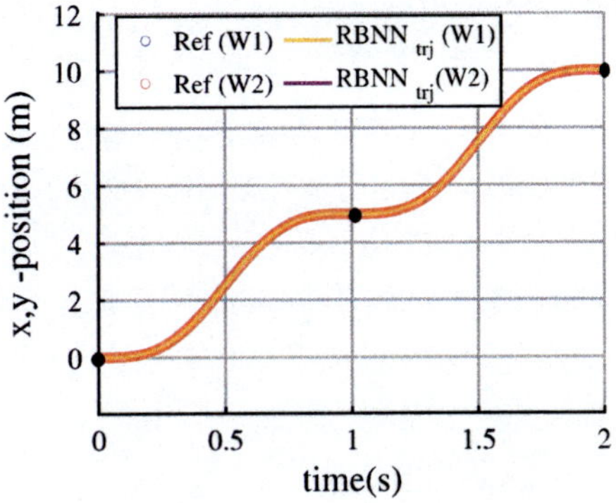

Fig. 5 RBNN trained trajectory

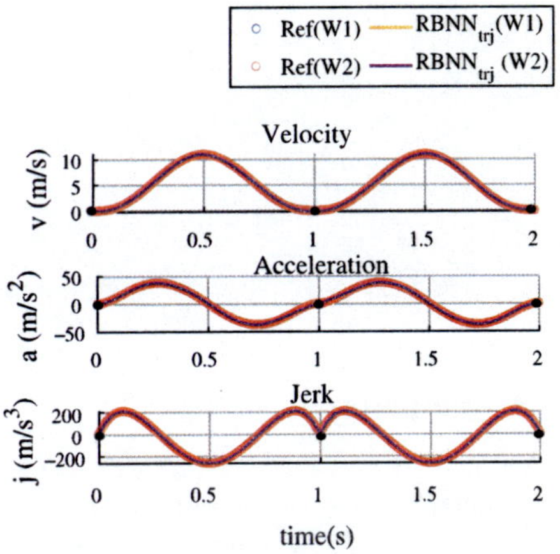

Fig. 6 RBNN trained velocity, acceleration, and jerk

Fig. 6. At each waypoint, the velocity, acceleration, and jerk boundary limitations are maintained, as seen in the graph. In the RBNN-based minimal snap trajectory, the acceleration fluctuates smoothly between the waypoints, but in the classic method, the acceleration approaches zero and lingers a little longer around the waypoint transitions. Figure 7 depicts a minimal snap trajectory in the 2D plane trained using a radial basis neural network. A closed trajectory is generated for $t = 0, 1, 2, 3,$ and 4 using the waypoints (2, 2, 10), (3, 3, 10), (4, 2, 10), (3, 1, 10), and (2, 2, 10). Figure 8 depicts the minimum snap trajectory generated using four waypoints in the 3D plane.

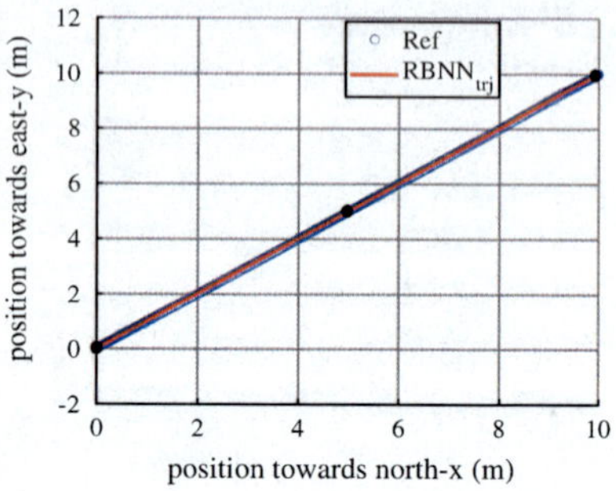

Fig. 7 RBNN trained 2D minimum snap trajectory

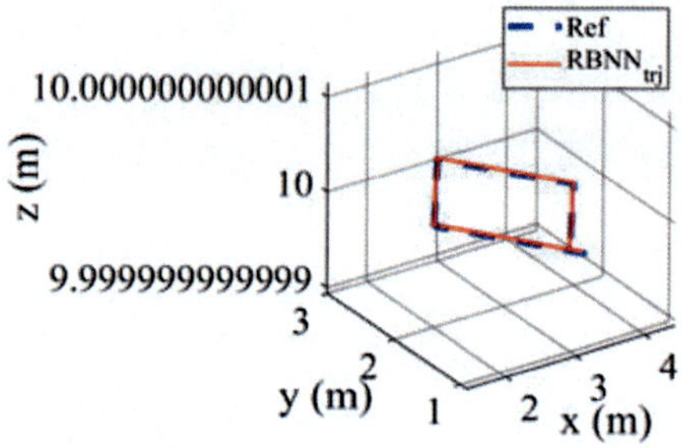

Fig. 8 RBNN trained 4 waypoint closed trajectory

5 Conclusion

This paper focused on developing an RBNN-based minimum snap trajectory for a quadrotor UAV. The waypoint trajectory is created using the minimum snap optimal criteria. The generated polynomial trajectory is fed into a radial basis neural network and trained. RBNN's excellent function approximation capability is applied for the trajectory approximation. The acceleration profile of the RBNN trained trajectory demonstrates improved performance of smooth transition between waypoints as opposed to a little discontinuity nature in the conventional methods.

References

1. Ayhan B, Kwan C (2018) Time-constrained extremal trajectory design for fixed-wing unmanned aerial vehicles in steady wind. J Guid Cont Dynam 41(7)
2. Techy L, Woolsey CA (2009) Minimum-time path planning for unmanned aerial vehicles in steady uniform winds. J Guid Cont Dynam 32(6)
3. Hu B, Mishra S (2019) Time-optimal trajectory generation for landing a quadrotor onto a moving platform. IEEE/ASME Trans Mech 24(2):2019
4. Dong W, Ding Y, Huang J, Zhu X, Ding H (2017) An efficient approach of time-optimal trajectory generation for the fully autonomous navigation of the quadrotor. J Dynam Syst Measurem Cont 139(6)

5. Liang X, Fang Y, Sun N, Lin H (2018) Dynamics analysis and time-optimal motion planning for unmanned quadrotor transportation systems. Mechatronics (50)
6. Silva JP, Wagter CD, de Croon G (2018) Quadrotor thrustVectoring control with time and jerk optimal trajectory planning in constant wind fields. Unmanned Syst 06(01)
7. Yu J, Cai Z, Wang Y (2014) Minimum jerk trajectory generation of a quadrotor based on the differential flatness. In: Proceedings of 2014 IEEE Chinese guidance, navigation and control conference, IEEE
8. Mellinger D, Kumar V (2011) Minimum snap trajectory generation and control for quadrotors. In: 2011 IEEE international conference on robotics and automation. IEEE
9. De Almeida MM, Moghe R, Akella M (2019) Real-time minimum snap trajectory generation for quadcopters: algorithm speed-up through machine learning. In: 2019 International conference on robotics and automation (ICRA). IEEE
10. Tang G, Sun W, Hauser K (2017) Learning trajectories for real-time optimal control of quadrotors. In: 2018 IEEE/RSJ international conference on intelligent robots and systems (IROS). IEEE J Solids Struct 14(6)
11. Richter C, Bry A, Roy N (2011) Polynomial trajectory planning for aggressive quadrotor flight in dense indoor environments. Springer Tracts in Advanced Robotics, Springer International Publishing, 2016, Engineering, vol 14
12. Lee C-C, Chung P-C, Tsai J-R, Chang C-I (1999) Robust radial basis function neural networks. IEEE Trans Syst Man Cybernet Part B (Cybernet) 29(6)
13. Miranda-Colorado R, Aguilar LT, Herrero-Brito JE (2018) Reduction of power consumption on quadrotor vehicles via trajectory design and a controller-gains tuning stage. Aerospace Sci Technol 78
14. Park J, Sandberg IW (1991) Universal approximation using radial-basis-function networks, Neural Comput 3(2)

REMICARE—Medicine Intake Tracker and Healthcare Assistant

Gulbakshi Dharmale, Pratiksha Shirsath, Abhishek Shinde, Vishwajeet Sawant, and Aditi Chougule

Abstract An automated reminder mechanism is built in this Android-based application. It emphasizes the contact between doctors and patients. Patients can set a reminder to remind them when it is time to take their medicine. Multiple medications and timings, including date, time, and medicine description, can be programmed into the reminder by using image processing. Patients will be notified through a message within the system, as preferred by the patients. They have the option of looking for a doctor for assistance. In this COVID-19 pandemic situation where nurses have to remind the patients in the hospitals to take their medications, our application can be useful, alerting the patient every time of the day when he/she has to take the medicine and in what amounts. Also, all the necessary tests report and prescriptions can be saved on the cloud for later use. Patients will be provided with doctor contact information based on their availability. Also, patients will be notified of the expiry date of the medicine, and the former history of the medicines can be stored for further reference. The proposed system prioritizes a good user interface and easy navigation. Image processing will be accurate and efficient with the help of powerful CNN-RNN-CTC algorithm. It also emphasizes on a secure storage of the user's data with the help of the RSA algorithm for encryption and the gravitational search algorithm for secure cloud access. We attempted to create a Medical Reminder System that is cost-effective, time-saving, and promotes medication adherence.

Keywords Android-based application · Reminder system · Medication adherence · Image processing · COVID-19 · CNN-RNN-CTC algorithm · Cloud · RSA algorithm

---

G. Dharmale (✉) · P. Shirsath · A. Shinde · V. Sawant · A. Chougule
Department of Information Technology, Pimpri Chinchwad College of Engineering, Pune, India
e-mail: gulbakshi.dharmale@pccoepune.org

V. K. Gunjan and J. M. Zurada (eds.), *Proceedings of 3rd International Conference on Recent Trends in Machine Learning, IoT, Smart Cities and Applications*, Lecture Notes in Networks and Systems 540, https://doi.org/10.1007/978-981-19-6088-8_25

1 Introduction

Teachers, students, businesspeople, housewives, and children all fall into the group of patients, and we all have a rigorous schedule. Responsibilities and stress are abundant in today's world. As a result, humans are susceptible to a variety of ailments, and it is our responsibility to keep ourselves fit and healthy. If the patient is at home, someone may look after him or her, but if the patient is not at home, is out of the city or state, it is difficult for family members to contact and remind them of their dose schedules on a regular basis. We rely entirely on devices, particularly smartphones, in our rapidly growing and technologically reliant lives. As a result, we will be able to make greater use of technology and make it more beneficial to us. According to one study, every ten percent increase in medication adherence cuts healthcare expenses by up to 29%. Walgreens discovered that every 1% increase in adherence saved around $50 in healthcare costs in another major research involving hundreds of thousands of consumers. Patients with chronic diseases such as hypertension or diabetes who followed their prescriptions exactly as recommended saved an average of $4000–$8000 per year on healthcare expenses. In another study conducted by a big firm, it was shown that enhancing adherence increased employee productivity by $18,000 per year. With proper drug therapy, 89,000 early deaths among hypertension patients might be avoided each year.

The following are examples of common behaviors:

1. One out of every two persons misses a dosage. One out of every three people forgets if they took their medication.
2. One out of every four people did not receive a refill on time.
3. Within 30 days, one out of every five Medicare patients is readmitted to the hospital. Half of them are rated non-compliant.
4. Between 41 and 59% of mentally ill people take their medicine just seldom or not at all. This adds to societal problems such as unemployment, homelessness, and suicide.
5. Nearly half of patients who get a prescription for a chronic ailment like cardiovascular disease stop taking it within the first year, according to CVS Pharmacy.
6. According to CVS Pharmacy/Caremark research, 68% of physicians want to be notified if their patients become non-adherent.

Some apps have been developed to address these issues, although they are not yet perfected [1].

2 Related Work

Medication systems have been built on a variety of platforms and concepts. Despite the fact that the use of healthcare-related apps is on the rise, there are a number of issues with their functionality. MyTherapy is a medicine tracking app. This software not only keeps track of your medicine use, but it also keeps track of your mood and overall wellness [2]. Another is Groove Health app. This app's built-in artificial intelligence engine assists you in better understanding the medicine you are taking by answering any medication or health inquiries you may have. You may learn more about your medicine on the app, set up personalized reminders to help you remain on track, and share your success with friends, family members, or caregivers [3]. Round Health also keeps track of your prescription history, so you can keep track of how many pills you have taken and how many you have missed [4]. Pill Reminder app will send you the notification until you mark the medication as taken. Existing reminder systems have several limitations. Here are a few examples: They do not provide any facility about checking and tracking expiry date of the medication. There is no facility of scanning the prescription and adding the medication name automatically. Also, existing systems do not have facilities to store past medical history of patients and former prescriptions. Also, the health-related documents such as maternity documents, several scan reports are not stored in any of the systems. Many of the systems offered need the buying of specialized hardware. Some systems feature a default alarm tone that users are unable to adjust.

2.1 Literature Survey on Different Techniques of OCR

Image processing and character recognition will be used to extract medicine names from the prescription, which will be handy for the user to operate. Various character recognition techniques include template matching, structural techniques, statistical techniques, neural networks, etc.

Steps of character recognition are as shown in Fig. 1.

The paper of Offline Handwritten Character Recognition Using Neural Network uses Fourier Descriptors that are used to train ANN model and then SVM is used to classify it. This model provided 96% accuracy and is used in many real-world applications [5].

In Optical Character Recognition Technique Algorithms, the proposed system uses neural networks with backpropagation to train the model. This model had an accuracy of 95% during real-time use [6].

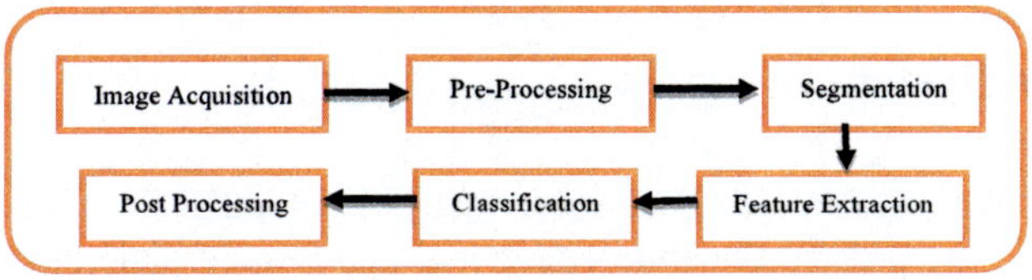

Fig. 1 Block diagram of character recognition [17]

Handwritten Cursive English Text Recognition Using Deep CNN-RNN-based CT uses CRNN-CTC algorithm. This model has better accuracy, so used in applications like pattern recognition, processing bank checks, etc. [7].

2.2 *Literature Survey on Different Techniques of Storage and Security Data*

The paper by Samir V. Zanjal mainly focuses on the inclusion of IoT in the smart home system. The system is implemented with the use of sensing element and a wireless module which should be secure [8].

The paper Secure Use of Cloud Storage of Data on Smartphones using Atomic AES on ARM Architectures mainly focuses on the use of AES algorithm and ARM architecture with the cloud storage for mobile application. ARM architecture allows the use of the mobile application on any mobile device [9].

Advanced Encryption Standard (AES): The Advanced Encryption Standard is an encryption method which encrypts the sensitive electronic data. This is a block cipher that consists of 128 bits.

Atomic AES: Atomic AES execution on ARM structures can get information more proficiently and quicker as proposed in this work as far as encryption time, decoding time, and throughput.

ARM architecture: Advanced RISC Machine or Acorn RISC Machine is the architecture with various figuring models set to be utilized in various conditions. This architecture makes it possible for the application to be used on a variety of devices.

The paper by Soloman Babatunde Olaleye mainly focuses on the use of Gravitational Search Algorithm with AES encryption of data. The use of the Gravitational Search Algorithm adds added security to the system by securing the connection to the cloud [10].

Gravitational Search Algorithm: In 2009 whose methodology depends on Newton's law of gravitational energy and movement.

Figure 2 describes the entire process of Atomic AES encryption on how plain text is converted to cipher text.

Table 1 describes the existing applications and highlights their main techniques and outcomes.

3 Proposed System

In our system, we propose a digital solution by developing an android application which can store personal healthcare records and remind patients to take their medicines on time. In our proposed system, users can scan and upload the image of the prescription which will be stored on the cloud. After scanning the prescription,

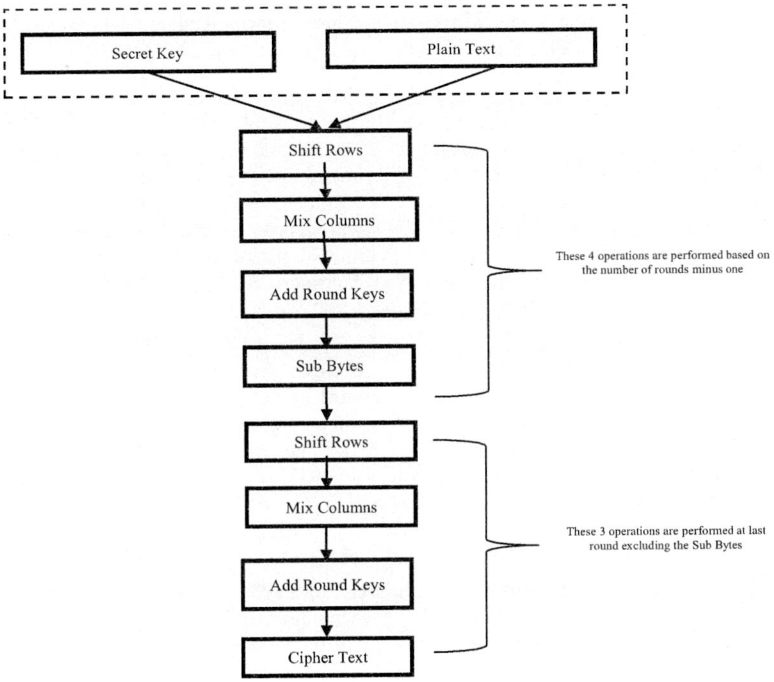

Fig. 2 Atomic AES encryption

the name of the medicine will be extracted for which the user can set a reminder. Users can also add extra meds in case he needs a reminder for his other medications. Also, he can enter the expiry date manually, and this application will send an alert message to the user before the expiry date. In addition to this, our application will also provide doctor's assistance for patients having trivial doubts such as change in dose or medicine, proper time to take the particular medicine, which medicine can cause some side effects when taken with a particular food item. Patients can also enter their height and weight if they want to measure their BMI. Storing medical history is also necessary for every individual. For example, a woman can store her maternity records. It will also be beneficial for patients suffering from some chronic disease. Doctors can refer to previous records during every visit.

(a) The proposed system is built on the Android operating system, and it uses a notification and automated alarm ringing mechanism to remind users to take their medications on time.
(b) For the scanning of prescriptions, we are going to use the CNN-RNN-CTC algorithm which has proven better results than other algorithms compared in the literature survey.
(c) Patients can also store health records on their cloud and can retrieve in times of need.

Table 1 Literature survey on different existing applications of medicine reminder systems

S. No	Title	Methodology	Techniques	Main outcomes
1	A Review on Medicine Reminder and Adherence System [11]	Studying various types of medicine reminder apps	The capsule with a tiny digestible antenna and a microchip. Sensor necklace that will record the exact time of pill taken. Alarm facility	Medication review in improving the knowledge and adherence of existing applications. Provides alarm facility and keeps the record of pills taken
2	Medication reminder APPs to improve medication adherence in coronary heart disease [12]	Using frequently available medication reminder applications in improving fidelity to patients with CHD	Use of eight-item Morisky Medication Adherence for report generation	Medication Adherence-based using eight-item question Morisky. Counts no. of pills missed by patients. Measures cholesterol level
3	A Smartphone-based Medication Self-management System with Real-time Medication Monitoring [13]	A smartphone-based medication self-management system (SMSS) that will be used to bring off all of the necessary functions	Use of SMSS technique with real-time medication monitoring	System supports omnipresent medication self-management using phone. Uses smartphones to store medical history
4	Current Trends in Electronic Medication Reminders for Self-Care [14]	To gather and examine publications related to electronic medication reminders used for medication adherence in self-care settings	Use of PubMed and MeSH for searching the publication	Mobile phones and other electronic devices are used for reminder messages. Improves medication adherence
5	Smart Phone Based Medicine In-take Scheduler, Reminder and Monitor [15]	Scheduling algorithms OMAT/ODAT are developed for efficient and intake of medicines	According to the priority of the medications, the OMAT algorithm schedules them one at a time based on the present priority of those medicines toward the patient	To provide a faster and efficient approach of intake of medicines by developing and using different scheduling algorithms

(continued)

Table 1 (continued)

S. No	Title	Methodology	Techniques	Main outcomes
6	Voice Based Medicine Reminder Alert Application for Elder People [16]	Voice-based reminders and color-based button options for communication	Text-to-voice conversion is based on the data entered by the user and then it sets the reminder	To set reminders using voice and providing color-based buttons so that patients can communicate with their family members easily in case of device failure

(d) Patients' sensitive data will be well secured with the help of Advanced Encryption Standard (AES) algorithm combined with the Gravitational Search Algorithm (GSA).
(e) Chat-with-doctor is a feature of our app which provides doctor assistance for minor uncertainties of patients.

This application can be useful if the patient requires something or has some questions. This application can remind the patient to take his meds on time even after he is discharged from the hospital. The user has the ability to communicate with their doctors. Everything can be updated by doctors. For example, if a specific medicine's dosage needs to be adjusted. It is also feasible to upload and update the patient's food chart. If the patient has any questions, he can contact the doctor and also provide health and progress updates using the app. If a patient is or was suffering from a disease or ailment, the doctor can consult the patient's previous medical records saved on the cloud and treat him or her accordingly. Figures 3, 4, and 5 represent system overview and different modules in the proposed REMICARE system.

Figure 3 depicts the overview of the app. The information submitted by the patient into the system comprises the scanned image of prescription from which the name of

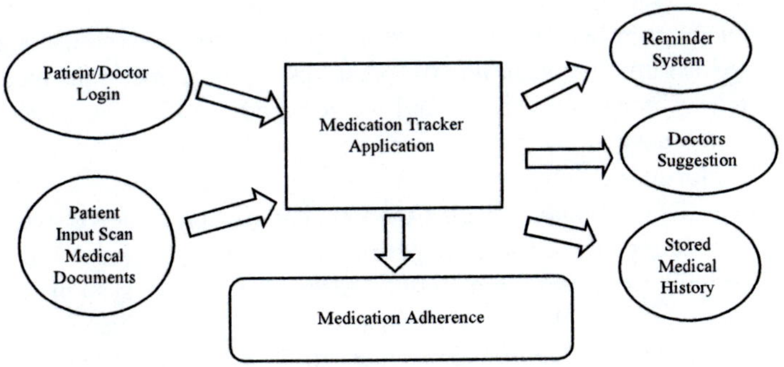

Fig. 3 System overview of REMICARE

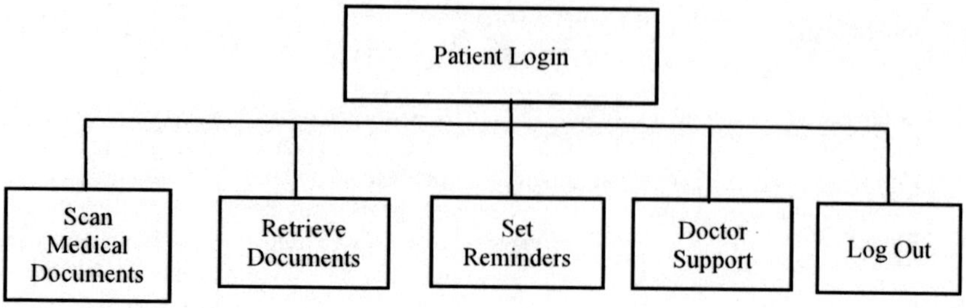

Fig. 4 Patient module

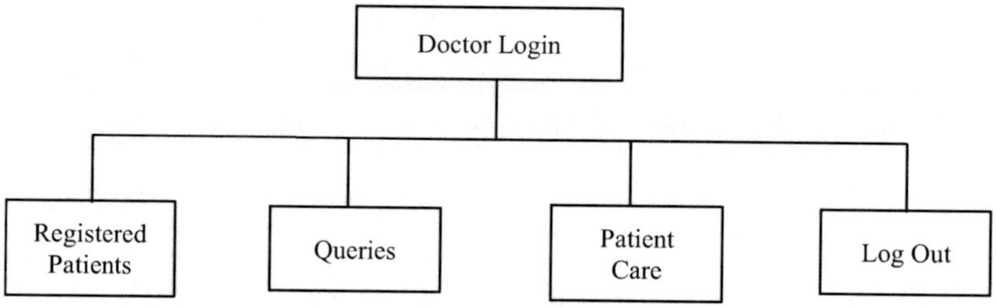

Fig. 5 Doctor module

the medication will be extracted. Also, users can set the reminder cycle as prescribed by the doctor or reminder can be set automatically by entering the dosage cycle. Users can also upload the past medical history and other health-related documents for quick and easy retrieval for further reference. Doctors can also login into the application to give the suggestions and to answer the queries given by the patients. Also, doctors can change the dosage cycle for the particular patient. The system's output focuses on "Medication Adherence." Medication Adherence refers to whether patients take their medicines as directed and whether they continue to take a medication after it has been given.

Figure 4 depicts the patient login module. After login into the application, the patient/user will be able to see the different options, such as scan the prescriptions and required documents, set reminder, doctor support, retrieve stored documents, and an option to log out. After scanning the prescription, the name of the medication will be extracted by image processing and the remainder will be set according to the dosage cycle. After that, patient will receive timely reminders for taking proper medications. Doctor support option is helpful for the treatment of various contagious diseases like COVID-19. Doctor can chat with the patient and can give suggestions, and also, they can change the dosages as per the requirements.

Figure 5 highlights the doctor login module. After login into the system, the doctor can see the registered patients list. Doctor can answer the queries given in the queries section. Also, a doctor can give the suggestions to the particular patient, and also, he/she can change the dosage cycle in the Patient Care section.

3.1 Image Processing

During the study of existing algorithms on character recognition, we found that deep CNN-RNN based on CTC provided better results. Character recognition using CNN-RNN is the electronic conversion of handwritten text pictures into machine-editable text. The network architecture called convolutional and recurrent neural network (CRNN) is shown in Fig. 6. It uses CNN layers, followed by LSTM layers and a final CTC layer [18].

Following are the steps involved in that process:

1. **Pre-processing**: Pre-processing is a basic phase of character recognition in which a set of operations are applied to scanned input pictures. This is where noise is filtered and smoothed which would otherwise complicate the recognition process. Finally, the image was scaled and sent to the model.
2. **Feature Extraction using CNN**: Convolutional layer is the first layer which extracts key features from an input image. By learning the features with the help of small squares of input file, this layer maintains the connection between pixels. After this, the image goes through different intermediate layers like maxpooling layer, normalization layer, ReLU layer, and finally the output layer.
3. **Information Gathering**: RNN is used to capture the shape of data accurately in a sequence through which relevant information propagates. Well-built and strong training characteristics are provided by RNN due to its ability to propagate information through long distances.

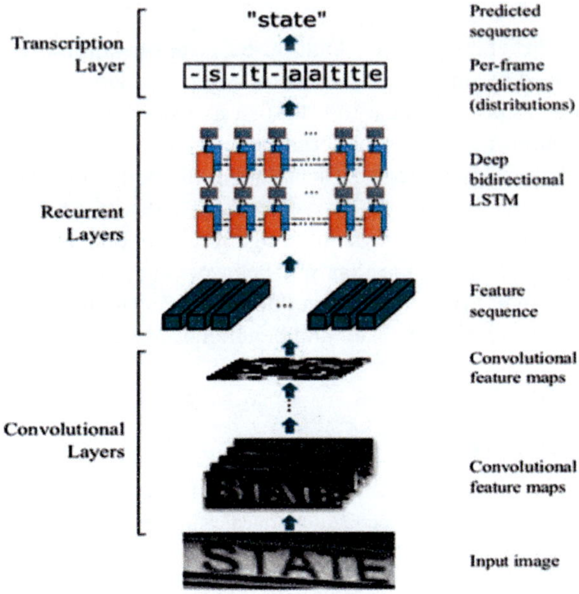

Fig. 6 Network architecture: convolutional and recurrent neural network (CRNN)-CTC

4. **Character Decoding**: Traditional handwritten character recognition methods tend to segment the input word image into sub-words and characters to identify it individually.

Therefore, segmentation plays a crucial role in such techniques, and every segment is considered as an observation where probability of need is calculated.

However, the chosen method aims to solve the problem without segmentation with the help of the potent CTC method which is a softmax layer following an RNN that allows sequence labeling without segmenting the input.

The Connectionist Temporal Classification method has demonstrated proficiency in handwriting recognition, to train deep neural networks, etc.

3.2 Secure Storage

In the application, the Advanced Encryption Standard is used for encryption of the data which is to be stored on the cloud. This enables the mobile data to be safe in both rest and transit states. Secure connection with cloud storage with the use of the Gravitational Search Algorithm will maintain confidentiality and integrity of the persons' sensitive data.

4 Conclusion

This paper proposes an Android-based application for medication reminders that can be used by many users and doctors for the medical purpose. In the study presented, an attempt was made to establish a system that is cost-effective, easy to use, and increases medication adherence.

Patients will receive a schedule of medicine intake times, as well as a description of the medication, the start and finish dates of the medicine, notice, an automatic alarm ringing system, and navigation. The planned reminder will not recommend any prescription that has not been prescribed by a doctor, ensuring the patient's safety and avoiding incorrect doses. The system will make the interaction between doctor and patient easier and beneficial.

References

1. https://www.pillsy.com/articles/medication-adherence-stats/20
2. https://www.mytherapyapp.com/
3. https://www.groovehealthrx.com/
4. https://www.imedicalapps.com/2016/11/round-health-app-personal-assistant-pill-taking/

5. Gupta A, Srivastava M, Mahanta C (2011) Offline handwritten character recognition using neural networks. In: ICCAIE 2011—2011 IEEE conference on computer applications and industrial electronics. https://doi.org/10.1109/ICCAIE.2011.6162113
6. Venkata Rao N, Sastry ASCS, Chakravarthy ASN, Kalyan Chakravarthy P (2016) Optical character recognition technique algorithms. J Theor Appl Inform Technol. 2005-2015 JATIT & LLS
7. Vyavahare S, Sagade M, Hajari K, Surwase S (202) Handwritten cursive english text recognition using deep CNN-RNN based CT. Int J Future Gener Commun Netw 13(2s):564–569
8. Zanjala SV, Talmaleb GR (2015) Medicine reminder and monitoring system for secure health using IOT. In: International conference on information security and privacy (ICISP 2015), 11–12 December 2015, Nagpur, India
9. Olaleye SB, Kant S (2018) Secure use of cloud storage of data on smartphones using atomic AES on ARM architectures. Int J Appl Eng Res 13(5)
10. Olaleye SB (2021) Security of sensitive data on android smartphones using cloud storage with reference to gravitational search algorithm. Int J Comput Sci Mob Comput 10(3)
11. Bagyalakshmi, Adhitya S, Youvashree (2019) A review on medicine reminder and adherence system. Int J Adv Res Electr Electron Instrum Eng 8(Special Issue 1). ISO 3297: 2007 Certified Organization
12. Santo K, Chow CK, Thiagalingam A et al (2017) Medication reminder APPs to improve medication adherence in Coronary Heart Disease (MedApp-CHD) study: a randomised controlled trial protocol, BMJ Open 7:e017540. https://doi.org/10.1136/bmjopen-2017-017540
13. Hayakawa M, Uchimura Y, Omae K, Waki K, Fujita H, Ohe K (2013) A smartphone-based medication selfmanagement system with real-time medication monitoring. Appl Clin Inf 4:37–52
14. Fang KY, Maeder AJ, Bjering H Current trends in electronic medication reminders for self-care. The promise of new technologies in an age of new health challenges. Series Stud Health Technol Inform 231:31–41. IOS Press Ebook. https://doi.org/10.3233/978-1-61499-712-2-31
15. Zao JK, Wang M-Y, Tsai P, Liu JWS Smart phone based medicine in-take scheduler, reminder and Moni. 978-1-4244-6376-31101$26.00 ©2010 IEEE
16. Mehala M, Viji Gripsy J (2017) Voice based medicine reminder alert application for elder people. Int J Rec Technol Eng (IJRTE) 8(6). ISSN 2277-3878, March 2020. https://doi.org/10.35940/ijrte.F7731.038620
17. Ameta D, Mudaliar K, Patel P (2015) Medication reminder and healthcare—an android application. Int J Managing Public Sector Inform Commun Technol 6:39–48. https://doi.org/10.5121/ijmpict.2015.6204.21
18. Shi B, Bai X, Yao C (Jul 2015) An end-to-end trainable neural network for image-based sequence recognition and its application to scene text recognition. arXiv:1507.05717v1 [cs.CV] 21

Detection of Fraudulent Credit Card Transactions in Real Time Using SparkML and Kafka

V. Ashwin, Vishal Menon, A. M. Devagopal, P. A. Nived, and J. Udayan Divya

Abstract As a result of the rapid expansion and development of E-Commerce, the use of credit card for online shopping and purchases has increased dramatically, and this rapid increase has led to a surge in fraudulent credit cards. Now being the most popular mode of payment in both offline and online purchases, credit card fraud has become a major menace. With the recent advancement in artificial intelligence, machine learning, data mining, sequence alignment, genetic programming, and fuzzy logic, innovative ways for detecting various credit card fraudulent activities have evolved. In our work, we have utilized SparkML and Scikit-Learn to develop and train several machine learning models for distinguishing fraudulent and legitimate transactions. We have also employed several data preprocessing techniques, such as class imbalance removal, which were implemented using various Spark packages. These machine learning models are then utilized to make predictions on Kafka-generated real-time data streams. Finally, we have used Streamlit in order to build an interface for displaying these predictions in real time.

1 Introduction

Credit card fraud transaction is gaining much popularity worldwide. Almost every company which offers credit card services is investing huge amount of money to overcome such fraudulent activities; unfortunately, the advance growth of technologies every year has also resulted in new fraud techniques. The historical activities of card holder provide good insights about the user as fraud or genuine. Furthermore, those behavioral patterns can be extracted while building algorithms for the models. In our work, we tried to improve the model accuracy by using ten different machine learning models (logistic regression, Gaussian Naive Bayes, random forest, K-neighbors, quadratic discriminant analysis, linear discriminant analysis, support vector classification, decision tree, AdaBoost, XgBoost), and instead of doing batch processing,

V. Ashwin · V. Menon · A. M. Devagopal · P. A. Nived · J. Udayan Divya (✉)
Department of Computer Science and Engineering, Amrita Vishwa Vidyapeetham, Amritapuri, India
e-mail: divyaudayanj@am.amrita.edu

V. K. Gunjan and J. M. Zurada (eds.), *Proceedings of 3rd International Conference on Recent Trends in Machine Learning, IoT, Smart Cities and Applications*, Lecture Notes in Networks and Systems 540, https://doi.org/10.1007/978-981-19-6088-8_26

we have done it live streaming. We have also done certain data preprocessing steps in order to achieve better accuracy for results. Since transaction predictions are handled in millions in we need to use distributed frameworks to derive such large-scale transactions and identify a transaction is genuine or fraud. We have used spark as a distributed framework that is similar to hadoop. We used PySpark for implementing our entire project. We have also implemented our entire work on Streamlit which is a Python open-source framework, which helps to create Web apps in a shorter period of time.

2 Literature Review

In [1], the authors have identified the problem of fraudulent credit card transactions and have applied data science and machine learning to explore and solve the problem. This paper focuses on the preprocessing as well as deployment of the data using multiple anomaly detection algorithms. The authors have only provided a high-level explanation of the main ML algorithms they have used and have not much explained about the performances of each algorithms. The paper also does not explain about the model inference as a result this work cannot be directly used for real-world applications.

The authors of [2] have used convolutional neural networks for detecting fraudulent transactions. They also have used methods such as simultaneous minority oversampling technique (SMOTE) to overcome the issue of imbalanced data. Their work is divided into two sections in which the first one focuses on training the model using the existing dataset with historical transactions, and the second face involves making predictions using the trained model. They have also focused more on cleaning and processing the dataset, and this involves feature selection, SMOTE, and feature transformation. The CNN implemented has a LeNet backbone architecture, and its performance was evaluated.

The authors of [3] conduct a thorough analysis of numerous Kafka settings and performance metrics in order to help users avoid bottlenecks, reach their full potential, and eventually exploit certain best practices for effective stream processing. In [4], the authors deliver solution to the Higgs Boson classification problem. They have used machine learning techniques such as logistic regression, decision tree, random forest, and gradient boosted tree.

The authors of [5] coin an innovative method which deals with credit card fraud detection. Research conducted by Fu et al. [6] focuses on fraud detection using traditional CNN model, in which some special patterns of fraudulent behaviors are captured from labeled data. The authors of [7] have used long short-term memory (LSTM) for sequence classification problem, which involves the fraud detection problem. They have also integrated some feature aggregation strategy along with their work [8].

In [9], the authors depend on bias of the training data which is generated as a result of the skewed distribution. In [10], the researchers combined both random forest and

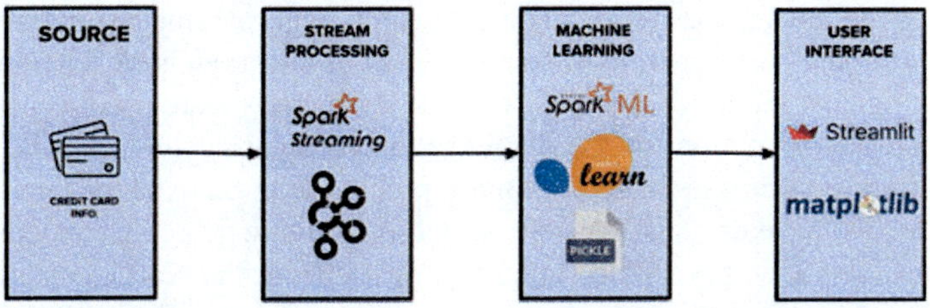

Fig. 1 System architecture

neural networks to form a new ensemble method which is capable of predicting with high accuracy and confidence.

In [11, 12], numerous supervised and unsupervised machine learning algorithms have been explored to solve the problem of fraudulent credit card transactions in a highly imbalanced dataset. From these works, it was observed that unsupervised machine learning algorithms generally perform better for classification and handles skewness more efficiently.

In [13], the authors have explored the smart grid which is an automated system that uses groups of sensors embedded into power grid systems. Modern technologies can be utilized to collect big data from such systems. The authors have presented Spark as a central platform for storing and performing big data analytics on smart grid for various applications.

In [14], the authors have introduced a general pipeline for of big data. While dealing with real-time data streams, for performing analytical operations, different tools and services such as Hadoop ecosystem, Apache Kafka, and Spark need to be used. The general pipeline introduced by the authors can be really useful while deploying big data analytic programs in the real world.

In [15], the authors have presented an architecture for dealing with live-stream data associated with IoT devices used for healthcare applications. The proposed method is fault tolerant and feasible for many applications.

In [16], the authors explore the vulnerability in IoT-based automated systems. They present a cloud-based service for monitoring the huge amount of data generated and transferred between devices and keeping the systems secure (Fig. 1).

3 System Architecture

The model used for classifying fraudulent credit card transactions is trained using the dataset which are preprocessed using SparkSQL modules. Various data preprocessing techniques such as removing outlier data, removing row vectors with NaN or missing values, remove irrelevant character values in the dataset, and extraction of important features from the dataset using principal component analysis and low rank approximations were done to improve the quality of the results obtained by our trained model during prediction. The gridsearch hyperparameter tuning method was used to find the best hyperparameters for each of our models. After training all the models, the best model parameters from all the models were used to determine the most accurate prediction for the newly streamed input vectors. All the trained models will be stored in the form of a binary pickle file to be used for later prediction purposes. We have used the Apache Kafka software distribution to mock a data stream in our local machines. The streams retrieved from the Kafka producer will be used as real-time data which is required for prediction by Kafka consumers. The machine learning pipeline is also integrated in the Kafka consumer script where the data retrieved by the Kafka consumer is converted into a NumPy one dimensional array format to be accepted by the models saved in pickle formats. The Streamlit open-source framework is utilized for creating the graphical user interface of our streaming application. Streamlit widgets utility has been used created custom UI elements for the real-time application. We have used the Matplotlib Python framework for plotting the pie graph plots of the results in real time inside our streaming application. Further, information regarding the real-time streaming pipeline design is explained in the methodology section.

4 Methodology

4.1 Dataset

The dataset contains transactions made by credit cards during the month of September 2013 by European cardholders. This dataset presents transactions that occurred in two days, where 492 fraudulent transactions were recorded out of 284,807 transactions. The dataset is highly unbalanced, the positive class (frauds) account for 0.172% of all transactions. We have performed a number of preprocessing techniques to reduce this class imbalance, and the dataset thus obtained is shown here. The dataset contains labels V1, V2, .. etc., representing the features obtained using principal component analysis, and the original features are kept hidden.

4.2 Data Prepossessing

A common issue with various real-life datasets is associated with class imbalance. Here, we observed that more than 99% of the training dataset contains genuine transaction records, and the rest less than 1% constitutes fraudulent ones. Hence, the model cannot be properly trained and is usually biased toward the majority category while making the predictions. To overcome this issue, firstly, we have used SMOTE, which is a statistical technique for increasing the number of cases in a dataset in a balanced way. Here, we use this technique to randomly duplicate examples in the minority class, thereby increasing their overall frequency. Another method is undersampling which is used for randomly deleting examples in the majority class and basically is the opposite of oversampling. Both techniques were used to balance the dataset consisting of 50% of each of the classes, and the trained model was able to achieve better results.

4.3 Machine Learning Models

In order to perform detection of fraudulent credit card transactions, the 10 different machine learning models were trained and used for estimating the final prediction. Each models provided different accuracy in different scenarios; hence in our real-time implementation, we have done a statistical analysis after getting the prediction from each models and have taken the prediction which is more statistically relevant (having the greatest mode). The following machine learning models were used in this project.

Logistic Regression Logistic regression comes under the class of predictive algorithms. In this, we use independent variables called X (features of our models) to predict categorical dependent variables called the target variable Y. This statistical model uses logistic functions to represent the conditional probability. Using binary regression, we calculate the conditional probability of Y, i.e., the dependent variable for the given independent feature variable X. $P(Y = 0|X)$ and $P(Y = 1|X)$ are two ways to write it. $P(Y|X)$ is approximated by a sigmoid function applied to a linear combination of input features.

Gaussian Naive Bayes Gaussian Naive Bayes works under the principle of Bayes theorem. It comes under supervised learning techniques and is a basic but strong approach for predictive modeling. The Naive Bayes approach is simple to comprehend. When there are a lot of data points, Naive Bayes has a better accuracy and speed. The Gaussian Naive Bayes is a Naive Bayes variation that accepts continuous data and assumes that each class is regularly distributed.

Random Forest Aggregation of a large number of decision trees results in the generation of random forest. Class which has the most number of votes will determine the overall prediction of the model. They rectify their faults, thus resulting in much better results. Random forest's core notion is that a large number of substantially uncorrelated models (decision trees) acting as a committee will surpass any one of the constituent models individually. They rectify their faults, thus resulting in much better results.

K-Neighbors K-nearest neighbor is a supervised greedy algorithm that is greatly used for conducting discriminant analysis where there is a lack of known probability densities of potential parametric estimates, thus making them difficult to estimate. One of the major disadvantages of using the KNN algorithm is that the algorithm becomes exceedingly slow as the size of the training set increases. The primary principle of KNN is that it calculates the distances between a query and all of the instances in the data and then picks the number of examples (K) closest to the query and voting for the most frequent label (in the case of classification) or averaging the labels.

Quadratic Discriminant Analysis Quadratic discriminant analysis is kindred to linear discriminant analysis, whereas mean and covariance are relaxed as equal for all the classes.

Support Vector Classification Support vector machine algorithms aim to derive the hyperplane on an N-dimensional feature vector that can classify the data points efficiently. SVM takes care of multiclass support in accordance with the one–one scheme.

Decision Tree In order to solve classification and regression problems, decision trees are used which is a non-parametric supervised learning method. A model is created using simple decision rules which are obtained using data features.

AdaBoost AdaBoost (adaptive boosting) belongs to the class of ensemble methods in machine learning. Decision trees with one level, or decision trees with only one split (decision stumps), are the most popular algorithm used with AdaBoost. This algorithm uses an iterative approach where the model gives equal weights to all the data points in the decision tree in the first iteration. Then, it assigns higher values for weights to data points that are misclassified. The points which are assigned higher weights are given higher priority in the next iteration. This process continues until a low error value is achieved.

Gradient Boost Gradient boosting machine learning method has varied applications ranging from classification to regression. Like AdaBoost, this also belongs to the class of ensemble models where the technique returns a prediction model which is a combination of weak predictors. Similar to AdaBoost, gradient boost also commonly uses decision trees as its basic prediction models. The algorithmic design for gradient boosting is similar to other boosting algorithms as it also uses an iterative approach, but the only point of difference from the other model is the fact that gradient boosting approach allows optimization of differentiable loss function, this makes gradient boosting more effective than other boosting and decision tree approaches like random forest.

Linear Discriminant Analysis Linear discriminant analysis or normal discriminant analysis or discriminant function analysis is a dimensionality reduction technique which is often employed for supervised classification problems. It is used to represent group differences, such as separating two or more classes. It is used to project higher-dimensional features onto a lower-dimensional space.

4.4 Kafka

We have used the Apache Kafka distribution to mock a data stream in our local machines. We have created a Kafka producer that reads data from CSV files and converter the data into Python dictionaries (key-value pairs) where the time stamp from the data is used as the dictionary key, and the other features from the data which is used for model prediction are stored as the values. These dictionaries are published

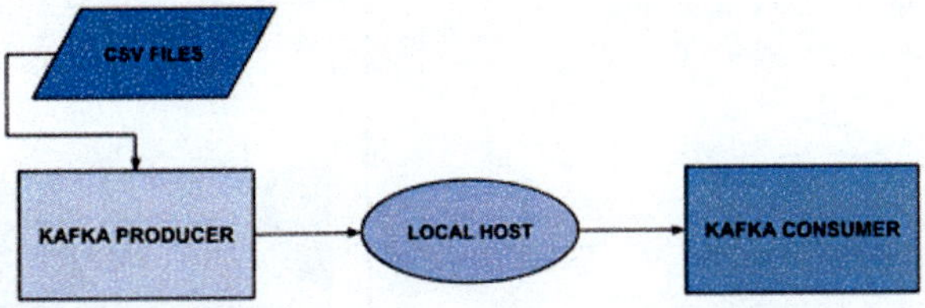

Fig. 2 Kafka workflow for retrieving mock data

into the local host using a TCP socket. The delay between the published messages can be adjusted at the Kafka producer.

Next, we use a Kafka consumer to retrieve the data from the local host using a socket connection. We have generated a Kafka consumer integrated with machine learning pipelines which helps in prediction using the real-time mock data that we retrieve using the socket. The consumer first reads the data from the host and then converter the bytes data into a NumPy array which could be used for prediction. The feature values from the NumPy array are then used for prediction using the model pickle files. We have also integrated the Kafka consumer with Streamlit functionalities to provide a Web interface for the consumer. This Web interface allows us to visualize the prediction results from the real-time mock data in a seamless and interactive way. The architecture of the designed Kafka framework is provided in Fig. 2.

4.5 Streamlit Visualizations

The entire streaming application has been built using the Streamlit framework. Streamlit is an open-source framework for creating frontend and backend pipelines for data centric applications. We built custom UI elements for the Streamlit application using the inbuilt widgets module provided by the framework. The pie plots built using matplotlib are frequently updated in the Streamlit environment in real time with a regular time delay of two seconds with the help of the Streamlit media module.

5 Results

With the use of various data preprocessing techniques like synthetic minority oversampling technique (SMOTE), undersampling, and oversampling, we have successfully balanced out the original dataset. The class distribution in the new dataset is graphically represented in Fig. 3.

We have implemented and trained ten different machine learning models to solve this classification problem, and Table 1 represents their performances based on the

Fig. 3 Class distribution of the preprocessed dataset

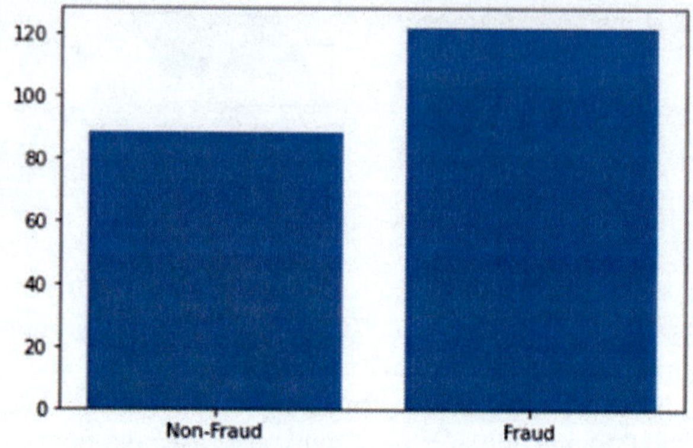

Table 1 Errors made by each model

Machine learning model	Error obtained
Decision tree	120
Logistic regression	200
KNN	150
Linear SVM	450
Polynomial SVM	75
Random forest	60

errors made by each of the model. As seen in the graph, random forest algorithm has the least amount of errors while testing on the validation set. We have observed that all of the used algorithms have performed well with relatively good accuracy, hence to make our system more robust, our final prediction is calculated by taking the mode of all the predictions made by each of the ten different models, and this way even if a single model fails for some particular test case, it will not affect the final predictions made by the system.

We have saved our trained models as pickle files which are then used for live inference. We have used Kafka for generating the mock real-time data.

Figure 4 represents the command line output on starting the Kafka producer. The messages generated from the CSV data and published to the local host are printed out in the terminal in each iteration. We can clearly see that the messages are generated in a Python dictionary format where the message timestamp acts as the dictionary key, and the feature values of the message act as the dictionary values.

Figure 5 shows the Web app interface that is activated when we run the Kafka consumer. The Web app interface has a input button which when clicked starts the prediction process for the live data.

When the button is pressed, the Kafka consumer starts retrieving the messages from the local host using socket connection. The retrieved message is then used for prediction using the stored ML model pickle files. The count of the predicted

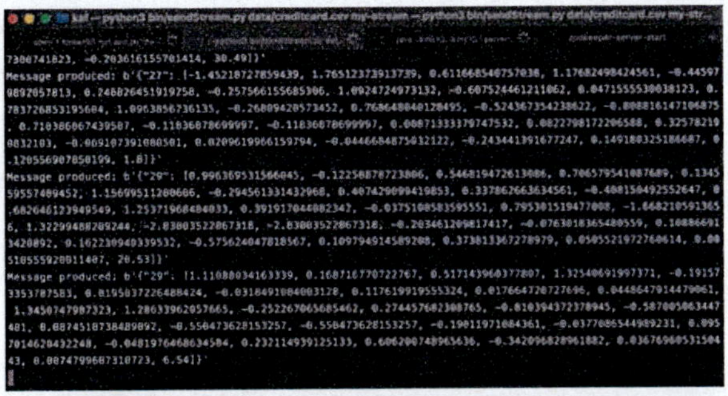

Fig. 4 Command line output when Kafka producer is executed

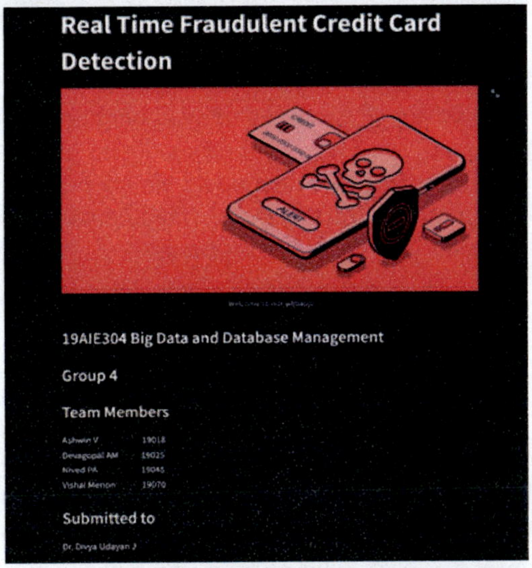

Fig. 5 Streaming application home page

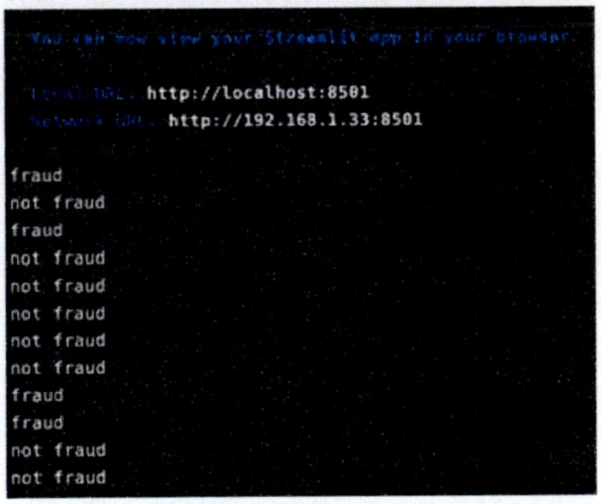

Fig. 6 Streamlit application with Kafka consumer script executed with real-time outputs displayed on terminal

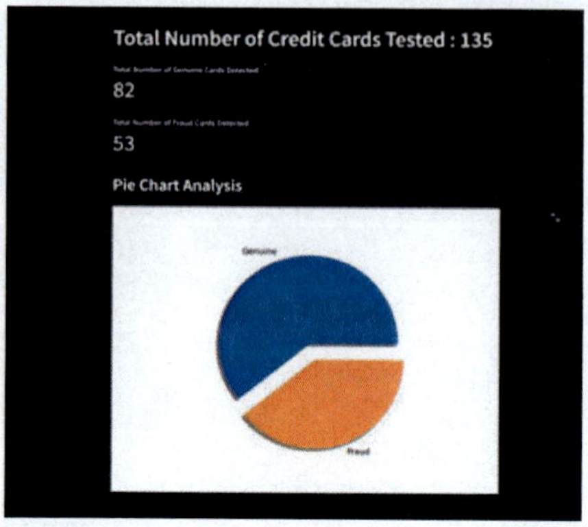

Fig. 7 Real-time outputs being displayed and updated in real-time in the streaming application

fraudulent credit cards can be seen in both the command line Fig. 6 and in the Web app Fig. 7. The command line provides a simple one line output of the predicted class, and the Web application provides a more intuitive visualization (pie chart) with other details like the total number of fraudulent, non-fraudulent credit cards detected, and total number of credit cards tested.

6 Conclusion

The recent increase in the credit cards used for online purchases has caused a tremendous explosion in credit card fraud. As credit cards become the most popular method of payment for both online and offline purchases, incidents of credit card fraud are on the rise. In this paper, we have successfully created a real-time credit card fraud detection system using Apache Spark, Kafka, and machine learning algorithms. We have used Spark SOL for data processing and balancing out the unbalanced data which was used for machine learning. SparkML and Scikit-learn libraries were used to train 10 different models for the prediction of fraudulent credit cards. The trained models were stored as pickle files in order to be used for real-time prediction. A Kafka producer was created to generate mock real-time data from a CSV file and publish the mock real-time data into the localhost. We have also created a Streamlit application that acts as a Kafka consumer and retrieved real-time data to produce prediction results from the trained ML models. The Streamlit application provides an interactive Web interface to visualize and analyze the prediction results in a seamless and interactive way. Our interactive Web interface would analyze the count of the fraudulent credit cards detected and provide us a graphical visualization (pie chart) for the same.

Through this project, we have successfully established the applications of Kafka in real-time generating and streaming of data and the use of Spark in data processing and training machine learning models for prediction applications. The designed fraudulent credit card detection system can have varied applications in reinforcing security in online banking, purchases, transactions, productivity, etc.

References

1. Maniraj S, Saini A, Ahmed S, Sarkar S (2019) Credit card fraud detection using machine learning and data science. Int J Eng Res 8(09)
2. Modi K, Dayma R (2017) Int J Adv Res Eng Sci Technol Trans 4(8)
3. Le Noac'H P, Costan A, Bougé L (2017) A performance evaluation of apache kafka in support of big data streaming applications. In: 2017 IEEE international conference on big data (big data). IEEE, pp 4803–4806
4. Higgs boson discovery using machine learning methods with pyspark. Proced Comput Sci 170:1141–1146 (2020). https://doi.org/10.1016/j.procs.2020.03.053, https://www.sciencedirect.com/science/article/pii/S1877050920304907, the 11th international conference on ambient systems, networks and technologies (ANT)/the 3rd international conference on emerging data and industry 4.0 (EDI40)/affiliated workshops
5. Quah JT, Sriganesh M (2008) Real-time credit card fraud detection using computational intelligence. Expert Syst Appl 35(4):1721–1732
6. Fu K, Cheng D, Tu Y, Zhang L (2016) Credit card fraud detection using convolutional neural networks. In: International conference on neural information processing. Springer, pp 483–490
7. Xuan S, Liu G, Li Z, Zheng L, Wang S, Jiang C (2018) Random forest for credit card fraud detection. In: 2018 IEEE 15th international conference on networking, sensing and control (ICNSC). IEEE, pp 1–6
8. Randhawa K, Loo CK, Seera M, Lim CP, Nandi AK (2018) Credit card fraud detection using adaboost and majority voting. IEEE Access 6:14277–14284
9. Kim MJ, Kim TS (2002) A neural classifier with fraud density map for effective credit card fraud detection. In: International conference on intelligent data engineering and automated learning. Springer, pp 378–383
10. Sohony I, Pratap R, Nambiar U (2018) Ensemble learning for credit card fraud detection. In: Proceedings of the ACM India joint international conference on data science and management of data, pp 289–294
11. Rajeev H, Devi U (2022) Detection of credit card fraud using isolation forest algorithm. Pervasive computing and social networking. Springer Singapore, Singapore, pp 23–34
12. Sriram Sasank JVV, Sahith GR, Abhinav K, Belwal M (2019) Credit card fraud detection using various classification and sampling techniques: a comparative study. In: 2019 international conference on communication and electronics systems (ICCES), pp 1713–1718
13. Shyam R (2015) Bharathi GHSKSPPSK. Apache spark a big data analytics platform for smart grid 21:171–178
14. Rengarajan K, Menon VK (2020) Generalizing streaming pipeline design for big data. In: Machine intelligence and signal processing. Springer Singapore, pp 149–160
15. Fahad PMP (2018) Prediction of human health using machine learning and big data, pp 0195–0199. https://doi.org/10.1109/ICCSP.2018.8524352
16. Rakesh N (2016) Performance analysis of anomaly detection of different iot datasets using cloud micro services, pp 1–5. https://doi.org/10.1109/INVENTIVE.2016.7830157

Real-Time Face Detection and Face Recognition: Study of Approaches

Siddhartha Singh Bhadauriya, Sachin Kushwaha, and Shweta Meena

Abstract This study demonstrates various face detection and recognition techniques which have been studied till now and compares them on basis of their merits and demerits, discusses their methodologies of working and put forward a core idea of how face detection is done, mentioning about very basic term, so that any person who is not that good in technology can understand it and dive deep into this field. Some of the face detection techniques we will look into will be geometric-based face detection, feature-based face detection, and Haar-like feature based-face detection, giving special emphasis on Haar-like features-based face detection. In face recognition techniques, we looked into the lazy learner's face recognition approach, neural networks face recognition approach, and holistic face recognition approaches. Objective of the study is also to demonstrate about preparing a model which detects faces in a real-time environment. Face detection, nowadays, is the most primary check in any security system. So, automation in detecting faces will prove helpful. This model, rather than any other model, works on real-time data provided. Our model works on the fundamentals of the K-nearest neighbors algorithm, Haar cascade classifier (an object detection technique), and OpenCV (an open-source python library for computer vision, machine learning, and image processing).

Keywords K-nearest neighbor · Principal component analysis · Open-source computer vision · Artificial neural network · Rectified linear activation unit · Automated teller machine

1 Introduction

In this COVID era and digital age, where everything is virtual, one needs to think more about the security of their phones, tablets, because there is increase in case of cyber threats, as well as in ATMs, there are increase in cases of thefts, and we were in so much need of face recognition system. Whenever we have to do Know Your

S. Singh Bhadauriya (✉) · S. Kushwaha · S. Meena
Delhi Technological University, Delhi, India
e-mail: siddharthabhadauriya@gmail.com

V. K. Gunjan and J. M. Zurada (eds.), *Proceedings of 3rd International Conference on Recent Trends in Machine Learning, IoT, Smart Cities and Applications*, Lecture Notes in Networks and Systems 540, https://doi.org/10.1007/978-981-19-6088-8_27

Client (KYC), and there were increased cases of militancy so our armed forces and security agencies were in need of an advanced face recognition technique, because sometimes the photographs of militants in the records of our security service are so old and it is really tough to recognize them, so there was increased demand in this field, we took this opportunity and started our research on this project where we thought to go through various face detection and face recognition techniques, and with this aim, we continued working on this and also build a model which detects faces and also recognizes the face of persons in real time. In this study, we looked into various techniques of face detection and face recognition; we discussed their merits and demerits and came up with a project on it.

2 Related Work

We went through various approaches of solving the face detection and recognition problem, proposed by researchers, comparing them on the basis of their merits and demerits and getting enough idea of various approaches in the form of tables.

2.1 *Procedure for Face Detection*

Step-wise procedure of face detection looks like this

(a) Take a snap of face
(b) Detecting human face using the below-proposed algorithms
(c) Extract the region of interest using rectangle around the face.

2.2 *Various Face Detection Techniques*

For face recognition, we have to first understand how we can detect faces. For that, there are many approaches, we will discuss them and understand how they function. Their merits and demerits are illustrated in Table 1.

From the perspective of a common man, there can be some human knowledge-based approach, where a human eye detects a face, that is nose and it is above the mouth of a person, and they also know that a normal human face has two eyes, two ears, one nose, and one mouth.

Feature-based face detection: Feature-based face detection uses the concept of spatial features (i.e., features occupying space), grouping them using the grey level constraints as we have images already in grey format [1]. Yow et al. [2] proposed the idea of facial features (such as different components of the human face, i.e., eyes, ears, mouth, nose) and image features (such as edges, face contours, and corners inside the

Table 1 Merits and demerits of some existing methods

Approach	Pros	Cons
Feature-based Face Detection	1. High Accuracy 2. Takes less Implementation Time	1. Takes Much learning Time
Geometric-based Face Detection	1. Effectiveness is more 2. Execution time is less	1. False detection rate is high 2. Less precise
Haar-like feature-based Face Detection	1. It has much better feature extraction part 2. False detection rate is low	1. Time taking approach 2. Not simple to implement

image). After extracting points of interest, we implemented a probabilistic approach to check the probability of that image as a face. Template matching approach can also be used, where we make a parameterized template of a face, then find the correlation between the template and the faces in the video frame, thus detecting a face [3].

Geometry-based face detection: Ghimire et al. [4] suggested us the idea of geometric-based facial detection, which is as crystal clear concept of looking into the geometry of different face parts, such as eye, ear, nose, eyebrows such as eye is oval in shape, then nose in quite linear and geometry of mouth is identified differently, so that is mostly based on detecting the shape of various body parts and then detecting a face.

Ghimire et al. [4] also proposed the idea of appearance-based face detection, which can detect faces in various different expressions, when they express their emotions such as laughing, crying, shouting, smiling, and other appearances. People get confused with the idea of geometry and appearance-based face detection, because they seem similar, but, appearance-based detection depends both on the geometry of the different components of human face and their location, because whenever any person make some expressions, different facial components reach to different positions and their geometry also gets, somewhat altered [5].

Haar-like features-based face detection: Haar-like features are of very much importance in object detection techniques. Even Viola–Jones algorithm used Haar-like rectangle features and altered their approach to use Haar-like waves. It works on the principle of five rectangles, which can be slightly rotated over the entire face and looks for five regions of image that are up-down, left–right, horizontally middle, vertically middle, and diagonally [6] as shown in Fig. 1. In this approach, the dark and light regions generally work on the fact that the area belonging to eyes in image is generally darker than the cheeks, whereas the region of nose is quite blazing. To detect a face, when an up-down rectangle will scan over the region of eye and cheek, then it will see similarity with its pattern, and similarly, other rectangles will have

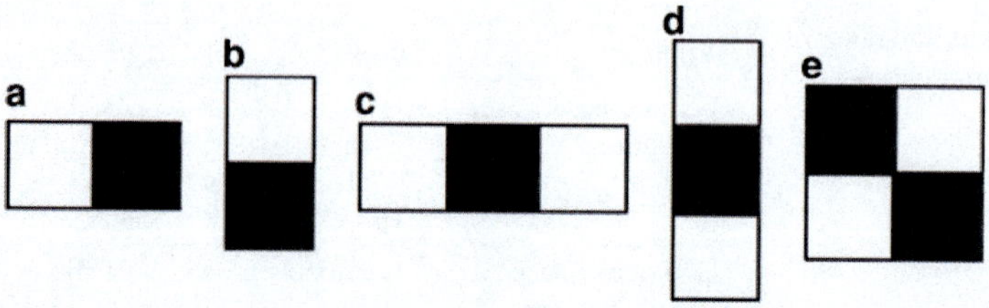

Fig. 1 Various Haar-like features **a** left–right, **b** up-down, **c** horizontally middle, **d** vertically middle, **e** diagonally via research gate commons. *Source* https://www.researchgate.net/profile/Ryan-Kastner/publication/272640190/figure/fig3/AS:391776583077898@1470418224940/Basic-group-of-Haar-like-features-in-the-OpenCV-library-includes-these-five-variations_W640.jpg

their zones to find similarity and if every rectangle satisfies their condition, then our face will get detected.

Working with these feature windows, and using them with 24*24 base window involves so much computational complexity. To get rid of this issue, concept of integral image came, which says, inside any feature rectangle, we don't need to calculate the sum of all pixels, we just have to calculate the sum of corner values of rectangle, and this improves the performance time of this approach.

Ahmad et al. [7] mentioned that using AdaBoost classifiers with Haar-like features gives the best face detection rate. It works on the principle of keeping only those features that anyhow lead to increase in the accuracy of AdaBoost. To obtain this, it makes a combination of weak classifiers which forms a strong one.

2.3 *Various Face Recognition Techniques*

After successfully learning about face detection, we will discuss some of the face recognition techniques whose merits and demerits are illustrated in Table 2.

Lazy Learning Approach: KNN algorithm, a lazy learning approach, works on classification of images based on looking on its K-nearest surrounding, Though it

Table 2 Merits and demerits of face recognition approaches

Approach	Pros	Cons
Lazy Learners Approach (KNN)	1. More accurate [8] 2. Requires no training time	1. Takes time in classification 2. No reusability of parameter values
Neural Networks (ANN)	1. Less accurate than KNN 2. Reusability of parameter values	1. Requires so many hyperparameters 2. Training is time expensive
Comprehensive Approach (PCA)	1. Works well with feature extraction	1. More sensitive to variations in shape of face

is put in lazy learner category due to almost no training involved, if we talk about accuracy, then it is quite a good algorithm, finding value of K in KNN is a very tedious task, as k being less, chances of noise will be more and increasing k will result in more calculation complexity of the algorithm, so we need to work on finding optimal value of k for our dataset [9].

For finding the neighbors, we will go with calculating Euclidean distance, which is:

$$d = \sqrt{((x_1 - x_2)^2 + (y_1 - y_2)^2} \tag{1}$$

Based on the above equation, we will get k-nearest neighbors, then computing probabilities for these neighbors based on their similarity with test data, and based on the probabilities, it will finally label test data to one of the nearest data points having the highest probability and the same process is repeated for every test data, we can further improve performance of our model if rather than taking a single set of neighbors, we take multiple sets of different neighbors as proposed by Guo [9].

Human Brain Approach (Neural Network)
This approach is based on the principle of human brain, that is as human brain consists of neurons. In this approach, known as ANN we make our image vector as the input layer and then we add some hidden layers whose task is to find some definite pattern for images and then classify the image that it belongs to which person [10].

This task is not as simple as it looks, extracting all the useful features from the image and then passing out useful information to the next layer, makes work easier for them, but it's a quite time-consuming task and complicated too.

To fetch out important features from the hidden layers, we also use the activation function in between that can be Sigmoid, tanh, ReLU, Softmax, etc. It helps in extracting useful linear and nonlinear patterns from the input layer and then carrying them forward to the output layer efficiently [11]. Sigmoid and tanh are nonlinear functions, ReLU is a partially linear function, and softmax is an activation function for multiclass classification [12].

Activation functions and options we can use for this study:

In layman terms, Sigmoid function is also considered as squashing function, which limits the output in range of 0 and 1. Its equation is written below:

$$S(x) = 1/(1 + e^{-x}) \tag{2}$$

'Tanh' activation function is similar to sigmoid, but comparatively better than that due to its range, which is -1 to 1. Equation is:

$$(e^x - e^{-x})/\left(e^x + e^{-x}\right) \tag{3}$$

'ReLU' keeps the value positive as $\text{ReLU}(x) = \max(0, x)$ so, it considers negative values as 0. In equation form:

$$R(x) = \max(0, x) \tag{4}$$

Softmax activation function is a probability-based activation function, where the winner class takes all, thus helpful in multi-class classifications [13].

In neural network detection, assignment of weight to input layer vectors plays a vital role in determining the outcome so to improve the value of weight we also propagate in reverse direction so that it can learn from its mistakes and change the value of weight accordingly. This process of propagating backward is called backward propagation. It is not a lazy algorithm like KNN [11].

Comprehensive Approach: This approach, as the name suggests, looks for the entire face rather than looking for different features or components of the human face. All methods we can discuss under this category follow the principles of PCA. In this approach, we create eigenfaces first, in which we derive a feature vector out of the input image in the form of a covariance matrix and train our model based on this. When we are provided with some new face for recognition, this approach looks into the weight of stored images and compare with all of them, and this process is not deterministic, it gives percentage to different stored images, and the image whose covariance matrix weight is more comparable, that is considered as the output [10].

3 Experiment Design

In this section, we will look into the experimental setup designed for solving the face detection and recognition problem. While preparing this model, we used the techniques which were the best among the above-mentioned techniques. Initially, the tool and techniques used in experiment are explained in brief followed by description of dataset and model build. Then phasewise description of experiment is given.

3.1 Learning About OpenCV

It is very easy for computer devices to understand the content of arrays, strings, but when it comes to images, it seems a quite tougher task. Image processing focuses on that, thus helping computers to actually understand the content of images. Intel worked on this and came with an open-source library OpenCV, which helps in image filtering (i.e., image modification or enhancement), object tracking aiming at pointing out single or multiple objects within a frame [14].

3.2 Learning About Haar Cascades

In Haar Cascades, we have different xml files for different part detections such as for eye, we have haarcascade_eye.xml and for frontal face, we have haarcascade_frontalface_default.xml. As a very starting step, we will have to import these files first, then we will focus on getting the x-coordinate, y-coordinate, width (w), and height (h) of the detected feature using detectMultiScale function. Once we are finished doing this, we will make two numpy arrays out of this, one being roi_gray and one being roi_color [15–17].

We created roi_gray, because we will pass the gray color version to the detectMultiScale function using which we got required coordinates [17]. Here, roi_color will be the numpy array created using the actual RGB scale while roi_gray is the grayscale image which we will use in our code aiming for faster processing. Immediately after this, we got four features so now we will loop through the features found out and will create a rectangle around the face and eye [15, 17].

3.3 Dataset for Model

This model, being based on a real-time environment, aims to collect its data directly in the form of images and name of person in the index. We used OpenCV to capture images and then used the cv2.rectangle() method to mark the rectangle over the entire face [16] as shown in Fig. 2a, b. This method is basically being provided with four parameters, one being the image itself, second being the starting and end coordinates of our required rectangle, third being the color of the rectangle (which is in rgb format), and last being the thickness of the rectangle boundaries in pixel [17, 18].

Once capturing images, we save this as our data in the format of a numpy array file starting as (name.py).

3.4 Building Model

We can summarize the whole process of face recognition in the following phases described below:

Phase 1: Data Collection

In this phase, we used OpenCV, which started with testing our camera and checking if it is taking pictures clearly without any picture quality issue. After checking it, we made a subdirectory named xyz, where we will store the photographs in numpy form of all the photographs gathered for training purposes, with the name of the person. Flowchart representation of the whole process is shown in Fig. 3.

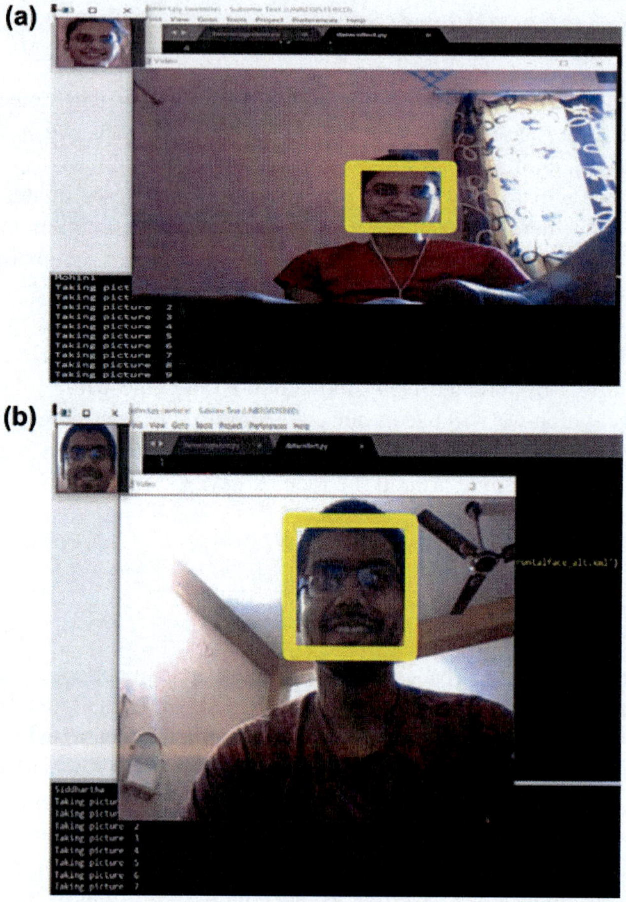

Fig. 2 **a**, **b** Data collection with image and name

Fig. 3 Data storing process

Now, to distinguish among different photographs, we saved the photographs with the name of that person, so that while detecting faces, our models return same name as given to it in input, important thing, we did is, we stored 30 photographs per person

Fig. 4 Face recognition approach

so that when we apply KNN, it checks with all those input numpy arrays of different persons.

Phase 2: Training Phase
In this model, we will use a pre-trained classifier for face, eyes, and smile which are already in OpenCV, and they can be easily obtained by haarcascades directory.

OpenCV provides us with a cascade classifier which does the work for us and fetches the models from wherever it is in Github; after that, we wrote a method which will accept an image and perform further detection, and this all is done after converting the images to grayscale format.

Then we use the GrayScale data with the model imported. It produces a list of coordinates for all the faces found in the photograph. We iterate on each of the entries in the coordinate list, it contains the x- and the y-coordinates and the width and the height. Then we create a rectangle around the faces and eyes with these values [6].

Phase 3: Face Recognition
As our dataset is almost ready with some images, we will start framing our face recognition model, where we will use the KNN algorithm to find out the most optimal face matching throughout our dataset to the face visible in the webcam. After testing on a few K values, we took our K to be 5 and then, from the test image, we will compute distances for each prestored numpy array along with the labels and distance used here is purely Euclidean distance [9] mentioned in Eq. (1).

After that, we will find out the label which occurred most times in the predicted label, and depending on those values we simply deflected the change in our already running webcam and showed the name of the detected person in the frame. Flowchart representation of the whole process is shown in Fig. 4.

4 Results and Discussion

In this section, we will look on the results obtained from the model and also discuss what other changes we can do to improve our model with possible limitations. In Fig. 5, we can clearly see that two of the images are being shown to the model via webcam and it is able to recognize both of them. Firstly, it is detecting both the faces, and secondly, it is also recognizing them together, and this is how our model was supposed to work.

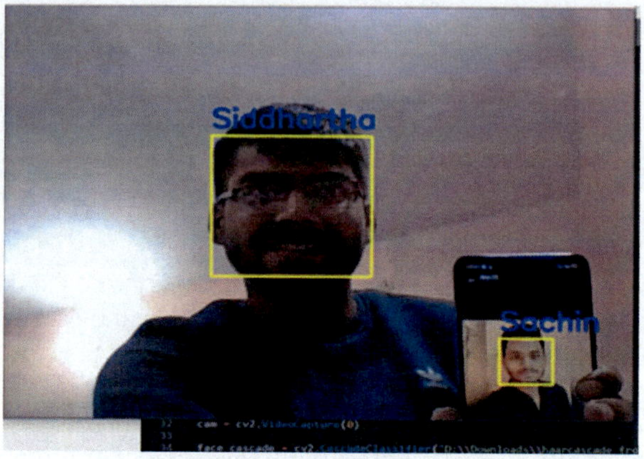

Fig. 5 Result showing face detection and recognition

This model though sometimes can show some mismatch in face detection, because as we can clearly see that every time, we create numpy array for our stored face, so there is a possibility that numpy arrays show some common nature for the image of two different people; it mostly happens when persons are look alike, e.g., Twins.

Our idea behind using KNN was simply that we knew we are dealing with numpy arrays, so when we will get our testing face matching with some input it will simply give us the label of face matching having the most probability.

An important aspect of this model is that if we get more than one face in front of the webcam, it will clearly identify them as objects and if images are trained previously, then it will successfully remember them.

Panjaitan et al. [8] showed in his study that KNN is more accurate when compared to ANN, for face recognition.

4.1 Limitations

Following are the scenarios where model might not perform well in detection and recognition process:

1. Condition when there is insufficient light while capturing image.
2. Condition in which a person's face is not properly exposed to the camera.
3. Condition, when the image is captured from the image of the person, and image quality is poor.
4. Conditions where two persons almost look alike, e.g., twins.

5 Conclusion

Winding up our study, we went through various face detection technologies such as knowledge-based face detection, feature-based face detection, geometry-based face detection, appearance-based face detection, and then Haar-like features-based face detection, compared them on basis of their merits and demerits and found that techniques of Haar cascade classifier with some modifications work better.

After that we looked out for face recognition techniques like lazy learners, neural networks and holistic approach and did comparative analysis. In the lazy learners approach, we worked on the k-nearest neighbors approach. In neural networks, we looked at ANN and multilayer neural networks, and then in a holistic-based face recognition approach, we looked at how we can use PCA for face detection. Going through all this, we build a model which detects face and it can even detect multifaces inside a frame and then recognize the face if it is stored in the database.

This study is of so much importance to humankind, because in this digital age, where everything is available online, privacy of digital gadgets is the utmost priority because that may contain very sensitive, personal, and important data. So, looking into the seriousness of this issue, this model is of so much value in current time and needs to be explored more so that there are no security-related threats further.

6 Future Work

After concluding our model, we see some scope of improvement in our model, which is listed below:

1. To improve accuracy of our model, we can further use techniques of deep learning to enhance the accuracy.
2. Rather than working on converting images to grayscale, we can simply work on the original image, which though may be more computational but may give better results.
3. Using multisets of K-nearest neighbors might improve the accuracy of the model.
4. Using some more xml files from our Haar cascade classifier such as smile classification can further improve the accuracy of our model.

References

1. Manjunath BS, Chellappa R, Malsburg CVD (1992) A feature based approach to face recognition. In: Proceedings 1992 IEEE computer society conference on computer vision and pattern recognition, pp 373–378. https://doi.org/10.1109/CVPR.1992.223162
2. Yow KC, Cipolla R (1997) Feature-based human face detection. Image Vis Comput 15(9):713–735. https://doi.org/10.1016/s0262-8856(97)00003-6

3. Dwivedi D (2018) Face detection for beginners. https://towardsdatascience.com/face-detection-for-beginners-e58e8f21aad9
4. Ghimire D, Lee J (2013) Geometric feature-based facial expression recognition in image sequences using multi-class AdaBoost and support vector machines. Sensors 7714–7734. https://doi.org/10.3390/s130607714
5. Wolf L Face recognition, geometric vs. appearance-based. In: Li SZ, Jain A (eds) Encyclopedia of biometrics. Springer, Boston, MA. https://doi.org/10.1007/978-0-387-73003-5_92
6. Adeshina SO, Ibrahim H, Teoh SS, Hoo SC (2021) Custom face classification model for classroom using Haar-like and LBP features with their performance comparisons. Electronics 10(2):102. https://doi.org/10.3390/electronics10020102
7. Ahmad F, Najam A, Ahmed Z (2012) Image-based face detection and recognition: "state of the art". Int J Comput Sci Issues 9(6). No 1, ISSN (Online) 1694-0814. https://doi.org/10.3390/s21165605
8. Panjaitan C, Silaban A, Napitupulu M, Simatupang JW (2018) Comparison K-nearest neighbors (K-NN) and artificial neural network (ANN) in real time entrants recognition. In: International seminar on research of information technology and intelligent systems (ISRITI), pp 1–4. https://doi.org/10.1109/ISRITI.2018.8864366
9. Guo G, Wang H, Bell D, Bi Y, Greer K (2003) KNN model-based approach in classification. In: Lecture notes in computer science (including subseries lecture notes in artificial intelligence and lecture notes in bioinformatics), vol 2888, pp 986–996https://doi.org/10.1007/978-3-540-39964-3_62
10. Zafaruddin GM, Fadewar HS (2014) Face recognition: a holistic approach review. In: International conference on contemporary computing and informatics (IC3I), pp 175–178. https://doi.org/10.1109/IC3I.2014.7019610
11. Mahanta J (2017) Introduction to neural networks, advantages and applications. https://towardsdatascience.com/introduction-to-neural-networks-advantages-and-applications-96851bd1a207
12. Sharma S (2017) Activation functions in neural networks. https://towardsdatascience.com/activation-functions-neural-networks-1cbd9f8d91d6
13. Browniee J (2020) Softmax activation function with python. https://machinelearningmastery.com/softmax-activation-function-with-python/
14. Naveenkumar M, Ayyasamy V (2016) OpenCV for computer vision applications. In: Proceedings of national conference on big data and cloud computing (NCBDC'15), pp 52–56. https://www.researchgate.net/publication/301590571_OpenCV_for_Computer_Vision_Applications
15. Khan T (2019) Computer vision—detecting objects using Haar cascade classifier. https://towardsdatascience.com/computer-vision-detecting-objects-using-haar-cascade-classifier-4585472829a9
16. Rajnis (2019) Python OpenCV. cv2.rectangle() method. https://www.geeksforgeeks.org/python-opencv-cv2-rectangle-method/
17. Saha P, Bhattacharjee D, De BK, Nasipuri M (2015) An approach to detect the Region of Interest of expressive face images. Procedia Comput Sci 46(Icict 2014):1739–1746. https://doi.org/10.1016/j.procs.2015.02.123
18. Ansari A (2017) Opencv Python program for Face Detection. https://www.geeksforgeeks.org/opencv-python-program-face-detection/
19. Chandra MLR, Kumar, BV, Sureshbabu B (2017) IoT enabled home with smart security. In: 2017 international conference on energy, communication, data analytics and soft computing, ICECDS 2017, pp 1193–1197. https://doi.org/10.1109/ICECDS.2017.8389630

ASIC Implementation of AI Edge Network on Chip (NoC) at 28 nm Technology Node and Its Various Timing Optimization Techniques at Each Stage

Riya Soni, Chandni Patel, and Nilesh Ranpura

Abstract People nowadays are habituated by instant communication at the same time; mobility is also demanded. Communication from anywhere and anytime requires good networking chip installed in the device. Networking on chip requires huge on-chip communication and over a fraction of time. Traditional network on chip (NoC) architecture interconnects all the routers via a bus. This topology was good approach with single processor. Now a days, multiprocessor network on chip (MPNoC) is much preferred. This type of networking chip operates on high frequency. ASIC design of such a high-speed chip becomes challenging in terms of timing closure. Timing closure is essential part of ASIC design flow. Hence, it is needed to be taken care of at each stage of physical designing of the chip. Here in this paper, physical design of such a networking chip is discussed with various optimization technique used at each stage. It is intended to reduce worst negative slack (WNS). Network on chip's basic architecture is also discussed here in this work.

Keywords Network on chip (NoC) · Multiprocessor network on chip (MPNoC) · Worst negative slack (WNS) · Application specific integrated circuit (ASIC)

1 Introduction

With increase in number of transistors in smaller area, the cell density increases. Current bus network on chip architecture cannot respond to millions of transistors on chip [1]. Traditional bus-based communication topology lacks for scalability and

R. Soni (✉)
Department of Electronics and Communication, Dharmsinh Desai University, Nadiad, Gujarat, India
e-mail: 20mepos001@ddu.ac.in

C. Patel · N. Ranpura
Einfochips, Ahmedabad, Gujarat, India
e-mail: chandni.patel@einfochips.com

N. Ranpura
e-mail: nilesh.ranpura@einfochips.com

V. K. Gunjan and J. M. Zurada (eds.), *Proceedings of 3rd International Conference on Recent Trends in Machine Learning, IoT, Smart Cities and Applications*, Lecture Notes in Networks and Systems 540, https://doi.org/10.1007/978-981-19-6088-8_28

predictability. They are not capable to cope up with the increasing requirements of future SoCs in terms of performance, power, and timing closure [1]. On-chip busses can serve a limited number of routers, and beyond that, performance degrades due to the bus parasitic capacitance [2]. Therefore, network on chip's ASIC design is challenging. To overcome this, mesh network is much preferred. With mesh topology, multicore design becomes more efficient as more number of routers can communicate at a time.

2 Network on Chip (NOC) Architecture

Network on chip architecture provides infrastructure to all the routers communicating to each other [3]. Topology means how nodes are connected to each other [4]. Topology of the nodes impacts on latency and bandwidth [4]. Traditional bus-based architecture is being replaced by mesh topology. In mesh topology as every routers are directly connected to each other, no centralization is required for all the routers to communicate. Hence, packets can easily and quickly travel through this kind of topology if the route is smaller. Beyond this, NOC must support to wide bandwidth and quality of services [5]. Because of this, timing closures are challenging in NOC. In mesh topology-based network architecture, the interconnection of resources and switches is independent of size of the network [3] which is a drawback of traditional bus-based architecture. In bus-based architecture with increase in size of the network, net length increases which increase parasites resulting in increase in net delays. This can be eliminated by mesh topology. Here, all the routers are connected to each other via switches using channel as a mesh or Manhattan like structure as shown in Fig. 1.

3 Physical Design Flow

The purpose of physical design is to transform an idea in form of RTL into silicon [7]. With advancement in VLSI technology, it is possible to incorporate thousands of transistors on a single silicon chip [7]. Following are the steps involved in physical design as shown in Fig. 2. Synthesis step acts as a perquisite for physical design. Floorplan decides the dimensions of the chip, and macros are placed inside the core. In power planning, power mesh is created to give VDD and VSS to the standard cell. In placement stage, all the standard cells are placed in the core. At CTS stage, a balanced tree is generated to give clock pulse to the entire sequential cell. All the physical routing is done at the routing stage. Here, signal and clock signals are routed at this stage.

Fig. 1 Mesh topology-based network on chip architecture [1]

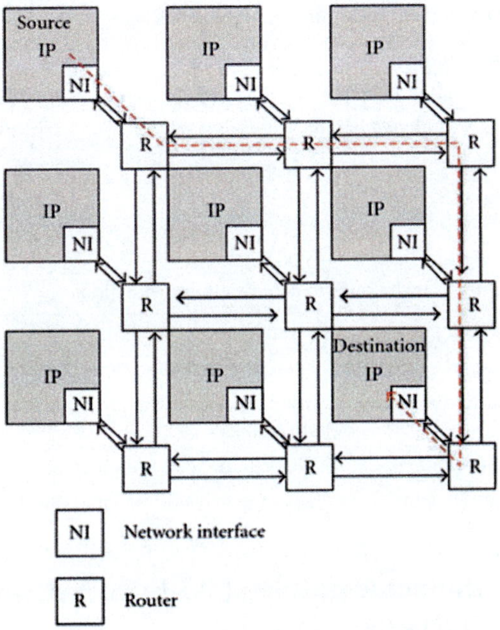

Fig. 2 Physical design flow

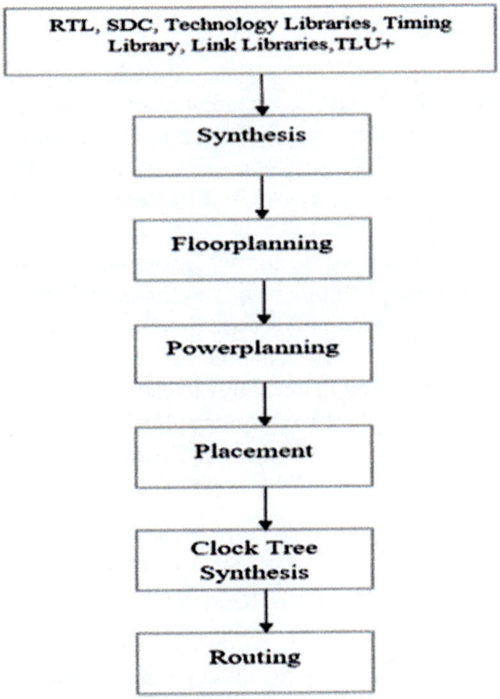

Table 1 Design configuration of Block-5

Operating voltage	0.81 V
Operating frequency	0.78 GHz
Core width and height	1167*972 μm^2
Macros	6
Total I/O ports	982

Table 2 Techniques to achieve timing violations

DFA analysis	Virtual connection of macros with other macros and I/O pads
Magnet placement	Fix object acts as a magnet and rest all
Buffer insertion	Insert a buffer in the path which reduces the transition time and net length also reduces
Gate cloning	Clock gating cell is duplicated so as to reduce fan-out
Pin swapping	Swaps the nets of the input pins to reduce slack

4 Implementation of AI Edge Networking Chip at 28 NM Technology Node

In this work, physical designing of a design called ARP-Block-5 is implemented. Chip designing briefly glances upon three aspects: performance, power, and area. To achieve these, various techniques are applied during entire physical design flow at every stage. Here, timing violation was taken into consideration and tried to reduce further at each stage. Some specification of Block-5 is listed in Table 1.

Various Techniques Used to Reduce Timing Violation at Different Stages of Physical Design

Timing closure occurs when timing estimations computed during synthesis do not match with timing estimates computed from the layout after routing [6]. Various optimization techniques were used to be optimize timing at different stages of physical design flow. A brief about those techniques is as shown in Table 2. Out of these techniques listed below, DFA is used at floorplan stage to place macros in the core area. Magnet placement is done after floor planning and before the actual placement. Buffers are inserted in the WNS path even at placement stage, and it can also be a helpful technique at CTS stage.

5 Simulation and Results

For simulation, here, technology-dependent netlist (.v) which was obtained from synthesis was taken into consideration. Here, with all the libraries required for the physical design flow, i.e., technology file (.tf), physical library (.lef), timing library

(.lib), constraints file, etc., were added to initialize floor plan. Here, ICC tool is used for physical design.

Operating voltage of Block-5 is 0.81 V. Its operating frequency is 0.78 GHz. Here, core width is 1167 μm, and its height is 972 μm. Its aspect ratio is 0.83. It has six macros in the design. All the macros are placed using data flow analysis as discussed earlier. There pin orientation is also taken into consideration while placing the macros. Floor planning of Block-5 is as shown in Fig. 3.

As this floor plan is for one of the block of networking chip and not at a chip level, placement blockage channel is created for feedthrough path in the middle of core area so that minimum path delay is generated after placement and routing stage. At this stage, tap cell was added to avoid latch up problem, and end cap cells were added to protect the standard cells while manufacturing. For power planning, M9 is used for power supply, and M5 is used for ground.

At placement stage, congestion-driven placement is chosen for implementation. Here, the macros above the blockage were chosen for magnet placement, so all the standard cell talking to those macros will sit near by the macros. Buffers are also placed near I/O ports. Psynopt is used to refine critical path to improve WNS path for setup time improvement. At this stage, without optimization, the WNS was 360 Pico second which is reduced to 160 Pico second after optimization.

At clock tree synthesis (CTS) stage, various clock optimizations were used to optimize the setup time improvement. At this stage, a balanced clock tree is designed having target latency and skew. For skew optimization, firstly, balance virtual tree is generated, and later on, its skew is optimized in second phase. Psynopt is executed to improve timing violation. At the end, this clock tree is routed. At this stage, without optimization was 310 Pico second which was 100 Pico second after applying all the optimizations. Balanced H-tree at CTS stage is as shown in Fig. 4.

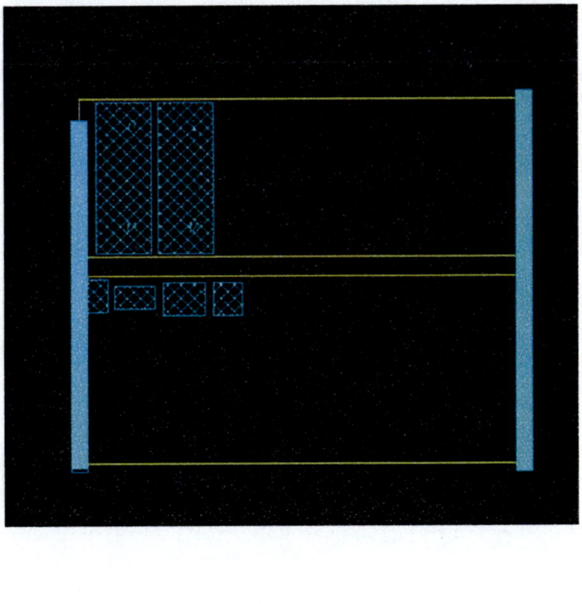

Fig. 3 Floor planning of Block-5

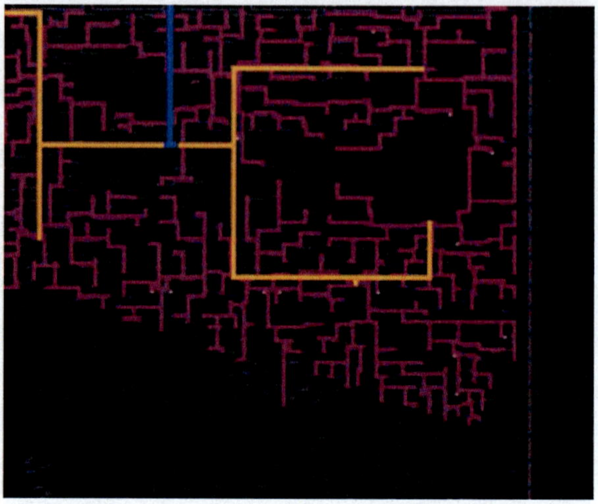

Fig. 4 Clock tree synthesis of Block-5

Table 3 Comparison of WNS at each stages

Placement	360	160
CTS	310	100
Routing	320	120

After CTS, routing of Block-5 is done. Here, congestion-driven routing is taken into consideration. As a result, there is no congestion in the design. Hence, WNS at this stage is 120 Pico second.

As discussed above, by using techniques like magnet placement before placing actual cells at placement stage, WNS was reduced at placement stage. We can additionally reduce WNS at this stage by switching HVT cells by LVT cells. At CTS stage we tried to insert buffers at critical path which in turn had reduced the WNS. We can go for clock cloning at this step for further reduction of WNS at this stage. A brief comparison of WNs at each stage is as shown in Table 3.

6 Conclusion

In ASIC design, timing closure is one of the most crucial part of the physical design flow. Here, we tried to reduce worst negative slack in the design using various techniques as discussed above. At placement stage, we manage to reduce WNS by 55%. At CTS stage, it has reached 100 picosecond which is 67% of the unoptimized WNS. Finally, at routing, we had obtained 120 picosecond which is reduced by 63%. Hence, WNS at each stage is optimized to improve setup and hold violations at each stages.

Acknowledgements I take this opportunity to express my deep sense of gratitude to Einfochips, Ahmedabad, as an institute, which makes its facilities available to their candidates doing their project work. My sincere thanks also goes to my guide Mr. Rahul Mehta, for there exemplary guidance, monitoring, and constant encouragement throughout the term work. Their blessing, help, and guidance given by them time to time were very important to me, and I shall carry them a long way in the journey of life on which I am about to embark.

References

1. Tsai W-C, Lan Y-C, Hu Y-H, Chen S-J (2012) Networks-on-chip: architectures, design methodologies, and case studies
2. Lauwereins R, Madsen J (eds) (2010) Design, automation, and test 105 in Europe
3. Kumar S, Jantsch A, Soininen J-P, Forsell M, Millberg M, Öberg J, Tiensyrjä K, Hemani A (2002) A network on chip architecture and design methodology
4. Chawade SD, Gaikwad MA, Patrikar RM (2012) Review of XY routing algorithm for network-on-chip architecture
5. Stensgaard MB, Spars J (2011) ReNoC: a network-on-chip architecture with recon- figurable topology
6. Gosti W, Khatri SP, Sangiovanni-Vicentelli AL (2001) Addressing the timing closure problem by integrating logic optimization and placement
7. Alpert CJ, Mehta DP, Sapatnekar SS (2008) Handbook of algorithms for physical design automation

3D IC Integration Using Blockchain

R. Radeep Krishna, P. Sivakumar, C. G. Abraham, and K. M. Sreedivya

Abstract Three-dimensional (3D) integrated circuit (IC) contains multiple layers with active devices, to intensely improve the chip performance, device density, and functionality facilitates various industrial applications. However, power and security in 3D IC is still poor due to inefficient placement and routing. To overwhelm these issues, we proposed BlockSec-3DIC method in 3D IC environment. The proposed work has four consecutive phases such as netlist protection, Trojan detection, placement and routing, and dynamic IC validation. In first process, we design a 3D IC with netlist preparation that is synthesized from the HDL codes, for that we proposed four Q-curve algorithm. All the transactions are hashed and stored the blockchain using secure hashing algorithm 3 (SHA-3). In second process, the netlists are analyzed for detecting Trojans by considering gate-level and chip-level features using bi-featured gated recurrent neural network (BF-GRU). In third process, we proposed jellyfish search algorithm (JSA) for optimal placement by considering wire length, average distance, and power consumption. And multi-weighted sum model (MSWM) is proposed for optimal routing by considering distance and thermal factor. Trojan-free netlist placement and routing is performed for enhancing the performance and reducing the power consumption. In final process, we proposed hidden Markov model-based validation (HMM-Valid) for validating the trustworthiness of the IC using the signature. The simulation is conducted using MatlabR2020a that evaluates the performance in terms of attack detection rate, netlist change rate, power consumption, time consumption, area, and wire length.

R. Radeep Krishna (✉) · C. G. Abraham
Mangalam College of Engineering, Ettumanoor, India
e-mail: radeep.krishna@mangalam.in

C. G. Abraham
e-mail: abraham.cg@mangalam.in

P. Sivakumar
Kalasalingam Academy of Research and Education, Krishnankoil, India
e-mail: siva@klu.ac.in

K. M. Sreedivya
Carmel College of Engineering and Technology, Punnapra, India
e-mail: sreedivya@carmelcet.in

V. K. Gunjan and J. M. Zurada (eds.), *Proceedings of 3rd International Conference on Recent Trends in Machine Learning, IoT, Smart Cities and Applications*, Lecture Notes in Networks and Systems 540, https://doi.org/10.1007/978-981-19-6088-8_29

Keyword 3-dimensional integrated circuit (IC) · Netlist protection. · Trojan detection · SHA-3 · Blockchain · Four Q-curve algorithm

1 Introduction

3D integrated chip (IC) design and fabrication is the emerging technology that has widespread applications over in Internet of Things (IoT) [1]. A 3D IC consists of multiple dies in which each die has multiple small components. These IC components are connected through silicon vias (TSV) [2]; the general structure of 3D IC is depicted in Fig. 1.

In general, four aspects of 3D IC integration are important those are partitioning, placement, routing, and TSV fault detection. In the absence of these processes, the IC performance will be degraded due to thermal heat effect, high power consumption, etc. [3–5]. However, lack of security aspect is another major challenging issue which needs better attention [6]. When it comes to security, 3D IC is subjected to many security threats [7]. A 3D IC has to cross many stages such as design, synthesis, placement, and routing and fabrication [8]. In this procedure, attackers' involvement can be encountered at any stage. Specifically, netlist which is the bas for IC fabrication is more vulnerable to many threats [9]. Important threats involved in 3D IC integration are,

- Hardware Trojans
- Malicious Chip Insertion

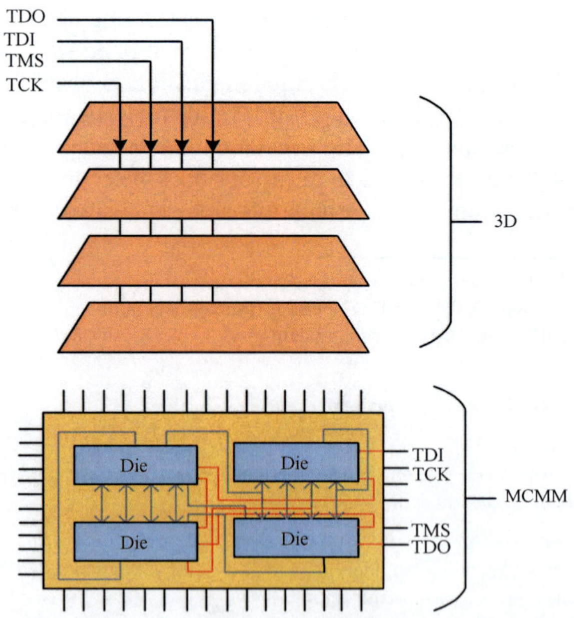

Fig. 1 General structure of 3D IC

- Gate-level Attacks
- Power Analysis Attacks.

Mainly, the aim of most of the attacks is to degrade the chip performance by increasing thermal heat of the chip or to inject the malfunction to the ICs [10]. Thus, it is necessary to overcome the attacks and vulnerabilities in 3D ICs in order to improve the performance of chips [11]. In security, provisioning many methods has been presented. Security provisioning and attack detection is performed in two methods including pre-silicon and post-silicon [12]. Pre-silicon models use the netlist analysis while post-silicon methods involve with golden model. In pre-silicon methods, some works introduce statistical methods to detect the threats [13]. For Trojan detection, machine learning (ML) and deep learning (DL) methods have been presented. Mainly, the netlist analysis involves with many features like gate information, pin information, I/O level, and so on [14].

The signals in the circuit were guarded in order to identify the abnormal events in the circuit. For this purpose, the additional hardware components were added to the circuit which affected the spacing in the IC [15]. The severe attacks such as brute-force attacks were mitigated by adapting the anomaly detection system in 3D IC which monitored the data streams and effectively detected the attack patterns without any additional hardware modification in the circuit. Further, the classification of anomalies was carried out, but the attack detection rate consisted of high false alarms which affected the reliability of these approaches [16]. Few methods implemented cluster-based detection of Trojan in the IC. The probability-based technique was deployed in the clusters to detect the malicious logic. These methods were stated to be less complex than other anomaly detection algorithm but were effective in detecting only less number of attacks and were also unable to detect the variants of known attacks [17]. The machine learning classifiers and decision trees were extensive used to classify the netlist into attack and attack free based on the gate-level features. These approaches were found to possess increased accuracy in detecting the attacks, but detection of attacks based on only gate-level features is not efficient and affects the reliability of the detection [18–20]. Still, accurate Trojan detection and end-to-end security provisioning is challenging due to lack of feature analysis, key distribution, and so on.

1.1 Research Aim and Objectives

The main aim of this research work is to assure high-level security for 3D ICs without affecting chip performance. To do so, this work is planned to focus on netlist protection, distributed integrity management, and Trojan detection processes. Additionally, the performance of the chip is managed as better through optimal placement and routing. This research work is applicable for many applications where 3D ICs play vital role. In particular, this research work focuses on IoT devices which are the significant elements of the IoT applications like smart city, smart health care, and so on. Recently, most of the research works focused on cost-effective and thermal-effective 3D IC designing methodologies. However, here, security of 3D ICs is not yet concentrated widely, but there is huge need for improving security. In 3D ICs, several attacks including hardware Trojans are initiated by the attackers in order to inject malfunction into the legitimate chips. On the other hand, presence of Trojans is the main cause for sudden thermal increase in the chip which relatively damages the chip. When it comes to security provisioning, four major problems are still not addressed. Those are, (i) Hardware Trojans are detected after manufacturing (i.e.,) post-silicon method which is not feasible for real-time since each application includes lots of ICs. Maintaining golden model for all chips is not feasible. (ii) Managing trust at each level of IC manufacturing is difficult since multiple parties involved in this flow. Securing netlist also requires separate key distribution scheme. (iii) Conventional machine learning methods are not suitable for Trojan detection because the algorithm needs to analyze more number of features at chip level and gate level. (iv) Majorly, mitigation techniques focus on removing chip components or redesigning the chip, but it is time-consuming process, and the Trojan impact is not fully mitigated.

The main research challenges involved in 3D IC are listed as follows:

- Poor accuracy in Trojan detection due to lack of important features
- High detection time due to ineffectual algorithms
- Insecure netlist distribution over foundries
- Improper integrity management.

The first and foremost objective of this work is to detect Trojans accurately through deep netlist analysis. The sub-objectives are listed as follows:

- To provide end-to-end security for 3D ICs design
- To protect netlist through optimal key distributions
- To perform optimal placement and routing in 3D ICs
- To validate the ICs by learning unique features.

1.2 Research Contribution

The integrity of the netlist is a significant factor to be considered in order to protect the circuit from being degraded. The proposed BlockSec-3DIC model addresses this issue by implementing blockchain technology. The major contributions of our proposed model are as follows:

- The integrity of the netlist data is protected by implementing four Q-curve algorithm-based digital signature, and storing of data is carried out in blocks in a hashed manner. The SHA-3 hashing algorithm is incorporated for this process.
- The hardware Trojan in the circuit is detected in order to overcome the vulnerabilities in the circuit. This is performed by BF-GRU in which both the gate-level and chip-level features are considered.

- The placement of IC block in the circuit is done optimally to achieve improved performance of the 3D IC. The JSA is implemented for this purpose which considered the minimization of power consumption. The route selection of signals is performed by implementing MWSM which considers both distance and thermal factor of the circuit.
- The dynamic validation of IC is executed in which the verification of signature of the chips is performed before fabrication using HMM-valid thereby improving the security of the IC.

1.3 Paper Organization

The rest of this paper is organized as follows: Sect. 2 explains the related work published in Trojan detection in 3D IC. Section 3 describes the major problem statements which are presented in 3D IC environment. Section 4 presents the proposed methodologies, pseudocode, and mathematical representation of 3D IC. Section 5 describes the experimental results of the proposed work and validates the efficiency of this work by comparing with various existing models. Section 6 represents the conclusion of this research, and it also includes the future work.

2 Literature Survey

Ghosh et al. [21] propose a technique to assure security for integrated circuits. In this work, counterfeit ICs insertion problem is concentrated. In order to resolve this threat, an automated defective pin detection methodology is proposed by using deep learning approach. For this purpose, a convolutional neural network (CNN) is proposed to detect the affected pins of ICs. Here, the depth map process is introduced to identify bent pins while pin texture analysis introduced to identify corroded pins. This paper finalizes that using CNN improves the attack detection in terms of accuracy. However, this work increases cost and only identifies defected pins which are ineffectual. Guin et al. [22] propose a robust security methodology for protecting ICs. The considered issues are unauthorized overproduction of ICs, sale of out-of-specification/rejected ICs, and so on. Mainly, this work focuses on preventing piracy of intellectual property. Here, a novel secure cell design is implemented for security infrastructure. This work highlights that this design has features like attack resistance, uniform key distribution, structural test capability without key, and post-silicon validation and debug capability. This work first scans the netlist; then, a secure netlist is generated by inserting secure cells in the netlist. At the end, the secure netlist is generated which further improves the security level. Although this work improves security, still it requires additional pin to assure the features which is inefficient. A multi-layer hardware Trojan detection framework is proposed for IoT IC chips [23]. Mainly, this paper focuses on system on-chip (SoC) security. In this work, both register transfer-level and gate-level security. A security layer called RG-Secure perception layer is introduced for feature analysis. The main aim of this perception is to extract and analyze the features from netlist. The RG-Secure combines third-party IP trusted design strategy and scan-chain netlist feature analysis processes. The high-dimensional circuit feature information is processed by the lightweight gradient lifting algorithm (light GBM) for feature analysis. However, Trojan detection accuracy is low since several important features are not extracted by light GBM. Also, light GBM has high time and computational complexity.

Random and clustered through silicon vias (TSV) defects are the main issues in 3D ICs [24]. In this paper, a pre-bond testing is proposed for detecting random and clustered TSV defects. Mainly, the test time minimization is focused in this work. Recursive bi-partitioning is the method used to achieve this goal. A heuristic approach is proposed for random TSV defect detection. For clustered defects, correlation factor is computed between each TSV pairs in order to identify the defected TSV. Overall more than 50% defected TSVs are identified accurately in both random/clustered TSVs. This paper highlights that this work minimizes the time consumption even in the worst-case scenarios. Limitation: In general, heuristic approaches are unable to provide optimum results which degrade the detection accuracy. Utyamishev et al. [25] introduce a security mechanism for protecting 3D IC from power attacks. The core idea of this work is to analyze the power variations exploited in the ICs. For this analysis, statistical methods are proposed. Through on-chip voltage variations, the attacker probe connected to the IC is detected. This work adjusts the density of on-chip sensors and exploits the sparse analysis to determine the attackers. In this work, the prediction error is minimized to 2% without increasing the complexity. However, the statistical method is not suitable for performing deep analysis on IC attacks. In paper [26] proposes a methodology for enhancing security of 3D ICs. In particular, this paper focuses on modifications made on third-party IP cores and commercial off-the-shelf (COTS) chips. The aim of this work is to achieve security provisioning for 3D ICs without maintaining a golden model (i.e.,) this work aims to prevent modifications when no golden model is available.

For that, a reverse engineering finite state machine (REFSM) helps end-users reconstruct a high-level description of the control logic. This logic is enabled from the flattened netlist. The proposed methodology recovers circuit control logic form. However, complexity and time consumption are high in this work to implement reverse engineering. Also, XOR-based key generation is insecure.

In paper [27], IP protection methodology is proposed to defend against reverse engineering issues. This paper uses the camouflaged gates which are specially designed to protect overall chip. The camouflaged gates are capable of performing different operations without affecting chip functionality. Here, four different kinds of camouflaged cells are introduced with optimum design. There are four main types of attacks: circuit partition-based, IC testing-based, brute-force-based, and satisfiability-based. For each attack, corresponding countermeasures are also introduced and named as clique-based camouflaging; CamoPerturb, AND-tree camouflaging, and equivalent class-based camouflaging. Yan et al. [28] introduce camouflaging which is the lay-out-level technique is proposed to protect 3D ICs. A fine-grained monolithic inter-tier vias (MIVs) is proposed to support ultra-high density device integration. A camouflaged SIMON which is the lightweight block cipher is introduced in this work to protect the 3D ICs. After deployment of camouflages, a chip-level analysis is exposed. Before that, the chip is processed with placement and routing procedures. Overall, this work reduces the cell-level power overhead. However, the power reduction and security is achieved in the cost of higher time consumption which is ineffectual. The optimal placement of TSVs in 3D ICs was executed by using the multi-objective genetic algorithm [29]. The objective of this paper was to achieve optimal placement by considering the significant parameters such as separation, congestion, power density, and wire length. Initially, the population is generated, and the binary tournament selection was implemented to select the chromosomes based on distance and rank. Further, the offspring population was generated by applying crossover and mutation. Then, the placement of TSVs was done by implementing non-dominant sorting. However, the time complexity of proposed optimization algorithm was huge which affected the placement process.

A novel zone-based routing strategy in 3D IC was proposed in [30]. The objective of this research work is to achieve scalability in number of IPs and optimal placement of nodes in 3D IC. The complexity of conventional routing strategies was reduced by implementing a lightweight zone-based routing strategy. Several zone strategies such as, layer zone, tree zone, regional zone, and local zone were addressed. The effective load balancing mechanism was proposed which considered the hop count. The scalability is further improved by the use of bridge routers. An efficient placement strategy based on power consumption for 3D IC was proposed in [31]. The objective of this research work is to reduce the power consumption and latency in the 3D IC. The portioning was executed by implementing clock tree synthesis (CTS). The processes involved in this approach are partitioning, placement of TSVs and nodes, and CTS. The power rate was estimated based on available space and power consumed, and the integration flower pollination algorithm with electro search was implemented for the effective placement. However, the security of 3D IC was not considered which will also affect the performance. The security of 3D IC against hardware Trojan attack was proposed by using game theoretic-based approach [32].

The major objective of this research work is to formulate active defense strategies against the Trojan attack. For this purpose, the non-cooperative game was implemented between the malicious manufacturers and defense designers. The fuzzy-based payoff modeling was performed to achieve the dynamic defense measures against the malicious hardware Trojan designers. Both the defense designers and malicious manufacturers were acted as selfish nodes to achieve the objective. The proposed method was found to provide resistance against the hardware Trojan attack.

3 Problem Statement

3D IC integration and fabrication is the significant research issue in recent times since all of Internet of Things (IoT) applications rely on this. The major results are explained detailed in below. In this work, 3D IC security is focused to mitigate the issues explored in trusted and untrusted foundries. Patnaik et al. [33] proposed split manufacturing (SM) method to obscure ICs from malicious entities. The major problems of this work are discussed as follows:

- Although SM method preserves security, there exists number of challenges such as (i) assembling of split dies that are manufactured from different foundries become risk and (ii) signal integration and performance of the IC becomes poor.
- Finally, placement and routing is performed in a random manner which is ineffectual. Also, manufacturing each die at different foundry without knowledge of other die affects the IC performance.
- SM method only ensures that a foundry (can be trusted or untrusted) cannot have the overall knowledge of the netlist. But, the Trojan can be injected or inserted in any partition of the netlist. Such Trojans are not detected and mitigated in this work.

Zhang et al. [34] propose an attack resilience method by using intrinsic noise. This paper aims to utilize the properties of the power distribution network (PDN). The PDN is responsible to supply power for all tiers of the 3D IC die. The main problems of this research are listed as follows,

- Increasing noise level for security provisioning affects the performance of 3D IC chip since the noise level increase affects the signal propagation
- Also, noise imputation results in higher thermal increase which damages the chip and leads to replacement
- Lack of netlist security increases the probability of Trojan injection and insertion.

Zaraee et al. [35] present an IC authentication scheme to protect the chips. This paper mainly focuses on mitigation of hardware Trojans injected during manufacturing stage of the ICs.

The main problems of this research are discussed as follows,

- In general, the uniqueness of gates in an IC varies with the thermal increase, noise imputation, and other factors. In this case, it is not possible to accurately identify the authenticity of an IC. Lack of deep analysis on the IC data increases vulnerability and affects the authenticity accuracy of the work.
- This work tends to identify the threats injected in the fabrication level through identification of location of the logic gates. However, the placement and routing depends upon the fabrication procedure which can be easily modified.
- Here, the validation is performed only after chip fabrication. If the chip is not authenticated, then the IC is not taken for further processes which are ineffectual.

Dong et al. [7] propose a machine learning-based approach for hardware Trojan detection in the chips that are used in Internet of Things. This work focuses on pre-silicon method which doesn't require any gold chips for Trojan detection. The major problems of this research are discussed as follows:

- XGBoost algorithm struggles to extract 56 features in all netlist since it performs sequential process, and the accuracy is low due to lack of deep analysis. Complexity and time consumption is also high since it needs to build several decision trees.
- Although 56 features are computed and Trojan list is detected, still timely detection is challenging. Since the IC must pass through a number of levels before fabrication, any one of those levels could be impacted. Thus, lack of trusted fabrication increases.

3.1 Research Solutions

The above mentioned problems are overcome by our proposed BlockSec-3DIC model by providing the following solutions:

- Optimal placement is performed by jellyfish search optimization based on multiple metrics which increases the performance of the chip
- Security and IC chip's reliability is assured by incorporating blockchain-based integrity management and netlist encryption by four Q-curve encryption
- Thermal issue is resolved by Trojan-free netlist generation by BF-GRU and optimal placement by using jellyfish search optimization
- Identify the Trojans by deep analysis using BF-GRU and validation is performed by HMM-based signature generation.

4 Proposed Work

In this research work, we have presented a novel BlockSec-3DIC design for achieving joint objective of security provisioning and performance enhancement. The proposed BlockSec-3DIC design involves with four major processes. Figure 2 illustrates the system model of the proposed work in an elaborative manner. In the proposed work, end-to-end integrity is attained by integrating the designing process with blockchain technology. That can be elaborated in below.

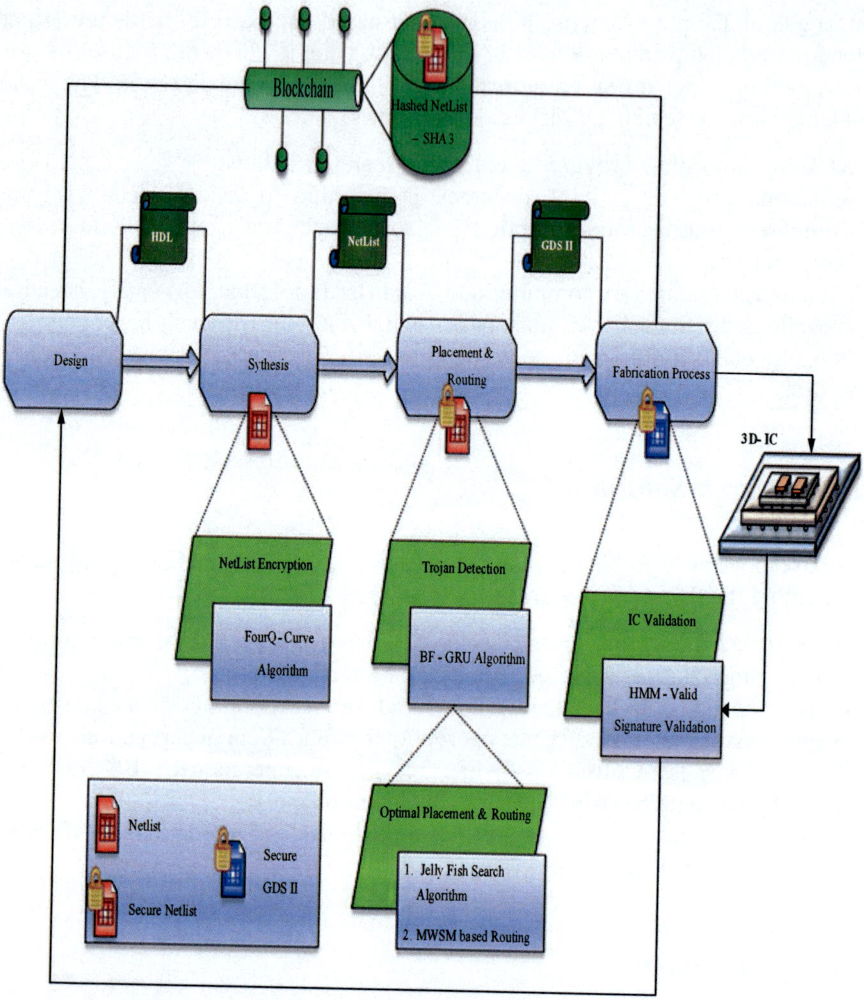

Fig. 2 BolckSec-3D IC system model

4.1 Netlist Protection

3D IC designing starts with netlist preparation which is synthesized from the HDL codes. Netlist includes connection of capacitors and resistors. The netlist is synthesized and signed by using four Q-curve algorithm. The four Q-curve is the lightweight cryptosystem that minimizes the hidden time costs involved in netlist protection. It performs 4–5 times faster than the elliptic curve cryptography. Four Q-curve is defined as follows,

$$\sum_{1=0}^{3} T_m = 0, \quad m = 1, 2, 3 \tag{1}$$

where the equation represents the functionalities of four Q-curve algorithm and $m = 1$, 2, 3 represents security, speed, and simplicity, respectively. When compared to ECC, it provides higher security with 128-bit level from a 4-dimensional decomposition which reduces the functions in ECC.

$$F(P)/C = -Y^2 + Z^2 = 1 + dy^2z^2 \tag{2}$$

where C represents as Edwards curve and P represents prime number, and the value of d is -1734398985892729835729857293759 8. i + 348598393853456456. In our work, we used fast arithmetic modulo prime method, i.e., $P = 2^{127} - 1$. And this research used extended twisted Edwards curve to refer the variables (y, z) over C. The tuple of the curve is represented as (y, z, s, t).

$$s \neq 0, Y = \frac{y}{s}, Z = \frac{z}{s}, t = \frac{yz}{s} \tag{3}$$

The secret key is generated by blockchain using this method. In each stage, the information is maintained in blockchain as the function of hash values. For hash generation, secure hashing algorithm 3 (SHA-3) is proposed. The signature key is also stored in the blockchain. In next stage, the key must be derived from the blockchain for access netlist. As the netlist is securely transmitted and access is restricted, the netlist modification issue is overwhelmed in this work

```
Pseudocode: SHA-3 algorithm
Input: Signed netlist
Output: Create hash
A[x]=b[x,0]⊕b[x,1]⊕b[x,2]⊕b[x,3]⊕b[x,4]
C[x]=A[x-1]⊕ROTATE(A[x+1],1)
b[x,y]=b[x,y]⊕C[x]
D[y,2x+3y]=ROTATE(b[x,y],s[x,y])
b[x,y]=U[x,y]⊕((NOT U[x+1,y]AND U[x+2,y])
b[0,0]=b[0,0]⊕rc
```

where pseudocode represents the SHA-3 algorithm, $s[x, y]$ represents rotation offset, and rc represents round constant and the shift operation moves from position i to $i + s$ in the modulo size 64 hash. And the hash format of signed netlist is stored in the blockchain which improves the security of netlists.

4.2 Trojan Detection

From the netlist, the placement and routing processes are implemented. Before that, the netlist is further analyzed to detect Trojans whose features are described in Table 1. For analyzing gate-level and chip-level features, we proposed a novel bi-featured gated recurrent neural network (BF-GRU) which extracts all important features from the netlist and detects the Trojans through deep analysis. In this phase, the netlist is assured without any Trojans.

In our work, we proposed BF-GRU for extracting the gate-level and chip-level features from the netlist for detecting hardware Trojans. Based on these features, we classified the features into two classes such as Trojan and without Trojans. Although BF-GRU and LSTM are more comparable, BF-GRU uses an update gate in its structure, which also has input and forget gate. The features are extracted using BF-GRU which are described as follows:

$$R(t) = \delta(w_R y(t) + v_R g(h-1) + a_R) \tag{4}$$

Table 1 Trojan features and descriptions

Features of Trojans	Descriptions
in-flipflop-x	The count of flip-flops is upgraded to x level from the input of the network
fan-in-x	The count of logic gates is fanins upgraded to x level from the network
out-flipflop-x	The count of flip-flops is upgraded to x level from the output of the network
in-multiplexer-x	The count of multiplexers upgraded to x level from the input of the network
in-loop-x	The count of loops upgraded to x level
out-multiplexer-x	The count of multiplexers upgraded to x level from the output of the network
out-loop-x	The count of loops upgraded to x level
in-const-x	The count of constants is upgraded to x level from the input of the network
in- nearest-pin	The small level to main level input from the network
out-const-x	The count of constants is upgraded to x level from the output of the network
out-nearest-pout	The small level to main level output from the network
{in. out}-nearest multiplexer	The small level to any level flip-flop from output of input of the network
{in, out}-nearest-flip flop	The small level to any level flip-flop from output or input of the network

$$U(t) = \delta(w_U y(t) + v_U g(h-1) + a_U) \tag{5}$$

$$g(t) = (1 - U(t))^{\circ} g(t-1) + U(t)^{\circ} g(t) \tag{6}$$

$$g(t) = \delta_g = \left(w_h y(t) + v_g (r(t)^{\circ} g(t-1)) + a_g\right) \tag{7}$$

where *R*(*t*) is represented as reset gate and *U*(*t*) represents as update gate and *w*, *v* represent as parameters, and δ_g represents as hyberbolic tangent and δ represents as sigmoid function. After construct the BF-GRU structure, the training method for GRU is determined for detecting hardware Trojan attack and classifies into two classes such as Trojan and without Trojan. The overall flow of Trojan detection process is presented in Fig. 3, and the algorithm of BF-GRU is provided below.

Algorithm: Trojan detection using BF-GRU
Input: Chip and gate-level features f{f1,f2,...,fn}
Output: Trojan detection
Initialization set of f()
Repeat
BF-GRU training
For i=1:I do
Select optimal features from f()
Classify {Trojan, and without Trojan}
End

4.3 Placement and Routing

In the Trojan-free netlist, placement and routing is performed to assure high-level performance and power reduction. For optimal placement, we present a new jellyfish search algorithm (JSA) is proposed. The placement algorithm places the components on the die based on wire length, average distance, and power consumption. The power consumption is estimated from the available space and number of cells which can be formulated as,

$$P_c = \sum_{k=1}^{n} C_n + S_r + P_d \tag{8}$$

where P_d, S_r, C_n denote the dissipated power, available space, and number of cells, respectively. The reason choosing (JSA) is that it has better convergence in achieving the objectives than other meta-heuristics algorithm such as whale optimization algorithm (WOA), genetic algorithm (GA), and particle swarm optimization (PSO). The best location for the placement of IC block is determined by manipulating the search

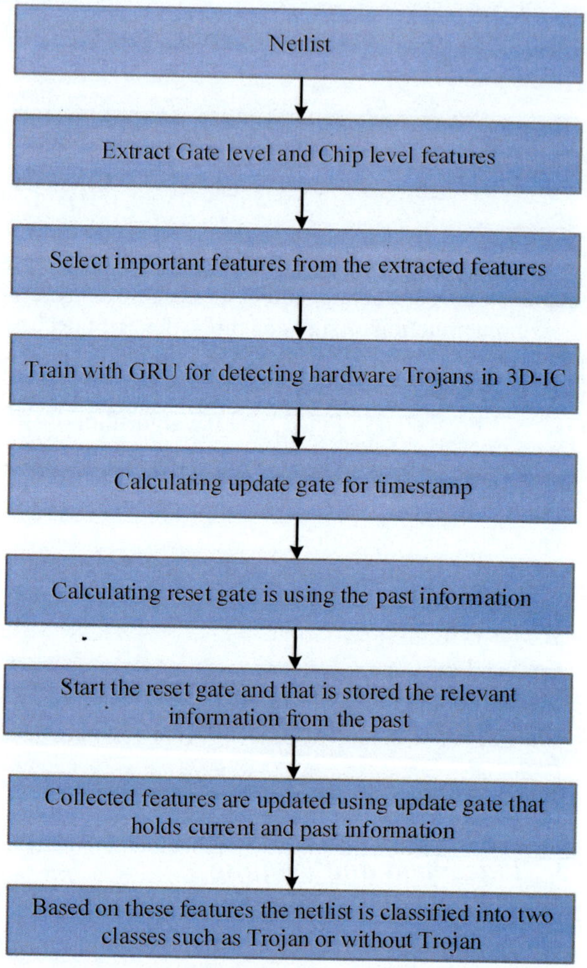

Fig. 3 Flowchart of hardware Trojan detection

strategy of jellyfish. The placement of die is formulated as,

$$Dp = \frac{1}{P_l}\sum Dp_i = \frac{1}{P_l}\sum\left(lc^* - \text{att}_f lc_i\right) = lc^* - \text{att}_f \mu \tag{9}$$

where P_l denotes the population of placement location, lc^* denotes the best location at time t, att_f represents the attention factor, and μ denotes the mean of all the placement locations. The variation between best location and mean location can be formulated as,

$$Vl = \rho \times \sigma \times r^f(0, 1) \tag{10}$$

where the likelihood area is formed as $\pm\sigma\rho$ in which σ is denoted as standard deviation and can be formulated as,

$$\sigma = \text{rand}^{\lambda}(0, 1) \times \alpha \tag{11}$$

Then,

$$Vl = \rho \times r^{f}(0, 1) \times r^{\lambda}(0, 1) \times \alpha \tag{12}$$

This is rewritten as

$$Vl = \rho \times r(0, 1) \times \alpha \tag{13}$$

where

$$a_t = \rho \times r(0, 1) \tag{14}$$

Hence,

$$Dp = lc^{*} - \rho \times r(0, 1) \times \sigma \tag{15}$$

The current position of jellyfish is expressed as

$$lc_i(t + 1) = lc_i(t) + r(0, 1) \times Dp \tag{16}$$

The proposed JSA algorithm performs optimal placement if die into two types, namely active search and passive search. In passive search, the location is determined only within the likelihood area whereas in active search the determination of location is carried out globally in the 3D IC, The updation function of passive search is expressed as,

$$lc_i(t + 1) = lc_i(t) + \delta \times r(0, 1) \times (U_l - L_l) \tag{17}$$

where the notations U_l and L_l denote the upper and lower limit of the likelihood area, respectively. The movement of jellyfish is determined by the motion coefficient δ which should be greater than zero. The search space is exploited which is expressed as,

$$s = lc_i(t + 1) - lc_i(t) \tag{18}$$

This can be rewritten as,

$$lc_i(t + 1) = s + lc_i(t) \tag{19}$$

The control mechanism is implemented to switch between the passive and active search. The threshold time is considered before which the passive search is carried out and after which the active search is started. The control function C_m is formulated as,

$$C_m = \left|\left(1 - \frac{t}{m_{\text{iterations}}}\right) \times (2 \times r(0, 1) - 1)\right| \tag{20}$$

The logistical map is utilized to perform initialization of placement process which can be expressed as,

$$lc_{i+1} = Pl_i(1 - lc_i) \tag{21}$$

The boundary condition for the optimal placement of IC block is represented as,

$$\begin{cases} lc'_{i,\text{dim}} = \left(lc_{i,\text{dim}} - U_{l,\text{dim}}\right) + L_l(\text{dim}), \text{ if } lc_{i,\text{dim}} > U_{l,\text{dim}} \\ lc'_{i,\text{dim}} = \left(lc_{i,\text{dim}} - L_{l,\text{dim}}\right) + U_l(\text{dim}), \text{ if } lc_{i,\text{dim}} < L_{l,\text{dim}} \end{cases} \tag{22}$$

where the location of placement in the circuit is denoted as $lc_{i,\text{dim}}$ and updated location is represented as $lc'_{i,\text{dim}}$. The pseudo-code for the proposed placement algorithm is presented below

```
Pseudo code: JSA algorithm
Begin
Declare objective function
Initialize search space, m_iterations, P_l
Initialize jellyfish using logistic map
Compute available locations
Find the jellyfish at best location
Initialize time: t=1
Repeat
For i=1: P_l do
Formulate C_m using eqn (20)
If C_m ≥ 0.5
location determination is carried out by eqn (9)
updation of new location is performed by eqn(10)
Else jellyfish follows swarm
If r(0, 1)>(1-C_m): passive search is exhibited
updation of new location is performed by eqn(17)
Else: active search is exhibited
Compute new location using eqn(19)
End if
End if
Update boundary condition
Update current location
```

```
End for i
t=t+1
Until t>, m_iterations
Optimal placement of die is performed
End
```

Once the die is placed, then optimal routing is enabled among IC components. For optimal routing, we proposed a novel multi-weighted sum model (MWSM) that selects optimal route based on distance and thermal factor. These factors are considered as criteria, and the routing of signal in the circuit is executed based on the summation of weights acquired by these criteria which is formulated as,

$$R_p^s = \sum_{q=1}^{n} w_q r_{pq}, \quad \text{for } p = 1, 2, \ldots, m \tag{23}$$

The R_p^s, for every route, is computed, and the route with the highest R_p^s is selected as the optimal route to perform the transmission and reception of signals between the die. Involvement of optimal placement and routing improves the performance of proposed IC design.

4.4 Dynamic IC Validation

After placement and routing, the IC fabrication is initiated. We test the trustworthiness of the IC through signature validation.

As the properties of an IC varies with time, we present a novel hidden Markov model-based validation (HMM-Valid) procedure.

The proposed method learns the unique pattern of the IC in chip and gate levels. The IC validation is performed based on the HMM probability. Procedure of HMM is defined as follows,

Set of n states $\{s_1, s_2, \ldots, s_n\}$ and state transition matrix is $M = \{a_{ij}\}$, where a_{ij} represents transition probability from state s_i to s_j.

$$a_{ij} = Pb(s_j \text{ at } T + 1 | s_i \text{ at } T) \quad 1 \leq i, j \leq n \tag{24}$$

Pair of n discrete symbols $\{d_1, d_2, \ldots, d_n\}$. Observation probability of matrix $\text{Mat} = \{b_{ij}\}$ where b_{ij} represents the generating symbol probability v_n from state s_j, and the initial distribution of probability for the states is represent as follows,

$$\Pi = \{\rho_j\}, j = 1, 2, 3, \ldots, n, \quad \rho_j = Pb(s_j \text{ at } T = 1) \tag{25}$$

The signature validation overall probability is denoted as follows,

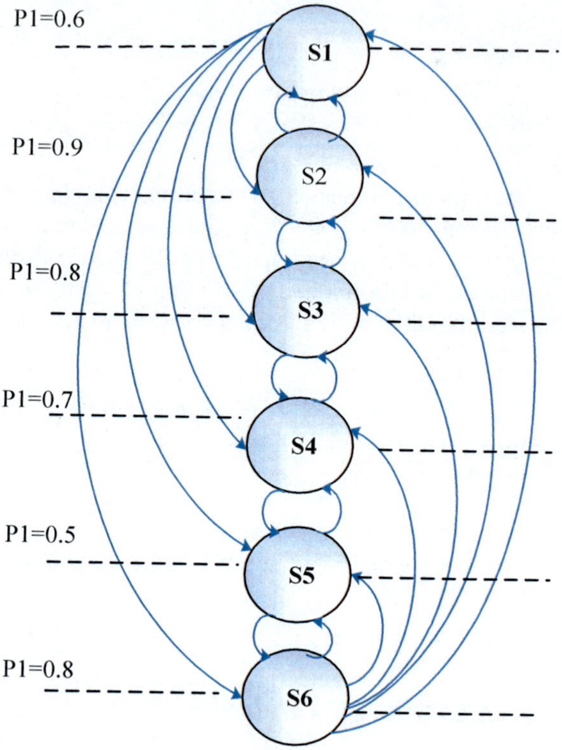

Fig. 4 Hidden Markov model

$$Pb\left(S_{ij}|S_{i1}, S_{i2}, \ldots, S_{ij-1}\right) = Pb(S_{ij}|S_{ij-1}) \tag{26}$$

Each state of hidden Markov model is generated randomly one of M visible states $\{v_1, v_2, \ldots, v_n\}$.

Figure 4 represents the HMM model which has six states, and the states are not permitted to transfer back to the existing states. The matric format of HMM is presented in Fig. 5. At a time instant, the model stays and visits in a one state. At time $T + 1$, the model stayed in the same states or moved to another states according the probability. This process is continued to reach a final state at time T. By this way, the IC signature validation is performed. Then, this signature is dynamically validated through HMM model. This phase assures that the IC is secure and not yet modified by any intermediate phase.

5 Experimental Study

In this section, the experimentation of proposed BlockSec-3DIC model was carried out which comprises of three sub-sections, namely simulation study, comparative

	c1	c1	c1	c1	c1	c1
c1	0.5	0.1	0.3	0.4	0.5	0.6
c1	0	0.8	0.9	0.5	0.2	0.1
c1	0	0	0.2	0.3	0.4	0.9
c1	0	0	0	0.6	0.9	0.3
c1	0	0	0	0	0.7	0.5
c1	0	0	0	0	0	0.8

Fig. 5 Matrix format of HMM

analysis, and research summary, implemented to evaluate the effectiveness of our model.

5.1 Simulation Study

The proposed BlockSec-3DIC model is simulated in MATLAB R2020a simulation tool. The system configurations required for the purpose of simulation are described in Table 2. The IBM dataset is used for the purpose of simulation which facilitates the implementation of netlist protection, Trojan protection, placement and routing of nodes and TSVs, and dynamic validation of ICs.

The IBM ISPD06 dataset comprises the details of both hard and soft macros which weigh more than 1500 block in size.

The details in the dataset include fixed and movable macros, netlist, area, cells, and circuit names in which the floor placement details are attained. The format of the dataset is presented in Table 3.

Table 2 System configuration

Hardware specifications	Processor	Pentium dual core and above
	RAM	8 GB
	Hard disk	60 GB
Software specifications	Simulation tool	MATLAB R2020a
	OS	Window 10

Table 3 ISPD 06 benchmark

Circuit name	# Cell	Area	#nets	Movable macros	Fixed macros
Ibm01	221,024	5.6	5673	0	435
Ibm02	245,475	10.7	8236	0	546
Ibm03	455,647	11.6	10,265	0	575
Ibm04	526,363	8.9	13,456	0	348
Ibm05	295,521	12.5	9395	0	624
Ibm06	573,453	6.7	15,083	0	718
Ibm07	1,012,343	8.9	16,454	0	691
Ibm08	2,123,678	4.7	18,324	0	1328
Ibm09	875,583	5.3	26,345	3421	637
Ibm10	347,821	12.0	23,487	0	234
Ibm11	456,291	0.6	25,438	0	532
Ibm12	561,255	5.9	34,563	72	725
Ibm12	567,342	2.4	34,756	0	834
Ibm12	346,573	5.3	65,732	0	659
Ibm12	423,573	7.1	83,245	2453	275
Ibm12	238,754	6.8	23,973	0	328

5.2 *Comparative Analysis*

In this sub-section, the evaluation of proposed BlockSec-3DIC model is carried out by comparing with existing works such as GLV-IC (Gate-level validation of integrated circuits with structured-illumination read-out of embedded optical signatures) and SM-3DIC (Split manufacturing for 3D ICs and obfuscation of vertical interconnects) in terms of performance metrics such as attack detection rate, Netlist change rate, power consumption, time consumption, area, and wire length.

5.2.1 Impact of Attack Detection Rate

The attack detection rate is referred as the rate at which the detection of hardware Trojan attack on the 3D IC is carried out. Figure 6 depicts the comparison of attack detection rate of the proposed BlockSec-3DIC model and other existing works with respect to number of iterations. The attack detection rate increases with increase in the number of iterations.

The attack detection rate of the proposed BlockSec-3DIC model is high due to the consideration of both gate-level and chip-level features for the purpose of detection by using BF-GRU. The efficiency of the proposed algorithm high than the existing works which results in high attack detection rate.

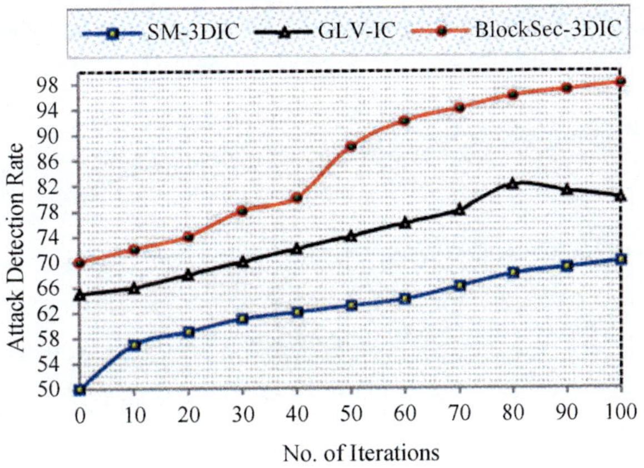

Fig. 6 Number of iterations versus attack detection rate

For instance, when the iterations range from 60 to 70, the attack detection rate of the proposed approach is 94% whereas the existing approach possess detection rate only up to 78%. The inefficiency of the existing approaches is due to the lack of consideration of all the significant features necessary for the purpose of attack detection.

5.2.2 Impact of Netlist Change Rate

The netlist change rate is referred as the rate of change of data in the netlist which is caused due to inefficient encryption of the input netlist data. Figure 7 illustrates the comparison of proposed model with that of other existing approaches in terms of netlist change rate with respect to number of iterations. From the figure, it is clear that the rate of change of netlist decreases with increase in number of iterations. The proposed BlockSec-3DIC model possess low rate of change of netlist than other existing approaches due to the protection of netlist by implementing four Q-curve algorithm to generate signature, and further, the integrity of netlist is protected by integrating the concept of blockchain in which the netlist is stored in blocks and hashed by using the SHA-3 algorithm.

The existing approaches lack in proper protection of integrity of netlist resulting in high rate of change of netlist. For instance, the netlist change rate of the proposed model is below 10% when the iterations reach 100 whereas the existing approaches possess up to 30% of change in netlist.

5.2.3 Impact of Power Consumption

The consumption of power is an important metric to be considered for the evaluation of performance of an approach. The power consumption can be formulated as,

Fig. 7 Number of iterations versus netlist change rate

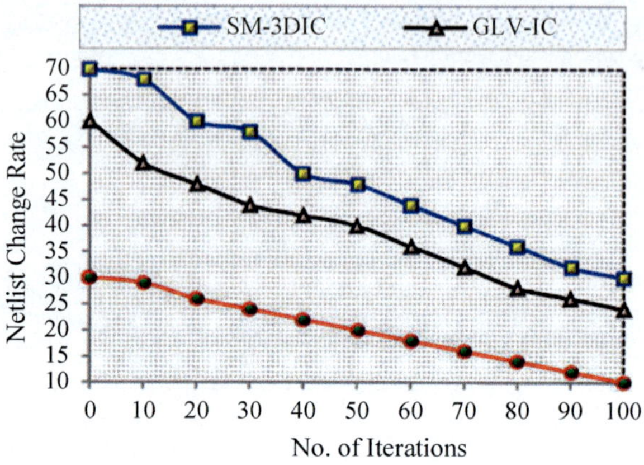

$$P_c = \sum_{k=1}^{n} \frac{P(C_k, f_n)}{\tau} \tag{27}$$

where k, C_k, f_n, τ denote the number of iterations, cell, function, and time, respectively. The comparison of proposed BlockSec-3DIC model approach and other existing approaches in terms of power consumption with respect to number of cells is presented in Fig. 8.

From Fig. 8, it is clear that the consumption of power increases with increase in the number of cells. The power consumption of our proposed model is low compared to other existing approaches due to the proper placement of die in the 3D IC. The optimal placement of die is carried out by implementing jellyfish search algorithm based on the significant parameters such as wire length and power consumption. Further, the proper routing strategy also contributes to reduction of power consumption in 3D IC. The existing approaches implemented random placement of die which

Fig. 8 Number of cells versus power consumption

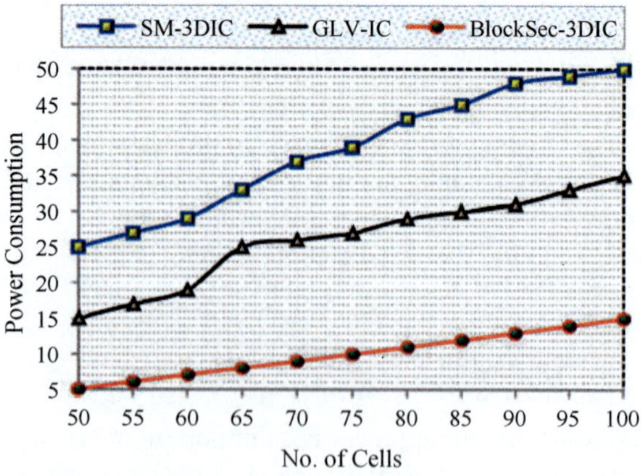

increased the power consumption and reduced the performance. For instance, the power consumption of our proposed BlockSec-3DIC model is 10 mW even when the number of cells reaches 100 whereas the power consumption of the existing approaches is near 48 mW which is 4 times greater than that of our BlockSec-3DIC model.

5.2.4 Impact of Time Consumption

The time consumption is a significant metric in validating the transmission efficiency of the circuit. The time consumption refers to the summation of time taken for transmitting the signal from the source and the time taken for reception of signal by the receiver. The time consumption can be expressed as

$$T = \sum T_{\text{st}} + T_{\text{sr}} \tag{28}$$

Figure 9 depicts the comparison of proposed approach and other existing works in terms of time consumption with respect to number of cells. The time consumption increases with increase in the number of cells in the circuit. The time consumption of our proposed model is less than other existing approaches due to the optimal routing of signals in the circuit based on the parameters such as distance and thermal factor. The selection of optimal route is carried out by using multi-weighted sum model in which the weights of each route based on the parameters are analyzed. The existing approaches provided inefficient routing thereby resulting in increased time consumption. For instance, the time consumption of proposed BlockSec-3DIC model is 6 ms when the number of cells increased to 100, whereas the existing approaches possess high consumption of time to about 27 ms.

Fig. 9 Number of cells versus time consumption

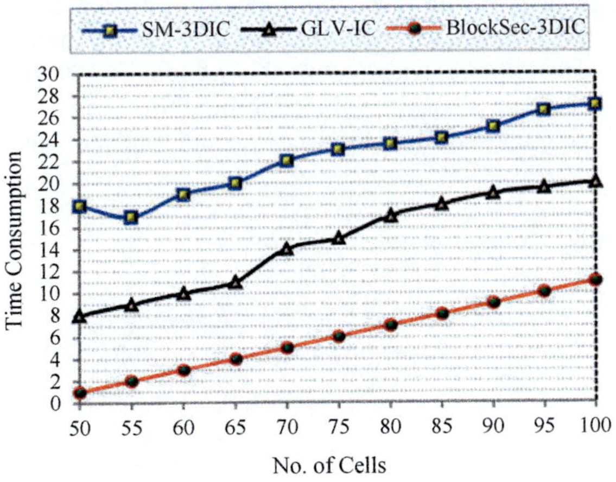

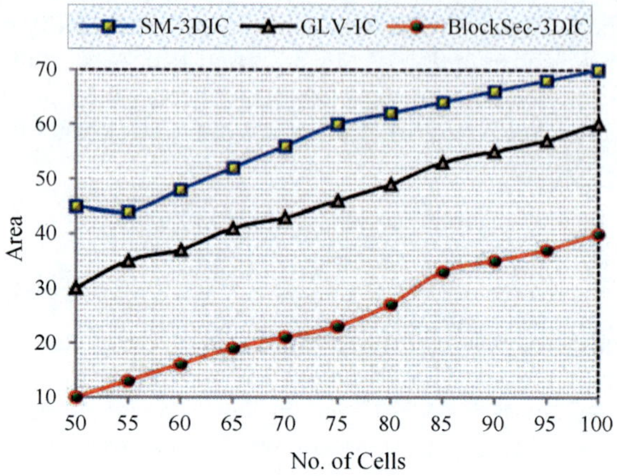

Fig. 10 Number of cells versus area

5.2.5 Impact of Area

The area is referred as the measure of compactness of the circuit. The lesser the area the higher will be the performance of the circuit. It is an important metric to evaluate the performance of 3D IC. The area can be represented as

$$A_r = H_c \times W_c \tag{29}$$

Where W_c and H_c denote width and height of the chip, respectively. Figure 10 illustrates the comparison of our proposed BlockSec-3DIC model and other existing approaches in terms of area with respect to number of cells. The area increases gradually with increase in the number of cells. The area consumed by our proposed model is lower than the existing approaches due to the consideration of wire length in placing the die. The adjustment of wire length and average distance reduces the area of the chip in an effective manner. For instance, the proposed model consumes 40 μm^2 of area when the number of cells is 100, whereas the existing approaches consume up to 70 μm^2 of area. From this, it can be concluded that the performance of our proposed BlockSec-3DIC model is better than that of the existing approaches.

5.2.6 Impact of Wire Length

The wire length is a significant metric which influences the area factor. The wire length of set of cells in a particular layer can be computed as,

$$W_l = \sum_{k=1}^{N} l(k_c) \tag{30}$$

where k denotes the layer and $l(k_c)$ represents the wire length of the cells in kth layer. The value of $l(k_c)$ is computed using the Euclidean distance. The proposed model is compared with other models in terms of wire length with respect to number of cells as presented in Fig. 11 The wire length increases with increase in the number of cells. The length of the wire should be optimal in order to achieve efficient consumption of area. The proposed BlockSec-3DIC model possesses less length of wire than other existing approaches due to the consideration of distance during the placement of die. For instance, the wire length of proposed model is 50 μm when the number of cells is 100, whereas the existing approaches possess wire length of about 80 μm. The increased length of wire by the existing approaches is due to inefficient placement strategies.

The numerical comparison of the proposed BlockSec-3DIC model and existing models is presented in Table 4. From this table, it can be concluded that our model is secure and outperforms the existing approaches in terms of attack detection rate, netlist change rate, power consumption, time consumption, area, and wire length.

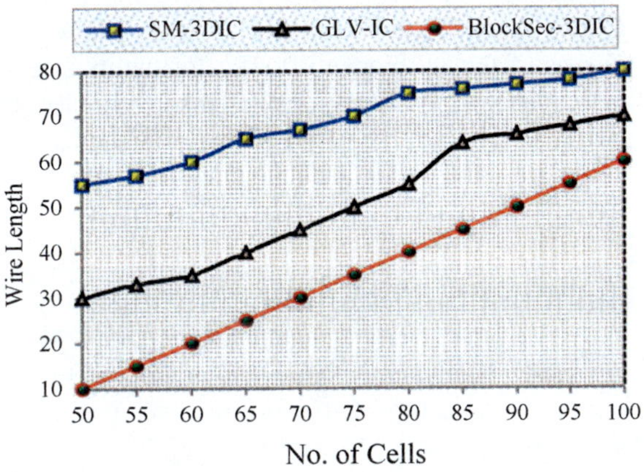

Fig. 11 Number of cells versus wire length

Table 4 Numerical analysis

Performance metrics	Proposed versus existing approaches			
	Scenario	SM-3DIC	GLV-IC	BlockSec-3DIC
Attack detection rate	No. of iterations	62 ± 2	71 ± 3	82 ± 1
Netlist change rate	No. of cells	48 ± 2	39 ± 2	20 ± 1
Power consumption	No. of cells	38 ± 0.5	26 ± 0.3	10 ± 0.1
Time consumption	No. of cells	22 ± 0.05	14 ± 0.02	6 ± 0.01
Area	No. of cells	57 ± 5	46 ± 3	24 ± 2
Wire length	No. of cells	69 ± 5	50 ± 4	35 ± 1

5.3 Security Analysis

The integrity of netlist should be protected to achieve efficacy in the performance of 3D IC. The presence of Trojans in the netlist results in malfunctioning of the chip. Additionally, there are several attacks which contribute to the performance degradation of the 3D IC. The proposed BlockSec-3DIC model mitigates these attacks which are mentioned as follows,

- **Brute force attack—**in this type of attack, the attacker captures the netlist to configure the contents in order to guess the required configurations. This attack is mitigated by performing netlist protection using four Q-curve algorithm and detection of hardware Trojan using BF-GRU in which both gate-level and chip-level features are used for detection purpose.
- **Wire length tracing—**in this type of attack, the comparison of wire lengths to find the normal wire length will be carried in order to manipulate the wire length for incorrect adversarial configurations. This attack is mitigated by dynamic validation of IC before fabricating which is performed by using HMM-valid.
- **Configuration prediction attack—**the attackers perform guessing of configurations of the IC blocks in order to manipulate the configurations to launch a malicious activity. This type of attack is mitigated by performing dynamic validation of IC in which the large number of IC blocks is optimally placed with random configuration which makes the guessing hard.

6 Conclusion and Future Work

In this paper, BlockSec-3DIC model is proposed for deep Trojan detection by validating the IC signature. Our work resolves the issues which are mentioned in the related work to detecting Trojans in 3D IC. The main aim of this research is to provide a security and reduces power consumption in 3D IC. Signature-based netlist protection is proposed for enhancing security for that we proposed four Q-curve algorithm. By utilising the SHA-3 hashing algorithm, 3D IC security is increased, and Trojan attacks are mitigated. All the transactions are stored in the blockchain for enhancing security. Trojans are detected by extracting the chip and gate-level features from the netlist using BF-GRU algorithm. For optimal placement, we proposed jellyfish search algorithm (JSA) by considering wire length, average distance, and power consumption. And multi-weighted sum model (MWSM) is proposed for optimal routing by considering distance and thermal factor. Finally, the trustworthiness of the IC is validated through signature using hidden Markov model (HMM) algorithm. The proposed model is experimented in MatlabR2020a, and our works provide high security and detect Trojans accurately in the 3D IC.

References

1. Reshma K, Priyatharishini M, Nirmala Devi M (2019) Hardware Trojan detection using deep learning technique. In: Soft computing and signal processing. Springer, Singapore, pp 671–680. https://doi.org/10.1007/978-981-13-3393-4-68
2. Sivakumar P, Pandiaraj K, JeyaPrakash K (2020) Optimization of thermal aware multilevel routing for 3D IC. Analog Integr Circ Sig Process 103:131–142. https://doi.org/10.1007/s10470-019-01513-y
3. Liakos KG, Georgakilas GK, Moustakidis S, Sklavos N, Plessas FC (2020) Conventional and machine learning approaches as countermeasures against hardware Trojan attacks. Microprocess Microsyst 79:103295. https://doi.org/10.1016/j.micpro.2020.103295
4. Ghandali S, Moos T, Moradi A, Paar C (2020) Side-channel hardware Trojan for provably-secure SCA-protected implementations. IEEE Trans Very Large Scale Integr (VLSI) Syst 28:1435–1448. https://doi.org/10.1109/TVLSI.2020.2982473
5. Yan C, Salman E (2018) Physical design of monolithic 3D ICs with applications to hardware security. In: Proceedings of the government microcircuit applications and critical technology conference
6. Nair RKR, Pothiraj S, Nagarajan R (2020) P2CTS-3D IC: an efficient placement aware partitioning and clock tree synthesis in 3D-integrated circuits. Appl Sci Eng Prog 13(4):377–393. https://doi.org/10.14416//j.asep.2020.09.001
7. Dong C, Chen J, Guo W, Zou J (2019) A machine-learning-based hardware-Trojan detection approach for chips in the Internet of Things. Int J Distrib Sens Netw 15(12):1550147719888098. https://doi.org/10.1177/1550147719888098
8. Cao K, Zhou J, Wei T, Chen M, Hu S, Li K (2019) A survey of optimization techniques for thermal-aware 3D processors. J Syst Architect 97:397–415. https://doi.org/10.1016/j.sysarc.2019.01.003
9. Lee JK, Park SY, Kim YH, Lee JC, Lee SH, Lee CH, Sung YH (2019) Three-dimensional integrated circuit (3D-IC) package using fan-out technology. In: 2019 IEEE 69th electronic components and technology conference (ECTC). IEEE, pp 35–40. https://doi.org/10.1109/ECTC.2019.00013
10. Medhat D, Dessouky M, Khalil D (2021) A programmable checker for automated 2.5 D/3D IC latch-up verification and hot junctions detection. Microelectron Reliab 124:114310
11. Zhang Z, Dofe J, Yellu P, Yu Q (2020) Comprehensive analysis on hardware Trojans in 3D ICs: characterization and experimental impact assessment. SN Comput Sci 1(4):1–13. https://doi.org/10.1007/s42979-020-00220-0
12. Dong C, Xu Y, Liu X, Zhang F, He G, Chen Y (2020) Hardware Trojans in chips: a survey for detection and prevention. Sensors 20(18):5165. https://doi.org/10.3390/s20185165
13. Zuo J, Zhang J, Yuen C, Jiang W, Luo W (2016) Energy efficient user association for cloud radio access networks. IEEE Access 4:2429–2438. https://doi.org/10.1109/ACCESS.2016.2566338
14. Zhang Z, Yu Q (2020) Invariance checking based Trojan detection method for three-dimensional integrated circuits. In: 2020 IEEE international symposium on circuits and systems (ISCAS), pp 1–5. https://doi.org/10.1109/ISCAS45731.2020.9180583
15. Hou Y, He H, Shamsi K, Jin Y, Wu D, Wu H (2018) On-chip analog Trojan detection framework for microprocessor trustworthiness. IEEE Trans Comput Aided Des Integr Circuits Syst 38(10):1820–1830. https://doi.org/10.1109/TCAD.2018.2864246
16. Elnaggar R, Chakrabarty K, Tahoori MB (2019) Hardware Trojan detection using change point-based anomaly detection techniques. IEEE Trans Very Large Scale Integr (VLSI) Syst 27(12):2706–2719. https://doi.org/10.1109/TVLSI.2019.2925807
17. Lu R, Shen H, Feng Z, Li H, Zhao W, Li X (2020) HTDet: a clustering method using information entropy for hardware Trojan detection. Tsinghua Sci Technol 26(1):48–61. https://doi.org/10.26599/TST.2019.9010047

18. Chen J, Dong C, Zhang F, He G (2019, August) A hardware-Trojans detection approach based on eXtreme gradient boosting. In: 2019 IEEE 2nd international conference on computer and communication engineering technology (CCET). IEEE, pp 69–73. https://doi.org/10.1109/CCET48361.2019.8988946
19. Hu T, Wu L, Zhang X, Yin Y, Yang Y (2019) Hardware Trojan detection combine with machine learning: an SVM-based detection approach. In: 2019 IEEE 13th international conference on anti-counterfeiting, security, and identification (ASID). IEEE, pp 202–206. https://doi.org/10.1109/ICASID.2019.8924992
20. Ren Z, Alqahtani A, Bagherzadeh N, Lee J (2020) Thermal TSV optimization and hierarchical floorplanning for 3-D integrated circuits. IEEE Trans Compon Packag Manuf Technol 10(4):599–610. https://doi.org/10.1109/TCPMT.2020.2970382
21. Ghosh P, Bhattacharya A, Forte D, Chakraborty RS (2019) Automated defective pin detection for recycled microelectronics identification. J Hardware Syst Sec 3(3):250–260. https://doi.org/10.1007/S41635-019-00069-7
22. Guin U, Zhou Z, Singh A (2018) Robust design-for-security architecture for enabling trust in IC manufacturing and test. IEEE Trans Very Large Scale Integr (VLSI) Syst 26(5):818–830. https://doi.org/10.1109/TVLSI.2018.2797019
23. Dong C, He G, Liu X, Yang Y, Guo W (2019) A multi-layer hardware Trojan protection framework for IoT chips. IEEE Access 7:23628–23639. https://doi.org/10.1109/ACCESS.2019.2896479
24. Maity DK, Roy SK, Giri C (2019) Identification of random/clustered TSV defects in 3D IC during pre-bond testing. J Electron Test 35(5):741–759. https://doi.org/10.1007/s10836-019-05824-w
25. Utyamishev D, Partin-Vaisband I (2018) Real-time detection of power analysis attacks by machine learning of power supply variations on-chip. IEEE Trans Comput Aided Des Integr Circuits Syst 39(1):45–55. https://doi.org/10.1109/TCAD.2018.2883971
26. Meade T, Zhang S, Jin Y (2017) IP protection through gate-level netlist security enhancement. Integration 58:563–570. https://doi.org/10.1016/j.vlsi.2016.10.014
27. Wang XY, Zhou Q, Cai YC, Qu G (2018) Spear and shield: evolution of integrated circuit camouflaging. J Comput Sci Technol 33(1):42–57. https://doi.org/10.1007/s11390-018-1807-6
28. Yan C, Dofe J, Kontak S, Yu Q, Salman E (2017) Hardware-efficient logic camouflaging for monolithic 3-D ICs. IEEE Trans Circuits Syst II Express Briefs 65(6):799–803. https://doi.org/10.1109/TCSII.2017.2749523
29. Saha D, Sur-Kolay S (2019) Guided GA-based multi-objective optimization of placement and assignment of TSVs in 3-D ICs. IEEE Trans Very Large Scale Integr (VLSI) Syst 27(8):1742–1750. https://doi.org/10.1109/TVLSI.2019.2908087
30. Bose A, Ghosal P (2020) A low latency energy efficient BFT based 3D NoC design with zone based routing strategy. J Syst Architect 108:101738. https://doi.org/10.1016/j.sysarc.2020.101738
31. Nayak DK, Banna S, Samal SK, Lim SK (2015) Power, performance, and cost comparisons of monolithic 3D ICs and TSV-based 3D ICs. In: 2015 IEEE SOI-3D-subthreshold microelectronics technology unified conference (S3S). IEEE, pp 1–2. https://doi.org/10.1109/S3S.2015.7333538
32. Yang D, Gao C, Huang J (2020) Dynamic game for strategy selection in hardware Trojan attack and defense. IEEE Access 8:213094–213103. https://doi.org/10.1109/ACCESS.2020.3040395
33. Patnaik S, Ashraf M, Sinanoglu O, Knechtel J (2019) A modern approach to IP protection and Trojan prevention: split manufacturing for 3D ICs and obfuscation of vertical interconnects. IEEE Trans Emerg Top Comput 9(4):1815–1834. https://doi.org/10.1109/TETC.2019.2933572
34. Zhang Z, Dofe J, Yu Q (2020) Improving power analysis attack resistance using intrinsic noise in 3D ICs. Integration 73:30–42. https://doi.org/10.1016/j.vlsi.2020.02.007
35. Zaraee N, Zhou B, Vigil K, Shahjamali MM, Joshi A, Ünlü MS (2020) Gate-level validation of integrated circuits with structured-illumination read-out of embedded optical signatures. IEEE Access 8:70900–70912. https://doi.org/10.1109/ACCESS.2020.2987088

Fire Detection Through Surveillance Videos Using Deep Learning in Real-World Applications

Kasindula Sai Padmakar, Samineni Sathvik Chowdary, Pavuluri Kumar Jaswanth, Gadiraju Aravind, and J. Divya Udayan

Abstract Unearthing fire and smoke in sighted landscapes is challenging because of the great range of texture and color. To address this issue, several fire and smoke picture taxonomy systems have been proposed; however, the majority of them depend on either rule-based approaches or hand-crafted attributes. The technologies that support fire and smoke detection systems are important to ensuring and providing maximum performance in today's surveillance situations. To overcome these limits, describe a unique technique based on a deep learning approach that employs a convolutional neural network with expanded convolutions. Fire may cause significant loss of life and property. Evaluated our method by training and testing it with a custom-built dataset of fire and smoke photos obtained from the Internet and manually classified. In terms of performance, our approach was compared against solutions based on well-known snipping architectures. The CNN used to recognize the activity of video abnormality and if any fire or smoke image was captured while analyzing frame by frame video then it was stored in the firebase database. If any fire frame is captured in the dataset then it gives the alert message to the android like notification. In terms of classification performance and complexity, our results suggest that our approach outperforms others.

K. Sai Padmakar · S. Sathvik Chowdary · P. Kumar Jaswanth · G. Aravind · J. Divya Udayan (✉)
Department of Computer Science and Engineering, Amrita Vishwa Vidyapeetham, Coimbatore, India
e-mail: divyaudayanj@am.amrita.edu

K. Sai Padmakar
e-mail: ksaipadmakar@am.students.amrita.edu

S. Sathvik Chowdary
e-mail: sschowdary@am.students.amrita.edu

P. Kumar Jaswanth
e-mail: pkumarjaswanth@am.students.amrita.edu

G. Aravind
e-mail: garavind@am.students.amrita.edu

V. K. Gunjan and J. M. Zurada (eds.), *Proceedings of 3rd International Conference on Recent Trends in Machine Learning, IoT, Smart Cities and Applications*, Lecture Notes in Networks and Systems 540, https://doi.org/10.1007/978-981-19-6088-8_30

Keywords Fire detection · Real-world applications · Deep learning · Smoke detection · Video analysis · Frame by frame detecting · Classification

1 Introduction

Early detection of fires and smoke is important so that you can intervene in time to avoid major damage. The main drawback of such sensors is that they can usually detect fires and flames only near the installation site. This limits its effectiveness over a wide range. These systems, detailed below, employ photographs or videos acquired by cameras that may be used both indoors and outdoors to detect fires and smoke.

The key characteristics of smoke and fire can vary widely due to various parameters such as formation, texture, color, etc., and it is difficult for such algorithms to detect fire and smoke with high accuracy in real photographs. Another method is to recognize and extract essential color and smoke information from images and videos using deep learning algorithms. The motivation for the current biggest strategy is to perceive photographic smoke and fire pits, but as mentioned above, they have some drawbacks.

To address these limitations, this study provides a unique deep CNN version for classifying fireplace and smoke photos. This technique can classify all smoke and hearth pixels at the same time and has several advantages, surpassing the various popularity of CNN mode, which detects hearth and smoke in the photographs seen today. There is currently no publicly available dataset for recognizing smoke or fire images, as far as we are aware. Therefore, to train the network and test the technology, we created our dataset using fire and smoke photos. Because the CNN model requires a large number of images for proper training, various data improvement techniques are used to generate additional training examples from a small amount of the original photo of the data benchmark.

2 Literature Survey

Chen et al. [1] designed a smoke detection system based on video. To assess whether each pixel depicts smoke, color-based static decision rules and diffusion-based dynamic decision rules are utilized. Toreyin et al. [2] offer another approach to fire and flame detection that combines motion and color cue data with edge blur and flicker capabilities. Muller et al. [3] created a video-based optical flow method, which is an essentially optimal mass transfer model that uses the direction and amplitude of flow to distinguish between movements that are caused by fire and those that are not caused by fire. Bugarovich et al. [4] proposed a computer vision-based approach to estimate fire size using GIS-based improved reality. Other methods have been projected, such as developing a rule-based system based on the photo's fire and smoke's color, shape, and temporal aspects [5]. The goal of the technique described above is to create rule-based algorithms that rely on hand-crafted qualities or specialized knowledge. Zhang [6] for example, demonstrated how to spot flames in forest photographs using two linked deep CNNs. The global image classifier assesses the full picture first in this technique. For smoke detection, Tao et al. [7] suggested a deep CNN model influenced by AlexNet [8] (GAN) [9] to generate fresh training images that improves network performance while addressing the problem of overfitting.

3 Dataset Collection

The learning of millions of parameters is common in deep learning techniques to picture recognition. Furthermore, this procedure necessitates the shooting of a huge number of photographs. As far as we know, there are no publicly available data standards, including smoke and fire records. Dataset is compiled and contains 31 videos from different situations. This collection contains 14 fire videos and 17 non-fire movies. One such sample dataset has been shown in Fig. 1.

The dataset is difficult and large, making it suitable for experimentation including burning objects and film of mountains with smoke and clouds made datasets difficult with both color-based and motion-based fire detection technologies. This is one of the reasons to choose this dataset for testing.

Fig. 1 Sample dataset of fire and smoke images

4 Proposed Methodology

CNN is a deep learning system inspired by the generation of superior visual mechanisms of perception. Its typical architecture has been shown in Fig. 2. Compared to those of established functions, CNNs are widely used in image classification and achieve encouraging classification accuracy in large data sets. To create feature maps, multiple cores of different sizes are applied to the input data during the convolution process. Sub sampling or grouping, which chooses the greatest activations within a small neighborhood, is applied to these feature maps. The fully connected layer is another key stage in the CNN pipeline, where high-level abstractions are modeled from the input data. During the training phase, the weights of the neurons are learned and changed to better reflect the input data.

Extensively the convolutional neural network (CNN) identifies fires early in CCTV surveillance footage as depicted in Fig. 3. The following are the article's major contributions:

Step 1: Deep learning (DL) for this focuses and offers a low-cost CNN framework for fire detection in video surveillance because previous manual engineering techniques have drawbacks.
Step 2: The video was converted frame by frame to analyze the accurate image of smoke or fire that occurred.
Step 3: Stimulated by the transfer learning approach, we qualified and fine-tuned a form with planning CNN for fire detection, which effectively dominated traditional fire detection schemes.
Step 4: The CNN algorithm is used to analyze frame activity and send data to the firebase database.
Step 5: The database sends a notification to the android mobile for the registered user in a particular organization/institution. Hence, our scheme is more suitable for early fire detection during surveillance to avoid huge fire disasters.

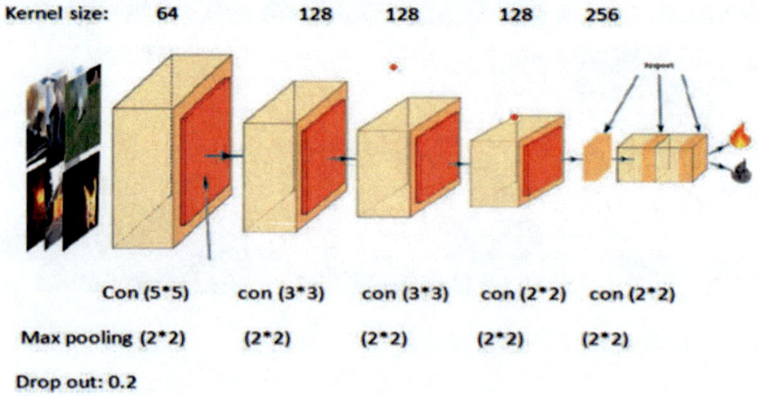

Fig. 2 Typical deep learning-based CNN architecture

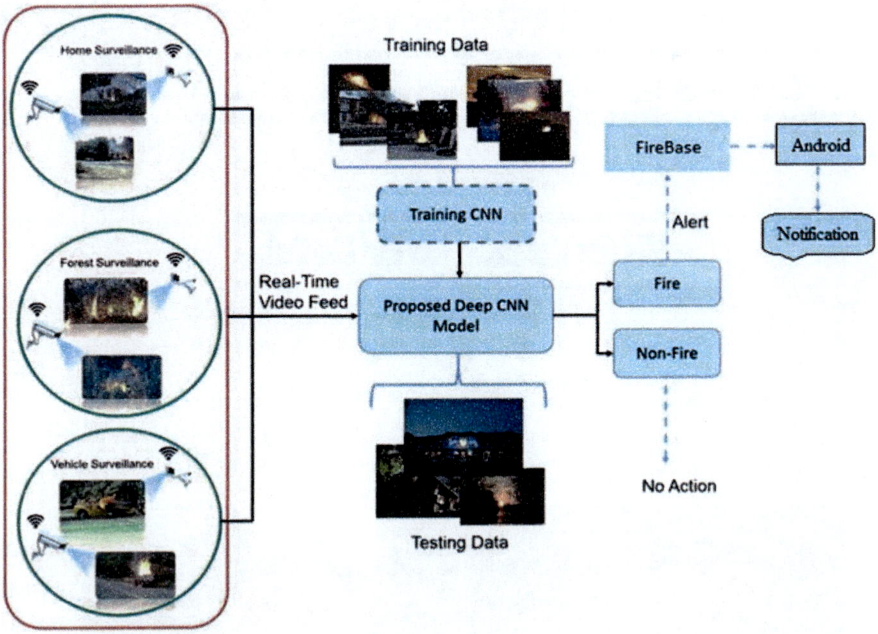

Fig. 3 Fire detection in surveillance videos using deep CNN frame by frame video detection

5 Frame by Frame Video Detection

While the academic community agrees that deep learning architectures learn deep features automatically from raw data, it takes some work to train multiple models with different parameters to get the best answer to the problem (Fig. 4). To do this, we train a large number of models with different parameter values based on the collected training data, their quality, and the type of problem.

The test picture is input and routed through the target model's architecture to obtain the inference. The probability for two firing and non-firing classes is output. The final label of stiff motion is determined by the maximum probability score between the two classes. For fire detection in surveillance video, another comparable technique based on movement and color characteristics is presented.

The fire integrated the qualities of form, color, and movement to boost accuracy even further, resulting in a multi-expert framework for real-time fire identification. Although the technique has dominated contemporary fire detection algorithms, it can yet be improved. Furthermore, the false alarm rate remains significant and can be decreased further (Fig. 5).

Computational Complexity is a term that refers to the difficulty of computing something. The suggested system saves the obtained visuals in the Firebase data set and sends a notice to the Android mobile phone with a label display in the case of a

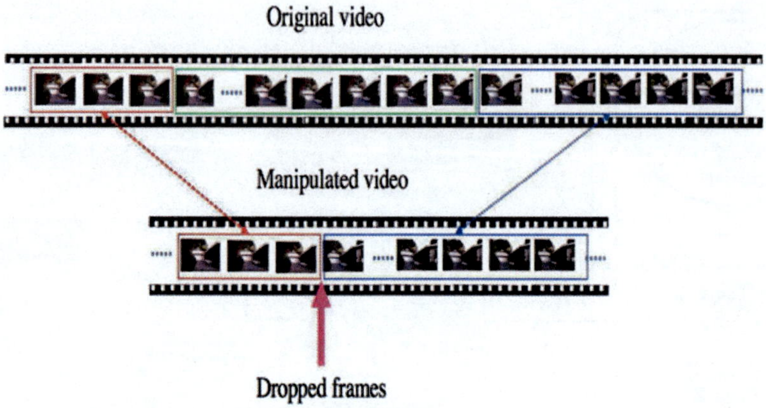

Fig. 4 Frame work of video analyzing process

Fig. 5 Sample frame from a video of the dataset

fire. As a result of this incentive, fire detection algorithms with lower computational costs, fewer false alarms, and higher precision are needed.

6 Results and Discussion

The suggested improved deep learning approach was evaluated and compared to current techniques in this paper and proposed a CNN technique which is used to recognize the activity of video frame matching. In terms of sensitivity, specificity,

Table 1 Analysis of the proposed and current machine learning algorithms' performance

Parameters	CNN (%)	SVM (%)	ANN (%)
Sensitivity	93	73	81.5
Specificity	90	78.8	74.9
Accuracy	98.4	63.4	74

accuracy, and fire detection, the equations below have been determined in comparison to current SVM and ANN classifier detection in terms of sensitivity, specificity, and accuracy of fire detection and sending the alert to the notification to the android Sensitivity, specificity, and accuracy are statistical parameters that are to be considered.

$$\text{Specificity} = \text{TN TN} + \text{FP} * 100$$
$$\text{Sensitivity} = \text{TP TP} + \text{FN} * 100$$
$$\text{Accuracy} = \text{TP} + \text{TN TP} + \text{FN} + \text{TN} + \text{FP} * 100$$

where

TP specifies the True Positive,
FP denotes the False Positive,
TN can indicate the True Negative,
FN represents the False Negative.

The suggested CNN approach is utilized to accurately identify fire alarms, as indicated in the experimental results in Table 1, and the chart shows the contrast.

In contrast to current approaches SVM and ANN classifier, the suggested method has a sensitivity level of 93%, a specificity level of 90%, and an accuracy level of 98.4%, as shown in Table 1and Fig. 6.

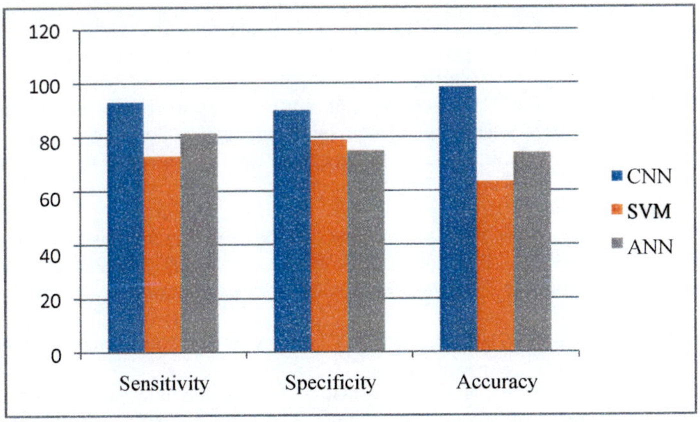

Fig. 6 Comparison graph for current DL algorithms versus proposed approach

Similarly, the total proportion of test data records that are accurately analyzed by classifier algorithms represents the video activity recognition of the supplied test dataset. Specificity and Sensitivity are alternatives for the metric of exactness used to assess fire detection performance.

7 Conclusion

The newly improved processing capabilities of smart devices have shown promise in monitoring systems for the detection of various abnormal occurrences such as fires, accidents, and other crises. If not managed quickly, fire is one of the most deadly hazards that can result in severe losses. This emphasizes the need of putting in place early fire detection systems. As a result, we provide a low-cost fire detection CNN architecture for surveillance films in this work. When compared to cutting-edge technology, fire monitoring systems, the proposed framework strikes a balance between fire detection accuracy and computation cost while also lowering false alarms.

References

1. Chen T-H, Huang S-F, Yin Y-H, Ye Y-T (2006) The smoke detection for early fire-alarming system base on video processing. Multimedia Sig Process. https://doi.org/10.1109/IIH-MSP.2006.265033
2. Töreyin BU, Güdükbay U, Enis Cetin A (2006) Computer vision-based method for real-time fire and flame detection. Pattern Recogn. https://doi.org/10.1016/j.patrec.2005.06.015
3. Mueller M, Kolesov I, Karasev P, Tannenbaum A Optical flow estimation for flame detection in videos. IEEE Trans Image Process. https://doi.org/10.1109/TCSVT.2015.2392531
4. Bugaric M, Jakovcevic T, Stipanicev D (2015) Computer vision-based measurement of wildfire smoke dynamics. Adv Electr Comput Eng 15(1):55–62. https://doi.org/10.4316/AECE.2015.01008
5. Çelik T, Ozkaramanlı H, Demirel H (2007) Fire and smoke detection without sensors: image processing based approach. In: 2007 15th European signal processing conference, pp 1794–1798
6. Zhang Q, Xu L, Xu J, Guo H (2016) Deep convolutional neural networks for forest fire detection. In: Proceedings of the 2016 international forum on management
7. Krizhevsky A, Sutskever I, Hinton GE (2012) Imagenet classification with deep convolutional neural networks. In: Advances in neural information processing systems, pp 1097–1105. https://doi.org/10.1145/3065386
8. Tao C, Zhang J, Wang P (2016) Smoke detection based on deep convolutional neural networks. In: International conference on industrial informatics—computing technology, intelligent technology, industrial information integration (ICIICII). https://doi.org/10.1109/ICIICII.2016.0045
9. Goodfellow, Xu B, Warde-Farley D, Bengio Y (2014) Generative adversarial nets. In: Advances in neural information processing systems, pp 2672–2680
10. Yin Z, Yuan F, Xia X, Shi J (2017) A deep normalization and convolutional neural network for image smoke detection. IEEE Access. https://doi.org/10.1109/ACCESS.2017.2747399
11. Sridhar P, Sathiya RR (2021) Computer vision based early electrical fire-detection in video surveillance oriented for building environment. J Phys Conf Ser

12. Sridhar P, Sathiya RR (2021) Real time fire detection and localization in video sequences using deep learning framework for smart building. J Phys Conf Ser
13. Sridhar P, Parameswaran L, Senthil Kumar T (2018) An efficient rule based algorithm for fire detection onreal time videos. In: Proceedings of the first international conference on intelligent computing
14. Thanga Manickam M, Yogesh M, Sridhar P, Thangavel K, Parameswaran L (2020) Video-based fire detection by transforming to optimal color space. In: Smys S, Tavares J, Balas V, Iliyasu A (eds) Computational vision and bio-inspired computing. ICCVBIC 2019. Advances in intelligent systems and computing, vol 1108
15. Pai ML, Varsha KS, Arya R (2018) Application of artificial neural networks and genetic algorithm for the prediction of forest fire danger in Kerala. In: Intelligent systems design and applications, vol 2. Springer International Publishing, Cham
16. Subathra P, Ramachandran R, Rajeev DC, Krishnan SG (2015) Deep learning—an overview. Int J Appl Eng Res

A Survey on Battery State of Charge and State of Health Estimation Using Machine Learning and Deep Learning Techniques

Mandeddu Sudhakar Reddy and M. Monisha

Abstract For long-lasting electric vehicles, accurate health evaluation and lifetime prediction of lithium-ion batteries are critical. Early diagnosis of poor performance allows for prompt battery system maintenance. This lowers operating expenses and lessens the risk of accidents and malfunctions. The rise of "Big Data" analytics and related statistical/computational technologies has sparked interest in data-driven battery health estimates. In this paper, we review several articles to highlight their achievability and also environmentally friendly in production with health of battery in reality. We distinguish how machine learning and deep learning algorithms helpful in estimating SOC and SOH of Li-ion battery that are utilized in durable electric vehicles. In addition, we explained the basics of battery, cells, types of battery along with its characteristics were analyzed. Moreover, we summarized the state-of-art table comprises techniques used, which state of estimation either SOH or SOC, metrics used by various machine learning and deep learning algorithms, and discussed their benefits too.

Keywords Machine learning (ML) · State of health (SOH) · Deep learning (DL) · State of charge (SOC) · Long short-term memory (LSTM)

1 Introduction

A battery is a chemical device that stores electrical energy in the form of chemicals and then turns that stored chemical energy into direct current (DC) electric energy through an electrochemical reaction. In 1800, an Italian physicist named Alessandro Volta invented the first battery. An electric current is used to move electrons from one substance to another (called electrodes) in an electrochemical reaction in a battery.

M. Sudhakar Reddy (✉) · M. Monisha
Department of ECE, VISTAS, Chennai, India
e-mail: sudhakar.mandeddu@gmail.com

M. Monisha
e-mail: monisha.se@velsuniv.ac.in

V. K. Gunjan and J. M. Zurada (eds.), *Proceedings of 3rd International Conference on Recent Trends in Machine Learning, IoT, Smart Cities and Applications*, Lecture Notes in Networks and Systems 540, https://doi.org/10.1007/978-981-19-6088-8_31

Then we are discussing about how cell and battery are collide with each other. Despite the fact that specific word series is frequently utilized, the fundamental ECU suitably energy saved space is known as a group. As previously stated, group defines essential ECU which produces electrical energy by converting chemical energy. A cell comprises of three primary components: two electrodes and electrolyte, as well as terminals, a separator, and a container in its most basic form. When it comes to electrodes, there are two sorts of electrodes: anode and cathode. The negative electrode is known as the anode that drops electrons to the peripheral path also oxidized in chemical response. The positive electrode, conversely, cathode represents electrons which are oxidized internally. This absorbs particles of electron as of everlasting circuit and is reduced in chemical response. As a result, the electrochemical oxidation–reduction reaction is responsible in power transfer series. The electrolyte is the cell's third most critical component. Between the two electrodes, an electrolyte functions as a channel for charge conversion represents as ions. As a result, electrolyte is also known as an Ionic Conductor which has ionic conductivity. A battery is typically made up of one or more "cells" that are electrically connected in series or parallel to give the required voltage and current levels depicted in Fig. 1.

The main objective of this survey is described as follows:

- To discuss about basics of batteries, cells and the working of battery as well.
- To review on several articles to make familiar about battery types along with its characteristics.
- To know how machine learning techniques helpful in estimation on battery SOC and SOH.
- To recognize how deep learning algorithms appropriate in evaluation on battery SOC and SOH.

Fig. 1 Configuration of battery

- To made comparison of several research articles related with battery to highlight the techniques utilized for battery SOH and SOC estimation.

The main contribution of this survey is to make familiar about the basic concepts of battery, cells, how battery works in electronic vehicles. Moreover, this reviewed work helpful for current researchers who are working in electronic vehicle field using modern techniques especially ML and DL algorithms. The remaining paper work is summarized as follows: Sect. 2 explains the literature survey of several existing articles regarding battery SOH and SOC estimation. Section 3 describes the types of battery along with its specific characteristics of each battery. Section 4 introduced how machine learning techniques suitable for predicting battery SOH and SOC estimation. Section 5 briefly about how deep learning algorithms useful in estimation of battery SOH and SOC. Section 6 introduces the overall survey table which described the comparison of several research works comprises of techniques used, kind of battery used, whether SOC or SOH is used, and benefits are discussed here. Section 7 explains the conclusion part of battery SOC and SOH estimation of this survey.

2 Related Work

Caliwag et al. [1] proposed hybrid method comprises of both Vector Auto Regressive Moving Average along with LSTM for predicting SOC as well as battery O_v while electric vehicle is motivated beneath CVS-40 drive cycle. This proposed approach attains least RMSE in estimation of SOC for battery in motor cycle. Erlangga et al. [2] utilized dual Kalman filter approach for Lithium battery SOC and SOH estimation. Guo et al. [3] applied the properties of battery divergence protection as well as ohmic resistance for estimating battery health (SOH) estimation. Based on accuracy metrics, the Lithium-Ion Phosphate battery health was estimated. Under varying SOH, polarization resistance does not change appreciably. The ohmic resistance is the principal cause of the battery's internal resistance variation. As a result, it is recommended to estimate the SOH of cells using ohmic resistance, which is defined as follows:

$$\mathrm{SOH} = \frac{R_{\mathrm{EOL}} - R_{\mathrm{Now}}}{R_{\mathrm{EOL}} - R_N} \times 100\%$$

When the actual capacity is 80% of the rated capacity, R_{EOL} is the value of the ohmic resistance. When the battery is made in the factory, the ohmic resistance value is R_N. In the current condition, R_{now} is the ohmic resistance value. Chang et al. [4] reviewed several articles on SOC estimating numerical approaches for all rechargeable and non-rechargeable batteries. Fan et al. [5] introduced open circuit voltage based on affine projection algorithm for estimating SOH on Li-ion battery in e-vehicles. Noura et al. [6] reviewed many articles regarding techniques of SOH estimation which are most challenging part in hybrid electric vehicles. How et al. [7] reviewed that utilization with hybrid model comprises of both data driven and model based approach for predicting SOC in Li-ion battery. Li et al. [8] reviewed on battery heath and battery lifetime in Li-ion battery using data driven approach.

3 Types of Battery Along with Its Characteristics

- Non-rechargeable batteries known as Primary,
- Rechargeable batteries are called Secondary.

The above two sorts of series are the fundamental kinds, despite the fact that there are countless different categories among kinds of series be shown in Fig. 2.

3.1 Non-rechargeable Battery

A primary battery is a simple and convenient power source as range of transferable electrical equipment like illumination, pictures, wristwatch, playthings, radios, etc. They are of the "use it and throw it away" kind because that particles are not reviving. Non-rechargeable series often low-cost, less weight, compact, and easy to use, requires little or no maintenance. The majorities of single-cell crucial series are utilized in home appliances, cylindrical and have a single cell structure (although, it is very easy to produce them in different shapes and sizes). Zinc-carbon-based batteries were used in the 1940s, during World War II, and afterward, maximum competence 50 W hour per kilogram. Major improvement in series expertise transpired among

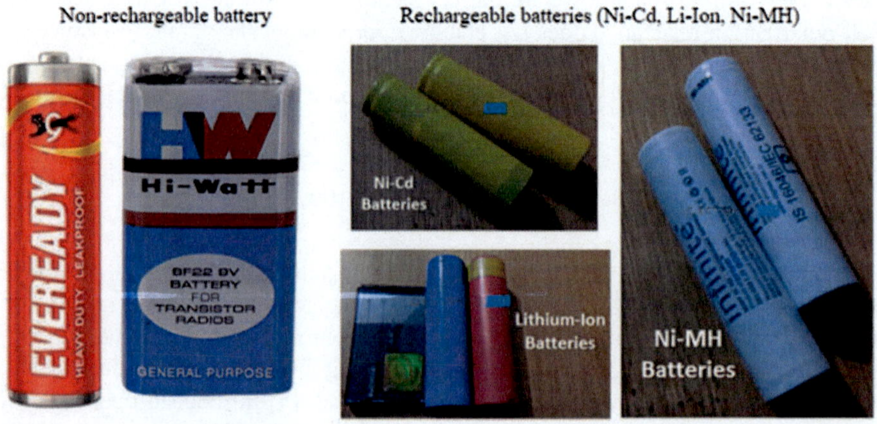

Fig. 2 Representation of rechargeable and non-rechargeable battery

1970–1990. The famous Zn-Al Mng O_2 series invented about this time, and they gradually superseded adult Zn-C types while dominant battery.

For the current period, Zn-MO also Cd–Mr O series utilized, as a result of environmental concerns over Mercury usage, these battery types were gradually phased out.

3.2 Rechargeable Battery

The resultant series is sometimes called boosting series since this may be repeatedly charged subsequent being released. By delivering the energy during ECC in the contradictory path of their release, the substance position of the group may "recharged" to their initial position. Rechargeable batteries are utilized in two dissimilar ways.

Way 1: Rechargeable series are effectively utilized as the device which store energy somewhere they are linked to main energy source electrically as well as charged by it too and hence providing energy when needed. The major real time example is UPS named as Uninterrupted Power Supplies.

Way 2: Another group of rechargeable battery is wherever battery is utilized and release as non-rechargeable battery. On one occasion, it is fully released rather than disposed it, the battery is re-energized with proper charging device. Such kind of procedure is applicable in real time appliances such as E-vehicles, mobiles and laptops.

3.3 Characteristics of Rechargeable and Non-rechargeable Batteries

Table 1 describes the kind of battery, whether primary or secondary batteries, along with its characteristics are explained.

4 SOC and SOH Estimation

Energy Management System is utilized for retaining the secure, consistent function of battery. This comprises of cell complementary, security to guarantee the function inside secure limits of battery, estimation of battery SOC and SOH. To ensure an accurate measure of a vehicle's remaining driving range as well as optimal battery pack balancing, a reliable state of charge estimation is essential. The SOC is similar to the fuel gauge found in gasoline-powered cars. SOC refers to the amount of charge left in the battery and is calculated as the ratio of the battery's residual capacity to its

Table 1 Description of kind of battery, whether it is rechargeable, non-rechargeable and its characteristics

Kind of battery	(Rechargeable/non-rechargeable)	Characteristics
Zinc-carbon	Non-rechargeable battery	Less cost, common, availability of several sizes
Mercury	Primary cell	Life time is long, very high capacity
Magnesium	Primary cell	Life time is long, very high capacity
Lithium	Non-rechargeable battery	Greater performance, very high power compactness
Silver/zinc	Non-rechargeable battery	Greater capacity, horizontal release and valuable
Lead-Acid (LA)	Rechargeable battery	Capacity ranges from 1 to 12,000 Ah
Nickel–Cadmium (Ni–Ca)	Rechargeable battery	Life time is very long, consistent and powerful, Discharge rate is high
Nickel Metal Hydrate battery	Rechargeable battery	Very high energy density
Lithium-Ion battery	Rechargeable battery	Very high energy density, high exact energy, life time cycle is longer

nominal capacity. The state of health (SOH) estimation, which has a value ranging from 0 to 100%, is commonly used to determine battery aging. It is a monetary value that does not correspond to a physical item and is expressed as a percentage. A SOH of 100% reflects battery health at the beginning of a battery's lifetime, when capacity is at its highest, and a SOH of 0% represents battery health at the end of the battery's lifetime, when capacity is at its lowest. The difference between a fully charged battery and the same battery in use is described by the state of charge of a battery. It has something to do with the amount of electricity left in the cell. It is calculated by dividing the remaining charge in the battery by the maximum charge that the battery can give. As shown below, it is expressed as a percentage.

$$\text{SOC} = \frac{(Q_0 + Q)}{Q_{\max}} \times 100\%$$

where Q_0 represents the initial charge of the battery, Q corresponds to the quantity of electricity.

The state of health (SoH) of a battery describes the difference between a researched battery and a new battery while also taking into account cell aging.

It's the proportion of a battery's maximum charge to its rated capacity. As shown below, it is expressed as a percentage.

$$\text{SOH} = \frac{Q_{\max}}{C_r} \times 100\%$$

whereas $Q_{\max}$ represents the maximum charge availability of the battery, C_r corresponds to the rated capacity.

5 How ML Algorithms Appropriate in SOH Estimation

Feng et al. [9] introduced partial charging segment based on support vector machine (SVM) for estimating online SOH on Lithium-ion battery. Also this work describes the institution of SVM model for new cell in battery with linear programming along with its parameter finding for all SOH via least square method. Feng et al. [10] detecting parameter inconsistency. Aloisio et al. [11] applied several machine learning algorithms like KNN, LDA, NB, Decision tree and Support vector machine for estimating SOH in Li-ion battery. Andre et al. [12] proposed SVM approach, dual Kalman filter method for estimating SOC and SOH on Li-ion battery. These two approaches were confirmed and validated via cell capacity in cycle form, cycle aging tests and capacity of battery. Man-Fai et al. [13] and Zou et al. [14] determined that basic machine learning algorithms appropriate for evaluating SOC and SOH in battery which are used in electric vehicles. Kim et al. [15] utilized reinforcement learning approach for predicting SOC for Li-ion battery. Using this reinforcement learning the features gets reduced by improved Kalman filter technique which is an iterative approach.

Roman et al. [16] built machine learning model based on pipeline for battery capacity prediction. For that battery capacity prediction, metric such as health of battery with 179 cells succession below several situations. The proposed pipeline based on ML algorithms evaluated SOH in linked with assurance gap via both parametric and non-parametric algorithms. Wang et al. [17] deduced that machine learning algorithms helpful in predicting SOC in Li-ion battery where some other algorithms were not supported during decoding because of non-linear problem. That SOC estimation in Li-ion battery was found in Hi-Fi range which means high quality and high capacity of battery too. Vidal et al. [18] reviewed several articles based on usage of machine learning technique for estimation of SOC and SOH in electric vehicle battery.

6 Deep Learning Techniques Suitable in Battery Health and Charge Estimation

Anjum et al. [19] designed deep learning-based neural network framework along with its hyper-parameter settings to efficiently evaluate SOC in electric vehicle battery cells. The layers in neural network and also neurons may diminish the computational

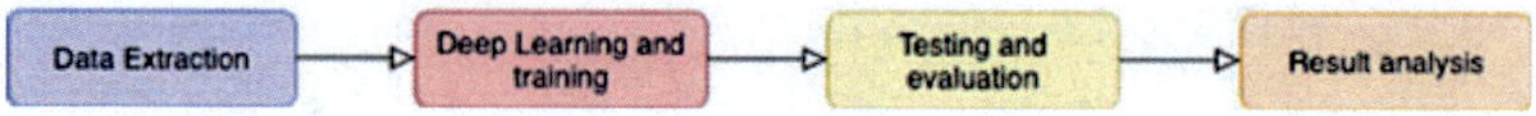

Fig. 3 Architecture of DL for SOC estimation proposed by [19]

costs, sources needed with no cooperation in performance. The framework of deep learning proposed by [19] for SOC estimation in E-vehicle battery is depicted in Fig. 3.

Venugopal et al. [20] built recurrent Neural Network for predicting SOC in battery especially Li-Nickel cobalt aluminum oxide battery. Liu et al. [21] introduced deep learning techniques and ultrasonic sensors for observing every states of battery. Metrics like Root Mean Square Error were estimated to predict the learning rate by the way states of battery was analyzed. Khan et al. [22] introduced NN for SOH estimation on lithium battery. Vidal et al. [23] proposed deep learning-based transfer learning especially LSTM-RNN approach for estimating SOC in Li-ion battery. Also, Kalman filter technique was utilized for testing, developing DL algorithms, and design the filter for every battery type. The structure of Li-ion battery proposed by [23] using LSTM-RNN is depicted in Fig. 4.

Gao et al. [24] proposed novel method by considering side reactions more lifetime of Li-ion corrupted battery for estimating SOC and SOH of battery. Zhang et al. [25] applied enhanced Radial Basis Function-based neural network model for SOC estimation of Li-ion battery pack.

Wei et al. [26] applied the integration of LSTM along with exogenous input neural network for estimating SOC in Li-ion battery which helps to solve the issues of gradient descent and gradient explosion by means of self-regressive approaches. The performance of the NN model was estimated through RSME evaluation by the way the Li-ion battery estimation was identified. The optimization NN method was utilized by Lipu et al. [27] for SOC assessment in Li-ion battery. The configuration is shown in Fig. 5.

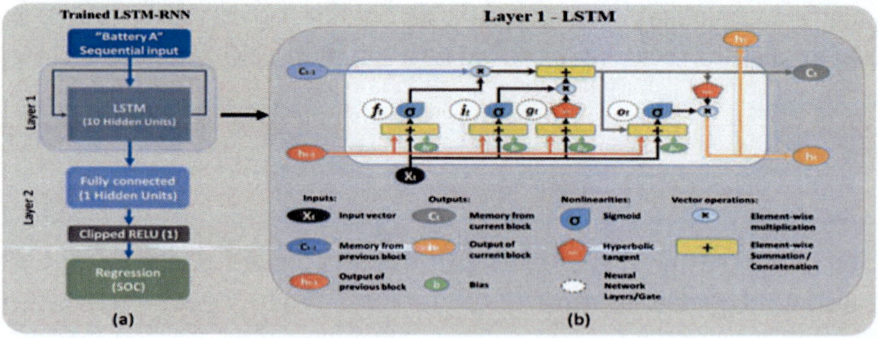

Fig. 4 Framework designed by [23] using hybrid method LSTM-RNN for SOC prediction

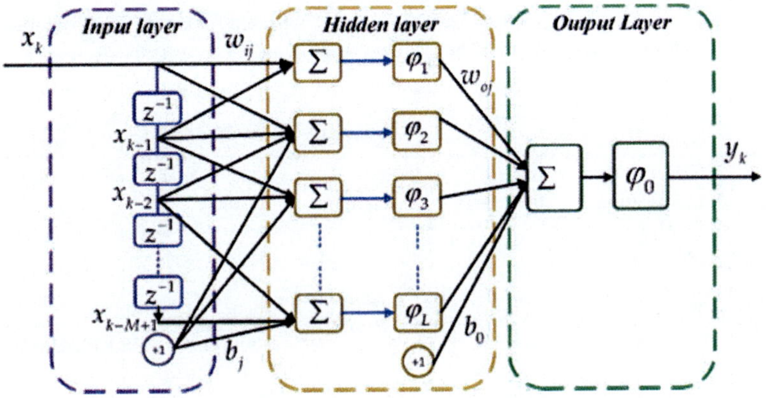

Fig. 5 The configuration on Neural Network based on time delay for estimation of SOC of Li-ion battery

Li et al. [28] proposed layers involved in Convolutional Neural Network method found the capacity of battery which is longer lifetime and fast too. This paper follows three phases such as sequence to image conversion, data separation and finally CNN design for estimating battery capacity. Also, based on RMSE value the model performance in finding capability of battery was estimated.

Shi et al. [29] applied hybrid approach includes deep learning based algorithm along with advanced Kalman filter method for finding SOC and SOH with highly optimized in Li-ion battery. Based on the findings of Mean square error, the performance of deep learning model was analyzed in SOC and SOH prediction for Li-ion battery.

7 Survey Table on Techniques Used, Battery Estimation, Advantages

In this section, we are discussing the techniques used, what kind of battery, battery estimated especially SOC/SOH, metrics used, and accuracy in percentage are summarized in Table 2.

8 Conclusion

In this survey, we reviewed several articles to highlight their achievability and also environmentally friendly in production with health of battery in reality. We discussed the basic concepts of battery, cells, and different kinds of battery such as rechargeable and non-rechargeable along with its characteristics. We surveyed how machine

Table 2 Overall survey table comprises of techniques used, what kind of battery, battery estimation by various researchers, metrics evaluated by each work and accuracy in percentage

Survey work	Techniques used	Kind of battery	Battery estimation (SOC/SOH)	Metrics found	Accuracy
Aloisio et al. [11]	Both classification (SVM, DT, KNN, NB) and Regression model	Li-ion batteries	SOH estimation	R^2 and Mean Absolute Error	91.5%
Zhang et al. [25]	RBF-NN	Li-ion battery pack	SOC estimation	Root Mean Square Error, Max Absolute Error	RMSE = 0.08
Erlangga et al. [2]	Coulomb counting method along with dual Kalman	Lithium battery	SOC and SOH estimation	–	Error below 1% only
Feng et al. [9]	Support Vector Machine	Li-ion battery	SOH estimation	Partial charging segments	Error less than 2%
Guo et al. [3]	Dual extended Kalman filtering approach	Electric vehicle battery	SOH estimation	Accuracy	Greater accuracy with less error
Andre et al. [12]	SVM with dual filter	Li-ion battery	SOC and SOH estimation	Open current voltage curve, pulse power method	Error less than 1%
Vidal et al. [23]	LSTM-RNN along with transfer learning	Li-ion battery	SOC estimation	Accuracy	Maximum accuracy
Fan et al. [5]	Open circuit voltage	Li-ion battery	SOH estimation	–	Capacity error less than %
Gao et al. [30]	Side reactions	Li-ion degraded battery	SOC and SOH estimation	Determining error	Estimating states of battery
Kim et al. [15]	Reinforcement learning	Li-ion battery	SOC estimation	Finding reward value	Error reduction
Wei et al. [26]	LSTM and NARX NN model	Li-ion battery	SOC estimation	RMSE	Less than 1%
Wang et al. [17]	Machine learning	Li-ion battery	SOC	Finding estimation error	Less than 1%
Roman et al. [16]	Machine learning pipeline	Li-ion battery	SOH estimation	RMSE	RMSE error is 0.45%

(continued)

Table 2 (continued)

Survey work	Techniques used	Kind of battery	Battery estimation (SOC/SOH)	Metrics found	Accuracy
Lipu et al. [27]	Neural Network	Li-ion battery	SOC estimation	RMSE	<1%
Man Fai et al. [13]	ML	E-vehicle battery	Both SOC and SOH	Throughput	High
Shi et al. [29]	Hybrid (DL + Kalman filter)	Li-ion battery	Both SOC and SOH	Error estimation	Less error rate 1.4%
Li et al. [28]	CNN	Li-ion battery	Estimation of battery capacity	RMSE	<0.02%

learning and deep learning-based algorithms appropriate in SOC and SOH estimation for batteries especially Li-ion battery which are utilized in electric vehicles. Moreover, the techniques used by several researchers, what kind of battery, estimation of state of health and state of charge, metrics used for finding the performance of the model, and finally accuracy percentage were summarized as survey table.

References

1. Caliwag AC, Lim W (2019) Hybrid VARMA and LSTM method for lithium-ion battery state-of-charge and output voltage forecasting in electric motorcycle applications. IEEE Access. https://doi.org/10.1109/ACCESS.2019.2914188
2. Erlangga G, Perwira A, Widyotriatmo A (2018) State of charge and state of health estimation of lithium battery using dual Kalman filter method. In: The 2018 international conference on signals and systems (ICSigSys). IEEE
3. Guo Q, Qu W, Deng H, Zhang X, Li Y, Wang X, Yan X (2017) Estimation of electric vehicle battery state of health based on relative state of health evaluation
4. Lin C-H, Wang C-M, Ho C-Y (2016) Implementation of state-of-charge and state-of-health estimation for lithium-ion batteries. In: IECON 2016—42nd annual conference of the IEEE industrial electronics society, pp 4790–4795. https://doi.org/10.1109/IECON.2016.7793672
5. Fan J et al (2019) IOP Conf Ser J Phys Conf Ser 1187:022014. IOP Publishing. https://doi.org/10.1088/1742-6596/1187/2/0220
6. Noura N, Boulon L, Jemeï S (2020) A review of battery state of health estimation methods: hybrid electric vehicle challenges. World Electric Veh J 11:66. https://doi.org/10.3390/wevj11040066
7. How DNT, Hannan MA, Hossain Lipu MS, Ker PJ (2016) State of charge estimation for lithium-ion batteries using model-based and data-driven methods: a review. IEEE Access 7:136116–136136. https://doi.org/10.1109/ACCESS.2019.2942213
8. Li Y, Liu K, Foley AM, Zülke A, Berecibar M, Nanini-Maury E, Van Mierlo J, Hoster HE (2019) Data-driven health estimation and lifetime prediction of lithium-ion batteries: a review. Renew Sustain Energ Rev 113:109254. https://doi.org/10.1016/j.rser.2019.109254
9. Feng X, Weng C, He X, Han X, Lu L, Ren D, Ouyang M (2019) Online state-of-health estimation for li-ion battery using partial charging segment based on support vector machine. IEEE Trans Veh Technol 1–1. https://doi.org/10.1109/tvt.2019.2927120

10. Feng F, Hu X, Hu L, Hu F, Li Y, Zhang L (2019) Propagation mechanisms and diagnosis of parameter inconsistency within li-ion battery packs. Renew Sustain Energ Rev 112:102–113
11. Aloisio D, Campobello G, Leonardi SG, Sergi F, Brunaccini G, Ferraro M, Antonucci V, Segreto A, Donato N (2020) A machine learning approach for evaluation of battery state of health. In: 24th IMEKO TC4 international symposium 22nd international workshop on ADC and DAC modelling and testing IMEKO TC-4 2020
12. Andre D, Appel C, Soczka-Guth T, Sauer DU (2013) Advanced mathematical methods of SOC and SOH estimation for lithium-ion batteries. J Power Sour 224:20–27. ISSN 0378-7753, https://doi.org/10.1016/j.jpowsour.2012.10.001
13. Ng M, Zhao J, Yan Q, Conduit G, Seh Z (2020) Predicting the state of charge and health of batteries using data-driven machine learning. Nat Mach Intell 2:161–170
14. Zou Y, Hu X, Ma H, Li SE (2015) Combined state of charge and state of health estimation over lithium-ion battery cell cycle lifespan for electric vehicles. J Power Sour 273:793–803. ISSN 0378-7753, https://doi.org/10.1016/j.jpowsour.2014.09.146
15. Kim M, Kim K, Kim J, Yu J, Han S (2018) State of charge estimation for lithium ion battery based on reinforcement learning. IFAC-PapersOnLine 51(28):404–408. https://doi.org/10.1016/j.ifacol.2018.11.736
16. Roman D, Saxena S, Robu V et al (2021) Machine learning pipeline for battery state-of-health estimation. Nat Mach Intell 3:447–456. https://doi.org/10.1038/s42256-021-00312-3
17. Wang W, Brady NW, Liao C, Fahmy YA, Chemali E, West A, Preindl M (2019) High-fidelity state-of-charge estimation of li-ion batteries using machine learning. arXiv:abs/1909.02448
18. Vidal C, Malysz P, Kollmeyer P, Emadi A (2020) Machine learning applied to electrified vehicle battery state of charge and state of health estimation: state-of-the-art. IEEE Access. https://doi.org/10.1109/ACCESS.2020.2980961
19. Anjum M, Asif M, Williams J (2021) Towards an optimal deep neural network for SOC estimation of electric-vehicle lithium-ion battery cells. In: Mporas I, Kourtessis P, Al-Habaibeh A, Asthana A, Vukovic V, Senior J (eds) Energy and sustainable futures. Springer proceedings in energy. Springer, Cham. https://doi.org/10.1007/978-3-030-63916-7_2
20. Venugopal P, Vigneswaran T, Sofana Reka S (2021) State of charge estimation of lithium batteries in electric vehicles using IndRNN. IETE J Res. https://doi.org/10.1080/03772063.2021.1906770
21. Liu C (2018) Battery states monitoring using deep learning and ultrasonic sensors. CS230, Deep learning, Winter 2018. Stanford University, CA
22. Khan N, Ullah FUM, Afnan, Ullah A, Lee MY, Baik SW (2021) Batteries state of health estimation via efficient neural networks with multiple channel charging profiles. IEEE Access 9:7797–7813. https://doi.org/10.1109/ACCESS.2020.3047732
23. Vidal C, Kollmeyer P, Chemali W, Emadi A (2019) Li-ion battery state of charge estimation using long short-term memory recurrent neural network with transfer learning
24. Gao Y et al (2018) Estimation of state-of-charge and state-of-health for lithium-ion degraded battery considering side reactions. J Electrochem Soc 165(16):A4018
25. Zhang L, Zheng M, Du D, Li Y, Fei M, Guo Y, Li K (2020) State-of-charge estimation of lithium-ion battery pack based on improved RBF neural networks. Hindawi Complexity 2020:10. Article ID 8840240, https://doi.org/10.1155/2020/8840240.
26. Wei M, Ye M, Li IB, Wang Q, Xu X (2020) State of charge estimation of lithium-ion batteries using LSTM and NARX neural networks. IEEE Access. https://doi.org/10.1109/ACCESS.2020.3031340
27. Hossain Lipu MS, Hannan MA, Hussain A, Ayob A, Saad MHM, Muttaqi KM (2020) State of charge estimation in lithium-ion batteries: a neural network optimization approach. Electronics 9:1546. https://doi.org/10.3390/electronics9091546
28. Li Y, Li K, Liu X, Zhang L (2020) Fast battery capacity estimation using convolutional neural networks. Trans Inst Meas Control 014233122096642. https://doi.org/10.1177/0142331220966425

29. Shi Y, Ahmad S, Tong Q, Lim TM, Wei Z, Ji D, Zhao J (2021) The optimization of state of charge and state of health estimation for lithium-ions battery using combined deep learning and Kalman filter methods. Int J Energ Res 45(7):11206–11230. https://doi.org/10.1002/er.6601
30. Gao T, Lu W (2021) Machine learning toward advanced energy storage devices and systems. iScience 24(1):101936. https://doi.org/10.1016/j.isci.2020.101936

Robust and Scalable Network Monitoring System Using Apache Spark

Rathinn Nair Melath, Srikar Rachabattuni, Sai Siddharth Cilamkoti, D. Ganesha Srinivas, Harshavardhana Chadaram, and J. Divya Udayan

Abstract Monitoring and restricting students' online activities have a lot of possibilities. A robust state-of-the-art network monitoring system and a proxy server architecture can be built to restrict access, monitor activity, reduce latency and increase throughput. In this paper, we have proposed a monitoring system and proxy servers that are built with scalable technologies. A prototype is built with four proxy servers and a central system that includes a Spark cluster and a Kafka cluster hosted on a single node with three Kafka brokers and one Topic with three partitions. The proposed architecture is very user-friendly, with both the API and the Dashboard hosted on the same server using application dispatching.

Keywords Apache Kafka · Apache Spark · K-mean clustering · NodeJs · Network monitoring · Proxy server

R. Nair Melath · S. Rachabattuni · S. Siddharth Cilamkoti · D. Ganesha Srinivas · H. Chadaram · J. Divya Udayan (✉)
Department of Computer Science and Engineering, Amrita Vishwa Vidyapeetham, Amritapuri, India
e-mail: divyaudayanj@am.amrita.edu

R. Nair Melath
e-mail: rathinnn@am.students.amrita.edu

S. Rachabattuni
e-mail: rachabattunisrikar@am.students.amrita.edu

S. Siddharth Cilamkoti
e-mail: saisiddharthc@am.students.amrita.edu

D. Ganesha Srinivas
e-mail: ganeshasrinivasd@am.students.amrita.edu

H. Chadaram
e-mail: charshavardhana@am.students.amrita.edu

V. K. Gunjan and J. M. Zurada (eds.), *Proceedings of 3rd International Conference on Recent Trends in Machine Learning, IoT, Smart Cities and Applications*, Lecture Notes in Networks and Systems 540, https://doi.org/10.1007/978-981-19-6088-8_32

1 Introduction

On March 24, India imposed a nationwide lockdown that included closing of universities and institutions in order to prevent the spread of the corona virus. COVID-19 is quickly spreading, causing a large number of people to be unable to attend schools and institutions. For now, it seems that the future is going to be adhered with online education, and there needs to be a proper system that can handle large number of students while keeping their online activity in check. The current systems are inefficient as they are centralized and can't scale well to handle the increase in number of students. The goal of this project is to improve the quality of education during the pandemic and in the future. The pandemic and the lockdown have had an impact on people's mental health all throughout the world. Many students are stressed out and depressed. Furthermore, not every student has equal access to digital tools. Not only that but also students with access to all of these devices are not concentrating on their education due to additional distractions such as social media: Youtube, Instagram, Facebook, and various computer games, among others. The sheer number of students learning online right now necessitates the usage of an incredibly efficient system. This is something that is being worked on as part of this project. This report investigates the options for monitoring and controlling students' online activity. It looks at a Robust Network monitoring system and proxy server architecture built to restrict access, monitor activity, reduce latency and increase throughput.

2 Literature Review

2.1 RAL—Reinforcement Active Learning for Network Traffic Monitoring and Analysis

Wassermann et al. [1] this paper addresses an efficient way of handling the data in a network arriving as a data stream and it also aids to comprehend the present condition of the application with limited data, and an example was also given on Network Attack Detection. This Paper is novel as it looks at Reinforcement Active Learning which is a stream-based, understanding the current state of the application being monitored using an active learning approach. This idea in the paper could be used as future scope for our project as well, because we have a similar structure in our architecture which is stream-based. This paper also helps us a way to solve the issue with labeling network monitoring samples for supervised learning using a stream-based active learning approach. They also have a benchmark test that pits them against a cutting-edge active learning algorithm as well as random sampling, on a public cyber-security dataset which is the MAWI dataset. They have also focused on mainly 2 types of attacks, which are net scan intrusions and flooding. The only limitation is that it would be difficult to apply this concept to other fields of network monitoring given that such a conversion was not mentioned. It is the research gap in this idea and could be included in the future scope.

2.2 *Design and Development of Centralized Squid Proxy Management System*

Mishra et al. [2] this paper addresses the complex and expensive proxy management solutions by implementing a centralized proxy management system. The project has proxy clusters of different sizes and they require a centralized management and monitoring system which is inspired from the paper. Here they have also described the architecture of the centralized squid proxy management solution. Centralized Squid Proxy Management System (CSPMS) was stated as the proposed solution. This solution provides a user-friendly Web Interface along with an automated proxy installer that can be used to manage the remote servers from a location which includes various technologies like cloning of groups, group-based management. The security risks that come with such architecture have not been discussed in the paper and should be looked at.

2.3 *An Efficient Network Monitoring and Management System*

Khan et al. [3] This paper describes the ideal characteristics of the network monitoring system, the automatic network monitoring system, RT ticketing system's importance, and its requirements for network management, as well as network monitoring by Nagios and its configuration. It also discusses the importance of efficient and fully automated network monitoring in large organizations such as universities, businesses and other industries for which manual network monitoring is difficult. Because large organizations have complex network topologies, manual network monitoring is a waste of time when it comes to identifying problem areas. This requires a synopsis of the network monitoring and management system that monitors and manages your network for you automatically. How Nagios sends an email or a text message to the network administrator when a node or service in the network has a problem. For managing the workflow, services are granted to manage hardware/software problems or clients, organizations typically require a management system. Furthermore, this paper has also stated an efficient and automated network monitoring system that alerts network administrators quickly when a problem occurs. Nagios is set up to create and monitor the entire topology of the network, as well as send notifications in the event of a state change anywhere on the network.

2.4 An Educational HTTP Proxy Server

Sysel and Doležal [4] the issue addressed in this paper is of an HTTP proxy server which is a multi-threaded HTTP proxy server that contains an embedded WWW server used for the GUI. This paper demonstrated that the developed educational HTTP proxy server is very effective software that can be used to give faster access to the resources, perform access control, log traffic and be effective in increasing its users' anonymity. The topics covered in this paper are as follows: how a proxy server enables the processing of more stringent requirements, HTTP proxy servers that are specialized, detecting and blocking malicious content, application of proxy server.

2.5 Network Monitoring System for High Speed Network Traffic

Kurt et al. [5] the problem which is addressed in this paper is the display of primary inspection of any platform used for network monitoring for advanced networks. This tool is responsible for doing complicated real-time analyses like app usage behavior and architecture planning. In the future, we can analyze the security and check the application's behavior. Furthermore we can keep an eye on the website if we decide to launch it in the future. A network monitoring system for the mobile proxy server was designed and implemented. It is common practice to divert requests to different proxy servers when visiting web servers in different countries in order to improve performance. In general, users do not know which proxy server provides the fastest network stats for any advanced level solution for a specific website.

3 Proposed Methodology

The architecture proposed is shown in Fig. 1. The Architecture can be divided into two parts each of which will be discussed in detail. Also a strategy for placing proxy servers will be discussed under Clustering.

3.1 Proxy Server System Design

Proxy servers are server applications that work as an intermediary separating end user clients and the destination server.

Proxy servers provide diverse functionality, security and privacy depending upon the use case. Proxy servers can be broadly classified into two types: Forward proxy server acts as a mediator between the server and the client, it is responsible for

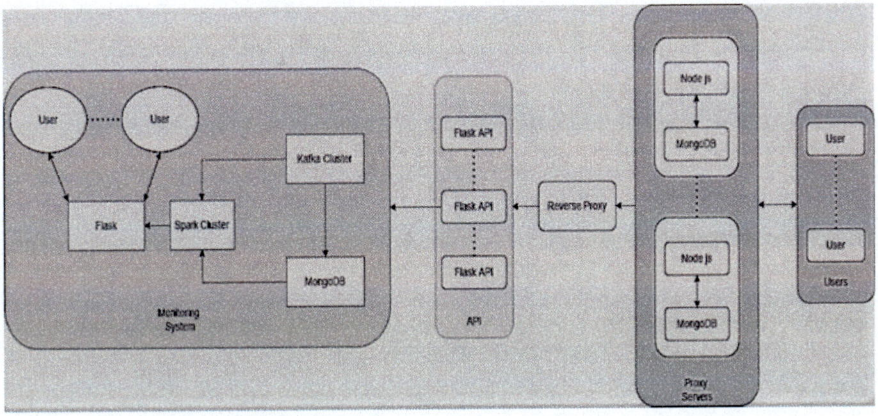

Fig. 1 Proposed system architecture

handling the outgoing requests also for acting on them after which it sends the request to the server. These types of proxy servers can also be used for caching and are helpful in improving the load time of web pages.

The most common type of proxy servers are forward proxies. Some examples of forward proxies are Virtual private networks and web content filters. The Forward proxy server implemented in this project can restrict web access for the client based on the requested hostname (acts as a web content filter). This server is implemented using socket programming in Node.js and the HTTP Connect Protocol. The HTTP Connect method starts a two-way communication with the requested resource and the Proxy Server then proxies the TCP Stream to and from the client. Reverse proxy server are placed between network and other servers in cases where a huge website might have multiple servers that collectively serve requests. Their main purpose is to improve the performance of websites and acts as a load balancer and is also used for caching. Load Balancing is the process of distributing the incoming requests between different servers which helps optimize response time. This project takes the help of Nginx as a reverse proxy.

The idea is that we use make use of both forward and reverse proxy servers to monitor and restrict activity of a set of users. The proxy server we developed is written using Nodejs. As shown in Fig. 2 each proxy server forms a cluster of users and serves them. The data is then sent to the central server. Node.js is an open-source JavaScript runtime environment which runs on the V8 engine and is used to execute JavaScript code outside a web browser. Unlike its counterpart php, Node.js is an asynchronous programming language. Node.js uses a single thread to attend all the client's requests. Figure 3 shows a basic workflow of the system.

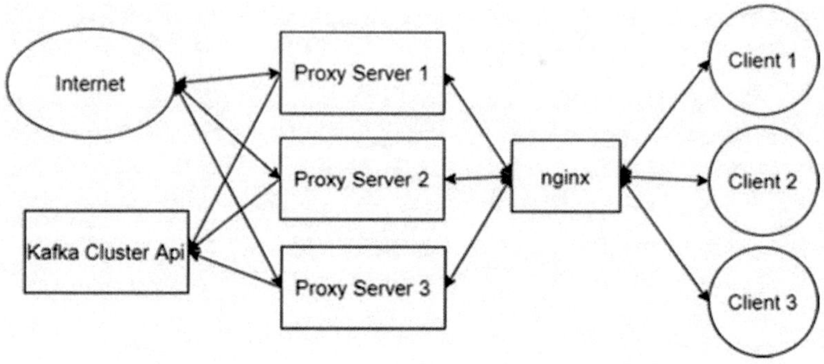

Fig. 2 Proxy server system management

3.2 Central Monitoring System

Data sent by the proxy servers can be difficult to process especially when the number of proxy servers increases. So, we need to develop the application using frameworks that are easily scalable. Spark Structured Streaming is both a fault-tolerant and scalable stream processing API which is built on top of the Spark SQL engine. Spark gives the freedom to express any streaming computations in the same way it is generally expressed on static data batch computation. The Spark SQL engine handles tasks like executing it both incrementally and continuously and also updating an aggregated result based on the streaming data that arrives continuously. Apache Spark is based on DAG and RDD's. Resilient Distributed Datasets (RDD) are data item collections which are partitioned and are stored on the workers' nodes of the spark cluster in-memory. Directed Acyclic Graph (DAG) can be thought as a sequence of computations which are performed on data where each edge is a transformation of the data and each node is an RDD partition. Apache Spark follows master/slave architecture by having a single master and any number of workers.

Structured Streaming can read from a few different types of input sources. Out of which one of the most reliable ones is Apache Kafka. We use Apache Kafka for processing stream data as it is a distributed data streaming platform in real-time. The Streaming data is stated as the data which is constantly created by large number of data sources that are normally received concurrently. Any streaming system would need to be able to receive this constant inflow of data, and proceed to process the received data in an efficient manner. Kafka is mainly used to create pipelines that can handle real-time streaming data. Combining storage, stream processing and messaging and allows storage and analysis of both real-time and historical data. A Broker is a Kafka server that runs in a Kafka Cluster. A Kafka cluster is generally made of one or more Kafka brokers.

The processes which push some records onto Kafka topics are known as producers. The processes which pull records from a Kafka topic are known as consumers. A Topic is an ordered log of records to which records are stored and published. Kafka

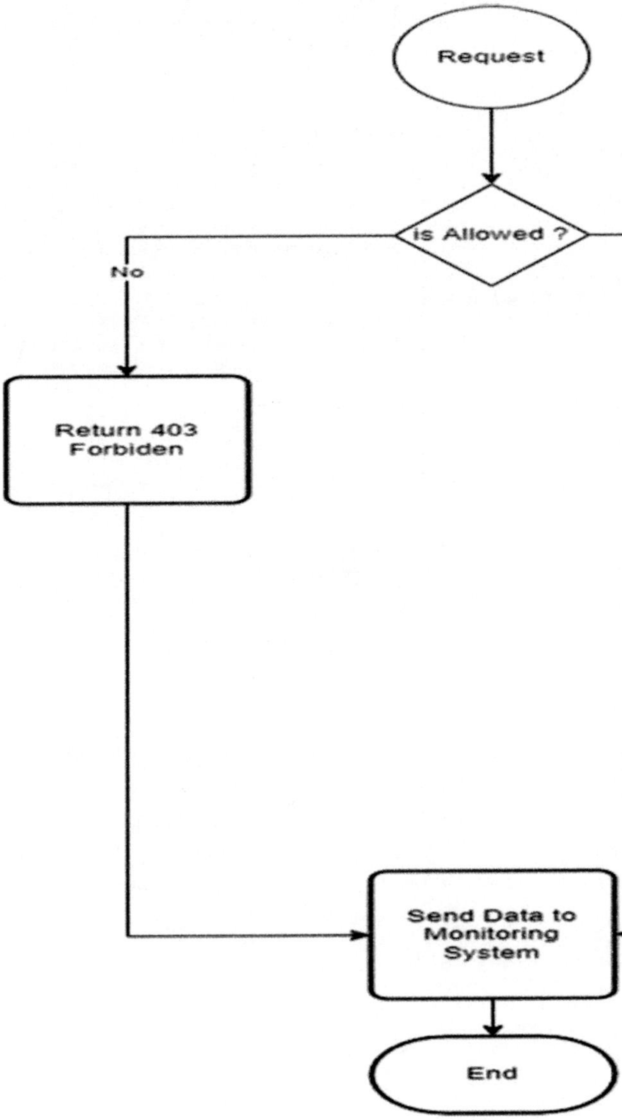

Fig. 3 Workflow of the proposed system

topics are split up into partitions that contain records. A unique offset is used to identify each record in a partition. Kafka allows a topic to have multiple partition logs, allowing for multiple consumers to read from a topic parallelly. Partitions allow topics to have the capability to be parallelized as the data is split into a topic distributed across multiple brokers.

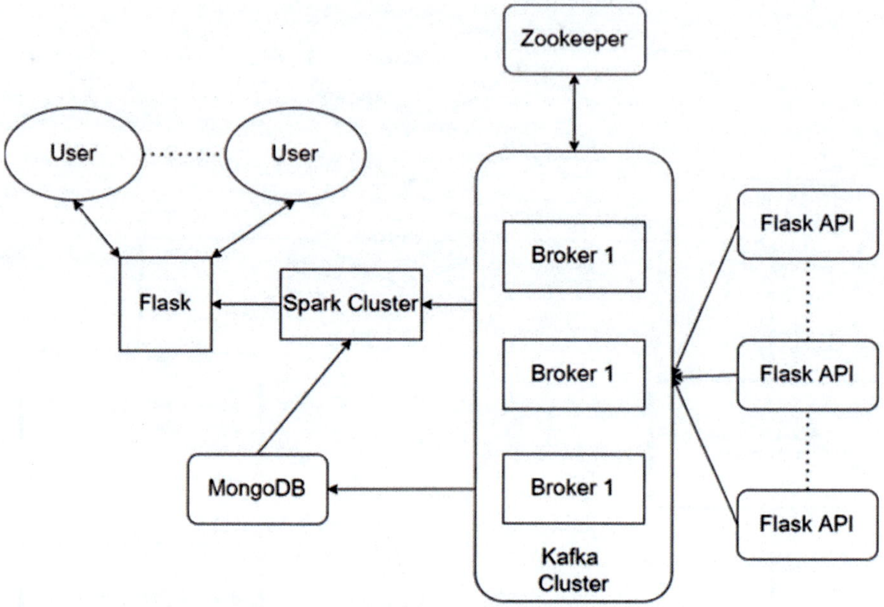

Fig. 4 Central monitoring system architecture

Kafka uses a binary protocol over TCP. All messages are size delimited and the protocol defines all APIs as request response message pairs. For serving the app we use Flask, Flask is a lightweight WSGI web application framework. An API was implemented to update the Proxy server Database centrally. The prototype was built on a single node spark and Kafka cluster with three brokers and one topic with three partitions with both the API and the Dashboard hosted on the same server using application dispatching. Figure 4 shows the architecture of the Central monitoring system.

3.3 Clustering

The word 'latency' means the delay. Network latency usually means the time taken for the packets in a network to reach its destination. Latency is usually measured as the time taken for the packets to reach its destination and return. It is also known as round trip delay. So, the round trip delay will have an important effect on the performance of the network. Proxy servers need to be clustered in a way that latency for the client is minimized and Application server throughput is maximized. The architecture achieves this by using k-means an unsupervised machine learning.

K-means clustering can be defined as a branch of partitioning methods in the clustering techniques. When given a set of n distant entities, the k-means clustering algorithm will try to partition the objects into k different clusters such that the inter-cluster similarity is lower and intra-cluster similarity is higher. k is an user specified parameter which is the number of clusters. The entities are defined with numeric attributes and any distance metric can be used to demarcate the clusters.

We have the cost function as:

$$J = \sum_{J=1}^{C} \sum_{k=1}^{N} X_k - \mu_j^2 P\omega_j | X_k$$

where c is the number of clusters. N is the number of data points, μ_j is the j'th centroid and X_K is the k'th data point. $P(\omega_j|X_k)$ is defined as

$$P\omega_j | X_k = \begin{cases} 1 \text{ arg min } X_k - \mu_j^2 \\ 0 \text{ else} \end{cases}$$

After solving for optimal value of μ_j we get

$$\mu_j = \frac{\sum_{x \in J} X}{|J|}$$

Figure 5 shows a flow diagram of the K-Means algorithm.

The System Architecture can be generalized as the following. Proxy Servers are clustered based on the K-means clustering and use Nginx as a load balancer. Each proxy server sends their log data to a central Kafka cluster. Spark Structured Streaming then reads from the Kafka Cluster and creates streaming data frames that are distributed. Computations and Transformation are then applied on the streaming data frames and written to an output sink. Flask then uses this data to serve the dashboard with the help of Dash and plotly. There is also an API implemented for centrally updating the proxy server database.

4 Results

Following Figs. 6, 7, 8, 9 and 10 are some of the results from the central monitoring system and a proxy server instance.

Figure 6 shows a Pie chart describing the share of HTTP method used. There are three types of HTTP method used in the network:

1. GET Method: Used to request resources
2. POST Method: Used to create/update a resource
3. CONNECT Method: Used to open a tunnel.

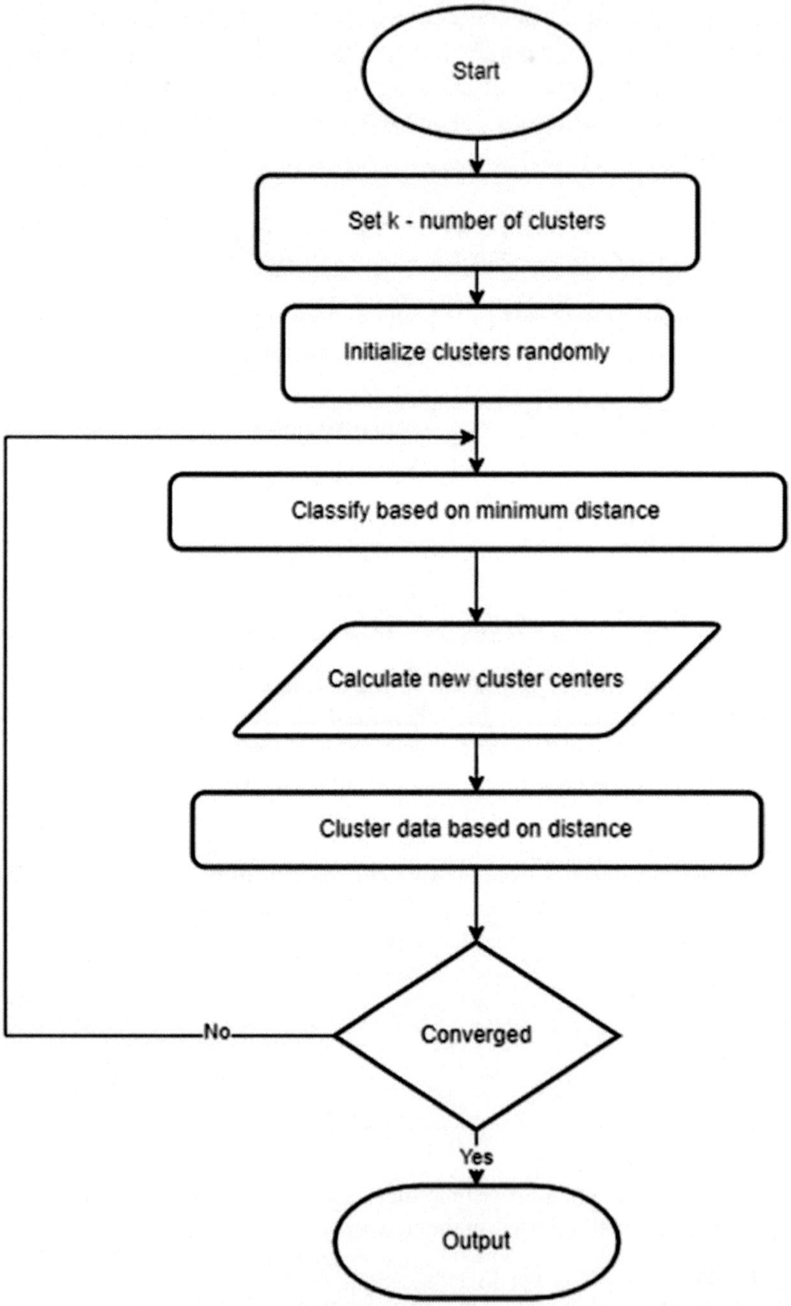

Fig. 5 K-means algorithm

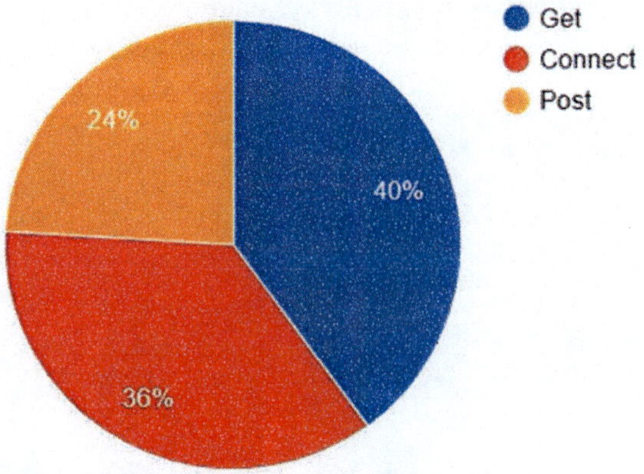

Fig. 6 HTTP method Pie chart

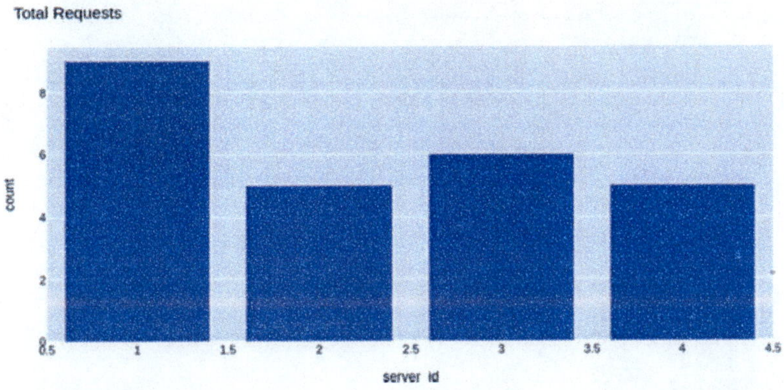

Fig. 7 Total requests bar graph

Figure 7 is a bar graph which shows the total number of requests for each proxy server. The X axis represents the server id and the Y axis represents the total count of requests received for that server.

Figure 8 is a bar graph showing the most requested websites in the network. The X axis represents the hostname (URL) and the Y axis represents the count of requests received for that hostname.

Figure 9 is a bar graph that shows the total number of blocked requests in the network. The X axis represents the server id and the Y axis represents the count of requests that were blocked for that server. Figure 10 shows a Pie chart describing the share of allowed and blocked requests.

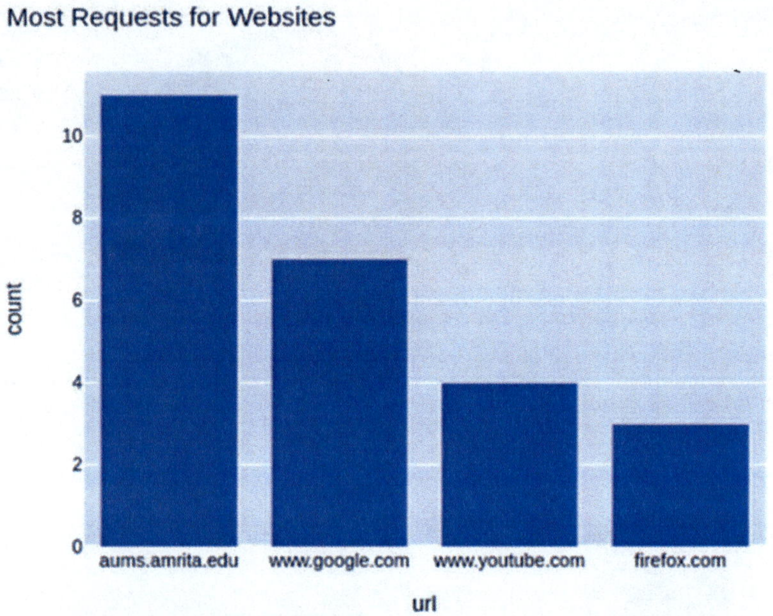

Fig. 8 Most popular websites in network

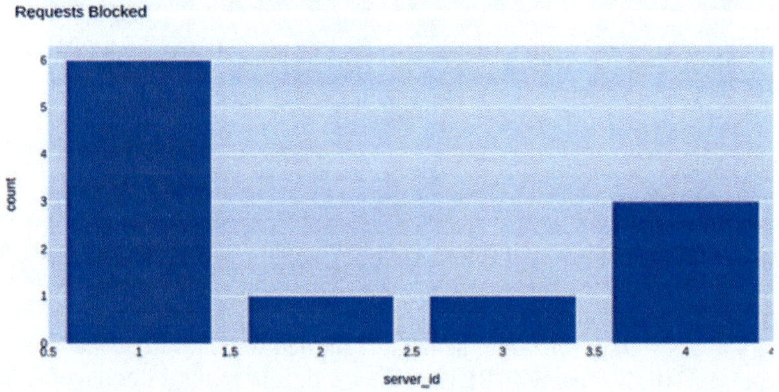

Fig. 9 Number of blocked requests for each server

5 Conclusion

The Project that we implemented can be used to monitor and restrict the network usage of the students. The Sheer number of students studying online right now demands an extremely efficient system to be of practical use. The Network monitoring system and proxy server architecture built is Robust, reduces latency and increases throughput. The prototype is built using 4 proxy servers and a central system with

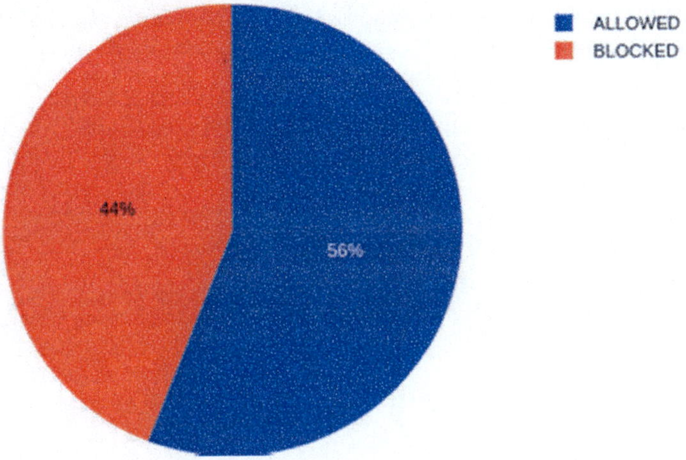

Fig. 10 Pie graph for requests blocked and allowed

spark cluster and Kafka cluster hosted on a single node with 3 Kafka brokers and one topic with three partitions with both the API and the Dashboard hosted on the same server using application dispatching.

References

1. Wassermann S, Cuvelier T, Casas P (2020) Ral-reinforcement active learning for network traffic monitoring and analysis. In: ACM SIGCOMM 2020 posters, demos, and student research competition
2. Mishra DD, Dhakwal V, Pathan S, Murthy C (2020) Design and development of centralized squid proxy management system. In: 2020 IEEE international conference on electronics, computing and communication technologies (CONECCT). IEEE, pp 1–6
3. Khan R, Khan SU, Zaheer R, Babar MI (2013) An efficient network monitoring and management system. Int J Inform Electron Eng 3(1):122–126
4. Sysel M, Doležal O (2014) An educational http proxy server. Procedia Eng 69:128–132
5. Kurt B, Zeydan E, Yabas U, Karatepe IA, Kurt GK, Cemgil AT (2016) A network monitoring system for high speed network traffic. In: 2016 13th annual IEEE international conference on sensing, communication, and Networking (SECON). IEEE, pp 1–3
6. Swetha K, Sathyadevan S, Bilna P (2015) Network data analysis using spark. In: Software engineering in intelligent systems: proceedings of the 4th computer science on-line conference 2015 (CSOC2015), vol 03, pp 253–259
7. Shyam R, Bharathi Ganesh HB, Sachin Kumar S, Poornachandran P, Soman KP (2015) Apache spark a big data analytics platform for smart grid. Procedia Technol 21:171–178
8. Menon VK et al (2019) A spark™ based client for synchrophasor data stream processing
9. Menon VK, Nair YC, Neethu PV, Soman KP (2016) Real time vehicular data analytics utilising bigdata platforms and cost effective ECU networks. Indian J Sci Technol 9:30

10. Kamal KUA, Sathyadevan S (2017) Intrusion detection system using big data framework. ARPN J Eng Appl Sci 12:3909–3913
11. Sanjanasri Jp, Menon VK, Soman Kp, Wolk K (2021) Mining bilingual word pairs from comparable corpus using apache spark framework. In: Proceedings of the 14th workshop on building and using comparable corpora (BUCC 2021), pp 2–7

Managing Data Protection and Privacy on Cloud

Satyavathi Divadari, J. Surya Prasad, and Prasad Honnavalli

Abstract When pandemic rose in 2020, people were fighting against COVID-19 virus and organizations had accelerated their digitization and cloud adoption rapidly (De et al. in Int J Inf Manag 55:102171, 2020 [1]) to meet the online based business during the lockdown. This chaos helped fraudsters and attackers taking advantage of the momentary lack of security controls and oversight. Federal Investigation Bureau (FBI) Internet Crime Compliant Center (IC3) 2020 reported highest number of complaints in 2020 (791 k +) compared to prior five years (298 k + in 2016), with peak losses reported ($4.2 Billion in 2020 compared to $1.5 Billion in 2016) (Internet Crime Complaint Center in Internet crime report. Federal Bureau of Investigation, Washington, D.C., 2020 [2]). Majority of these incidents were connected to financial fraud, identity fraud, and phishing for personally identifiable information (PII). Considering the severity and impact of personal data exposure over cloud and hybrid environment, this paper provides a brief overview of prior research and discuss technical solutions to protect data across heterogeneous environments and ensure privacy regulations.

Keywords Multi-cloud · Hybrid-cloud · Cyber security · Data privacy · Data breaches · Encryption · Data governance · Data discovery · Tokenization · Blockchain · Format preserving encryption · Regulations · COVID-19 · Pandemic

S. Divadari (✉)
Micro Focus India Pvt Ltd, PES University, Bangalore, India
e-mail: satyavathi.d@gmail.com

J. Surya Prasad · P. Honnavalli
PES University, Bangalore, India
e-mail: surya@pes.edu

P. Honnavalli
e-mail: prasadhhb@pes.edu

V. K. Gunjan and J. M. Zurada (eds.), *Proceedings of 3rd International Conference on Recent Trends in Machine Learning, IoT, Smart Cities and Applications*, Lecture Notes in Networks and Systems 540, https://doi.org/10.1007/978-981-19-6088-8_33

1 Introduction

With the increased cloud adoption by organizations, data proliferation is greater than ever with the low-cost availability of advanced compute and storage and always-on high-speed networks. Given such a large accumulation of data, sensitive and confidential information is getting added to the pile.

Protection of personally identifiable information (PII) and sensitive personal information (SPI) against leakage or exposure are essential for people who own the data as well as the organizations that are its custodians.

1.1 Impact of Data Leakage on Individuals

With the digitally scalable, always-on, and reliable cloud-based infrastructure, data is heavily accumulated on single place, hackers and other adversaries are breaching it, collecting the data, and trading it. The data includes patient health information, financial information, and other identity linked data on dark web [3, 4]. Financial institutions and insurance organizations use such data procured from authorized agents to decide on insurance policy issuance or sanctioning personal loans. While such claims sound like a science fiction movie to some, there are numerous proof points that confirm it is true. Figure 1 indicates an example of data monetization on dark web [4].

1.2 Impact of Data Leakage on Organizations

With the continued threat of hackers and data exposure, government and regulatory authorities across the world strengthened the data protection laws, that include European Global Data Protection Regulations (GDPR), California Consumer Privacy Act (CCPA), Personal Data Protection Bill (PCPB), and The Health Insurance Portability and Accountability Act (HIPAA). FTC enforced $5 Billion penalty and extended privacy restrictions on Facebook [5] in a historic resolution with a highest penalty as of 2019, and substantially strong obligations on organizations to enhance accountability of protecting data security and preserving end user privacy.

This case and similar instances demonstrate that enterprises are accountable to maintain strong security protection mechanisms and data privacy enforcement to maintain compliance with the regulatory requirements and stay abreast of continued cyber-attacks.

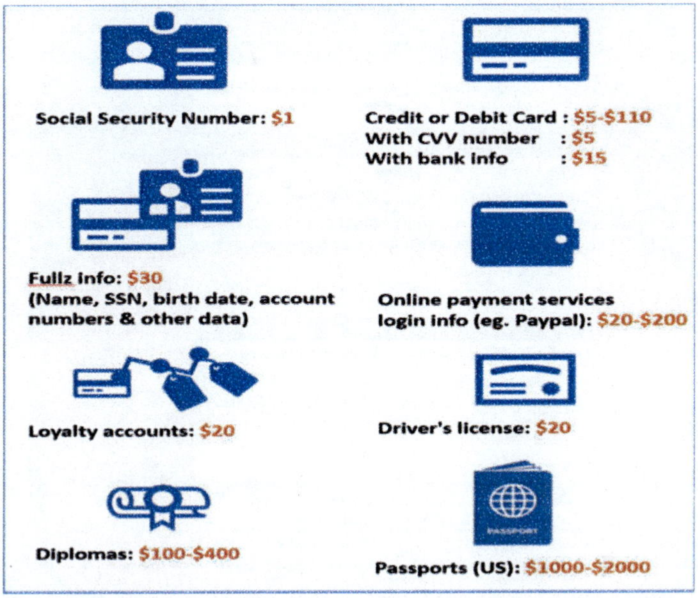

Fig. 1 Cost of leaked data on dark web

Data Discovery and Protection Methods

Organizations implement numerous safeguards and controls to discover and protect sensitive data against various types of cyber threats on the expanded threat landscape that include on-premises, cloud, hybrid-cloud, or multi-cloud.

1.3 Data Discovery

1.3.1 Identification of Sensitive Information

Deployment of privacy and security controls to comply with legislation begins with the identification of sensitive data that is being collected and processed across the enterprise.

In a simple excel or a database like the one in Fig. 2, it is easy to classify different fields as personal information (PII) or as sensitive personal information based on the standard definition of PII.

With the heaps of data being collected across cloud, and on-premises data centers, customer, and third-party storage, identifying and classifying critical and sensitive information is a complex challenge to handle. Unless one knows the data type/content, it is difficult to protect.

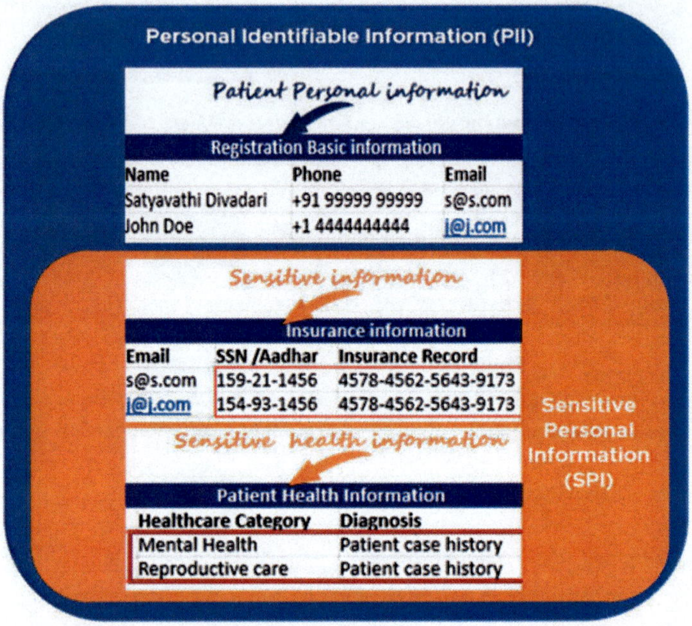

Fig. 2 Example of data classification

The above complexities can be handled by a centralized data governance solution that supports three key requirements, as shown in Fig. 3.

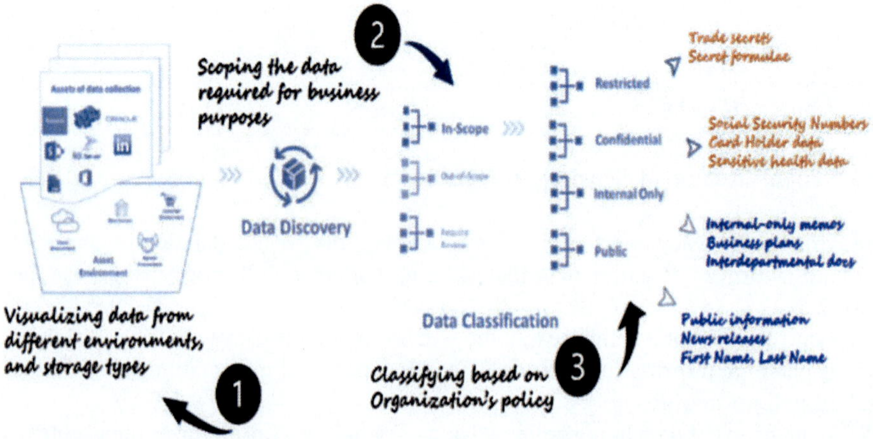

Fig. 3 Data discovery across different environments

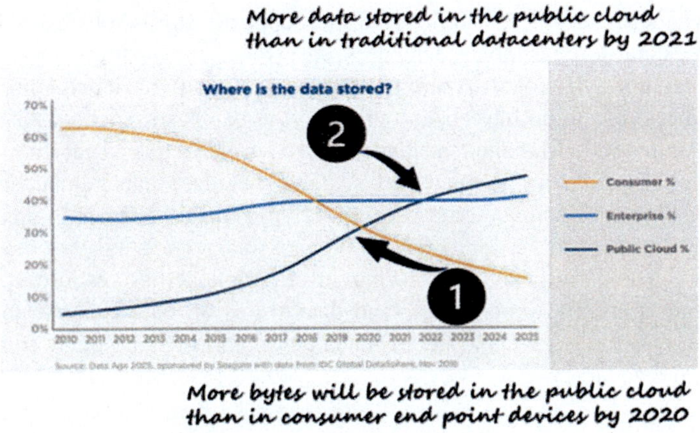

Fig. 4 Data storage trends on public cloud vs others

1. Visualize data elements on several heterogeneous storages located on-premises, cloud, and external environments to know and identify the critical data assets from others. Centralized data asset awareness is the key to protection and oversight.
2. Analyzing and scoping in the information relevant to current business requirements help in reducing the cost of storage by eliminating unnecessary data and archiving data required for long-term needs.
3. Classification of data assets based on the organization's policy helps in segregating the data that require higher protection than others.

1.3.2 Data Discovery on Popular Cloud Platforms

IDC Research in 2018 [6] predicted that customer workloads will increasingly migrate to cloud platforms. They predict that more data will be stored in the public cloud than in consumer endpoint devices by 2020, and storage on public cloud exceeds the storage on-premises by 2021 as shown in Fig. 4.

The above trend necessitates the need to study the mechanisms of how data discovery and classification can be carried out on popular public cloud platforms.

1.3.3 Data Discovery and Classification in the Cloud (AWS, Azure, GCP)

Amazon Web Services (AWS) Macie [7] discovers sensitive data among unstructured data that is stored in Amazon Simple Storage Service (S3) [8] as per definitions by data privacy regulations such as GDPR, PCI-DSS, and HIPAA. Macie can read several types of unstructured data in S3 buckets that include.txt,.json,.xml, Avro,.csv.tsv,.doc,.docx,.xls,.xlsx,.pdf,.tar,.zip,.gzip, and Parquet.

Amazon SageMaker [9], AWS Glue [10], and supporting tools help in data labeling the structured data stored in tables, databases, data stores, etc., on the AWS infrastructure platform. In the Microsoft Azure platform [11], sensitivity labels, information types, and discovery logic have been built into the databases such as Azure SQL Database, Azure SQL Managed Instance, and Azure Synapse Analytics. Google Data Catalog [12] offers a fully managed, scalable metadata management service to discover, classify, and manage data with underlying machine learning-based Data Analytics.

Scanning, discovering, and classifying information across resources such as consumer endpoints, file repositories, and data stores on data centers, and cloud storage is an enormous task and necessitates automation technologies to improve efficiencies and accuracy.

1.3.4 Data Discovery: Heterogeneous Environments

There are many third-party tools that help in data discovery and classification across heterogeneous environments mentioned above. Here is an example of a tool, File Analysis Suite from Micro Focus [13], which can scan across cloud, on-premises, public repositories, and tag the resources with data classification labels.

The mentioned technology enables connections to different cloud as well as on-premises environments such as Google Drive, Microsoft share point repositories, Local file servers, and other content repositories. It allows the user to define different data classification tags for example pay slip, credit card data, driving license, etc.

Once tags are created, data stored in different repositories are tagged and grouped under the respective classification groups, such as financial data and contact data. They are given ratings like partially sensitive, sensitive, highly sensitive, etc., to make it useful for the next level of the decision on protection controls. Figure 5 demonstrates an indication of tags.

2 Data Protection

While encryption is the first technique that instantly comes into mind when data protection controls are planned, it is just one of the methods. To maintain the strongest compliances like GDPR, comprehensive protection of data is needed. This paper recommends a five-step approach to comprehensively address data security and privacy concerns.

Step 1: Assessing the risk of the data

As per Federal Information Processing Standards (Fips) Publication 199 [14], Data classification levels are expected to be defined based on the potential data breach impact on the business in the event of security breach to Confidentiality, Integrity, or Availability. Hence, the risk exposure level is directly proportional to the sensitivity

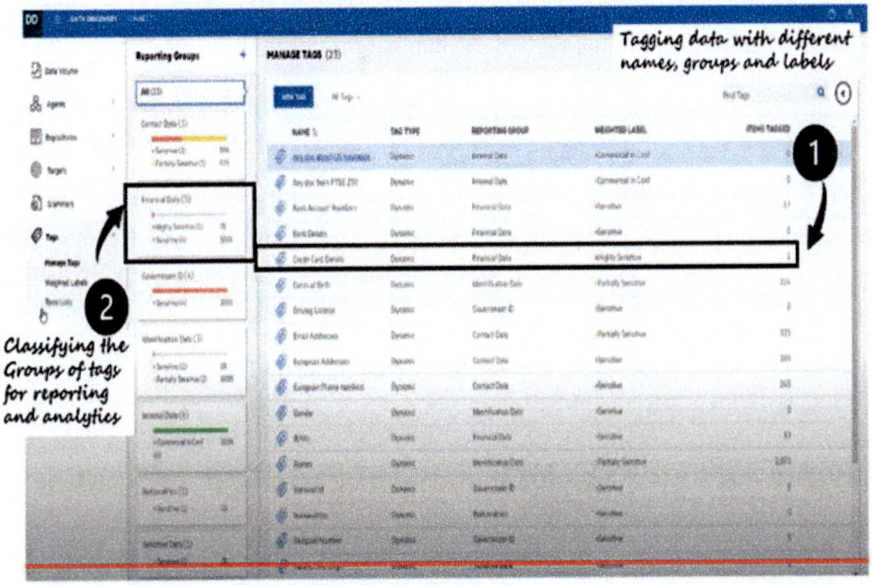

Fig. 5 Example of data discovery across hybrid-cloud

level of the data. Highly sensitive data means higher risk exposure to the organization, in the case of leakage or cyber-attack, and that requires equally stronger controls to protect and preserve the data.

Cost and efforts could be reduced considerably by performing risk assessment before control implementation.

Step 2: Enforcing Zero Trust principles to access data

With the extended use of remote working employees, network security devices no more act as a perimeter. Zero trust adoption calls for the least privilege principle, and access should be based on the risk level of the asset. Identity and Access Management acts as a new perimeter or a gateway to deliver the right access to the right assets. Additional authentication levels with fine-grained access controls are required to exercise privileged changes to higher risk or sensitive data.

Step 3: Encryption

High-risk data [15] is the prime target for attackers, as it gets them the financial advantage or recognition. Data encryption makes their efforts ineffective because they could not retrieve the data itself, but a concealed version. We will discuss different techniques of encryption in the next section while the data is at rest, on the move, or while it is in use.

Step 4: Backup and Archives

Important data assets [16] required regular backups to reduce the risk of accidental loss or disruption due to human error or technical misconfiguration or protect from ransomware attacks. Protecting archival data as per contractual or regulatory requirements by storing the data securely as per the classification is a complex problem to be handled.

Step 5: Secure Data deletion:

Data is deemed as securely deleted [17] from the system only when the data is made completely inaccessible or unusable to anyone that, including adversary or cyber attackers. Deletion of data securely is crucial to ensure that sensitive data will not land in hands of adversaries or hackers.

2.1 Encryption Methods for Cloud

Increased use of cloud and concerns around data privacy called for a requirement that was once a research question [18] posed by Rivest, Adleman, and Dertouzos in 1978, that is " Can computations be conducted on encrypted data, the need to decrypt it, while conserving the data integrity?"

After that intriguing question, several encryption methods were developed to meet the requirements of conducting analytics or permutations on the encrypted data. Format Preserving Encryption and Tokenization are two such approaches.

2.1.1 Format Preserving Encryption

Conventional encryption methods AES-CBC [19] impacts data structures, schemas, and applications design as the encrypted text format in length and structure changes completely. Whereas Format Preserving Encryption (FPE) preserves the data structure and database schemas as is. Thus, making the applications perform operations on the encrypted format at all times and improving the performance of the application processing. Figure 6 indicates comparison of traditional encryption vs FPE.

FPE uses the NIST-standard FF1(radix) mode of the Advanced Encryption Standard (AES) approved algorithm [20] that uses Format Preserving Feistel-based encryption.

Other FPE methods FF2 [21] and FF3[22] have certain concerns related to encryption methods. FF2 was never got approved. FF3 method was exploited by researchers with a cryptanalytic attack (2017) making the encryption unacceptable for general-purpose usage because the anticipated 128-bit security level was not met. In response to the attack, NIST reverted FF3 to FF1 (FF3-1) in a revised version of FPE in the 2019 release [23].

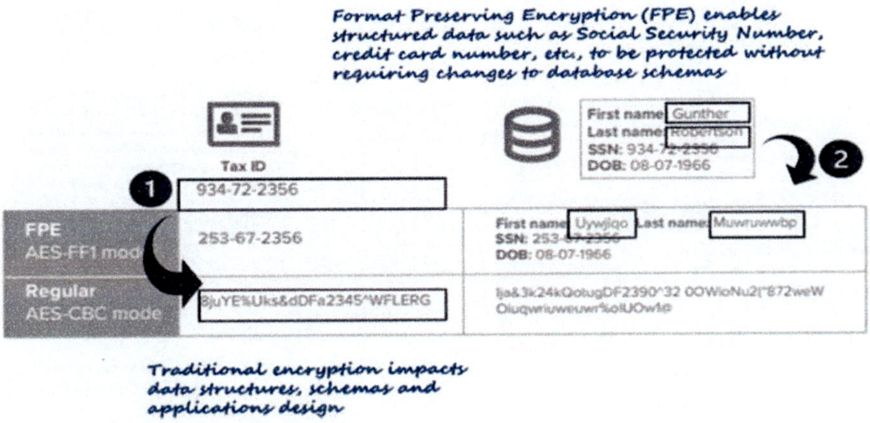

Fig. 6 Traditional encryption vs FPE

1. Data was split into 2 portions.
2. A keyed round function, with modular addition, applies on one data portion and it changes the other data portion.
3. The actions of these two portions were switched in the following iteration.

Ten iterations must be performed to complete the encryption.
Similar steps are applied in decryption.

1. Data is split into 2 parts. But the order of the round indices is reversed.
2. A keyed round function, with modular subtraction, is applied on one data portion and it changes the other data portion.
3. The actions of these two portions were switched in the following iteration.

Ten iterations must be performed to complete the Decryption.

The structure was explained in Fig. 7 for both encryption and decryption. Four iterations are listed in Fig. 7, but ten rounds are specified for FF1.

2.1.2 Business Uses of FF1 Mode of FPE

Sensitive data protection is the main usage of FPE. This includes protection of Payment Card Information, Bank Account details, Social Security Number, etc., encrypting the personally identifiable information (PII) such as these will make the information less vulnerable to focused attacks.

Field-level encryption of personally sensitive data eases the transaction by anonymizing the data in retail, health care, and financial applications that help in conducting analytics and transactions on the data without decrypting the data. As FPE preserves the original data format and length, systems that recognize a certain format treat that data as the original and carry on with the transactions on the encrypted data.

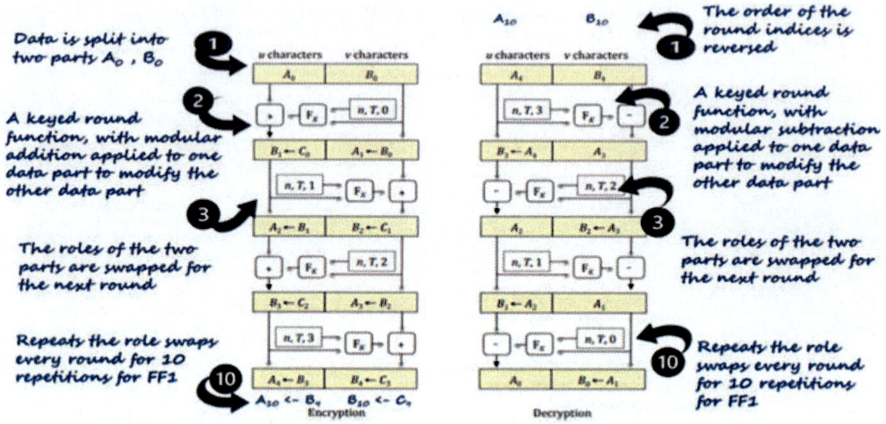

Fig. 7 Feistel structure for FPE FF1

Data remains encrypted unless there is a request from the data owner or other exceptional reasons. For example, a law enforcing authority is investigating a fraudulent case and is interested to know the financial details and sensitive information such as Aadhar, in that case, the ciphertext will be decrypted into plaintext by the authorized people after seeking the right permissions.

Multiple vendors offer FPE in their products and services, including Micro Focus Voltage SecureData, HashiCorp, Comforte, and Futurex.

2.1.3 Tokenization

To make the business transaction of physical assets easier, usage of digital substitutes of asset representations such as tokens or smart contracts are increasing rapidly [24].

Nowadays, gold, diamonds, real estate properties, and many more precious assets are getting tokenized. Considering the value of such transactions, addressing financial fraud, tax avoidance, and investor protection are the major concerns of the governments. To meet the regulatory compliance requirements and to protect the sensitive data, tokenization needs to be securely handled.

As per Payment Card Industry Data Security Standard (PCI-DSS) guidelines [25], a method by which a substitute value, called a "token," swaps the primary account number (PAN). The process of tokenization may or may not revert a token back to the original PAN. The token's security and reliability rely primarily on the unfeasibility of establishing the original PAN with the knowledge of just substitute value (i.e., token)."

To meet regulatory compliance requirements and to protect sensitive data, tokens are used as a substitute for original data. For example, a token represents a credit card number, and there is no way to associate the token to the card itself. However, it fulfills the purpose of completing the transaction whatever the worth the token was issued for.

Such solutions alleviate the responsibility of PCI compliance and privacy-related concerns, but they have other challenges. With the raised interest in the use of tokenization to meet compliance, different methods arrived to fulfill different business use cases.

Reversible tokens: Tokens that can be converted back to the original plain text. They are either cryptographically created ciphertexts or data maps using relational database functions. Format Preserving Encryption is one of the encryption methods used to create reversible tokens.

Irreversible tokens: Tokens can never be converted back to the original primary account number PAN. They are either authenticable or non-authenticable.

The authenticable token is like a hash function to authenticate whether the PAN was used in creating the token. This procedure cannot be reversed to reconstruct the PAN itself. Whereas non-authenticable can never be linked to a specific PAN, however, they could be linked to a consumer or account within the merchant. The most discussed tokenization method in recent times is blockchain-based tokenization.

2.1.4 Blockchain-Based Tokenization and Previous Research

Blockchain-based transactions, commonly knowns as ledgers or tokens, are immutable [26] with timestamps that evade tampering of the information. Timestamp monitoring and access trails offer traceability and risk of counterfeit transactions.

With such strong claims, the tokenization of valuable assets is getting implemented using blockchain technology [27]. Researchers analyzed its business use cases in real estate finance [28], renewable energy and green buildings [29], critical infrastructure such as energy microgrid transactions [30], smart city infrastructure with intelligent transportation [31, 32], pharmaceutical industries, and medical research [33–35].

2.1.5 Concerns and Future Research Possibilities

While blockchain technology seems to gain the robust attraction of adoption, the representation of virtual assets with tokens has been a highly mystifying matter [26].

The first point of concern by the asset owners is that token issuers control profits and cash flows connected to the token, thus diminishing the value of ownership of the associated assets. Another concern is about the separation of claim and ownership rights of tokens vs assets and interlinked components.

With the value and strength of the technology offered by blockchain, there are many business use cases with tokenization that could be powered by blockchain while addressing the dilemmas of usage, access, and ownership. We foresee a strong possibility to conduct future research in the adoption of blockchain-based tokenization.

3 Conclusions

Hyper-scale digital transformations and cloud adoptions are growing rapidly with the new trend of remote workers and digital businesses during the pandemic. With the increased business demand on the cloud, adversaries are attacking organizations with advanced threats and exploits are exposing sensitive data to the public. Governments are strengthening laws and regulations to enhance accountability on organizations for their security and privacy practices.

Cloud service providers are offering native tools to discover, classify, and protect the data and workloads on their platforms. Security companies are coming up with advanced technologies to address the hybrid and multi-cloud complexities to secure the data across the life cycle from discovery to destruction.

Regular data encryption techniques are not effective to provide protection while running data analytics on a cloud platform for banking, health care, and e-commerce industries. Field-level encoding technologies, for instance, Format Preserving Encryptions and Tokenization, preserve the data in original format, while running the transactions, support in managing the performance of applications. We propose to reassess the blockchain-based tokenization for various business purposes, along with strong process definitions around ownership and responsibilities of tokens and their associated assets.

References

1. De R, Pandey N, Pal A (2020) Impact of digital surge during Covid-19 pandemic: a viewpoint on research and practice. Int J Inf Manag 55:102171. https://doi.org/10.1016/j.ijinfomgt.2020.102171
2. Internet Crime Complaint Center (2020) Internet crime report. Washington, D.C: Federal Bureau of Investigation. Federal Burea of Investigation. Accessed on 18 Jan 2022 at https://www.ic3.gov/Media/PDF/AnnualReport/2020_IC3Report.pdf
3. Wilson E (2019) Disrupting dark web supply chains to protect precious data. Computer Fraud Secur 2019(4):6–9. https://doi.org/10.1016/S1361-3723(19)30039-9
4. Stack B (2022) Here's how much your personal information is selling for on the dark web. Experian. Haettu 4 (2018): 2021. Accessed on 18 Jan 2022 at https://www.courts.ca.gov/opinions/links/S248130-LINK1.PDF
5. Henderson JG (2019) FTC imposes $5 billion penalty and sweeping new privacy restrictions on facebook. Federal Trade Commission. https://www.ftc.gov/news-events/press-releases/2019/07/ftc-imposes-5-billion-penalty-sweeping-new-privacy-restrictions. Accessed on 8 Jan 2022
6. Reinsel D, Gantz J, Rydning J (2018) DATA AGE 2025: the digitization of the world from edge to core. An IDC White Paper–#US44413318, Sponsored by Seagate. https://www.seagate.com/files/www-content/our-story/trends/files/idc-seagate-dataage-whitepaper.pdf
7. Discovering sensitive data with Amazon Macie, Amazon Web Services. https://docs.aws.amazon.com/macie/latest/user/data-classification.html
8. Amazon Simple Storage Service User Guide, Amazon Web Services. https://docs.aws.amazon.com/AmazonS3/latest/userguide//Welcome.html
9. Amazon Sage Maker user guide, Amazon Web Services. https://docs.aws.amazon.com/sagemaker/latest/dg/whatis.html

10. Amazon Glue user guide, Amazon Web Services. https://docs.aws.amazon.com/glue/latest/dg/security.html
11. Microsoft Azure Data Discovery & Classification documentation. https://docs.microsoft.com/en-us/azure/azure-sql/database/data-discovery-and-classification-overview
12. Google Data Analytics Products, Data Catalog Overview. https://cloud.google.com/data-catalog
13. File Analysis Suite, Micro Focus documentation. https://www.microfocus.com/documentation/file-analysis-suite/3.6/
14. Federal Information Processing Standard (FIPS) 199, Standards for Security; Published. 1 March 2004
15. Tao Y, Lei Z, Ruxiang P, Fine-grained big data security method based on zero trust model. In: 2018 IEEE 24th international conference on parallel and distributed systems
16. Zhang J, Li H (2017) Research and implementation of a data backup and recovery system for important business areas. In: 2017 9th international conference on intelligent human-machine systems and cybernetics
17. Reardon J, Basin D, Capkun S (2013) SoK: secure data deletion. In: 2013 IEEE symposium on security and privacy
18. Vaikuntanathan V (2011) Computing blindfolded: new developments in fully homomorphic encryption. In: 2011 IEEE 52nd annual symposium on foundations of computer science, pp 5–16. https://doi.org/10.1109/FOCS.2011.98
19. Frankel S, Glenn R, Kelly S (2003) The AES-CBC cipher algorithm and its use with IPsec, IETF. https://www.ietf.org/rfc/rfc3602.txt
20. Bellare M, Ristenpart T, Rogaway P, Stegers T (2009) Format-preserving encryption. 295–312. https://doi.org/10.1007/978-3-642-05445-7_19
21. Bellare M, Rogaway P, Spies T (2010) The FFX mode of operation for format preserving encryption, Draft 1.1. http://csrc.nist.gov/groups/ST/toolkit/BCM/documents/proposedmodes/ffx/ffx-spec.pdf
22. Dworkin M, NIST Special Publication 800-38G (2016) Recommendation for block cipher, modes of operation: methods for format-preserving encryption. https://doi.org/10.6028/NIST.SP.800-38G
23. Dworkin M (2019) Recommendation for block cipher modes of operation, methods for format-preserving encryption, draft NIST special publication 800-38G Revision 1, February 2019. https://doi.org/10.6028/NIST.SP.800-38Gr1-draft
24. Weingärtner T, Tokenization of physical assets and the impact of IoT and AI. Lucerne University of Applied Sciences & Arts—School for Information Technology
25. Ferraiolo H, Chandramouli R, Ghadiali N, Mohler J, Shorter S, Guidelines for the authorization of personal identity verification card issuers (PCI) and derived PIV credential issuers (DPCI). NIST Publication, NIST.SP.800-79-2. https://nvlpubs.nist.gov/nistpubs/SpecialPublications/NIST.SP.800-79-2.pdf
26. All you need to know about tokenization, its benefits, challenges and future outlook. https://e27.co/academy-tokenization-benefitschallenges-future-outlook-20181109/. Accessed 8 Jan 2022
27. Chen Y (2018) Blockchain tokens and the potential democratization of entrepreneurship and innovation. Bus Horiz 61(4):567–575
28. Stein Smith S (2020) Stablecoins and the decentralized organization. In: Stein Smith S (ed) Blockchain, artificial intelligence and financial services. Springer, Cham
29. Uzsoki D (2019) Tokenization of infrastructure: a blockchain based solution to financing sustainable infrastructure. Int Instit Sustain Dev
30. Mengelkamp E, Gärttner J, Rock K, Kessler S, Orsini L, Weinhardt C (2018) Designing microgrid energy markets. A case study: the brooklyn microgrid. Appl Energy 210:870–880
31. Gong Y, Liao JH (2019) Blockchain technology and simulation case analysis to construct a big data platform for urban intelligent transportation. J Highway Transp Res Dev 13(4):77–87
32. Zhong BJ, Adriaens P (2020) Digital financing model for bridges in Washington State. In: International conference on transportation and development. Seattle, WA, pp 300–308

33. Pêgo A, et al. (2021) Blockchain and clinical data economics: the tokenization of clinical research in the EU. In: de Oliveira Rodrigues D (ed) Political and economic implications of blockchain technology in business and healthcare. IGI Global, pp 269–291. https://doi.org/10.4018/978-1-7998-7363-1.ch011
34. Agbo CC, Mahmoud QH, Eklund JM (2019) Blockchain technology in healthcare: a systematic review. Healthcare 7(2):56
35. Yaqoob I, Salah K, Jayaraman R et al (2021) Blockchain for healthcare data management: opportunities, challenges, and future recommendations. Neural Comput & Appl. https://doi.org/10.1007/s00521-020-05519-w

Vehicular Communication Strategy Using Machine Learning and Image Processing to Enhance Observations and Control on the Road Side Area

Kamlesh Gautam, V. K. Jain, Sourabh Singh Verma, and Sonali Vyas

Abstract The Vehicular Communication System (VCS) is a cooperative engagement of Vehicle Nodes with the required intervention of On-Board Unit (OBU), Cluster Head Unit (CHU), Road Side Unit (RSU), and Infrastructure Domain Unit (IDU). These all units provide exceptional performance with the help of Digital Fragment Processing (DFP), Machine Learning (ML), Internet of Things (IoT), and 5G technologies. However, during vehicular communication lot of data is required to transceive. This high volume of data traffic increases the unbearable load at each transceiver node, this increases data transmission delay. The delay affects the reliability of VCS, this needed to be resolved using some enhanced efficient strategy. Therefore a CHUS algorithm has been proposed in this article that follows the CHU strategy. The simulation results show that our proposed strategy algorithm performs better with existing strategies by enhancing the reliability of VCS by reducing average delay.

Keywords VCS · VANET · Vehicle communication · CHUS · RSU · Reliability enhancement

K. Gautam
Poornima College of Engineering, Jaipur, Rajasthan, India
e-mail: kamlesh@poornima.org

V. K. Jain
Sage University Bhopal, Bhopal, India

S. S. Verma (✉)
Manipal University Jaipur, Jaipur, India
e-mail: ssverma80@gmail.com

S. Vyas
University of Petroleum and Energy Studies, Dehradun, India

V. K. Gunjan and J. M. Zurada (eds.), *Proceedings of 3rd International Conference on Recent Trends in Machine Learning, IoT, Smart Cities and Applications*, Lecture Notes in Networks and Systems 540, https://doi.org/10.1007/978-981-19-6088-8_34

1 Introduction

The Vehicular Communication System consists mainly of four units namely, OBU, CHU, RSU and IDU. Out of these units, OBU is installed in the vehicle itself while CHU, RSU and IDU are installed on Road Side Area (RSA) [1]. In VCS, a CHU controls all OBUs within its communication range. Based on the data of many such CHUs, a respective RSU processes. The entire city VCS is handled by an IDU based on data from several such RSUs. The On-Board Unit of VCS consists of—OBU display, OBU keyboard, OBU speaker, Electronic License Plate (i.e., for administrative authentication/handling of transceiver data), OBU inside/outside sensors, memory, etc. The OBU sends and receives data to and from CHU. The CHU process data with help of its components in several steps like data generation, anomaly detection, feature vector creation, predictive models, fragment processing, noise reduction, quality improvement, data compression, coding, data storage, data communication, etc. The data communicator of CHU transceives data to the respective OBU as well as communicates with the data communicator of RSU. Then in RSU, the data from the data communicator is stored as historical data as well as it is fed to a pre-trained predictive model (like VGG-16, ResNet, etc.) for fragment data classification[2]. Then fragment data is used with feature vectors in another pre-trained predictor model, which is based on the co-occurrence matrix. The output of this model is then used to pass to the data communicator in form of an extracted frame with the expected labels. The data communicator transmits the data to the respective CHU along with passing it to the data communicator of IDU. The IDU process the data with the information updater along with the event handler, which checks if the predicted event has occurred or not. The outcome of the event handler is used for reinforcement learning for future prediction updating. After reinforcement learning, the data is stored as historical data, which is then processed by the data communicator of IDU for respective road area predictions.

2 Background

The Vehicular Communication System is employed in VANET, which is based on DSRC, WAVE and IEEE802.11 to enhance various applications related to safety, congestion, convenience, etc. As VANET also suffers from all the threats that were present in its parent, i.e, Mobile Ad-Hoc Networks[3, 4]. Due to this reason, Vehicular Communication systems required concerns to provide reliable, scalable and efficient communication between vehicle nodes to include various advanced applications [5]. These applications are related to a road accident, traffic congestion, traveling time, convenience purpose, transport purpose, commercial purpose, safety purpose, driver situation prediction (drowsiness and fatigue)[6], road area object and object speed detection, data set of road area criminal activity, and road area event prediction, etc. These types of advanced applications are now also possible in VCS with help of machine learning algorithms. To perform these types of applications, data always needs to be collected, controlled and transceived. But always, this data increases congestion in Control Channel (CC) and Service Channel (SC) [7]. In the VCS, if this congestion is decreased in the channel, then packet loss and delay will also decrease, network performance will increase and predictions of machine learning will become real-time predictions. That's why congestion has to be reduced to enhance the reliability of the Quality of Service (QoS) of VCS [8]. So far, two types of strategies have been used to reduce this congestion. The first is Close Loop Strategies (CLS) and the second is Open Loop Strategies (OLS) [9].

2.1 Close Loop Strategies (CLS)

In the CLS strategy of VCS, firstly feedback is taken about the channel conditions and to calculate that feedback it is seen how useful the channel is, how many messages are pending within the message queue and how long does it take for the channel to be occupied. On the behave of these feedback calculations, if it seems that congestion is increasing more than a limit inside the channel, then according to the CLS strategy, some solutions are used to reduce that congestion. These solutions are Rate-Based Solutions (RBS) [10], CSMA/CA-Based Solutions (CBS) [11] and Hybrid Solutions (HS) [12].

2.2 Open Loop Strategies (OLS)

In Vehicular Communication System the OLS is a strategy that prevents data congestion in the channel before congestion occurs. Because OLS strategy does not include feedback regarding the time of channel occupancy and channel load. OLS strategy works on the available data from the beginning and keeps reducing the data so that there is no unbearable congestion of data in the channel. For this data reduction Power-Based Solution (PBS) [13] and Prioritizing and Scheduling-Based Solution (P&S-BS) are used [14].

3 Proposed Strategy

Algorithm: Pseudocode of Proposed Cluster Head Unit Strategy (CHUS) Algorithm

Input: $C_a[i++]$ from CHU Sensors and Message Fragments from OBU Sensors.

Output: $[M_{F_{el}}\ [i_{FPS}++]]$ and $F_{el}\{OBU[f_N]\}$

1. Begin
   //Currently only CHU Sensors data $C_a[i++]$ is input
   STEP-1
2. $if C_a[i++] \neq 0$
3. Call to $YOLOAv5$
4. $YOLOAv5import = C_a[i++];$
5. $YOLOAv5detect = opt(c1), opt(c2), opt(c3);$
   Where
   $opt(c1) = C_a\{F_{el}(Obj_j)[f+++];$
   $opt(c2) = C_a\{F_{el}(Obj_{jk})[f+++];$
   $opt(c3) = S\{Obj_{jk}\};$
   Where $S\{Obj_{jk}\} = \Delta Obj_{jk}/\Delta t$
6. Then find out f_{VOA}, f_{RA} and $J = (f_{VOA}/f_{RA}) \times 100;$
7. $if(J < L)$
8. Case 1: P_m to OBUs and RSU = "$JAM\ Probability\ Low$";
9. $else$
10. $if(J < H)$
11. Case 2: P_m to OBUs and RSU = "$JAM\ Probability\ Medium$";
12. $else(J \geq H)$
13. Case 3: P_m to OBUs and RSU = "$JAM\ Probability\ High$";
14. $endif$
15. $endif$
16. $else$
17. P_m to Self = "$No\ Information\ from\ CHU\ Sensors$";
18. $endif$
    STEP-2
19. $if C_a\{F_{el}(Obj_{jk})[f+++]\} == C_a\{F_{el}(V_N)[f++]\}$ $in C_a\{F_{el}(V_N)[f+++]\}$
20. $if(S\{V_N\} \leq Th_L or S\{V_N\} \geq Th_{SV_r} or S\{V_N\} \geq Th_H)$
21. $if S\{V_N\} \cap RA_{tf}$
22. Call to $YOLOAA_b;$
23. $YOLOAA_b import = C_a\{F_{el}(V_N)[f++]\};$
24. $YOLOAA_b opt = F_{el}\{A_n[f++]\};$
25. Then $\left[M_{F_{el}(A_{n+1})}\right] - \left[M_{F_{el}(A_n)}\right] = \begin{bmatrix} E_{11} & \cdots & E_{1w} \\ \vdots & \ddots & \vdots \\ E_{\alpha 1} & \cdots & E_{\alpha z} \end{bmatrix};$
    Where $\beta = (\gamma/(\alpha + z)) \times 100;$
26. $if(\beta \geq \beta_p)$
27. $\left[M_{F_{el}(A_{n+1})}\right] = 0;$
28. $else$
29. $FPSimport = \left[M_{F_{el}(A_{n+1})}\right];$

$FPSopt = \left[M_{F_{el}}[i_{FPS} + ++]\right];$
$endif$
$else$
$C_a\{F_{el}(V_N)[f + +]\} = 0;$
$endif$
$else$
$C_a\left\{F_{el}\left(Obj_{jk}\right)[f + ++]\right\} = 0;$
P_m to RSU = "$NO\ EVENT$";
$endif$
$else$
$C_a\left\{F_{el}\left(Obj_{jk}\right)[f + ++]\right\} = 0;$
$endif$

In the urban areas, a lot of events occurred on road side areas regarding safety, vehicle congestion, accident, parking, toll-plaza, etc. Currently, all these events' observation and control are enhanced with help of VCS. But during this lot of data is required to effective. That increases the unbearable load at the transceiver nodes. By increasing this load, channel congestion duration, packet transmission delay and loss are also increased [15]. That affects directly the scalability and reliability of VCS with its real-time performance. This performance of VCS can also be enhanced by an efficient strategy. Therefore, a CHUS algorithm is proposed in this work. This strategy is called Cluster Head Unit Strategy (CHUS) because it employs CHU, that unit is installed on the side area of the road. This unit is capable of making the cluster of all the road area vehicles coming in its communication range. The CHU allows the vehicles and some electronic devices to communicate within this cluster, as well as provide some important possible predictions with help of OBU and RSU. CHU plays a major role in the data collection of this entire process. CHU gathers data in form of text and image fragments from OBU sensors of vehicles and itself sensors. The CHUS is applied to this gathered data with help of a machine learning-based CHUS algorithm.

According to the CHUS algorithm in the first stage, the OBU and CHU sensor data is categorized using the already available $YOLOAv5$ algorithm in form of three-level cases $opt(c1)$, $opt(c2)$ and $opt(c3)$. First level data includes a continuous fragment array ($C_a\{F_{el}(\text{Obj}_j)[f + ++]$) of features extracted–labeled jth type objects. Second level data includes a continuous fragment array of features extracted–labeled with the number of each type object. Third-level data includes the speed of all objects. Based on these cases data, this algorithm predicts the probability of jamming on road areas. In the second stage, only filtered and minimized data is sent to the Road Side Unit for the processing of RSU Machine Learning Algorithms. For this, the fragments in which any (V_N) is an object is filtered with the threshold value (Th) of the speed of the vehicles. Now out of these fragments with the help of an anomaly detection algorithm ($YOLOAA_b$), the anomaly identified fragments are selected. These fragments are further minimized based on fragments similarities. Then remaining data is processed to Fragment Processing System (FPS) for fragment quality handling. In this way the proposed algorithm minimizes the data, thereby reducing the load of data on the transmitting node and reducing the average delay of transmitting data.

4 Performance Evaluation

In this work SUMO and NS-3 are used for OBU mobility and OBU ad-hoc network, respectively. For the simulation urban environment is used to control congestion. In these simulators, the pink city road pattern is adopted as an urban scenario with four crossroads (chaupar) for performance evaluation. This work simulation uses CBS as a MAC layer transmission solution, IEEE 802.11p as a standard wireless protocol, Nakagami as obstacles modeling propagation model and Poisson Distribution (PD) to generate data. In this work, the size of fragments is 400 Bytes [12]. The importance value of fragments is decided according to the remaining deadline time of fragments. The distance between OBUs and CHU is defined according to the position of OBUs at CHU. In this work, the CHUS strategy is compared with three different Open Loop Strategies (i) D-FPAV [16], (ii) CABS [14] and (iii) NC-CC [13]. According to the first strategy, safety fragments are separated into event-driven and beacon fragments, and the relevance is checked with the distance from the sender. So that the channel uses level can be felt by each node. Based on this, transmission power is tuned with higher priority for event-driven fragments. In the second strategy to adjust dynamically the priority of beacon fragments according to channel congestion, beacon fragments include details of vehicle passion, velocity and direction. On behave of these a TDMA-based time slot is scheduled for beacon fragments of each road are vehicles. So that channel access delay and packet reception rate can be enhanced. In the third strategy, channel congestion is reduced by reducing beacon fragments. For this application layer transmission power is reduced with help of packet-level network coding. But in the proposed strategy a CHUS algorithm is used to reduce redundant fragments. So that the congestion of data can be reduced and the average delay can be reduced automatically due to its low as shown in Fig. 1 with the comparison of previously available strategies.

For the performance evaluation of the CHUS strategy following the performance, matrices are employed.

$$\text{Average Delay} = \frac{1}{T}\sum_{x=1}^{y}\frac{\lambda_x}{\rho_x - \lambda_x}$$

where

T = Total amount of fragment traffic occurrence in the fixed Time interval.
$x = 1, 2, 3, \ldots, y$ (Number of links in a network).
λ_x = Fragment arrival rate from xth link.
ρ_x = xth link Fragment delivery rate.

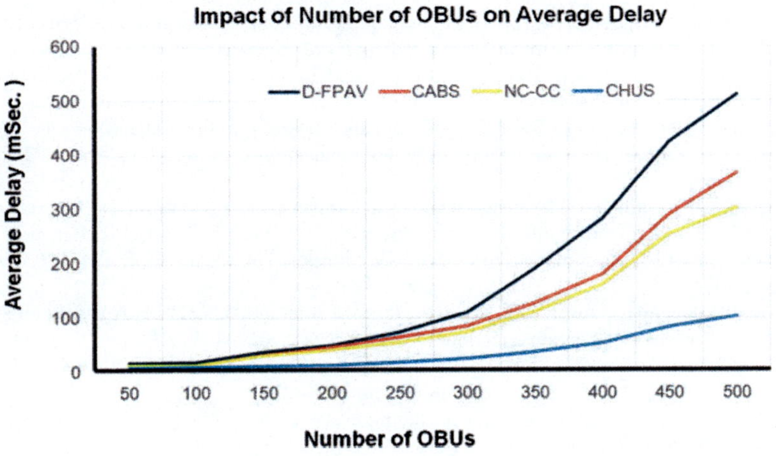

Fig. 1 Impact of the number of OBUs on average delay

A set of simulations was carried out to evaluate the impact of the number of OBUs, simulation time and OBUs density on the introduced performance metrics.

5 Conclusion

The article explores the necessary information about VCS and the background for the proposed work. The literature significantly concludes the problem of average delay in data transmission. To resolve this gap an algorithm CHUS has been proposed. This machine learning-based algorithm reduces redundant fragments from the transmission to a substantial level. By reducing this, the data congestion load on the transmitting node has reduced and the average delay is notably decreased. The proposed algorithm improves the reliability of the Vehicular Communication Systems in the Road side Area. The performance of this reliability enhancement is evaluated using the NS-3 simulator. The simulation result shows that the average delay of transmitting data has been substantially reduced from the channel using the proposed strategy algorithm as compared with other strategies.

References

1. Cooper C, Franklin D, Ros M, Safaei F, Abolhasan M (2017) A Comparative survey of VANET clustering techniques. IEEE Commun Surv Tutorials 19(1):657–681
2. Gautam K, Jain VK, Verma SS (2021) Enhancement in the reliability of vehicular communication system for road side area. In: International conference on new trends in applied mathematical sciences and its applications (ICNTAMSA-2021). Indonesia

3. Verma SS, Patel RB, Lenka SK (2016) Investigating varying time black hole attack on QoS over MANET. In: Satapathy S, Joshi A, Modi N, Pathak N (eds) Proceedings of international conference on ICT for sustainable development. Advances in Intelligent Systems and Computing, vol 408. Springer, Singapore
4. Verma SS, Patel RB, Lenka SK (2015) Analyzing varying rate flood attack on real flow in MANET and solution proposal: real flow dynamic queue (RFDQ). Int J Inf Commun Technol Inderscience 7
5. Karagiannis G, Altintas O, Ekici E, Heijenk G, Jarupan B, Lin K, Weil T (2011) Vehicular networking: a survey and tutorial on requirements, architectures, challenges, standards and solutions. In: IEEE communications surveys & tutorials, vol 13, no 4, pp 584–616, Fourth Quarter 2011
6. Ramzan M, Khan HU, Awan SM, Ismail A, Ilyas M, Mahmood A (2019) A survey on state-of-the-art drowsiness detection techniques. IEEE Access 7:61904–61919
7. Qian Y, Moayeri N (2008) Design of secure and application-oriented VANETs. In: VTC Spring 2008—IEEE vehicular technology conference, pp 2794–2799
8. Singh RK, Tyagi N (2012) Challenges of routing in vehicular ad-hoc networks: a survey. Int J Electron Commun Comput Eng 3(1):126–132
9. Shen X, Cheng X, Zhang R, Jiao B, Yang Y (2013) Distributed congestion control approaches for the IEEE 802.11p vehicular networks. IEEE Intell Transp Syst Mag 5(4):50–61. Winter 2013
10. Fei Y, Raymond Y, Sumit R, Jinyun Z (2011) Efficiency and reliability of one-hop broadcasting in vehicular ad hoc networks. IEEE J Select Areas Commun 29(1):151–160
11. Chih-Wei H, Chung-Hsien H, Huei-Ru T (2011) MAC channel congestion control mechanism in IEEE 802.11p/WAVE vehicle networks. In: 2011 IEEE vehicular technology conference (VTC Fall), pp 1–5
12. Taherkhani N, Pierre S (2016) Centralized and localized data congestion control strategy for vehicular ad hoc networks using a machine learning clustering algorithm. IEEE Trans Intell Transp Syst 17(11):3275–3285
13. Pratap SK, Abdelhakim H, Soumaya C (2014) Congestion control in vehicular networks using network coding. IEEE Int Conf Commun (ICC) 2014:2736–2741
14. Songnan B, Junho O, Jaeil J (2013) Context awareness beacon scheduling scheme for congestion control in vehicle to vehicle safety communication. Ad Hoc Netw 11(7):2049–2058
15. Sharma AK, Nandal A, Dhaka A, Dixit R (2020) A survey on machine learning based brain retrieval algorithms in medical image analysis. Health Technol 10:1359–1373
16. Mare TM, Paolo S, Hannes H (2006) Distributed fair transmit power adjustment for vehicular ad hoc networks. In: 2006 3rd annual IEEE communications society on sensor and ad hoc communications and networks, pp 479–488

Emotional Labour and Perceived Stress at Workplace—HR Analytics

G. V. Manoj Kumar, Brig Anuj Chawla, Kunjam Nageswara Rao, and G. Sita Ratnam

Abstract Human resource (HR) analytics is a technique for collecting and analysing human resource data to improve organisation performance. Emotional labour is the psychological work individuals perform in conforming to emotional display rules in a given environment. The perceived stress at the workplace and emotional labour at the workplace are known to have negative health consequences for employees and compromise the performance of organisations. The study with these metrics significantly influences decision-making for individuals and businesses. Data from serving officers with twenty years of experience from the Armed Forces and Civil Services revealed emotional labour and demand-resource work stress by both the services at varying levels. Multi-supervised machine learning techniques like logistic regression (LR), Naive Bayes (NB), support vector machine (SVM) and decision tree have are considered to build a model. The data collected from Armed Forces and Civil Service officers were used for features extraction. The dataset was loaded into the model to identify, classify and analyse the emotional labour and perceived workplace stress. Findings based on the display of emotional behaviour at the workplace and the perceived workplace stress by the employees were plotted. The model's performance was evaluated using a variety of machine learning techniques. This study is probably one of the few projects to report the correlation between emotional labour and stress at the workplace. The data revealed emotional labour, primarily as deep acting, in 76.1 and 72.2% of officers of the two services, respectively. The concept can be developed further to investigate the effects of occupational stress on health metrics and the changing priorities of the people in the wake of prevailing pandemic and rise in attritions rate.

G. V. Manoj Kumar (✉) · K. N. Rao
Department of Computer Science and Systems Engineering, Andhra University College of Engineering, Visakhapatnam, AP, India
e-mail: informmanoj@gmail.com

B. A. Chawla
National Defence College, Delhi, India

G. Sita Ratnam
Lendi Institute of Engineering and Technology, Jonnada, India

V. K. Gunjan and J. M. Zurada (eds.), *Proceedings of 3rd International Conference on Recent Trends in Machine Learning, IoT, Smart Cities and Applications*, Lecture Notes in Networks and Systems 540, https://doi.org/10.1007/978-981-19-6088-8_35

Keywords HR analysis · Emotion labour · Workplace stress · Perceived work stress · Sentimental analysis · Health and safety executive (HSE) indicator · Consequences of emotion labour · Human resource performance · Target value analysis · Predictive power score · Emotional labour · Key performance indicator · Organisational stress · Surface acting (SA) · Deep acting · Display emotion · Regulating emotion

1 Introduction

HR analytics is the science that unifies all data pertaining to human resource tasks such as recruiting, talent identification and best utilisation of human resources for the strategic execution of plans to benefit the organisation. Employees are a critical component of a profitable and productive business [1]. The HR team in the Indian scenario does not always consider the influence of emotional labour at the workplace in general. Emotions are a complex phenomenon, triggered mainly in response to interactions between individuals and the workplace environment, where individuals spend a significant part of their working hours. Surprisingly, research into organisational behaviour has long neglected the study of emotions at the workplace, which plays a paramount role. The organisation's management can visualise potential situations [2] that impact the organisational growth by periodic assessment of human resource stress factors through the emotional labour metrics to control the attrition rate and enhance the organisation's productivity. At the workplace, HR analytics and emotion labour can deduce human resources' functions [3], orchestration, and tactics based on data and data analysis.

2 The Literature Survey—Emotional Labour at the Workplace

The concept of emotional labour has changed over time. Ashforth and Humphrey [4] argued in 1993 that emotional labour was not always deleterious. They applied the concept of 'identity' to emotional labour, saying that it would have negative consequences only if individuals failed to identify with their assigned roles and emotions at work. The concept of emotional labour was first defined by Hochschild [5], an American sociologist, Morris and Feldman [6] added another layer to the concept of emotional labour in 1996 when they characterised it in terms of four dimensions: frequency, variability, conformity to display rules and emotional dissonance. They were the first to recognise both the individual and organisational components of emotional labour. Brotheridge and Grandey [7] classified emotional labour in 2002 into 'job-focused' and 'employee-focused' to reconcile the numerous perspectives on emotional labour. While the former took into account characteristics of the job such as intensity and variety of emotions expected to be displayed, duration of and frequency of social interactions and display rules of the organisation, the latter focussed on the way individual employees managed their emotions.

An Emotional Labour Scale (ELS) based on six dimensions was formulated based on Brotheridge and Grandey. Pugliesi[1], in 1999, discovered that organisational norms for emotion display were associated with occupational stress, psychological distress and job satisfaction. Schaubroeck and Jones found in 2000 that display regulations have a deleterious effect on bodily symptoms such as insomnia, fatigue and disease. Additionally, emotional labour has been connected to 'burnout'. Three characteristics of burnout include emotional exhaustion, depersonalisation and critical personal success. It is generally agreed that 'surface acting' is more likely to be related to burnout than other emotional labour forms. Zammuner and Gali have reported that deep acting, on the other hand, is thought to correlate positively with the sense of personal accomplishment and not lead to emotional exhaustion and depersonalisation.[2]

The present pandemic has imparted a few additional variables like family bonding, flexible timings and work from home. The situational factors become significant predictors of emotional labour; the workplace variables that affect emotional labour are employee autonomy, social support, the frequency and duration of interpersonal connections. Job autonomy is associated with job satisfaction [8].

3 Factors Affecting Emotional Labour

Workload and work intensity are also important factors that affect emotions at the workplace. The QPSNordic and the HSE (Health and Safety Executive, United Kingdom) indicator are two recent survey methods for assessing the psychosocial work environment. The United Kingdom's Health and Safety Executive (HSE) developed the HSE indicator tool [9] to measure perceived workplace stress using a 'management standards' approach. The HSE enables businesses to quantify felt workplace stress by comparing perceived stress to management criteria. Three supervised machine learning techniques are used to build a model for identifying, classifying and analysing Emotion Labour and perceived stress at the workplace. The techniques used in the proposed model is support vector machine (SVM), Naïve Bayes (NB), logistic regression (LR) and decision trees (D.T). The gathered data were pre-processed, and features were extracted manually. These features were fed as input to the proposed model, and efficiencies were compared. This study on the role of emotional labour at workplaces in the Indian defence environment [10] and the

[1] Pugliesi, K. (1999). The consequences of emotional labour: Effects on work stress, job satisfaction, and well-being. Motivation and Emotion, 23(2), 135–154.

[2] Zammuner, V.L. & Galli, C. (2005a). Wellbeing: Causes and consequences of emotion regulation in work settings. *International Review of Psychiatry, 17*(5), 355–364.

influence of present-day stress factors on human resources will aid in taking corrective actions if necessary for the welfare of the Human Resources and Organisational growth.

4 Dataset

The data were collected from defence officers with twenty plus years of experience through two different questioners for emotional labour and perceived stress at the workplace. In addition, personal interviews were also conducted to capture the data. Each response was awarded a numerical value based on the selection of choices [Never-1, Rarely: 2, Sometimes:3, Often: 4 and Always: 5]. A total of 12 and 35 attributes for four facets of emotion labour, six facets of perceived stress, respectively, are captured from 753 members. The data measure perceived workplace stress in six areas that are associated with poor health outcomes in employees, increased workplace absenteeism and reduced productivity for the organisation. These six areas are 'demands', 'control', 'support', 'relationships', 'role' and 'change' at the workplace. The four facets of Emotion Labour considered are 'Surface Acting', Deep Acting', 'Display of emotions' and 'Automatic Regulation of emotions'. The data were pre-processed for training. Univariate and multivariate visualisation plots were used to illustrate the data. The whisker plot was used to visualise data in the univariate form. The histograms were used to plot many variables. The histogram illustrates the distribution of numerical data in an approximate manner. Data validation is carried out utilising the K-fold cross-validation technique, which partitions data into training and testing sets. Cross-validation is a technique that uses resampling to evaluate machine learning models using small datasets. The model is trained using 80, 70, 60% of the data and tested with 20, 30 and 40% data, respectively, to ensure correctness and the model was validated using the sequence as explained in Fig. 1.

The training and comparing outputs from the pre-processed data were given in Table 1, and the correlation of the first phase of the interview is given in Table 2.

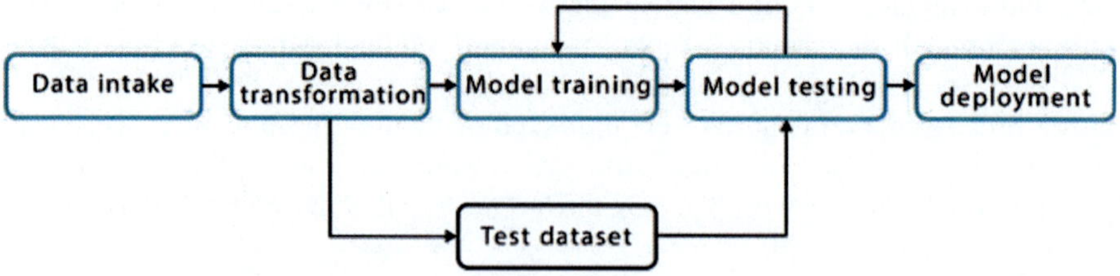

Fig. 1 Procedural process of machine learning

Table 1 Train and test accuracy of data

S. No.	ML method	Train accuracy	Test accuracy
1	Logistic regression	97.81	98.79
2	SVM	98.01	97.99
3	KNN	100	95.5
4	Decision tree	100	98.39

5 Results

The HSE questionnaire recommends analysis and interpretation of data in terms of the percentage of respondents reporting mean scores greater than 3 in the given population. The data revealed that the three areas of great perceived workplace stress for both civilians and armed forces are 'Demands', 'Change' and 'Control'. The employees of the armed force reported lesser stress compared to the other group coherent across all the six aspects of workplace stress. The differences were statically significantly.

Figures 2 and 3 depict the emotion labour facets and correlation, while Figs. 4 and 5 display results for perceived stress.

1. The data reveal that in officers of both the armed forces and civil services, a score for deep acting and display of emotions were higher than that of surface acting and automatic regulation of emotions. The officers of the two organisations have a comparable score on deep acting, as depicted in Fig. 2. It is evident that more than 70 of the respondents of both groups were seen to have deep acting at the workplace.
2. The study discovered that officers in the Armed Forces and the Civil Services were subjected to emotional labour on the job. Deep acting and a variety of emotional displays were shown to be the key contributors to emotional labour in both study groups.
3. In terms of perceived workplace stress, it is observed that the most often cited stressor by officers from both departments was a demand-resource imbalance. Additionally, officers of the Civil Services identified 'control' and 'change' as major occupational concerns.

6 Conclusions

The study was the first to examine and report a correlation between specific workplace-based stressors and facets of emotional labour in non-client focussed workplaces like the Armed Forces and the Civil Services.

The findings of the study also validate the selection and training strategies of both services. Since the majority of respondents were found to identify with their expected role and display deep acting as the predominant form of emotional labour, it can be

Table 2 Correlation of scores of the emotional labour questionnaire with scores on the HSE perceived workplace stress questionnaire

	Means	Std. dev	Demand	Control	Support	Relationship	Role	Change	SA	DA	ED	Auto-reg
Demand	23.2	5.2	1.000	0.582	0.597	0.666	0.386	0.602	−0.145	−0.294	0.157	−0.385
Control	19.8	4.9	0.582	1.000	0.803	0.626	0.485	0.723	−0.182	−0.034	0.347	−0.431
Support	33.0	6.3	0.597	0.803	1.000	0.727	0.529	0.833	−0.170	−0.153	0.333	−0.409
Relationship	15.2	2.8	0.666	0.626	0.727	1.000	0.384	0.617	−0.246	−0.235	0.224	−0.339
Role	22.0	3.1	0.386	0.485	0.529	0.384	1.000	0.439	−0.262	−0.115	0.283	−0.385
Change	9.7	2.7	0.602	0.723	0.833	0.617	0.439	1.000	−0.054	−0.157	0.258	−0.299
SA	8.5	2.7	−0.145	−0.182	−0.170	−0.246	−0.262	−0.054	1.000	−0.025	0.035	0.618
DA	10.7	1.8	−0.294	−0.034	−0.153	−0.235	−0.115	−0.157	0.025	1.000	0.253	0.117
ED	10.2	2.1	0.157	0.347	0.333	0.224	0.258	0.258	0.035	0.253	1.000	−0.295
Auto-reg	8.3	2.6	−0.385	−0.431	−0.409	−0.339	−0.385	−0.299	0.618	0.117	−0.295	1.000

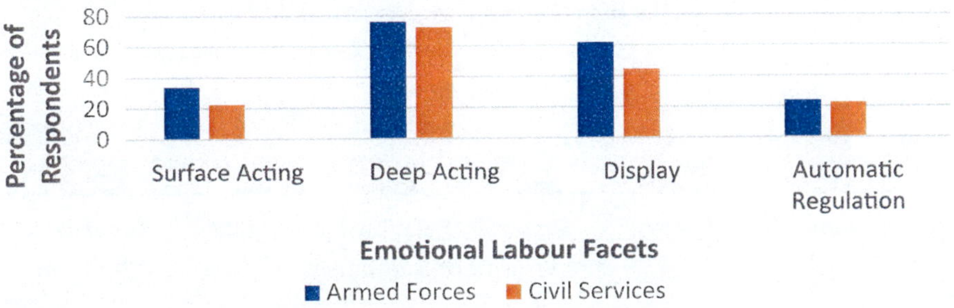

Fig. 2 Emotion labour facets in armed and civil services

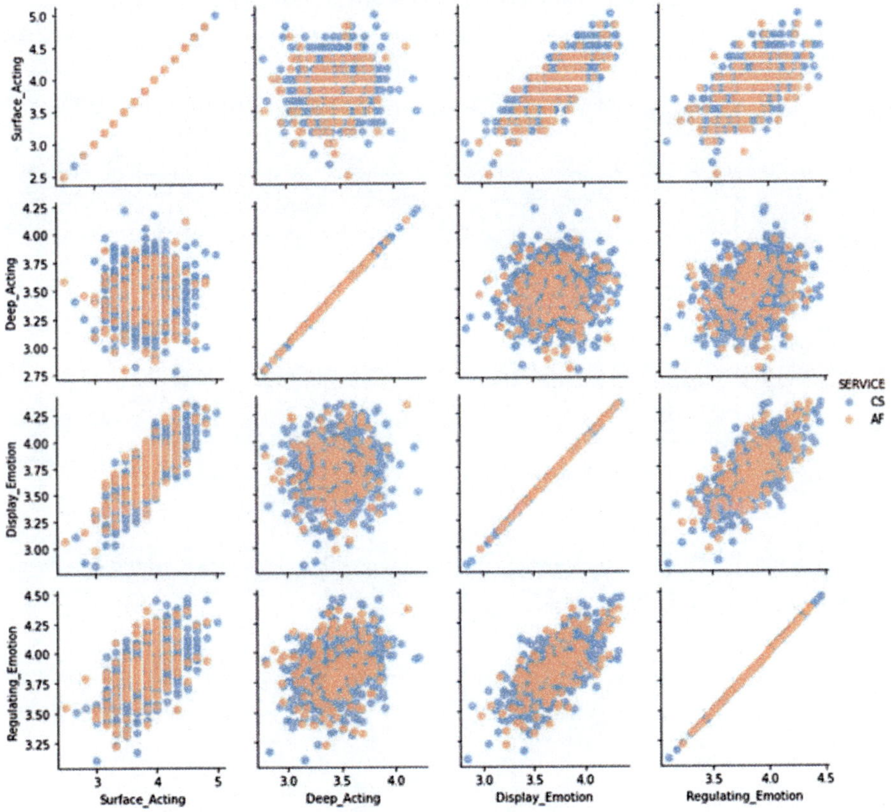

Fig. 3 Correlation of emotion labour parameters

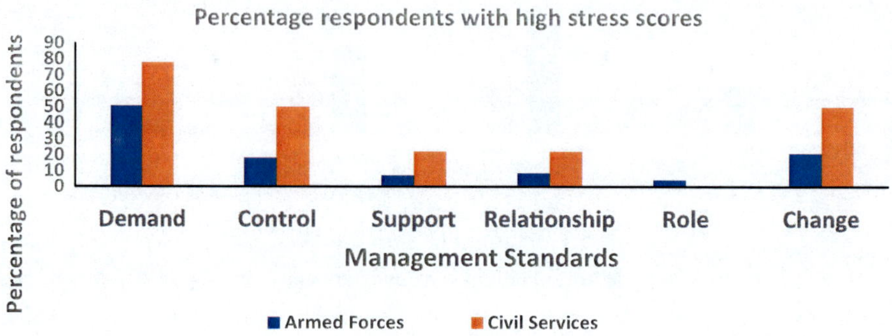

Fig. 4 Percentage of officers with mean high-stress facets on six management standards

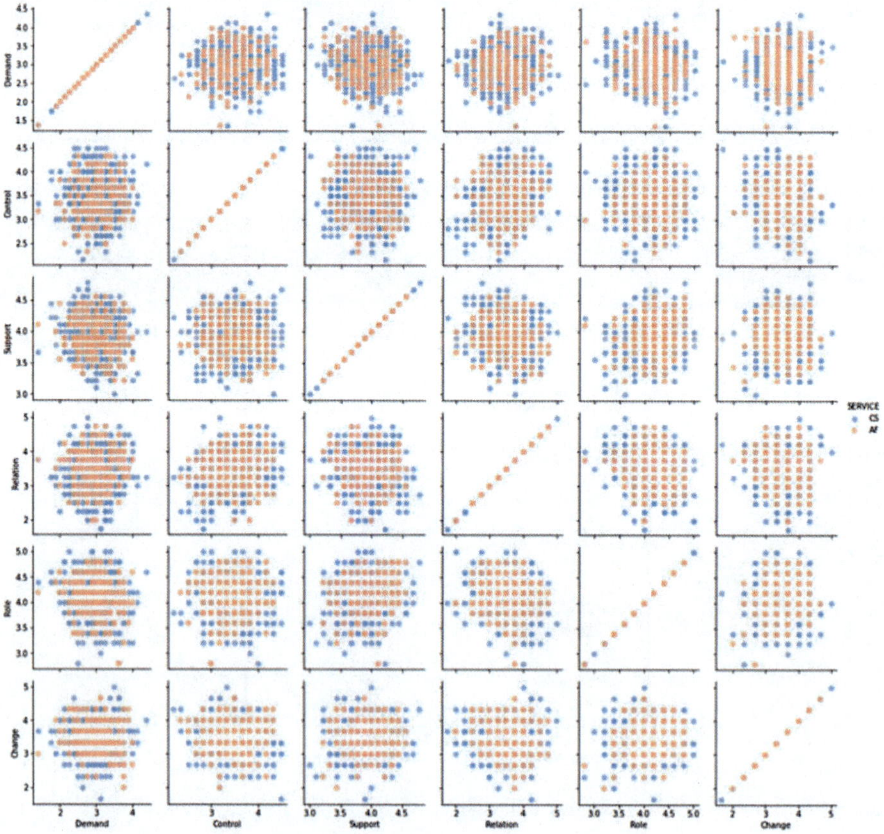

Fig. 5 Correlation of perceived work stress parameters

concluded that the selection process of both services is successful in identifying candidates best suited to serve in the respective organisations. The study can be further extended to understand the attrition rate amongst the employees in other sectors post-pandemic situations, which is a significant problem to be addressed by HR.

Bibliography

1. Sisodia DS (2017) Evaluation of machine learning models for employee chirn prediction. In: International conference on inventive computing and informatics
2. CCMBSDN, Brockett (2019) A system for analysis and remendiation of attrition. In: IEEE international conference on big data
3. VVRP, Padmaja KVNSB (2018) A machine learning approach for stress detection using a wireless physical activity tracker. Int J Mach Learn Comput 8(1)
4. HR, Ashrodorth BE (1993) Emotional labour in service: the influence of identity. Acad Manage Rev 18:88–115
5. HAR (1983) The managed heart. University of California Press, Berkely
6. JA, Morris FDC (1983) The dimentions, antecedents and consequences of emotional labour. Acad Manag 21:989–1010
7. Grandey B (2002) Emotional labour and burnout: comparing two perspectives on people work. J Vocat Behav 60:17–39
8. BM, Mehra R (2018) The effect of emotion labour on organisational commitments: an emperical evidence in Indian banking sector. Amity J Train Dev 3(1):12–21
9. RMCJCSK et al (2004) Management standards and work-related stress in UK: practical development. Work Stress 18:113–136
10. Chawala A (2021) Emotional labour and workplace stress in the Indian Armed Forces and civil services: an exploratory study. Research Thesis, New Delhi

Autonomous Traffic Sign Detection and Recognition in Real Time

N. Malarvizhi, Athreayasa Kalyan Jupudi, Manikanta Velpuri, and T. V. K. Dheeraj

Abstract Every day, over 400 traffic accidents occur in India, according to official statistics. Traffic sign recognition plays very important role in intelligent autonomous vehicles as well as in driver assistant system to disburden the driver. The main aim of this project is to reduce the cause of road accidents because of an inactive drivers on the road. We are developing a model which makes image classification and buzzer to avoid drivers distraction. In this model we are improving the performance to detect the Indian road signs to assist the drivers. And we will also reduce the over speed problems, they will follow traffic rules strictly by using this model. Our project will also help new learners to improve their driving experience. A skeleton for an Autonomous Traffic Sign Detection and Recognition (TSDR) in real time has been proposed. The proposed framework includes two parts: traffic sign detection and recognition of detected traffic signs. Future autonomous vehicles will benefit from this framework because it allows for the traffic sign detection and recognition in real time from the complex images that occur on the road. The proposed framework helps the drivers to improve road safety as well to help new learners to improve their driving experience also.

Keywords Artificial intelligence · Deep learning · Neural network · Tensorflow · Keras · Computer vision

N. Malarvizhi (✉) · A. K. Jupudi · M. Velpuri · T. V. K. Dheeraj
CSE, Vel Tech Rangarajan Dr Sagunthala R&D Institute of Science and Technology, Chennai, Tamilnadu, India
e-mail: drnmalarvizhi@veltech.edu.in

A. K. Jupudi
e-mail: vtu15933@veltech.edu.in

M. Velpuri
e-mail: vtu15955@veltech.edu.in

T. V. K. Dheeraj
e-mail: vtu15739@veltech.edu.in

V. K. Gunjan and J. M. Zurada (eds.), *Proceedings of 3rd International Conference on Recent Trends in Machine Learning, IoT, Smart Cities and Applications*, Lecture Notes in Networks and Systems 540, https://doi.org/10.1007/978-981-19-6088-8_36

1 Introduction

Machine learning algorithms are helping to build Advanced Driver Assistance Systems (ADAS) in the automotive sector, which are designed to develop the driver's protection and comfort. Automatic traffic sign recognition systems, such as Volvo's Road Sign Information (RSI) and Toyota's Road Sign Assist (RSA), have been on the market for a few years and are integral parts of modern ADASs. As a result, vehicle manufacturers are eager to take advantage of recent advances in deep learning, particularly deep Convolutional Neural Networks (CNNs), which have demonstrated state-of-the-art concert in a diversity of computer vision tasks such as image classification, localisation, and detection [1].

The growth of automobiles is outpacing the growth of the economy. Accidents and the number of people killed in car accidents. A lack of situation on routes such as express highways is responsible for the majority of accident deaths. Driving a car has become more and more common in people's lives. As a result, traffic safety is crucial. For independent vehicles, traffic signs are used for traffic notification, guideline steering, and the board of significant data [2]. These signs are intended to influence driving behaviour. Because of the massive increase in the number of street cars around the world, the number of street accidents has also increased significantly. Obliviousness of the street sign, impediment of the street sign, and interruption of the drivers are some of the major causes of disasters.

Traffic sign categorization is the process of automatically detecting traffic indicators such as speed limit signs, yield signs, merge signals, and so on. Expert systems rely significantly on traffic sign detection and identification, such as traffic assistance driving systems and automatic driving systems. It immediately helps both human and robot driving systems detect and recognise traffic signs. If we can recognise traffic signs automatically, we can build "smarter cars". Self-driving automobiles entail traffic sign recognition to efficiently read and understand the roadway. Likewise, in order to aid and protect drivers, "driver alert" systems in cars must recognise the road environment.

Furthermore, traffic signals ensure that road users follow specified regulations, reducing the chances of traffic offences. The usage of traffic signals also helps with route guidance. All road users, including automobiles and pedestrians, should prioritise road signals. For a variety of reasons, such as focus issues, tiredness, and sleep loss, traffic signs are ignored. Poor vision, the outside world's influence, and the surrounding surroundings are all elements that contribute to ignoring the warnings.

2 Literature Survey

The authors in [3] describe a revolutionary lightweight CNN architecture for traffic sign identification that does not require the use of a GPU. The authors' key hurdles in identifying traffic signals in real time circumstances include image distortion, speed, motion effect, noise, and faded hue of signs. The German Traffic Sign Recognition Benchmark and the Belgian Traffic Sign Classification dataset were utilised in the experiments. The authors in [4] suggested an efficient convolutional neural network for detecting minor traffic signs. The authors focussed on small object detection challenges and suggested an efficient convolutional neutral network for detecting small traffic signs, comparing accuracy to R-CNN and Faster R-CNN.

The paper [5] describes a CNN Ensemble-based traffic sign detection and recognition system. The suggested system is separated into detection and recognition modules, and it is tested on the Belgium and the German Traffic Sign Benchmark data sets. In the detection stage, photos of traffic signs are captured and an object is located from the image; in the identification step, a convolutional neural network assembly is used to provide a tag to the identified sign. The developers of [6] formed a system for detecting and recognising traffic signs. For traffic sign detection and recognition, the authors utilised a Convolutional Neural Network (CNN) with the mask RCNN.

There are some model that can detect the road signs and some more objects also, they use the techniques of image classification and object localisation. These models have some pros and cons that made us to improve the models further. Now the Tesla and MG cars also using this type of techniques to detect the objects on the roads and use some sensors like radar to position the objects in the correct location in the 3d space, but these models should need further improvement to get best results. The major problem in the existing models are specific to particular country or location, not suitable for the Indian road signs and also have high rate of occupation problem and noise that result to reduce in accuracy and the performance of the model.

On the Indian road sign data set, the proposed TSDR model will perform well. This model has a high level of accuracy, which means it will function well in the actual world. The proposed model will detect 32 different types of Indian road signs that we trained on. The model's overall accuracy is greater than 90%, which is sufficient for us to apply it in the actual world. Because it was trained on numerous categories, the suggested TSDR model makes it easier to predict Indian road signs. The model's accuracy is quite great, and it will also help to reduce traffic accidents.

3 Proposed TSDR Architecture

The suggested system uses the CNN algorithm to detect and recognise traffic signs. Pre-processing of the input is done before classification to remove noise and to reduce complexity. We used technologies like deep learning, computer vision, and python to build this application. The proposed TSDR model will be able to detect the 32 different road signs. We trained our model with 20,000 images which has been collected from German and Indian road sign data set. We used Keras to build the model as Keras provides vast number of features which enabled us to build state-of-the-model that will detect the road sign with accuracy of above 90%. Figures 1 and

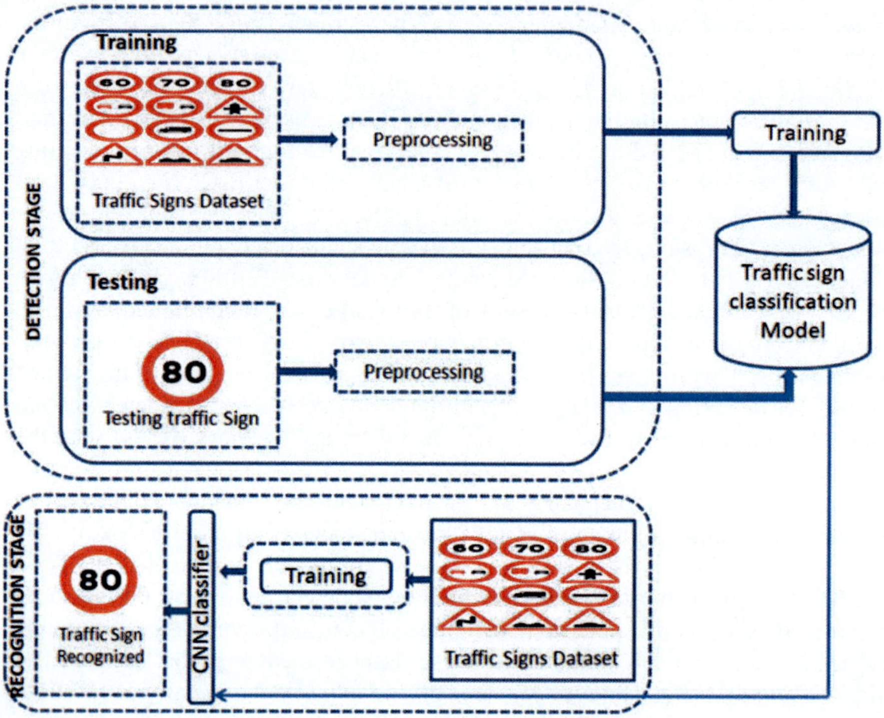

Fig. 1 TSDR architecture diagram

2 depict the General Architecture of TSDR and a detailed flow of processes in the design of TSDR, respectively.

The abstract working for the components in the process is given below:

Pi Camera: It is a compact light load camera that bolsters the pre-processing of image. It switches to the MIPI camera sequential interface convention when Pi is used. It's ordinarily utilised in picture handling, AI or in observation ventures; it's regularly utilised in examination rambles since the payload of camera is a smaller amount. Apart from these modules, Pi may also use standard USB webcams that are often used with PCs.

Grey Scaling: It is the method of transforming a picture to shades of grey from other colour spaces such as RGB and HSV. The detected input image by the Pi camera ranges from completely black to completely white.

Histogram Equalisation: The histogram of the input image is used to change the contrast of the image in this image processing technique. It equalises and improves the image's contrast by scattering out the most common pixel intensity values or stretching out the intensity range of the supplied image.

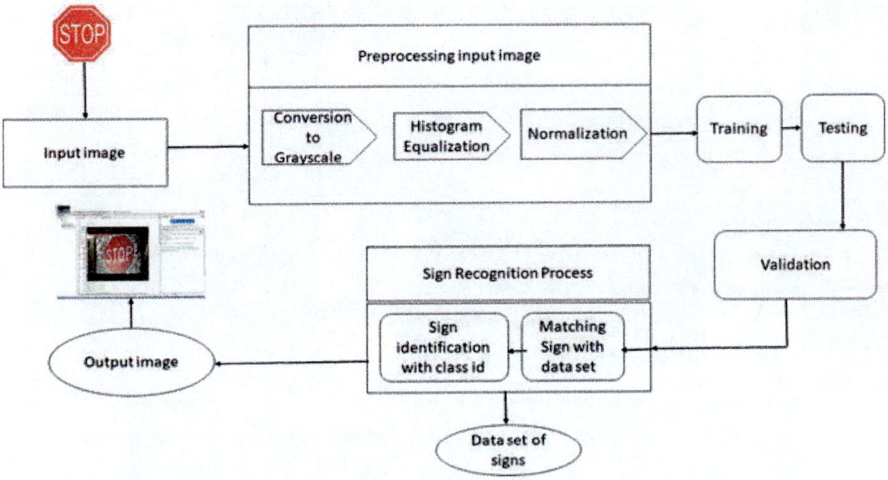

Fig. 2 Detailed flow of processes in TSDR

Image Validation: Image validation occurs in gateway server image processing threads. Image compliance collects the validation data so it can apply the specified data set of image validation rules, which allows a rule dataset to execute the same image test using different threshold criteria and validate.

Image Recognition: Speed restrictions, prohibited entrance, traffic signals, turn left or right, no passing of big trucks, and so on are all examples of traffic signs. The technique of determining which classes a traffic sign belongs to is known as traffic sign recognition.

Steps for the TSDR Model

(1) **Data collection from various sources**: The road sign data has been collected from the German and Indian Road sign data set. The data set consist of 20,000 images of different road signs. There are 32 different road signs that are similar to Indian road signs in 30 classes. So we made our prediction with the 32 different road signs as shown in Fig. 3.

(2) **Pre-processing of Data**: The images are resized into 32*32 pixels. The images are then converted into grey scale and histogram equalisation and then the images is given to the model for testing.

(3) **Splitting of training and testing data**: The imported data set of images is separated into the Training, Validation, and Testing. The distribution of training dataset as number of images used in each class of traffic sign is depicted in Fig. 4. The number of images taken for each is as follows:

 (a) For Training 16,472 images with 32×32 and 3 Layers
 (b) For Validation 4119 images with 32×32 and 3 Layers
 (c) For Testing 5148 images with 32×32 and 3 Layers

Fig. 3 Different road signs used in TSDR

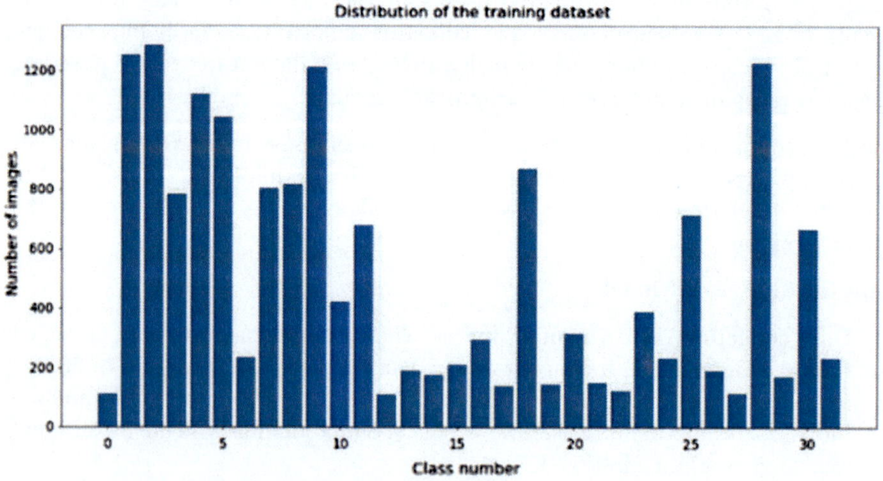

Fig. 4 Distribution of training dataset

(4) **Model creation and selection**: The model is trained with the training data for 10 times. The training images have divided into batches with 50 images each batch. The proposed model has 11 layers of network and 32 layers node will predict 32 different classes of road signs.

(5) **Validation of model on testing data**: Once the model is created, it has been validated by extracting the image path with corresponding labels. Next is to resize our images to 32 $\times$ 32 pixels to predict the model.

(6) **Deployment of the selected model**: The recognition network is trained on the cropped image part of the dataset to identify different road signs in an image.
(7) **Prediction of output**: The model can detect 32 different types of road signs which will make the system more efficient.

4 Results and Discussion

We used data from a German traffic road signs data set that was just uploaded on Kaggle. Kaggle is a platform for all data science and AI challenges. Our programme will be able to predict 32 different road sign kinds. The German road sign dataset has been reduced from its original size of 20,000 pictures. The first layer, also known as the input layer, is made up of 32×32 nodes (1024 nodes), which is the same as the image size. Layer 2 has 784 nodes and is also known as the Hidden Layer. The Relu linear activation unit is used to activate the neural network and perform the appropriate calculations. The rectified linear activation unit addresses the problem of vanishing gradient and allows us to design large deep neural networks. The disappearing gradient is a problem that completely damages the neural network's output. The convolution layer is then connected to the densely connected neural network via the flatten layer.

Figure 5a and b show the input design and the trained pickle model, respectively. The image will be delivered through the model to predict the output, and the model output will be provided along with the gathered image to a subsequent process to alert the driver with an LCD display and recognise the indications. For the purpose of demonstration we have used the Pycharm software and we have used python 3.7 for the computation purpose.

The model can detect 32 different types of road signs which will make the system more efficient. The model has the accuracy of 99%. The system will be easily integrated with other hardware equipment's. Figure 6a and b depict the "Ahead Only"

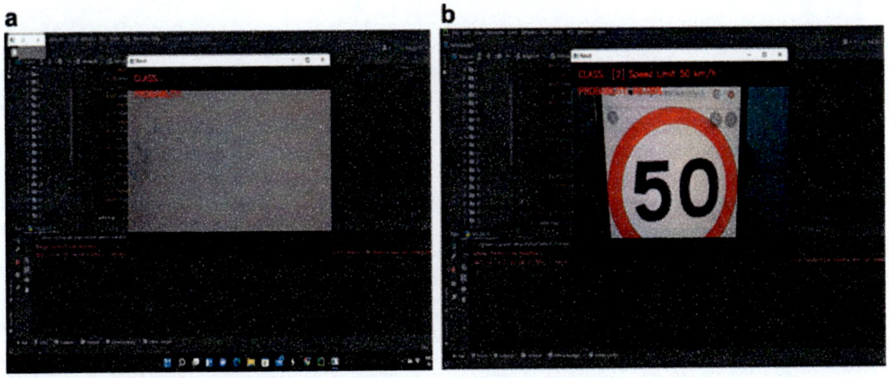

Fig. 5 **a** Input design. **b** Output GUI

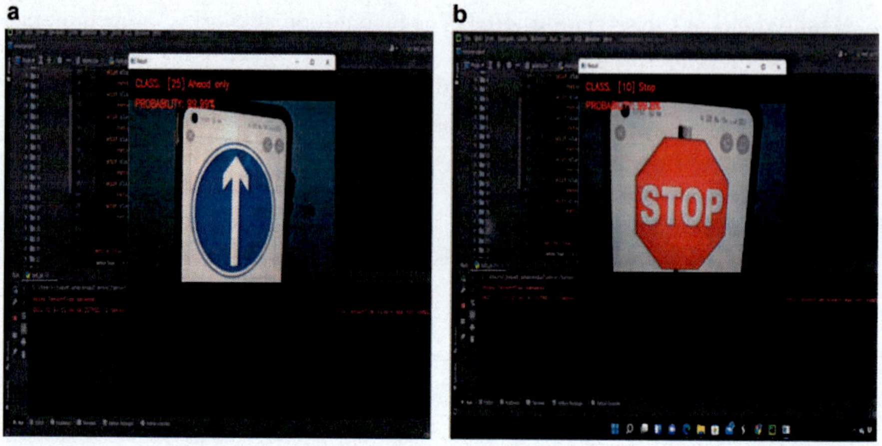

Fig.6 **a** Ahead only. **b** Stop sign

and "Stop sign". The model predicted the class of the road sign and the probability of the model.

Figure 7 shows the performance analysis of accuracy and loss both in training and validation. At 0th position the accuracy is zero and error is also zero. At 1st epoch the training accuracy has become 0.90 and validation accuracy is 0.95. At 2nd epoch the training loss has became 0.2 and validation loss became 0.01. After this epoch, the training and validation loss is stable and consistent. The loss we obtained by training this model is very good and suitable for all types of vehicles. We trained the model for 10 epochs. The final accuracy of the model is 0.9985 and error is 0.011.

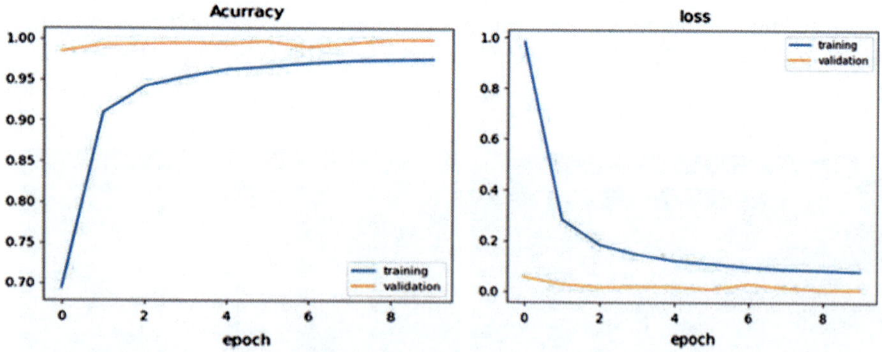

Fig. 7 Performance analysis of accuracy and loss

5 Conclusion and Future Enhancements

The suggested TSDR model can predict 32 various types of traffic signs, which is useful in real-world applications. To be used effectively, the model can be integrated with other equipment. The model has a precision of 0.996573984623 (best if it is near to 1). During the picture acquisition step, the images will be captured with a camera attached on the vehicle and the recognition will be done. When a traffic sign is detected, the system provides a voice alert. This model can be utilised in situations when accurate navigation is required. We can improve this model to make more predictions on new road signs with reduction in error by training the model with huge amount of data. The model can be further improved to make suitable for all type of vehicles.

References

1. He K, Zhang X, Ren S, Sun J (2015) Deep residual learning for image recognition. ArXiv preprint arXiv:1512.03385
2. Starkey NJ, Charlton SG (2020) Speedometer monitoring when driving with a speed warning system. Euro Transp Res Rev (ETRR) 12(1):1–12
3. Haque W, Arefin S, Shihavuddin A, Hasan M (2021) DeepThin: a novel lightweight CNN architecture for traffic sign recognition without GPU requirements. Expert Syst Appl 168:114481
4. Song S, Que Z, Hou J, Du S (2019) An efficient CNN for small traffic sign detection. J Syst Architect 97:269–277
5. Vennelakanti A, Shreya S, Rajendran R, Sarkar D (2019) Traffic sign recognition using a CNN ensemble. In: IEEE international conference on consumer electronics (ICCE), pp 1–4
6. Tabernik D, Skočaj D (2020) Deep learning for large-scale traffic-sign detection and recognition. IEEE Trans Intell Transp Syst 21(4):1427–1440

Human Posture Monitoring

Ashutosh Manorkar, Lavitra Misal, Kaustubh Narkhede, and Meera Arun Thorat

Abstract The problem this report sets out to serve is Human Posture Monitoring. Keeping good posture is very important because it helps in improving spine health. People who keep good posture are less likely to experience back and neck pain. The paper proposes a solution to these problems of fixing human posture. Performing different Asanas that can help fix posture is important. The posture monitoring while performing Asanas is important otherwise it could result in various problems. To keep these activities in check while performing them can yield better results in the long run. Keeping bad posture can result in some chronic problems as one grows old. This can be prevented in advance with practicing some daily Asanas, so the earlier the better. The objective is to monitor the posture of humans using technology like OpenCV and TensorFlow and make comparisons to see if the individual is following the right posture.

Keywords Machine learning · OpenCV · TensorFlow · Image processing · Video processing · Posture monitoring · Neural network · Deep learning

1 Introduction

With the advancement of technology and improvements to computers, it becomes better at interacting with humans in different ways. It leads to systems that can analyze humans in various different ways such as their posture, eye movement and facial changes. Recently, studies have focused on an individual's posture, which is critical for good health as a person should pay attention to his posture or else it might cause them problems later. There have been a lot of attempts at identifying the posture of humans and quite a few have succeeded but not completely as every image is different. This has been achieved using Deep Learning, in which the computer will analyze a data set, in the proposed case, an image, and once taught the important posture points in a person's body, should be able to later detect it in new images.

A. Manorkar (✉) · L. Misal · K. Narkhede · M. A. Thorat
Pimpri Chinchwad College of Engineering, Pune, India
e-mail: ashumanorkar@gmail.com

V. K. Gunjan and J. M. Zurada (eds.), *Proceedings of 3rd International Conference on Recent Trends in Machine Learning, IoT, Smart Cities and Applications*, Lecture Notes in Networks and Systems 540, https://doi.org/10.1007/978-981-19-6088-8_37

Deep learning is a more specific field of Machine Learning which takes inspiration from brain structure and function, which are basically neural networks. The goal of deep learning is to create artificial neural networks.

This field was something that was interesting and this project set out to achieve posture detection. In this case, the project uses an amazing library, the OpenCV library which was created to provide a general package to be used for computer vision apps and to make it easy to use machine perception in commercial products. Using a simple camera/webcam, it is possible to capture an individual and then obtain the main joints and points of the person in the image. After getting this input, the TensorFlow library can be used to train a neural network using a deep learning classifier. Inputting an image into the program will result in an image with the posture points drawn onto the image. The algorithm will mark individual parts such as the ears, chin, chest, hands and legs. The algorithm will then connect the points to form a human skeleton. This posture can then be used for various applications and also for comparing it to other postures.

2 Literature Survey

This study uses the bottom-up approach to find the individuals in the image. In this method, the algorithm first finds the distinctive points (joints) of each individual in the image and then creates a human skeleton from the points found. The paper [1] uses something the paper calls Part Affinity Fields (PAFs), which is basically a list of 2-dimensional fields of vectors that hold the co-ordinates and rotation of the extremities of every individual onto the image. The algorithm is made so that it jointly learns extremities' co-ordinates and their relation according to two forks in the network. In the program, it takes a 2-dimensional color image as an input to create 2-dimensional co-ordinates of points (joints) for every human in the image.

This paper discusses two methods to segment and retrieve the image data. This segmentation helps in sorting out individuals and their individual parts that the algorithm will work on to see if it is a point of interest (joint). The paper discusses how segmentation can be obtained by using gradient texture and feature space, and also unsupervised clustering. Segmentation is usually done so as to output a resemblance map that can be used for further analyzing. The paper [2] also uses bitmaps—which is a set of pixels that are directly related to the pixels on the monitor. These are also called raster images. This image of a bitmap is made from a set of pels, with each pel having some unique color. All this research on image processing is now being bundled up with machine learning and other technologies so that the computer is able to better analyze the different patterns of different objects.

This research paper [3] emphasizes more on the paucity of easily available methods for one person Human Pose Estimation which are both computationally efficient and effective in estimation precision. The technique of using the advantage of Conv-Nets and proposing an improved approach called Efficient Pose has helped a lot. The sole goal is to change Open-Pose into a family of scalable ConvNets for high precision and computationally efficient single-person pose estimation from 2-dimensional images. To gauge the performance and analyze the standard of the approach taken, the paper [3] deals with two separate qualified studies. First is gauging the Efficient Pose model by comparing it to the real Open-Pose model on one person Human Pose Estimation. Second, comparing the algorithm to the current best single-person Human Posture Estimation methods on the official MPII pose dataset challenge, buckling down to accuracy as a function of the amount of parameters. The given Efficient Pose models try to induce high computational efficiency, while bridging the gap in the availability of higher precision Human Pose Estimation networks.

The research paper [4] aims at increasing pose detection efficiency without performance degradation but still getting accurate results. The paper deals with identifying the real problem with pose estimation and its accuracy. The research paper has used the method where one could see basic Convulsion Neural Network foundation for the best human pose networks such as Hourglass are not efficient in creating small networks because there are more channels in each layer and it's quite difficult to train. To get away with these barriers, the paper [4] proposed an effective solution of coming up with the network and proposing a better method of training for small pose networks which uses distillation fashion. The method can be called Fast Pose Distillation (FPD). This technique helps analyze pose estimation in a more efficient manner by comparing it with the best alternatives. The so stated FFD methods are better and cheaper too. This does help in the long run. Observing smaller model sizes while also getting the same level of performance is quite a gateway.

The paper [5] has proposed a unique estimation approach that is an easily understandable model to serve complicated examples of position monitoring. The paper has compared single pose estimation with multi posture estimation and the paper showed that multi posture estimation required for calculations of postures of everyone in the location and is a considerably more demanding job because of occlusions, different vocalizations and communication between the groups of people. The technique has used a complex ResNet architecture and proposed a detailed information buildup for training. To encounter some difficult cases, the paper suggested a unique data development method that helps co-ordinate input postures and let the network learn to recognize the imprecise joints conjectures and rectify them.

This research paper provides general solutions for pose estimation as well as monitoring. They are straightforward but effective. The paper [6] has used a method where pose monitoring uses the same greedy matching technique. The paper has modified the technique and used an optical pose dispersion and resemblance measurement. The paper involves skip layer attribute concatenation. A particularity of the three methods proposed in this research paper is the three up sampling steps and also three levels of nonlinearity which are used to form feature maps having high-resolution and heat maps. Obtaining high-resolution feature maps is crucial, based on the foregoing facts and the outstanding performance of the baseline given in this study work.

This research paper suggests how for an organized prediction problem such as that of human posture detection, CPMs naturally recommend a structured framework that replenishes gradients as well as by putting intermediate supervision on the network at regular intervals, the network is guided to produce increasingly exact belief maps. The paper also proposes various training schemes of such a sequential prediction model. The paper proposed multi-staged architecture is totally distinguishable and as a result, back-propagation can be used to train the system from beginning to end. With the aim of getting long range interactions between parts, the frame of the network in each stage of the sequential forecasting framework is encouraged by the goal of achieving a colossal receptive field on the image as well as the belief maps. The paper [7] has proposed a technique to achieve best results on standard guidelines which have the MPII, LSP, and FLIC datasets, and examine the effects of jointly training a multi-staged architecture with frequent transitional supervision.

In this research paper [8], an Omni-Pose framework is used which guesses the location of people's joints contingent on contextual details because of multiple attribute representation in the network. The circumstantial approach that allows the network to take in details from the full frame, obviating the need for post-processing using statistical or geometric techniques. Moreover, the waterfall module allows for well shape detection, which in turn results in accurate assessment of occluded joints. The proposed Omni-Pose framework in this research paper escalates in performance by assimilating Gaussian heat map modulation that amplifies de-convolution activities in the multi-scale architecture for encoding and decoding for a more accurate depiction of joint identification and a reduction in network delayed feedback. The paper has proposed the unique. The Omni-Pose framework is a single-pass, fully trainable multi-scale technique that produces good multi-person posture detection results.

The first approach of Key-Point-based posture estimation for (transparent) 3 dimensional objects from RGB images is presented in this research paper [9]. There were numerous challenges, one of which was the absence of a large dataset for transparent 3D object position estimation from stereo pictures with clarified key-points. To overcome the challenges of data acquisition and elucidation, the paper has introduced an efficient method of apprehending and identifying RGB images for objects that are transparent. The research paper [9] proposes a deep model, Key-Pose that predicts 3-dimensional key-points on transparent objects using cropped stereo RGB inputs to address the difficulty of dependable depth. The model analyzes the depth implicitly by integrating details from the pair of images, and forecasting the 3-dimensional location of key-points.

This research report [10] explored the research line of sustaining high-resolution representations and the investigation of the high-resolution network (HR-Net) for a wide range of visual tasks, which was originally designed for human posture estimation. A HR-Net preserves high-resolution depiction by staging multi-scale fusions across aligned convolutions and associating high and low resolution aligned convolutions. Not only are the resulting high-resolution representations impenetrable, but they are also semantically precise. The representations from all the high to low resolution parallel convolutions, rather than just the high-resolution representations in the original HR-Net, were examined in this study work. This change results in a minor overhead and more impregnable high-resolution representation. HR-NetV2 is the name of the resulting network.

3 Proposed Work

This project mainly works on detection of human posture in images and their accuracy. It will work on algorithms to compare two postures—of different individuals and parameters where the comparison can be made. It aims to find a proper method which can compare the two postures of two individuals so that various inferences can be made and be used in applications such as correcting the postures of individuals doing exercises, Yogasanas and other activities. The aim is to first take images as an input and find individuals from the image. For this, the project plans to use a trained model which will be able to plot the points on the limbs of the individual. After these co-ordinates have been plotted then multiple co-ordinates of correct postures of different activities such as exercises and Yoga poses can be stored. This will create a dataset for the usage which can be used to compare more postures with the dataset. The next task the project aims to do is find out a method to compare two postures with another. The project plans to use the angles between the joints of the individual to compare with the angles of the joints of the other person. This comparison then can be used to tell the user how perfect their poses are and can also be used to tell them what changes they might have to do.

Module 1: Performance Evaluation of human detection in images with certain performance metrics.
Module 2: Implementation of the detection of humans and their postures in various images.
Module 3: Implementation of algorithm to compare two postures and also a certain correction percentage.

4 Result Analysis

Table 1 below shows the posture correctness value in a percentage format where the comparison of different Yogasanas where an image is taken as a reference and stored in the dataset and all other images are compared with it. The program depicts the percentage correctness value.

Figure 1 depicts the flow of executing the whole process of comparison between different Asanas. The key-points of the joints get detected and stored in the database and then used for calculating the percentage values. The case study for the above Asana is explained below.

4.1 Case Study

For case study the paper has proposed the use of "UtkataKonasana" also known as Sumo or Malasana pose and compared the correct posture with three images having

Table 1 Input posture correctness value

Sr. No	Reference image	Input Sr. No	Input image	Correctness (%)	Average correctness (%)	Yogasana whose correctness is tested
1	 Image.1	a	 Image.1a	83.571	78.141	Tadasana
		b	 Image.1b	75.714		
		c	 Image.1c	75.714		
2	 Image.2	a	 Image.2a	83.928	73.809	Mountain pose
		b	 Image.2b	61.785		

(continued)

Table 1 (continued)

Sr. No	Reference image	Input Sr. No	Input image	Correctness (%)	Average correctness (%)	Yogasana whose correctness is tested
		c	 Image.2c	75.714		
3	 Image.3	a	 Image.3a	70.0	76.157	Virabhadrasana
		b	 Image.3b	85.142		
		c	 Image.3c	73.333		
4	 Image.4	a	 Image.4a	84.285	76.428	Butterfly asana

(continued)

Table 1 (continued)

Sr. No	Reference image	Input Sr. No	Input image	Correctness (%)	Average correctness (%)	Yogasana whose correctness is tested
		b	 Image.4b	80.0		
		c	 Image.4c	65.0		
5	 Image.5	a	 Image.5a	61.632	53.826	Leg swing
		b	 Image.5b	47.346		
		c	 Image.5c	52.50		

(continued)

Table 1 (continued)

Sr. No	Reference image	Input Sr. No	Input image	Correctness (%)	Average correctness (%)	Yogasana whose correctness is tested
6	 Image.6	a	 Image.6a	78.571	74.999	Sukhasana
		b	 Image.6b	65.714		
		c	 Image.6c	80.714		
7	 Image.7	a	 Image.7a	62.857	49.364	Trikonasana
		b	 Image.7b	40.952		

(continued)

Table 1 (continued)

Sr. No	Reference image	Input Sr. No	Input image	Correctness (%)	Average correctness (%)	Yogasana whose correctness is tested
		c	 Image.7c	44.285		
8	 Image.8	a	 Image.8a	63.333	56.122	Vrikshasana
		b	 Image.8b	38.367		
		c	 Image.8c	66.666		
9	 Image.9	a	 Image.9a	91.785	80.901	Utkatakonasana

(continued)

Table 1 (continued)

Sr. No	Reference image	Input Sr. No	Input image	Correctness (%)	Average correctness (%)	Yogasana whose correctness is tested
		b	 Image.9b	70.204		
		c	 Image.9c	80.714		
10	 Image.10	a	 Image.10a	85.0	76.904	Vajrasana
		b	 Image.10b	74.285		
		c	 Image.10c	71.428		

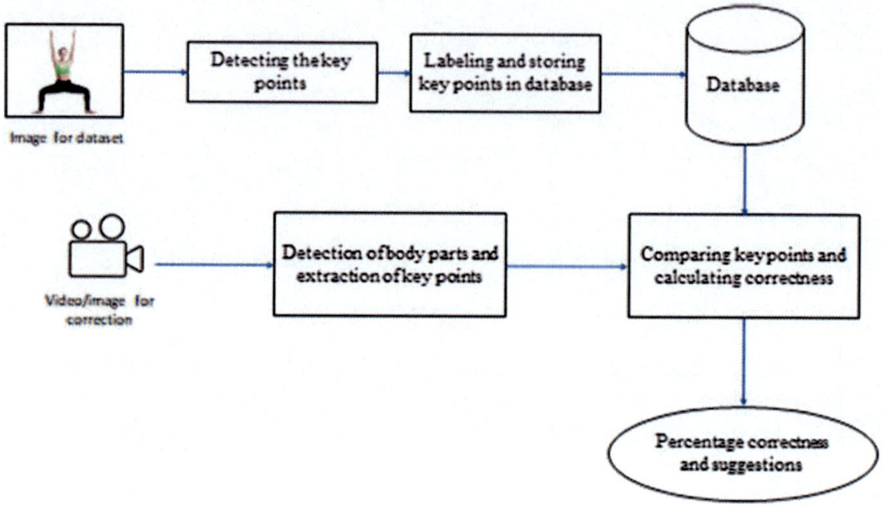

Fig. 1 Brief flow diagram

posture slightly incorrect. The program first finds the set of correct angles between the joints which will be used to compare other postures for correctness.

To extract the correct set of angles the image 1 in Fig. 2 is provided to the program and stored the angles. Then used image A, image B, image C to check correctness. Posture in image A has left leg turned outward from body, in image B along with an improper position of left and right leg there is also a slightly tilted hand position and in image C left leg and right hand are not in correct position. As the difference in posture of image C is very less and that in image B is high with respect to the correct pose, and results in a high correctness value for image A and a low correctness value of image B.

When the images were provided to the program which compared each angle recorded. It resulted in the correctness values as 91.785%, 70.204%, 80.714% for image A, B, C, respectively. These correctness values are what were expected earlier.

Fig. 2 UtkataKonasana correctness comparison

The table above has comparisons between different images—the images in the column "Reference Image" is the image that will be used for comparison. These are the images that the program analyzes and stores their data points and the angles between different joints for comparison. Each reference pose has been compared with three variations of the same pose as seen in the column "Input Image". The "Correctness" column shows how similar the input image is to the reference image in a percentage. This is calculated using the comparison algorithm in the program. The "Average Correctness" column shows the average correctness for each image. The final column "Yogasana whose correctness is tested" shows the name of the Yogasana that is being depicted by the images.

5 Future Scope

This project has a lot of future scope. Until now, only the backend analysis algorithm has been completed. In the future a frontend could be added to this project. This project could be used in lots of applications as mentioned before. The program can also suggest correctness to a user. It will be able to tell which part of the user's posture is wrong and can tell them how they could fix their posture. This can help a lot of people lead healthier lives by assisting them in their exercises, Yogasanas, physiotherapy and much more.

6 Conclusion

Human posture is imperative to maintaining a healthy lifestyle for any individual. With the increase of modernity all around the world, humans have gotten worse at maintaining postures as they spend most of their lives sitting down and slouching. With this, research has also increased on human posture and how it can be maintained using software. This project sets out to fix this problem and has reached its goal. This project can also be taken further and lots of improvements can be made to it so that it can interact with the user.

References

1. Realtime Multi-Person 2D Pose Estimation using Part Affinity Fields./* IEEE April 2017
2. Computer Vision and Image Processing: A Paper Review. Int J Artif Intell Res 2018
3. Efficient Pose: Scalable single-person pose estimation, Applied Intelligence, 6 Nov 2020
4. Zhan F, Zhu X, Ye M (2019) Fast human pose estimation. University of Electronic Science and Technology of China, 2 Apr 2019
5. Fieraru M, Khoreva A, Pishchulin L, Schiele B (2018) Learning to refine human pose estimation Apr 2018

6. Xiao B, Wu H, Wei Y (2018) Simple baselines for human pose estimation and tracking, 21 Aug 2018
7. Wei SE, Ramakrishna V, Kanade T, Sheikh Y (2016) Convolutional pose machines
8. Artacho B, Savakis A (2021) OmniPose: a multi-scale framework for multi-person pose estimation
9. Liu X, Jonschkowski R, Angelova A, Konolige K (2020) Keypose: multi-view 3D labeling and keypoint estimation for transparent objects, 18 May 2020
10. Sun K, Zhao Y, Jiang B, Cheng T, Xiao B, Liu D, Mu Y, Wang X, Liu W, Wang J (2019) High-resolution representations for labeling pixels and regions, 9 Apr 2019
11. Girdhar R, Gkioxari G, Torresani L, Paluri M, Tran D (2018) Detect-and-track: efficient pose estimation in videos. In: Proceedings of the IEEE conference on computer vision and pattern recognition, pp 350–359
12. He K, Gkioxari G, Doll´ar P, Girshick R (2017) Mask R-CNN. In: Computer vision (ICCV), 2017 IEEE International Conference on. IEEE, pp 2980–2988
13. He K, Zhang X, Ren S, Sun J (2017) Deep residual learning for image recognition. In: Ilg E, Mayer N, Saikia T, Keuper M, Dosovitskiy A, Brox T (eds) Proceedings of the IEEE conference on computer vision and pattern recognition. pp. 770–778 (2016). [14] Flownet 2.0 (2017) Evolution of optical flow estimation with deep networks. In: IEEE conference on computer vision and pattern recognition (CVPR), vol 2
14. Ioffe S, Szegedy C (2015) Batch normalization: accelerating deep network training by reducing internal covariate shift. In: International conference on machine learning, pp 448–456
15. Dai J, Li Y, He K, Sun J (2016) R-FCN: Object detection via region-based fully convolutional networks. In: NIPS
16. Jin S, Ma X, Han Z, Wu Y, Yang W, Liu W, Qian C, Ouyang W (2017) Towards multi-person pose tracking: Bottom-up and top-down methods. In: ICCV PoseTrack Workshop
17. Kingma DP, Ba J (2015) Adam: a method for stochastic optimization. ICLR
18. Krizhevsky A, Sutskever I., Hinton GE (2012) Imagenet classification with deep convolutional neural networks. In: Advances in neural information processing systems. 16 B. Xiao et al., pp 1097–1105
19. Lin TY, Maire M, Belongie S, Hays J, Perona P, Ramanan D, Doll´ar P, Zitnick CL (2014) Microsoft coco: common objects in context. In: European conference on computer vision. Springer, pp 740–755

Classification of Devanagari Characters Dataset Using Deep Learning Methods

S. Brunda, C. S. Soumya, D. P. Akarsha, C. Manjunatha, C. V. Nikshep, Samuel Roushan Kumar, and G. B. Nijaguna

Abstract Deep Learning is a growing set of approaches for extracting useful information and knowledge from large amounts of data. Deep Learning research and tools have focussed on commercial sector applications. Only a fewer Deep Learning research have focussed on scientific data. This paper highlights the Machine Learning techniques applied to handwriting character recognition using network models built by using Deep Learning Techniques. Handwriting Recognition is now getting attention for researchers to assist technology for visually impaired, blind and human-robots for business documents. We have experimented Devanagari Character database with two different architectures GoogLeNet and AlexNet to evaluate the performance and achieve highest accuracy.

1 Introduction

People learn to read and write at school, and as they become older, they gain more advanced reading and writing skills that may be printed or typed on a machine. The majority of people have no trouble reading light or thickly written texts. They are skilled at igitizeng distorted letters or expressions in text written in fancy font styles, broken, fractured text, or misspelt words based on context. The most important requirement for computers today is the ability to quickly interpret a book. For the last six decades, many academics have sought to replicate human reading with computers in order to discover efficient and effective approaches, but with little success. As a result, handwritten character recognition is a key research problem in the field of computer visibility and pattern capture. Character recognition can be used to igitise large amounts of data in a variety of commercial and scientific applications, including

S. Brunda · C. S. Soumya (✉) · D. P. Akarsha · C. Manjunatha · C. V. Nikshep · S. R. Kumar · G. B. Nijaguna
Nitte Meenakshi Institute of Technology, BSF Campus, Yelahanka, Bengaluru, Karnataka 560064, India
e-mail: soumya.cs@nmit.ac.in

V. K. Gunjan and J. M. Zurada (eds.), *Proceedings of 3rd International Conference on Recent Trends in Machine Learning, IoT, Smart Cities and Applications*, Lecture Notes in Networks and Systems 540, https://doi.org/10.1007/978-981-19-6088-8_38

passports, invoicing, office automation, automated mail sorting and check-out identification. To detect characters, an optical scanner is used to capture a copy of the original page, which is then saved as a bitmap. To identify written text from visuals to excite the letters, the bitmaps are processed using an Optical Character Recognition (OCR) method. Earlier OCR systems saved bitmaps that matched photographs based on certain fonts. Each OCR invention goes awry by attempting to match the real character. The primary aim of this research is to identify the individual letters using the GoogLeNet and AlexNet models by training the model and assessing its efficiency.

2 Related Work

This section gives a review of the literature on the various stages of the OCR system. During the past several years significant experimental in the identification of the characters of various scripts, particularly in different languages like English, Arabic, Chinese, etc. and other Indian scripts have been carried out. We have a short overview of literature survey relevant with foreign literature, Indian scripts given. In 1979 research began on automated identification on these Devanagari scripts. A systematic method for this Devanagri script identification described in IIT Kanpur Sinha's thesis [1]. OCR method for printed Devanagari was created by Pal and Chaudhuri [2]. The first solution for Devanagari scripts was done by B. B. Chaudhuri, U. Pal et al., of Indian statistic institute. Now years later the research into this field is booming. Sethi and Chatterjee [3] developed a structural technique to distinguishing handwritten Devanagari numbers. Right and left slants, as well as other primitives, are employed. To normalise each aspect, Bajaj has normalise three unique characteristics—(1) Compactness characteristics; (2) Instant characteristics of the right, left, upward downward profile curves (3) Descriptive component characteristics.

Hanmandlu [4] suggested the box technique for extracting information from handwritten numbers, which requires dividing number images into boxes spatially. Code and gradient-based characteristics are used to normalise Devanagari numbers. Kirsch directional edges cover distance transform, code, gradient and directional remoteness distribution.

For the Devanagari character recognition problem, we're looking at the deep belief network (DBN) [5]. Two stage structural groups based on structural properties are used to categorise complicated characters in the first step. Characteristics such as changed wavelet approximation, Euclidean distance, and pixel thickness features from mechanically classified and normalised characters are extracted and applied to three different neural networks in the second stage. Finally, if the recognition output is distinguishable from all neural networks with 43 adjusted ripple approximation properties, it is picked by a majority vote.

3 Architecture of the System

In the model, data is collected from the Internet source and an image is introduced. Additionally, the picture will be rescaled for future usage. The dataset is divided into two parts: training and testing. Following that, we consecrate the test set and chose X percent of our training dataset at random as the actual training set and the remaining (100 X) percent as the validation set, where X is a fixed percentage (say, 80%).The model is then iteratively trained and validated on these 2 distinct sets. Max-Pooling is a compression method that uses the grid's maximum pixel value to reduce the size of an image. This also helps to prevent overfitting and makes the model more general. Following model construction, it is necessary to train the model. We were capable enough to construct an efficient artificial convolutional neural network for image recognition (Fig. 1).

Finally, we constructed and trained the model using the training data. Model testing is done after the model has been trained. During this step, a test set of data is loaded. The model's real accuracy will be confirmed because it has never encountered this data set. Finally, the model may be used in real-world scenarios when it has been saved.

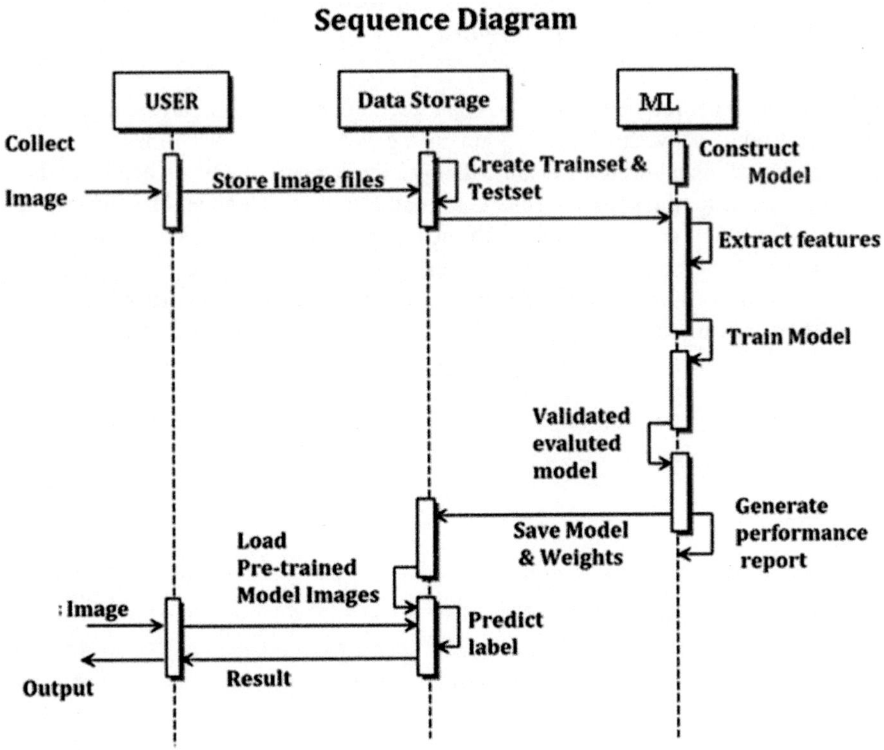

Fig. 1 Diagram of sequence

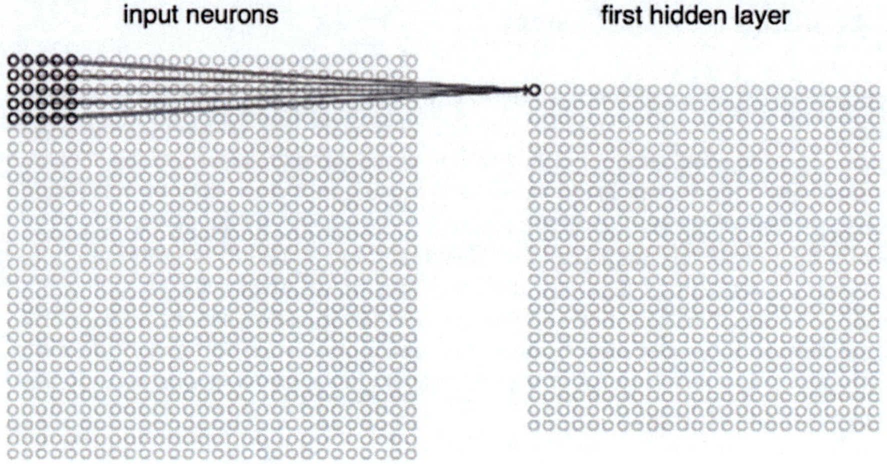

Fig. 2 Neurons layers

4 Technologies for Implementation

4.1 Convolutional Neural Network

A convolutional neural network (CNN) is particular design of artificial neural network, introduced by the Yann LeCun in the year 1988. CNN utilises certain characteristics of visual cortex. Assume that the input-matrix analysis begins at the top left of the image, the software selects a smaller matrix, which is referred to as a filter. The CNN then goes sideways with the contribution picture as a result of the filter. The job of the filter is to multiply values by the original pixel values. At the conclusion, a number is obtained. Because the picture in the left corner has been read, the filter advances one unit to the right, repeating the process. After applying the filter to all points, a matrix is generated which is less than the original matrix (Fig. 2).

4.2 AlexNet

AlexNet is a deep neural network created in 2012 by Alex Krizhevsky and colleagues. It was created for the purpose of classifying pictures for the Imagenet LSVRC-2010competition, where it obtained state-of-the-art results [6]. Additionally, it supported multiple GPU's. AlexNet is a deep convolutional network that was designed to handle large coloured pictures (224 × 224 × 3). It supported approximately 62 million trainable parameters in total.

4.3 *GoogLeNet*

At the ImageNet Large-Scale Visual Recognition Challenge 2014(ILSVRC 2014), GoogLeNet was built on a deep convolutional neural network architecture dubbed "Inception," which was responsible for defining the new state-of-the-art for classification and detection. Overfitting is an issue that may occur when a network is designed with numerous deep layers. The GoogleNet architecture was hence designed to mitigate the problem by allowing filters of various sizes to work on the same level. The network grows wider instead of deeper as a result of this approach.

5 Results and Analysis

In this last step, we have examined and evaluated the routine of our classification model on our prepared picture dataset. To assess the efficacy of our newly developed classification and to compare it to existing methods, we use accuracy to quantify classifier effectiveness (Fig. 3).

We measured success by measuring how many of the test set images were correctly categorised into their respective category bin out of all the categories. Our approach could be useful to apply to character recognition tasks when there are limited resources to train with. The work trains each model-on epochs to achieve maximal accuracy (Table 1) (Figs. 4 and 5).

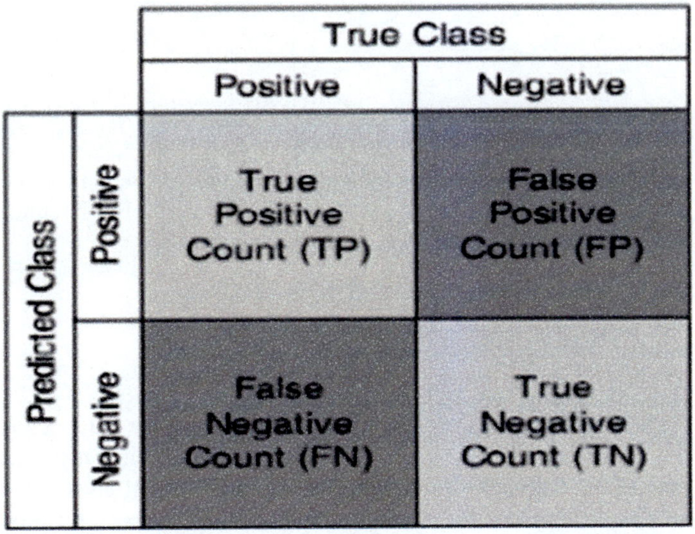

Fig. 3 Confusion Matrix accuracy

Table 1 Accuracy validation

Architecture	Epochs	Training accuracy	Validation accuracy
AlexNet	50	0.99	1.00
GoogLeNet	4	0.8196	0.8168

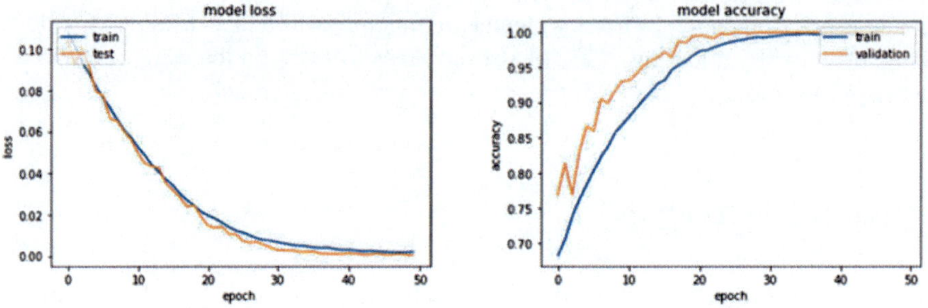

(a) The figure depicting the loss for the Alexnet Architecture

(b) The figure depicting the accuracy for the Alexnet Architecture

Fig. 4 Results of validation and training accuracy for AlexNet Architecture

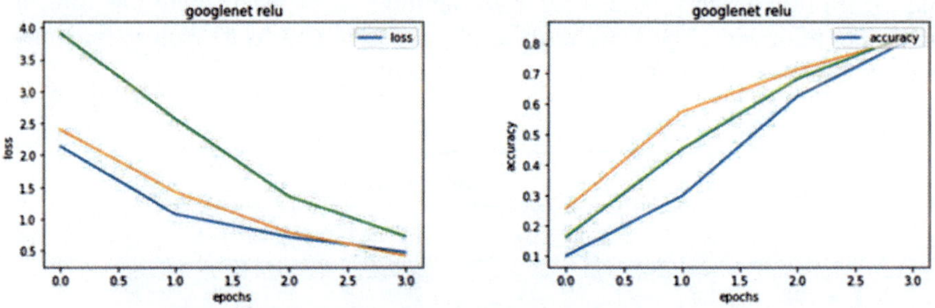

(a) The figure depicting the loss for the GoogLeNet Architecture

(b) The figure depicting the accuracy for the GoogLeNet Architecture

Fig. 5 Results of validation and training accuracy for GoogLeNet architecture

6 Conclusion and Future Work

Deep Learning is a well-known approach that has been put to the test in a number of fields, including computer and document analysis. To recognise unconstrained handwritten Devanagari characters, we employed the ResNet18, GoogLeNet and AlexNet Models. We were able to automatically detect and categorise the best qualities thanks to the Deep Neural Network.

- Out of all the potential categories, we counted how many of the test set images were correctly sorted into their appropriate category bins.
- Our method might be useful for character recognition difficulties when there are minimal resources to train with. Each model in the study is trained using epochs to achieve maximum accuracy.

References

1. Sinha RMK (1973) A syntactic pattern analysis system and its application to Devnagari script recognition. Ph.D. Thesis, Indian Institute of Technology, Kanpur, India
2. Chaudhuri BB, Pal U (1997) An OCR system for reading two Indian scripts: Bangla and Devanagari. In: Proceedings of the fourth international conference on document recognition, pp 1011–1015
3. Sethi IK, Chatterjee B (1977) Machine identification of Devanagari numerals written by hand. J Inst Electron Telecommun Eng 22:532–535
4. Sinha RMK, Mahabala H (1979) Machine recognition of Devanagari script. IEEE Trans Syst Man Cybern 9(8):435–441
5. Madhvanath S, Govindaraju V (1999) Local reference lines for handwrit ten word identification. Pattern Recogn
6. Impedove S, Ottaviano L, Occhinegro S (1991) Optical character recognition—a survey. Internal J Pattern Recogn Artif Intell 5(1–2):1–24

LightHARs: A Light Human Activity Recognition System

Darsh Dave and Randheer Bagi

Abstract Human activity recognition (HAR) is the method to identify an activity of a human by rendering images, video, and sensor data collectively. While performing HAR and working with images, the video dataset is not secure because these datasets reveal the user identity, but the sensor data do not. Therefore, in this paper, we proposed a cost-efficient deep neural network model i.e. light human activity recognition system (LightHARs) which takes less parameters and computational time as compared to existing HAR models. It outperformed existing models by achieving 4.5% more accuracy.

Keywords HAR · ResNet · VGG · LSTM · GRU

1 Introduction

With the emergence of deep learning, many real-life challenges are solved [1–6]. Human activity recognition is one of the most complex applications among that. However, various methods are proposed in the literature for human activity recognition. Many of them are using vision-based datasets and give better performance but are not cost-effective. Therefore, to mitigate computational cost, we proposed a sensor data-based HAR system. Here, the main challenge is to achieve high performance with a limited number of parameters so that the model is lighter in nature and cost-effective. In this paper, we introduced a cost-effective model that is light human action recognition system (LightHARs). It uses a lesser number of parameters compared to other HARs models. We utilize the WISDM lab activity prediction dataset to feed our LightHARs.

Randheer Bagi: Contributed equally to this work.

D. Dave (✉) · R. Bagi (✉)
Parul University, Vadodara, Gujarat, India
e-mail: davedarsh1999@gmail.com

R. Bagi
e-mail: randheer.bagi19507@paruluniversity.ac.in

V. K. Gunjan and J. M. Zurada (eds.), *Proceedings of 3rd International Conference on Recent Trends in Machine Learning, IoT, Smart Cities and Applications*, Lecture Notes in Networks and Systems 540, https://doi.org/10.1007/978-981-19-6088-8_39

In the organization of the paper, Sect. 2 covers the state-of-art methods in the field of HAR. In Sects. 3 and 4, we discussed the WISDM dataset and its preprocessing part *i.e.* how the division of dataset will take place for training and testing by giving them labels. Section 5 gives a detailed explanation of the proposed HRAs model architecture. At the end, we discuss the results in Sect. 6.

2 Background Work

In this section, we do a survey on existing work related to human action recognition. Initially, we focus on video and image data process models in deep learning. After that then, we analyze some hybrid models which focus on sensory data, and finally, we see the models which are more related to our proposed model.

Most of the work on image and video datasets is based on separating between the meanings of action and activity [7]. It relies on depth sensors to capture body movement and recognize human activity in 3D stereo and motion capture systems [8]. Authors in [9] perform HAR using a single hybrid deep learning model, which is evaluated on the challenging UCF Sports, UCF101, and KTH datasets. It is essential to consider the complexity of an activity recognition method when defining it. A tree-structured taxonomy is discovered in 2016, which is subdivided into several categories [10]. Some of the researchers focus on ontologies for human behavior recognition [11]. In [12], the authors use ResNet-50 and CPN as a backbone network to construct a multitask network for the task of recognizing human actions, whereas in [13] authors use exceptional CNN structure for human motion reputation and detection technique. Twin-stream CNN is used in [14], which is based on the fusion of sEMG power for activity recognition.

3 Dataset

Human motion recognition WISDM lab dataset [15] consists of records amassed through managed laboratory conditions with the rate of 20Fs. This dataset contains general 1,098,207 samples. Each sample is distributed in the 6 pass time which might be on foot, walking, upstairs, downstairs, sitting, and status, and their distribution is illustrated in Table 1. This dataset is collected using body acceleration, gravity acceleration, body angular speed, body angular acceleration.

4 Preprocessing

In this section, we will process raw data to be experiment ready. We have raw accelerometer data we convert them into x, y and z frequencies by applying the formula:

Table 1 Activity samples in dataset

Activity	Samples
Walking	424,400
Jogging	342,177
Upstairs	122,869
Downstairs	100,427
Sitting	59,939
Standing	48,395
Total	1,098,207

Table 2 Labeling of accelerometers data

X	*Y*	*Z*	Label
−0.6946377	12.680544	0.50395286	0
2.6	9.66	1.0351465	1
0.27240697	1.6480621	8.117727	2
0.000503	−0.099190	0.337933	3
−2.5606253	7.7772183	−0.0	4
−0.800225	0.267827	0.475569	5

$$X = 10 = 1\text{g} = 9.81(\text{m/s})^2$$

$$Y = X \times 1\text{g} = 9.81(\text{m/s})^2$$

$$Z = X \times 1\text{g} = 9.81(\text{m/s})^2$$

The result of these X, Y and Z is the average of the square roots of the sum of the values of each axis squared [16] which is calculated by:

$$\sqrt{xi^2 + yi^2 + zi^2}.$$

based on acceleration formula result data divide into $-x$, $+x$, $-y$, $+y$, $-z$, $+z$ which is illustrate in Fig. 1.

4.1 Labeling

In this section, we label the axis data in form of an integer value instead of a string. Therefore, LabelEncoder is used [17] to convert the labels into a numeric form, as in Table 2.

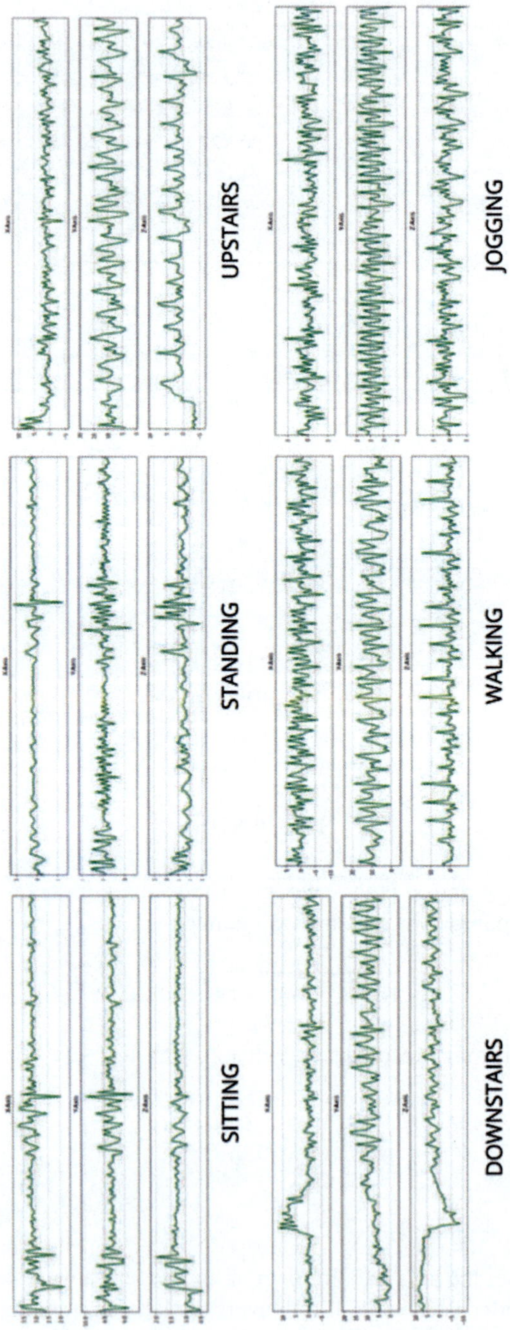

Fig. 1 Variations in frequencies from each activity over 80 ms

4.2 *Training and Testing Data*

In this section, we prepare training and testing data from our labeled dataset. For preparing training and testing data, we create X and Y array and scale both using formula:

$$Z = (X - u)/s.$$

For prediction using X, Y and Z, we create FrameSize and HopeSize using Fs which is 20 as per our dataset collection rate.

$$\text{Frame Size} = \text{Fs} \times 4, \quad \text{Hope Size} = \text{Fs} \times 2.$$

Based on above formula resulted FrameSize is 80 and HopeSize is 40. Add FrameSize with every axis value from range from 0th sample to total number of sample subtract with FrameSize and HopeSize. Make X and Y as 3D array by resize it with scaled X or scaled Y, FrameSize, N_features which is 3. Output of that process is X = (scaled_X, FrameSize, N_features) = (532, 80, 3) and Y = (Scaled_Y) = (532). Take X and Y array and split data for training and testing by ratio of 85 and 15%, we give X_train & Y_train85% and X_test and Y_test15%. Data sample which is split is X_train = (452, 80, 3), Y_train = (452) and X_test = (80, 80, 3) and Y_test = (80).

5 Proposed Section

In this section, we discuss the overview and detailed model architecture of proposed LightHARs.

5.1 *LightHARs Overview*

LightHARs has backbone of 2D convolution network architecture [18]. We feed the model with 6 training activities as we illustrated in Fig. 2. To feed data to our model, it must be reshaped in such a way that each person has multiple two-dimensional records which hold 80 times slices for each of three accelerometer readings. One record is associated with one label. The input layer is a vector with 240 elements and three hidden layers with 100 nodes each. All layers are fully connected, one additional layer upfront for reshaping the input into (80, 3) matrix and a softmax activation layer as the final layer.

5.2 Architecture of LightHARs

From Fig. 2 we observe that the input first pass to conv2D [19]

$$z1 = X \times f$$

In this formula X stands for input and f stands for filter. We set dropout rate 0.1 [20] using formula:

$$Er = \frac{1}{2}\left(t - \sum_{i=1}^{n} PiWiIi\right)^2 + \sum_{i=1}^{n} Pi\,(1 - Pi)\,W^2 i I^2 i,$$

and activate using $relu$ activation [21] *i.e.*,

$$R(z) = \begin{Bmatrix} z & Z > 0 \\ 0 & Z <= 0. \end{Bmatrix}$$

As per the observation from Fig. 2 connect two similar block with flatten layer sequentially, and pass to dense layer by using formula:

$$Y = \sum (w \times X) + b.$$

Now, we feed to the dropout layer with 0.4 rate. Finally, we apply sigmoid activation and perform prediction using:

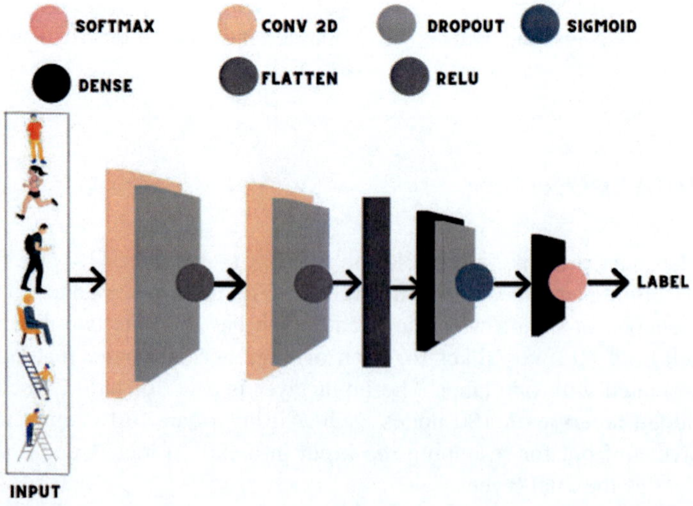

Fig. 2 Block diagram of LightHARs

$$S(x) = \frac{1}{1 + e^{-}x}$$

$$\sigma(Z) = \frac{e^{z}}{\sum_{j=1}^{K} e^{z} j}$$

6 Results and Discussion

In this section, we discuss our LightHARs's results and perform a comparison with state-of-art literature.

6.1 LightHARs Comparison with Existing Model

We compare LightHARs with GRU, ResNet, LSTM, Mobilenet, VGG for performance measure.

6.1.1 Comparison with Mobilenet

We observe that Mobilenet has more number parameters than LightHARs because of Mobilenet architecture which works on sigma gate and softmax activation function and LightHARs use only require layer and different activation for boosting layer performance. Mobilenet has more filters that cause more data loss that will reflect in training-testing time. Results are illustrated in Fig. 3 and Table 3.

6.1.2 Comparison with Resnet

ResNet has more parameters than LightHARs because the architecture of ResNet works with 50 number of blocks and softmax activation function and LightHARs

Table 3 Comparison of LightHARs with others in terms of performance, time, and parameters

Models	Recall (%)	Precision (%)	f-measure (%)	Parameters	Time
Mobilenet	66	66	66	4,264,232	47.76
ResNet	92	92	92	25,630,440	154.78
LSTM	87	87	87	103,465	41.21
GRU	87	87	87	54,201	41.01
LightHARs	95	95	95	322,550	10.81

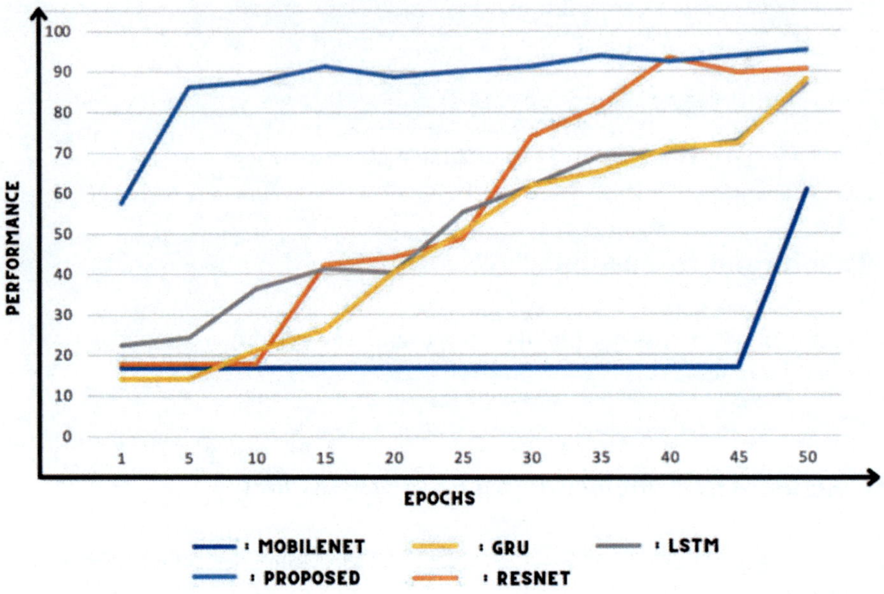

Fig. 3 Performance of models over epochs

use only require layer and different activation for boosting that layer performance. ResNet also causes more data loss in training-testing time as shown in Fig. 3 and Table 3.

6.1.3 Comparison with LSTM

LSTM has less amount of parameters than LightHARs because LSTM architecture work on short-term memory and less layer and LightHARs use only require layer and different activation for boosting that layer. LSTM architecture works well in the sequential type of data. Results are reflected in Fig. 3 and Table 3.

6.1.4 Comparison with GRU

GRU has less amount of parameters than LightHARs because GRU architecture works on 2D data and less layer with only *tanh* activation function and LightHARs use only require layer and different activation for boosting the performance of that layer and it uses 3D data. GRU takes more time in training-testing time because of its activation function *tanh*. GRU results are illustrated in Fig. 3 and Table 3.

7 Conclusion

In this work, a novel light framework for recognizing human activity is proposed. The proposed cost-effective LightHARs are reflected improved HAR accuracy. It shows a significant improvement of 4.5% in terms of performance and a minimum of 30 s less in training and testing time. LightHARs require less amount of parameters compared to existing models.

References

1. Bagi R, Dutta T, Nigam N, Verma D, Gupta HP (2021) Met-MLTS: leveraging smartphones for end-to-end spotting of multilingual oriented scene texts and traffic signs in adverse meteorological conditions. IEEE Trans ITS, pp 1–10
2. Bagi R, Mohanty S, Dutta T, Gupta HP (2020) Leveraging smart devices for scene text preserved image stylization: a deep gaming approach. IEEE MultiMedia 27(2):19–32
3. Bagi R, Dutta T, Gupta HP (2020) Cluttered textspotter: an end-to-end trainable light-weight scene text spotter for cluttered environment. IEEE Access 8:111433–111447
4. Bagi R, Dutta T (2021) Cost-effective and smart text sensing and spotting in blurry scene images using deep networks. IEEE Sens J 21(22):25307–25314
5. Bagi R, Dutta T, Gupta HP (2020) Deep learning architectures for computer vision applications: a study
6. Gupta A, Gupta HP, Biswas B, Dutta T (2021) A fault-tolerant early classification approach for human activities using multivariate time series. IEEE TMC 20(5):1747–1760
7. Turaga P, Chellappa R, Subrahmanian VS, Udrea O (2008) Machine recognition of human activities: a survey. IEEE Trans CSVT 18(11):1473–1488
8. Yang C, Wang L, Wang X, Mao S (2022) Environment adaptive RFID based 3D human pose tracking with a meta-learning approach. IEEE J Radio Freq Identif
9. Jaouedi N, Boujnah N, Bouhlel MS (2020) A new hybrid deep learning model for human action recognition. Journal King Saud Univ-Comput Inf Sci 32(4):447–453
10. Sarafianos N, Boteanu B, Ionescu B, Kakadiaris IA (2016) 3d human pose estimation: A review of the literature and analysis of covariates. Comput Vis Image Underst 152:1–20
11. Rodríguez ND, Cuéllar MP, Lilius J, Calvo-Flores MD (2014) A survey on ontologies for human behavior recognition. ACM Comput Surv (CSUR) 46(4):1–33
12. Lin Y, Chi W, Sun W, Liu S, Fan D (2020) Human action recognition algorithm based on improved ResNet and skeletal keypoints in single image. Math Prob Eng, vol 2020
13. Patil S, Prabhushetty KS (2022) A survey on human action recognition and detection techniques. In: ICT analysis and applications, pp 157–165. Springer, Heidelberg
14. Xu L, Zhang K, Yang G, Chu J (2022) Gesture recognition using dual-stream CNN based on fusion of semg energy kernel phase portrait and IMU amplitude image. Biomed Signal Process Control 73:103364
15. Kwapisz JR, Weiss GM, Moore SA (2011) Activity recognition using cell phone accelerometers. ACM SigKDD Explor Newslett 12(2):74–82
16. Park YS, Cho MH, Oh CS, Kang YJ (2022) Coherence-based sensor set expansion for optimal sensor placement in active road noise control. Mech Syst Signal Process 169:108788
17. Tajane K, Sheth S, Satale R, Tumbare T, Panchal O (2022) Breast cancer detection using machine learning algorithms. In: Intelligent sustainable systems, pp 347–355
18. Vafeiadis T, Kolokas N, Dimitriou N, Zacharaki A, Yildirim M, Selvi HG, Ioannidis D, Tzovaras D (2022) A comparison of 2DCNN network architectures and boosting techniques for regression-based textile whiteness estimation. Simul Model Pract Theory 114:102400

19. Feng L, Yang L, Liu S, Han C, Zhang Y, Zhu Z (2022) An efficient EEGNet processor design for portable EEG-based BCIs. Microelectr J, pp 105356
20. Yagli GM, Yang D, Srinivasan D (2022) Ensemble solar forecasting and post-processing using dropout neural network and information from neighboring satellite pixels. Renew Sustain Energy Rev 155:111909
21. Li GH, Sekine R, Nehra R, Gray RM, Ledezma L, Guo Q, Marandi A (2022) All-optical ultrafast ReLU function for energy-efficient nanophotonic deep learning. arXiv:2201.03787

Twitter Data Analysis to Enhance Malware Detection Using ML

Abhishek Singh, N. Sai Ganesh, G. Vamsidhar Reddy, A. Vishal Chandra, A. Harshith Varma, and J. Divya Udayan

Abstract In current times, we have seen an escalation of cellular or apps, consisting of useful, congenial apps, and additionally malicious apps (or malware). Detecting fraudulent applications is a challenging but critical task, as malicious apps may cause significant damage and financial losses to their users. The majority of strategies for detecting malware rely on characteristics acquired from the apps' code via static or dynamic analysis. Much zero-day malware software, on the other hand, avoids such mechanisms and enters the market. We recommend using social media statistics, specifically, Twitter, to supplement the statistics contained within the code and facilitate the detection of zero-day android malware apps. We recommend picking out tweets that mention android malware, primarily those that can contribute to the malware's spread. The notion is that clients who try to sell it and/or spread malware share the same characteristics as spammers. We utilized the Twitter Developer APIs to scan a huge number of tweets that had URLs that were similar to those found in android apps. The tweets were recorded in a MongoDB database, together with meta-statistics about their retweets/favorites and customers. The URLs found throughout the stream of tweets were matched with android apps using data gathered from the Google Play Store. Furthermore, utilizing a platform known as AndroZoo, which uses antivirus tools such as VirusTotal to detect malware, the apps identified in tweets that were linked to apps in the Google Play Store were classified as benign or malicious. Furthermore, Twitter users who post malware are being investigated to uncover tendencies similar to those seen in unsolicited mail, which might be used to identify zero-day malware.

Keywords Bot detection · Malware detection · AndroZoo · Tweepy · Machine learning · Google Play Store · Twitter · Data visualization · SparkML

A. Singh · N. Sai Ganesh · G. Vamsidhar Reddy · A. Vishal Chandra · A. Harshith Varma · J. Divya Udayan (✉)
Amrita Vishwa Vidyapeetham, Amritapuri, India
e-mail: divyaudayanj@am.amrita.edu

V. K. Gunjan and J. M. Zurada (eds.), *Proceedings of 3rd International Conference on Recent Trends in Machine Learning, IoT, Smart Cities and Applications*, Lecture Notes in Networks and Systems 540, https://doi.org/10.1007/978-981-19-6088-8_40

1 Introduction

We compared the URLs of android applications in this document using data from AndroZoo and Google Play Store. The AndroZoo VT detection number was used to compare and flag applications detected on any of the platforms. AndroZoo is a trustworthy interface for determining the application's precise name. As a result, the findings were incorporated into the analysis as basic reality. There were 30,950 links that matched the AndroZoo database. Among these, 9750 were found to be harmful, while the rest 21,200 were found to be innocuous. We also reviewed the metadata of its users and filtered tweets that contained information about harmful android apps. This method was used to look at user attributes and create a pattern based on abnormalities. On Twitter, users were not safeguarded or verified. Users who shared links to safe apps, on the other hand, had comparable data. As a result, we may deduce that the individual attributes derived from the user object alone are insufficient to forecast a user's personality-on Twitter, to be precise. This latter pattern may be used in conjunction with machine learning classifiers to identify spammers on Twitter. There are still a lot of apps that haven't been branded. These apps are either new to the market or have not piqued the public's interest; therefore, they go unnoticed. These might be sent to AndroZoo, VirusTotal, or other reputable antivirus companies to display the number of detection.

2 Literature Survey

In this selected research paper [1], we come to understand that the project had been performed using three linking techniques, which allows making connections between tweets and the apps that their references. Twitter data that are accurately linked to apps via HTTP links can be used to improve machine learning classifier performance across a wide range of common malware detection classifiers.

Preliminary findings in Section IV-B show that Twitter data are a beneficial dimension for android malware detection. Scaling up our methods to larger datasets can improve the accuracy of the model. While there is still work to be done in both the techniques for linking the tweets and the apps, as well as how to leverage the resulting linked Twitter data, our work, which is the first of its kind integrating social media data with Android malware detection, has shown potential for future research. The malware must not only be present in the app store but must also be downloaded and consumed by end users in order to be effective. Social media links are a popular way for information to spread to new devices. As a result, this approach enables a more comprehensive examination of android malware as a whole. Through the app's binary-based features, we can see not only what malware does on a device, but also how it attempts to spread via online social media.

We can identify three major avenues for future work. To begin, it would be beneficial to leverage Twitter data to a greater extent in order to gain a better perspective from the tweets that can be linked to apps. This could imply looking into more advanced methods for detecting Twitter spam. Crawling a graph of Twitter, users discussing android apps could also aid in the establishment of credible Webs. A second approach would be to use a threat model, similar to the one used by Twitter, to make it more resistant to potential data source poisoning attacks. Finally, it would be interesting to broaden the social media data sources beyond Twitter, possibly including blogs, other social networks, and so on.

In this selected research paper [2], a system for real-time malware alerting based on a set of tweets captured via Twitter APIs and analyzed using a Bayes Naive classifier following that, groups of tweets about the same topic, such as a new malware infection, are condensed in order to generate an alert. There have been tests performed in order to assess the system's performance, and results demonstrate the effectiveness of our implementation.

An intelligent system has been designed to analyze the Twitter stream in order to generate automatic alerts when news of malware attacks or other computer security threats spreads across the Internet. The use of dynamic length observation windows and dynamic threshold is key components of the architecture proposed, allowing the system to adapt its behavior to the volume of tweets captured at a given time. Furthermore, the use of some preprocessing steps and a Naive Bayes classifier helps to improve the alerting system's performance by filtering out irrelevant tweets. Future work can be expanded by providing alerts on generic events based 6 on an initial set of keywords. Furthermore, in order to increase the likelihood of detecting significant events, the set of keywords used to extract information from Twitter streaming should be expanded. We can investigate the use of more sophisticated techniques to estimate the trustworthiness of users and, as a result, their feedback. To that end, we can provide the system with a reputation management algorithm, such as those previously reported, to ensure the robustness of reputable systems against security attacks.

3 Datasets Description

In order to detect whether the given android app is malicious or not, we are using the AndroZoo dataset which is a popular dataset that is updated constantly with new apps, which contains the information about the link to the android app, how many antiviruses have detected the app as malicious out of the tens of different antivirus software it uses to scan the apps, and it also provides us many details about the app which are explained below.

As discussed above, the attributes in the dataset give us many insights about the apps present in the dataset [3]. Attributes available for each entry in the dataset are.

1. **apk_size**: the size of the apk.
2. **Sha256, md5, sha1**: different hashes of the apk.
3. **dex_size**: This attribute tells us the size of the classes.dex file.
4. **pkg_name**: This attribute tells us the name of the android package.

5. **vercode**: vercode tells us the version code of the android app which is mentioned in the manifest file.
6. **vt_detection**: This attribute tells us the number of antiviruses from VirusTotal which had detected the apk as malicious.
7. **vt_scan_date**: the latest date on which the scan had been done.
8. **markets**: This attribute tells us the names of the market in which AndroZoo has detected this app. Each market is separated by '|'.

Out of all these attributes, the attributes which we need to detect whether an app is malicious or benign are the pkg_name and the vt_detection. Using these two features, we can classify apps as malicious or benign.

4 Proposed Methodology

4.1 Data Collection

To get the necessary tweets and extract URLs from those tweets, we can make use of the Tweepy library which utilizes twitter's streaming API. By getting Twitter's developer credentials, we get access to Twitter's tweets, and Tweepy gives further options to filter the tweets and directly access metadata associated with the tweets without additional parsing. As the project is associated with finding tweets that are spreading links to malware applications, using Tweepy [4], we can filter the tweets based on some of the associated keywords (malware, Play Store, etc.).

To prevent any sort of disconnection (as client machines might not be able to keep up with the incoming data) while the tweet data are being streamed, we make use of Kafka as a message queue [5]. The producer takes the data and sends it to the topic of the tweet which is utilized by a consumer (to store data in MongoDB) and Logstash (part of the elk stack which is being used for data visualization).

On the consumer side, the data that are being received from the topic of the tweet is being added to the corresponding tweets collection in MongoDB. From the MongoDB database, we filter the tweets to find the tweets which have a non-empty URL associated with them [6, 7]. The entities.URLs.expanded_url field has the extracted URLs corresponding to the tweet and thus doesn't require additional parsing to find the URL in a tweet.

Now that we have the possible malware URLs which are being spread, we need to actually check which URLs are considered malware. For this, we make use of a dataset from AndroZoo (https://androzoo.uni.lu/lists) which consists of the package names of the app and the number of times it has been detected as malware by an anti-virus. Using the count of antiviruses detecting an app as malware, we filter out the harmful applications and do some processing on the package names.

As we have the URLs scraped from the tweets and the malicious URLs from AndroZoo, we compare both of them to find the actual malicious URLs which are being circulated on Twitter [8]. We are getting the tweet's data in real time, and thus, data have to be collected over several days to actually detect malicious URLs. But, due to the lack of sufficient hardware, we weren't able to get sufficient data, and thus, our outputs were not satisfactory. With access to better hardware, we would be able to actually detect and analyze the malicious URLs from Twitter.

4.2 Data Visualization

For real-time visualization of the Twitter data being used, we made use of the ELK stack. Here, Logstash is used for collecting the data transforms the data ingests from the tweets topic and stores the data in the Elasticsearch database. Elasticsearch then searches and analyzes the data obtained (Figs. 1 and 2).

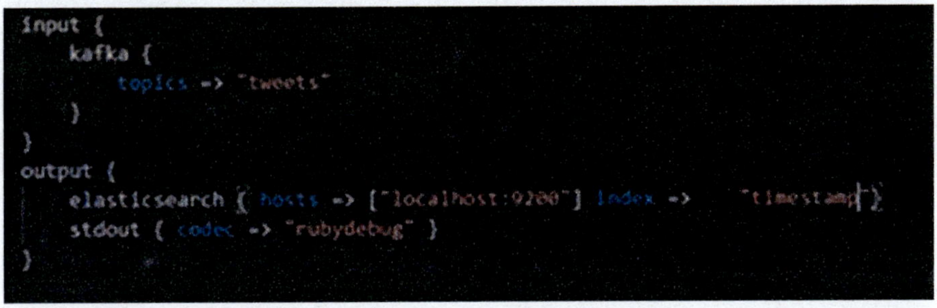

```
input {
    kafka {
        topics => "tweets"
    }
}
output {
    elasticsearch { hosts => ["localhost:9200"] index =>     "timestamp" }
    stdout { codec => "rubydebug" }
}
```

Fig. 1 Storing the data into the database

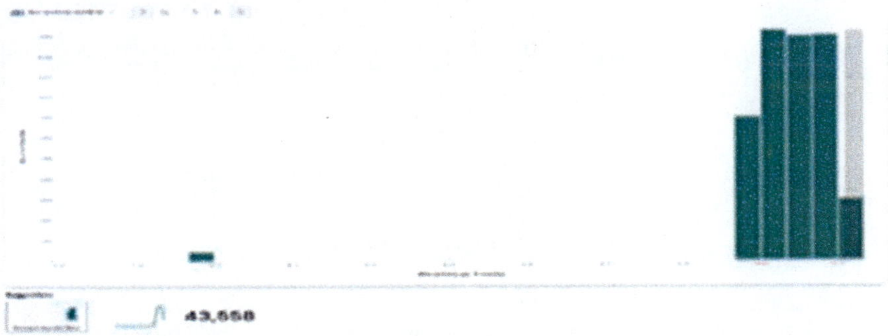

Fig. 2 Visualization of data using Kibana

```
_id: ObjectId("61b1b18ce153d8d4af8dbd52")
url: "/store/apps/details?id=com.niksoftware.snapseed"
```

Fig. 3 Format of the data stored in database initially

```
_id: ObjectId("61b1b18ce153d8d4af8dbd52")
url: "/store/apps/details?id=com.niksoftware.snapseed"
link: "snapseed.niksoftware.com"
```

Fig. 4 Format of the data stored in the database after processing

4.3 Scrapping Play Store

To see if there are any apps in the Google Play Store which are common to the malicious app found from the Twitter data, we have scrapped the Google Play Store by creating a crawler using Scrapy. The function of the crawler is to collect the links of the apps present in the Google Play Store [2]. We first start scraping from the base URL of Google Play Store apps. We find the apps present in the base URL, store them in the database, and follow those links; we repeat this process till the desired number of apps are scrapped (Fig. 3).

Format of the data stored in the database initially. In order to compare this data with the previously scrapped data from Twitter, we need the link of the app which can be obtained by reversing the URL field and modifying it which can be observed by comparing the link field, URL field in the image shown below [2]. In order to obtain this format, the database is processed (Fig. 4).

4.4 System Flow Chart

See Fig. 5.

4.5 Detecting Bots on Twitter Using Machine Learning

A Twitter bot is software that creates automatic messages, follows Twitter users, or serves as spam to lure people to click on the microblogging site Twitter. We will utilize machine learning techniques in this research to determine if a Twitter account is a bot or a human individual. We did a lot of feature engineering and feature extraction, and we were able to figure out if an account was a bot or not by looking at 20 different features [9]. We tried a few different algorithms before settling on our own, which gave us an AUC of over 95%.

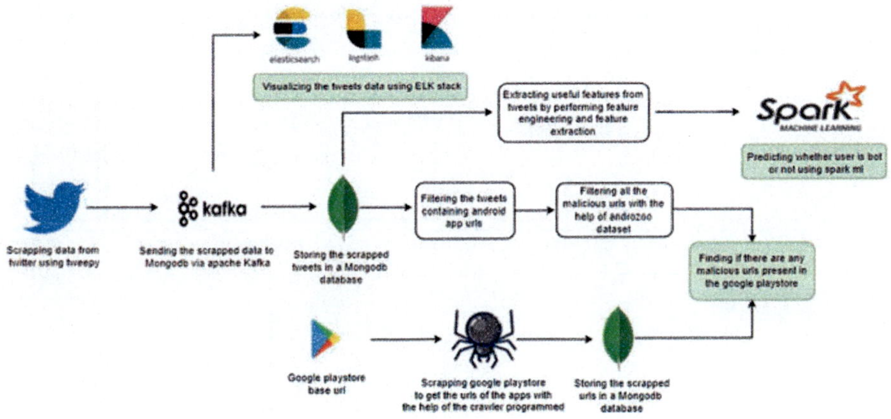

Fig. 5 Flowchart of the project

This machine learning project aims to determine whether or not a Twitter profile is a bot. Starting with feature extraction, we evaluated the test accuracy of multinomial Naive Bayes, decision trees, and random forest models [7, 10].

Finally, we chose random forest classifier for the task which gave us a test accuracy of over 90% in detecting bots (See Figs. 6 and 7).

Now that you understand what a bot is, the real question is whether you would recognize one if you came across one [7]. Some bots are more difficult to identify than others, depending on their level of stealth.

Bots on Twitter usually have the following characteristics:

- The founding date of many Twitter bots is quite recent.

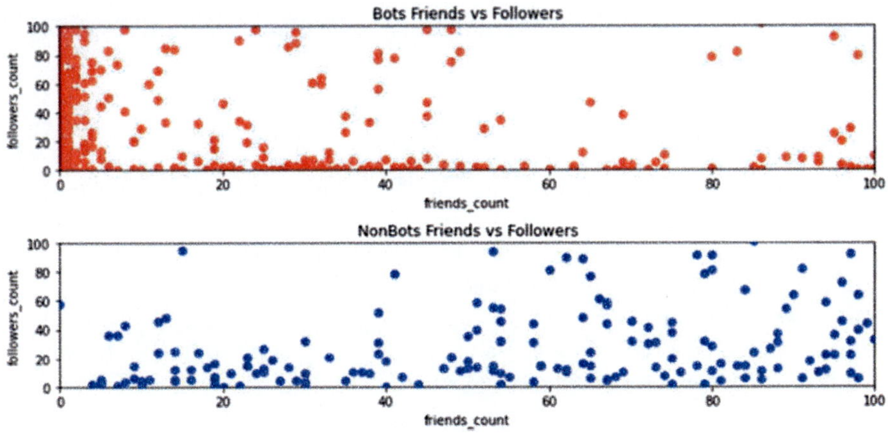

Fig. 6 Friends count versus followers count of bots and non-bots using seaborn regplot

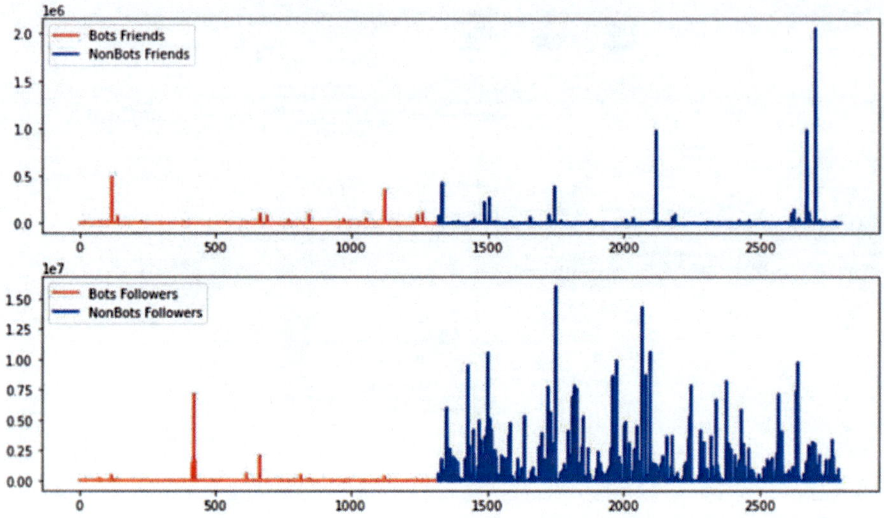

Fig. 7 Friends count versus the followers count of bots and non-bots using matplotlib

- Many bot user names contain digits, which might suggest that they were generated automatically. Rather than tweeting fresh stuff, the account mostly retweets it.
- The account's tweet frequency is far higher than that of a real user.
- Some bot accounts may be identified because they send a lot of tweets but only have a few followers, while others can be identified because they send a lot of tweets but only have a few followers.
- At around the same moment, many bots tweet the same stuff as other people.
- Automated activity can also be detected by short replies to other tweets.
- Bot's Twitter accounts are frequently connected with no biography or even a photo (Fig. 8).

5 Result

- The variables id, statuses count, default profile, default profile picture, and target variable do not correlate.
- There is a strong correlation between verified, listed_count, friends_count, followers_count, and the target variable.
- We cannot perform correlation for categorical attributes. So, we will take screen_name, name, description, status into feature engineering. While using verified, listed_count for feature extraction.

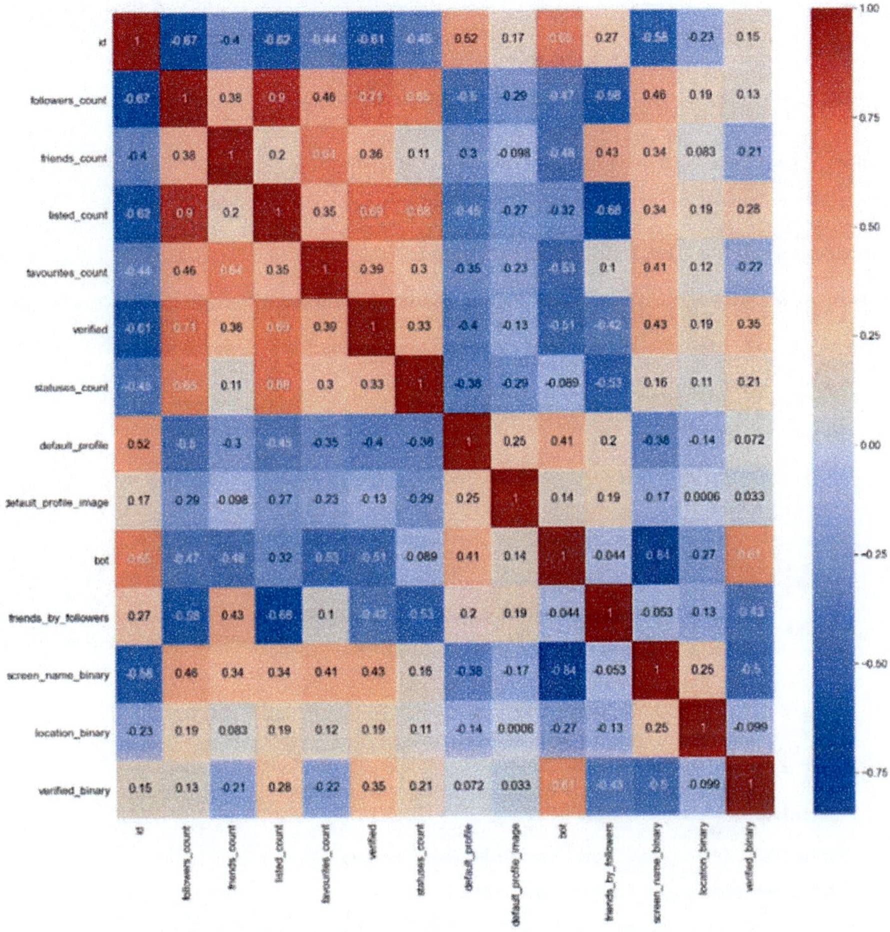

Fig. 8 Generated HeatMap (cool warm) between features using Spearman correlation

6 Model Performance Testing

See Figs. 9, 10, and 11

7 Results and Discussion

In this document, we compared the URLs of android applications from AndroZoo and the Google Play Store. The AndroZoo VT detection number was used to compare and flag applications found on any of the platforms. AndroZoo is a reliable platform

```
: recall = tp/(fn+tp)
  precision = tp/(fp+tp)
  f1 = (2*recall*precision)/(recall+precision)
  print("F1 Score is: ",f1)

  F1 Score is:  0.7853403141361258

: evaluator = BinaryClassificationEvaluator(rawPredictionCol='rawPrediction',labelCol='bot')
  print("The area under ROC for train set is {}".format(evaluator.evaluate(predict_train)))
  print("The area under ROC for test set is {}".format(evaluator.evaluate(predict_test)))

  The area under ROC for train set is 0.8299374078420743
  The area under ROC for test set is 0.8224646000765415
```

Fig. 9 Logistic regression model—binary classification evaluator and F1 score

```
: recall = tp/(fn+tp)
  precision = tp/(fp+tp)
  f1 = (2*recall*precision)/(recall+precision)
  print("F1 Score is: ",f1)

  F1 Score is:  0.8823529411764706

: evaluator = BinaryClassificationEvaluator(rawPredictionCol='rawPrediction',labelCol='bot')
  print("The area under ROC for train set is {}".format(evaluator.evaluate(predict_train)))
  print("The area under ROC for test set is {}".format(evaluator.evaluate(predict_test)))

  The area under ROC for train set is 0.9471908759892869
  The area under ROC for test set is 0.9270091848450059
```

Fig. 10 Random forest model—binary classification evaluator and F1 score

```
bots = 0
nonBots = 0
for key in user:
    if user[key][1]:
        bots += 1
    else:
        nonBots += 1
print("Total Bot percentage: ",bots*100/(bots+nonBots),"%")

Total Bot percentage:  8.175638434686753 %
```

Fig. 11 Total bot percentage of predicted Twitter data

for determining the exact name of the application. As a result, the observations were incorporated into the analysis as a fundamental truth (See Figs. 12, 13 and 14).

In addition, we filtered tweets containing information about malicious android apps and examined the metadata of its users. This method was used to investigate user characteristics and create a pattern based on anomalies. On Twitter, users were not protected or verified. However, similar data were discovered for users who shared links to innocuous apps [9].

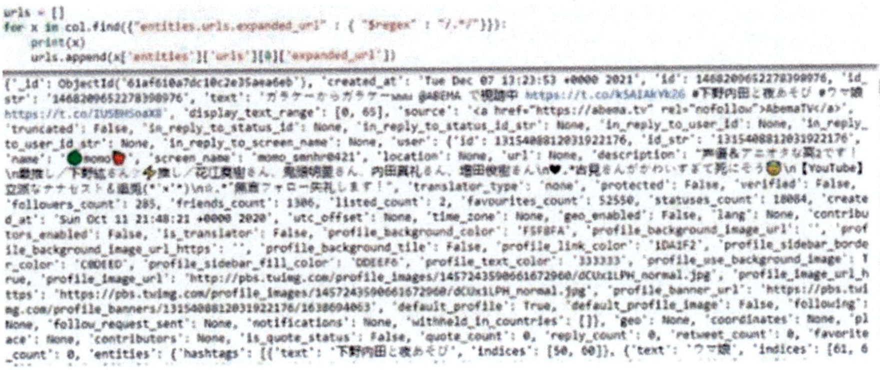

Fig. 12 URLs scraped from Twitter

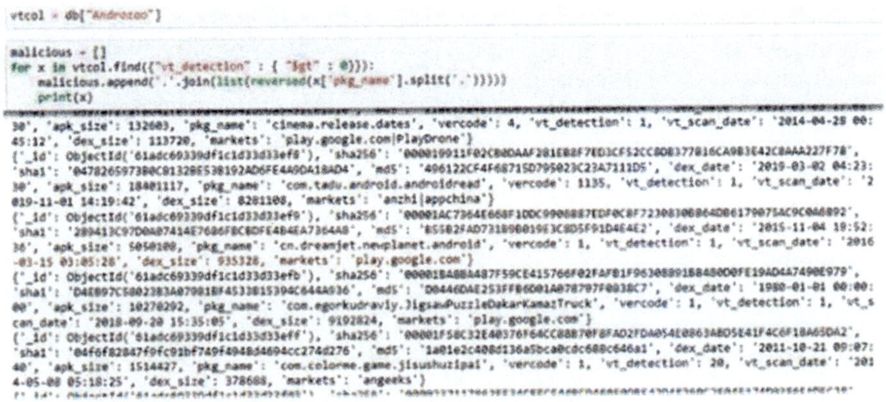

Fig. 13 Malicious applications are taken from AndroZoo

8 Conclusion

As a result, we can conclude that individual characteristics extracted from the user object alone cannot predict a user's nature [11]. This pattern can be used in conjunction with machine learning classifiers to determine whether Twitter users are spam, bots, or real. There are still a large number of unlabeled applications. These applications are either new to the market or have not piqued the public's interest, leaving them undiagnosed. These could be submitted to AndroZoo, VirusTotal, or other reputable antivirus vendors to show the number of detections.

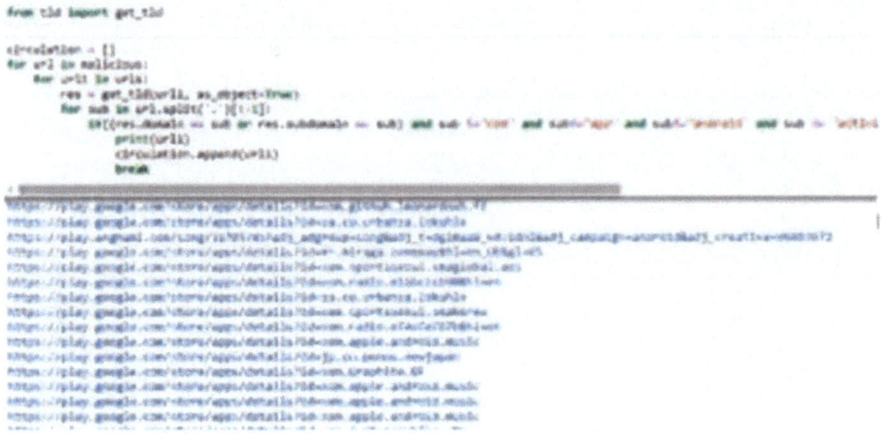

Fig. 14 Possible malicious URLs

Acknowledgements We express our gratitude to our beloved Chancellor and world-renowned humanitarian leader Shri. (Dr) Mata Amritanandamayi Devi (AMMA), for inspiration and motivation. We would like to thank the staff and faculty members of the department for providing immense support and suggestions to improve this paper.

References

1. DeLoach J, Caragea D (2017) Twitter-enhanced Android malware detection. In: 2017 IEEE international conference on big data (Big Data), pp 4648–4657
2. Concone F, De Paola A, Re GL, Morana M (2017) Twitter analysis for real-time malware discovery. AEIT Int Annual Conf 2017:1–6. https://doi.org/10.23919/AEIT.2017.8240551
3. Tang J (2016) Trendsmap a real-time us trends map for twitter. Master's thesis, Brown University
4. Roesslein J (2020) Hello tweepy
5. W3Techs. Androzoo, 2016. Last accessed 16 September 2017
6. Dhanya KA, Gireesh Kumar T (2019) Efficient android malware scanner using hybrid analysis. Int J Recent Technol Eng (IJRTE) 7(5S3). ISSN: 2277-3878
7. Shyam R, Bharathi Ganesh HB, Sachin Kumar S, Poornachandran P, Soman KP (2015) Apache spark a big data analytics platform for smart grid. Procedia Technol 21:171–178
8. Vinayakumar R, Soman KP, Poornachandran P, Kumar S (2018) Detecting android malware using long short-term memory (LSTM). J Intell Fuzzy Syst 34:1277–1288
9. Twitter (2020) Data dictionary
10. Radhika N, Vimalkumar K (2017) A big data framework for intrusion detection in smart grids using apache spark. In: International conference on advances in computing, communications and informatics, ICACCI 2017
11. Kumar PN, Balasubramanian P, Parambalath G, Mahesh E, Balasubramanian P (2019) Big data analytics: a trading strategy of NSE stocks using bollinger bands analysis. In: Data management, analytics and innovation, vol 869. Springer Singapore, Singapore, pp143–154

Prediction of Osteosarcoma Using Machine Learning Techniques

Devesh Kumar Srivastava, Aarushi Batta, Tanuj Gupta, and Aditi Shukla

Abstract Osteosarcoma is the name given to the most common bone malignancy which occurs in individuals of five to twenty-five years of age. This study focuses on finding out a prediction algorithm by comparing widely used algorithms on both feature, and image dataset. The aim is to find an algorithm with commendable accuracy which can be used in clinical settings. The work is done in two halves, i.e. on feature dataset and image dataset. The algorithms used for the feature dataset are Extra Trees Classifier, Convolutional Neural Network, XGBoost Classifier and DenseNet classifier. The concatenation of Viable and Necrotic Tumour gave the highest training accuracy of 96.22% followed by the group of Non-tumour versus Viable which gave the training accuracy of 94.56% compared to others. For the image dataset, Convolutional Neural Networks is used to predict Osteosarcoma. We managed to implement the model by utilizing trainable 2.5 million parameters of Convolutional Neural Network (CNN) which indicates that the model works on less complexity.

Keywords Osteosarcoma · Extra trees classifier · Convolutional Neural Network · XGBoost classifier · DenseNet classifier · CNN

1 Introduction

Whilst bone tumour is responsible for approximately 5–10% of new cancer cases in early stages, Osteosarcoma is the most familiar form of bone tumour. It has 5-year survival rate of about 60% [1]. The malignant is seen to be developed in individuals of 5–25 years of age. There are age peaks of occurrence amongst the patients, with a peak age of youngsters beneath age ten and teenagers at age ten-twenty [2]. Osteosarcoma is a form of mesenchymal tumour which often metalizes to the lung and peripheral bone. Therefore, metastatic ability is a key element in figuring out the analysis and diagnosis of osteosarcoma [3, 4]. The disease has an estimated incidence

D. K. Srivastava (✉) · A. Batta · T. Gupta · A. Shukla
SCIT, Manipal University, Jaipur, Rajasthan, India
e-mail: devesh988@yahoo.com

V. K. Gunjan and J. M. Zurada (eds.), *Proceedings of 3rd International Conference on Recent Trends in Machine Learning, IoT, Smart Cities and Applications*, Lecture Notes in Networks and Systems 540, https://doi.org/10.1007/978-981-19-6088-8_41

of 0.34/100,000 per year [5]. The symptoms and signs of osteosarcoma typically start with moderate bone pain which is generally localized, redness and heat on the place of the tumour. Patients feel growing pain, which frequently influences their motion and functions of joints. If the early segment of osteosarcoma is not treated, it is far predicted to look a huge variety of metastasis consisting of at lungs, different bones and smooth tissues. To diagnose this cancer, a patient has to go through series of medical tests including physical examinations and detailed health history. The presenting symptoms are similar to that of any other malignancy, but pain in more than one area can also additionally portend skeletal metastasis; therefore, they need to be properly investigated [6]. Although primarily the biopsy-based techniques can successfully find out the malignancy, obstacles in biopsies that are histologically guided and MRI scans have constrained the detecting capacity. Also, the coaching of histological specimens is very time consuming. For instance, a correct detection of osteosarcoma malignancy calls for coaching of at the very least fifty histology slides to symbolize an aircraft of a huge third-dimensional tumour [2]. The research focuses on finding out machine learning algorithms with noticeable high accuracy to aid the medical processes. The primary aim was to compare different machine learning algorithms on a publicly available feature dataset along with image dataset consisting of three categories namely, Non-tumours, Necrotic Tumours and Non-Viable Tumours.

2 Literature Review

The computer-aided era in biopic and radiological detection became possible from 2010 [7]. Extraordinary developments have been executed in clinical photographs, mostly because of the deep Convolutional Neural Networks (CNNs) and large scale datasets withinside the technological area [8]. This era has been extensively carried out to a lot of clinical photographs for the possible detection of various sicknesses, including chest X-ray pneumonia, pulmonary oedema, breast cancer, pulmonary fibrosis, gastric cancer and gastric endoscopic photographs for celiac sickness [9, 10]. Several research are primarily based on the concepts of deep learning have been identified as a current essential enhancement in histological photograph detection. However, maximum efforts of photograph detection are targeted on histological pictures of breast cancer. In 2017, Jongwon [11] did a pilot take a look on the histopathology of breast cancer, which achieves an AUC figure of 93% on microscopic biopsy images in classifying malignant or benign tumours. They display that transfer learning is possible and it is a pre skilled version which is beneficial in classifying the histological pictures. Recently, enhancements in the computing capacity, the development of deep learning algorithms and the advent of huge information have added growth in tertiary synthetic intelligence. Deep learning (DL) dependent via way of means of constructing a version to mimic the human mind, is one of the synthetic intelligence structures primarily based on neural networks. Deep learning strategies are presently taken into consideration kingdom of the artwork for the type of images [12] and implemented for a few fields of scientific pictures [13]. The neural community starts via way of means of simulating neural cells and seeing to simulate the human mind to the usage of a simulation version referred to as a perceptron. A multilayer perceptron is built arranging layers and making them with perceptron's where in all nodes within the version are completely connected, permitting the answer of extra complex difficulties. Artificial Intelligence era has been implemented within the area of thoracic imaging and has evolved withinside the following areas [14]: the detection of pulmonary nodules: differentiation amongst malignant and benign lesions; prognosis of diffused; lung diseases (i.e. retrieval device for same cases); and development of 3D-evaluation and photo quality (e.g. Pixel shine, a noise discount set of rules the usage of gadget getting to know; AlgoMedica, Sunnyvale, CA). Sun found that the maximum region beneath neath the curve (AUC) changed into 0.899 $\pm$ 0.018 was carried out by CNN, which turned out to be considerably better than the conventional CADx with the AUC = 0.848 $\pm$ 0.026 [14]. Performance evaluation on the most important public dataset, the Image Database Resource Initiative and The Lung Image Database Consortium, shows that the proposed technique has appreciably decreased the range of fake positives to 2.8 in keeping with test with a promising sensitivity of 95.6%. These effects reveal the importance of the technique in automated lung nodule classification and detection [15].

In a study by Arunchalam [16], the object reviews the first completed automatic device to evaluate necrotic and possible tumour in osteosarcoma, the usage of deep learning models and histological photographs. The intention is to label numerous areas of the tissue right into a possible tumour, a necrotic tumour and Non-tumour. They put into use both deep learning and Machine Learning (ML) models. The ensemble learning version completed an ordinary accuracy of 93.3% with class precise accuracy of 91.9% for Non-tumour, 95.3% for Possible Tumour and 92.7% for Necrotic Tumour. In mining algorithms and ML, the primary premise is that the schooling and capacity information ought to be withinside the identical area and distribution. The trouble arises whilst one does not have any sufficient access to information withinside the precise study domain. Hence, the basic parameters to train the deep learning model can be obtained from the pre-training networks and apply them to larger datasets of different domains. In those situations, knowledge-shifting considerably improves getting to know outputs if accomplished successfully even as minimizing high priced information labelling efforts [17].

3 Research Methodology

The dataset used was retrieved from the research of Arunachalam et al. the dataset contains both Haematoxylin and eosin (H&E) stained osteosarcoma histology pictures. The records changed into accrued through a group of scientific scientists at University of Texas Southwestern Medical Centre, Dallas. Archival samples for fifty affected individuals were dealt with at Children's medical centre, Dallas, during the period 1995–2015 and have been used to create this dataset. The pictures are labelled as Non-tumour, Viable Tumour and Necrosis in keeping with the main cancer's kind

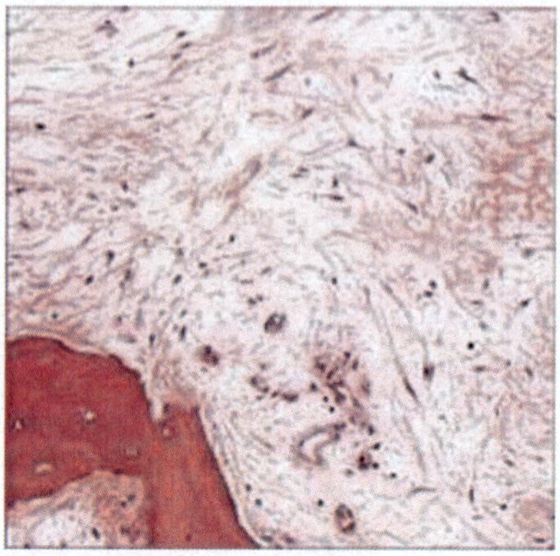

Fig. 1 Viable Tumour

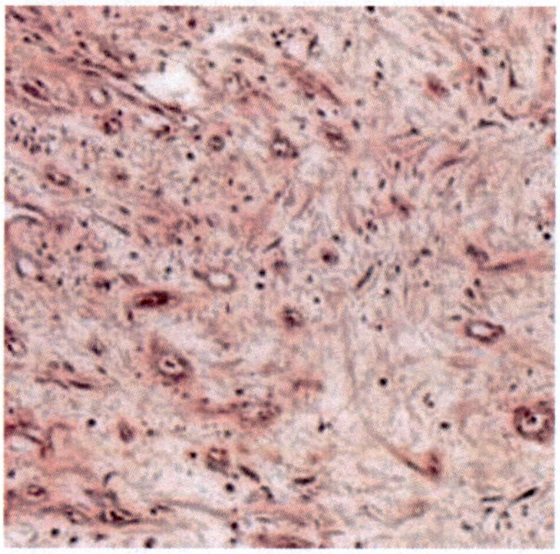

Fig. 2 Necrotic Tumour

in every photograph. Sample pictures are shown in Figs. 1, 2 and 3. The dataset includes 1144 images of length 1024 X 1024 at 10X decision with the subsequent distribution: 536 (47%) Non-tumour, 263 (23%) Necrotic Tumour and 345 (30%) Feasible Tumour tiles.

The feature dataset had 1144 rows and 69 columns. The distribution is shown in Fig. 4.

3.1 Pre-processing

The feature dataset of 1144 rows and 69 columns was split into training and testing with a ratio of 80% and 20%, respectively. After this step, we proceeded to do feature selection. For this, the columns with threshold >0.9 in the correlation matrix is dropped. This resulted in dropping of 48 columns in total. After this, there are 915 rows in train dataset and 229 rows in test dataset along with 20 columns in both. Initially for the image dataset, we have 1144 tiles, the images from original

Fig. 3 Non-tumour

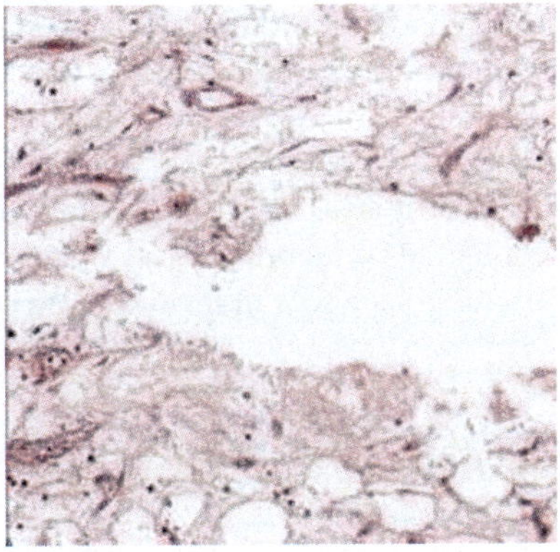

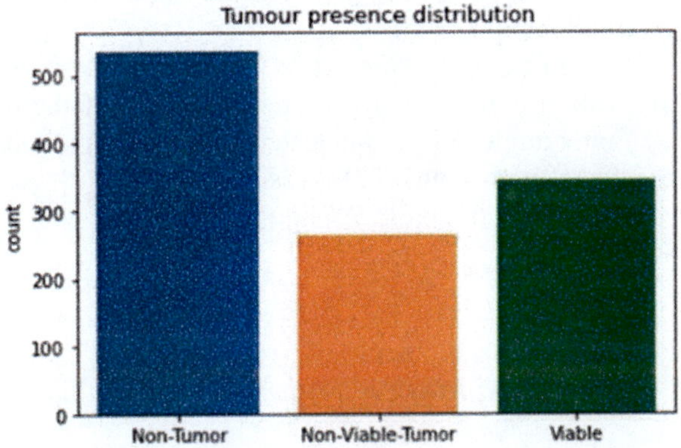

Fig. 4 Distribution of tumour in feature dataset

dataset are used for model training. The dataset is unbalanced and so for an unbiased and stable result, we have balanced the dataset with augmented images in all the three classes. The updated dataset has 1740 images in training dataset. The dataset is split up into training and testing with a ratio of 70% and 30%, respectively. For deep learning models the image dataset is randomly rotated and normalized, for all pertained models the width and height are reduced to the required size needed for the classification of input image and then normalized with mean and standard deviation.

4 Proposed Work

We started with applying different machine learning algorithms on the Feature Dataset namely: Extra Trees Classifier, Convolutional Neural Network, XGBoost Classifier and DenseNet classifier. Convolutional Neural Network is chosen for the application of Image Dataset. The dataset resulted in overfitting when it is applied to high complexity and run for epochs greater than 40. The model tended to overfit the training model very quickly which led to reduction in validation accuracy drastically. Therefore, it is important to maintain and monitor number of epochs and introduce dropout in the model to reduce the dimensionality issues encountered. The use of 2.5 million parameters offered an insight that the dataset could work on lesser complexity and still be able to offer good results. The use of a smaller number of parameters and reaching the similar results (as shown by the VGG19 model) depicts that the number of training instances in the dataset were not as high as suggested earlier.

4.1 Extra Trees Classifier

Extremely Randomized Tree Classifier collects and aggregates the results of various de-corrected decision trees produced during a "forest" and then produces an output of its classification result.

The Extra trees classifier gave the best results in terms of validation accuracy, whilst still not showing overfitting indicators of the training dataset. Extra Trees Classifier is implemented and tested with various numbers of decision trees present in the forest with both—0.9 threshold and 0.99 threshold datasets. The classifier gave its best result, i.e. 82.7 when implemented with 500 decision trees in the 0.9 threshold dataset.

4.2 Convolutional Neural Network (CNN)

Applications utilizing CNN with the image dataset gave the best result when implemented using MaxPooling and Dropout functions in the model. The necessary use of Dropout suggests that models are prone to overfitting and that the number of neurons in hidden layers need to be omitted. Furthermore, the implementation is sensitive in the number of epochs it is run for, suggesting that the implementation of the dataset is prone to further dimensionality issues. The layers of CNN are shown in Fig. 5.

The CNN model suggested in our application uses the Adam Optimizer and Categorical Cross-Entropy for its hyperparameters; the architecture of the model comprises 3 convolutional layers with the ReLu activation function, and 2 Maxpooling layers after which it is flattened and introduced to fully connected ReLu

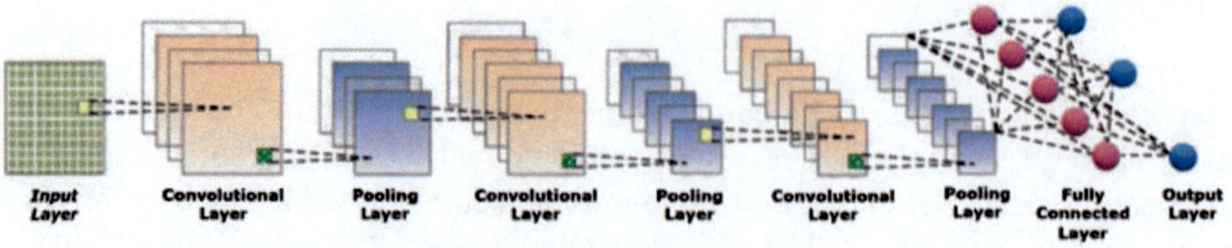

Fig. 5 Layers of Convolutional Neural Networks

layers. A dropout of 0.5 is also introduced prior to the last layer. VGG16: For the classification of the following image dataset, we have used VGG16 as defined in Fig. 5 as a feature extraction and then embedded our own classifier for result. We have to freeze the pre-trained network for the mode to behave as a feature extractor and then passed it through a classifier of three dense layer a ReLU layer, Batch Normalization followed by a dropout of 0.5 with the output layer of LogSoftMax as shown in Eq. 1. For training the given dataset we have used NLLLoss function as shown in Eq. 2 and Adam optimizer with initial learning rate of 0.05.

Training Algorithm

Input: Model, dataset, learning parameter

Output: The accuracy and training loss of the model on the given dataset

Step 1: For every epoch

Step 1.1: For input in batch size

Step 1.1.1: Training the model with the features of the dataset

Step 1.1.2: Forward propagation and computing the total loss via NLLLoss function

Step 1.1.3: Back propagation and update parameters by gradient descent

Step 1.2: Record the total loss and accuracy for every epoch

Step 2: End.

SoftMax Activation Function

$$\sigma(\vec{z})_i = \frac{e^{z_i}}{\sum_{j=1}^{K} e^{z_j}} \tag{1}$$

NLLLoss Function used

$$l(xy) = \left\{ \sum_{n=1}^{N} \frac{1}{\sum_{n=1}^{N} W_{y_n}} l_n \right. \tag{2}$$

Other classifiers are listed below are used for better investigation as better output. InceptionV3—We have tried different methods for feature extraction and the next one being InceptionV3.The model is a pre-trained on "ImageNet" dataset with trainable layers set to false. These layers will act as feature extraction. With under 25 million parameters, this model is also used in the research for leukaemia. The frozen layer here is followed by the same classifier as VGG16 and the results are compared. It is shown in Fig. 6.

DenseNet 121/201—We have also tried different versions of dense, this model passes the information obtained by each layer to all the subsequent layers, this can help in reducing channels and the network can be more compact. Each of the composition layer is passed through different sets of convolution and pooling layer. DenseNet

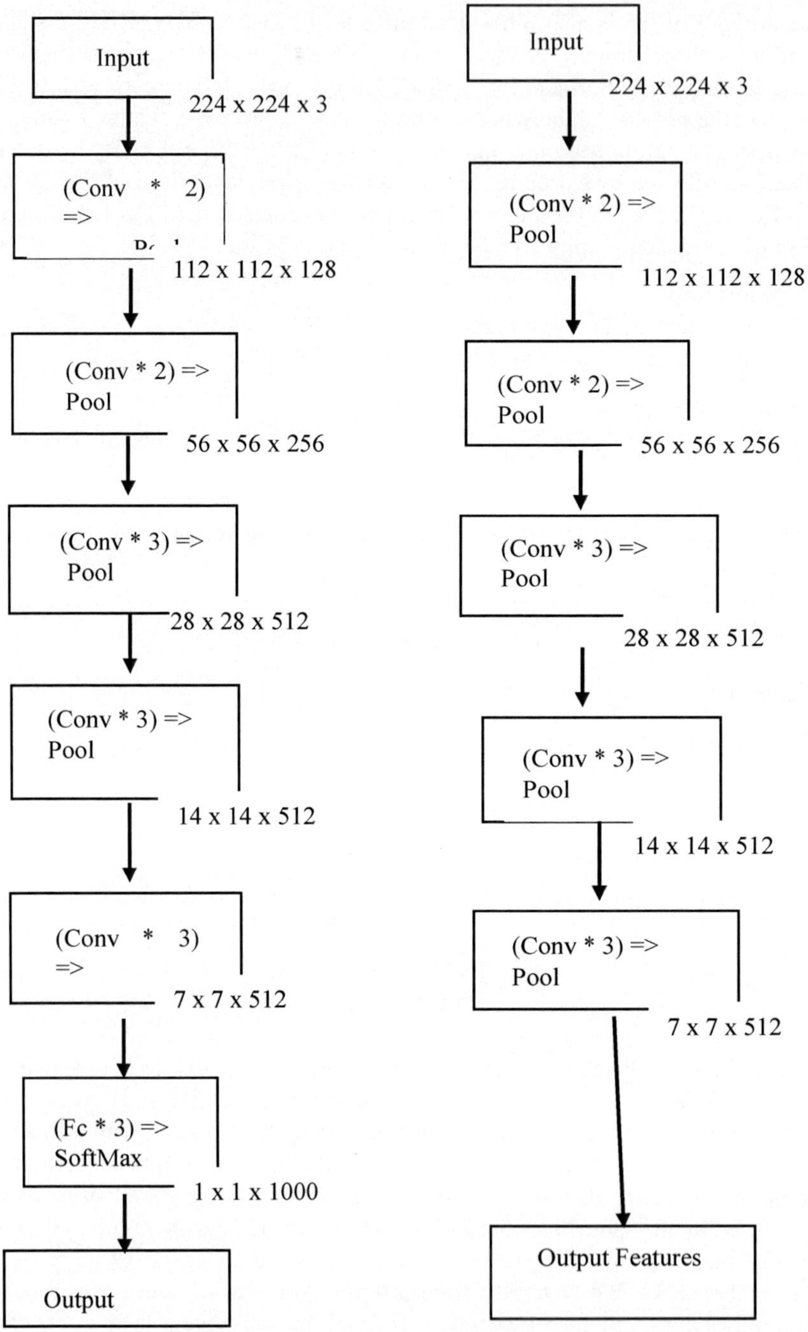

Fig. 6 Left is VGGNet, right VGGNet for feature extraction

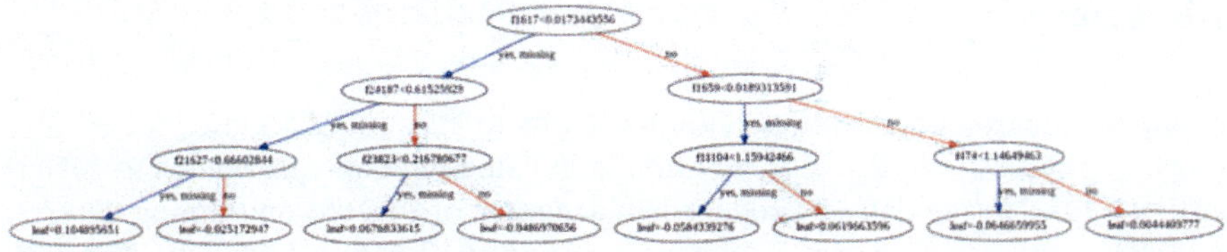

Fig. 7 XGBoost tree

help us in getting diverse features, strong gradient flow and maintains computational efficiency.

XGBoost Classifier—Instead of adding top layers of our classifier, we have used these feature extractor algorithms with XGBoost classifier instead, the result does not match up with the custom classifier but still is enough to be included in our study. XGBoost is an optimized parallel tree boosting algorithm that helps in fast computation. Boosting is a sequential learning method that combines weak learners to get improved accuracy. XGBoost paired used different feature algorithm. Algorithm provided good result. The computation time needed for the model is less than the time needed for the Custom classifier. As shown in Fig. 7, we plotted the tree generated by our XGBoost classifier. The tree is having a depth of 4 and the Number of trees selected for this graph is three Random Forest Classifier—Like XGBoost, Random Forest also combines a few decision trees on different sub samples and averages the results to improve the predictivity and decrease overfitting of the model.

The results we obtained are less than XGBoost and another custom classifier. We have combined random classifier with all the feature extractor algorithm and then evaluated the results.

4.3 Concatenation of Groups

Whilst performing multiclass classification, the possibility of data overfitting one another, due to the extra number of training instances needed to be accounted for, increases. Thus, it is important to perform binary classification between various groups individually as well as concatenated. We implemented all combinations possible for the dataset classifications and results showed the concatenation of Necrotic Tumour and Viable Tumour; and the concatenation of Non-tumour and Viable Tumour groups gave the best results. Furthermore, we also saw that the classification between Viable and Necrotic Tumour, as well as Non-tumour and Viable Tumour both gave the best outcomes.

5 Results

The implementation of Inception classifier led us to a Training accuracy of 94.7% and testing accuracy of 92.1%. In addition to this, the implementation of DenseNet classifier on the image dataset produced an accuracy of 91.76% on training dataset and 90.73% on testing dataset. The concatenation if Viable and Necrotic Tumour gave the highest training accuracy of 96.22% followed by the group of Non-tumour versus Viable which gave the training accuracy of 94.56%. The result is shown in Table 1.

In [16], the maximum accuracy achieved was that of Viable Tumour, i.e. 92.6%. The accuracy for Necrotic Tumour and Non-tumour were 91.5 and 89.5%, respectively. In comparison to this study, our model is accurate by 0.3% for Viable Tumour, 1.1% for Necrotic Tumour and 3.1% for Non-tumour.

The limitations of this model consist of the shortage of evaluation of model from the pathologists, albeit our model reaches a high performance, it is recommended that the model should always be used under the supervision of a pathologist. This method can capture new cases in clinical practises. Besides, the existent dataset will not indicate the long term histological images from the patients, therefore the generalization of our model could prove to be problematic to deal with the issue. It might be useful to be adopted in medical facilities and clinical settings to assess its performance.

Table 1 Results of selected models

S. No.	Result		
	Implementation	Training accuracy (%)	Testing accuracy (%)
1	Extra trees classifier	82.7	80.55
2	DenseNet classifier	91.76	90.73
3	Inception classifier	94.7	92.1
4	VGG16 classifier	87.7	83.5
5	DenseNet+XGBoost	86.4	83.4
6	DenseNet with RandomForest classifier	83.9	80.7
7	Inception with XGBoost classifier	84.15	83.72
8	Inception with RandomForest classifier	85.33	82.13
9	VGG16+XGBoost classifier	84.32	82.15
10	VGG16 with RandomForest classifier	81.32	80.67
11	Non-tumour versus Viable+Necrotic	94.51	89.68
12	Necrotic versus Non-tumour+Viable	92.00	89.09
13	Viable versus Necrotic	96.22	90.11
14	Non-tumour versus Viable	94.56	90.80

6 Conclusion

It is important to automate the prediction of cancer using computer-aided systems since the human accuracy goes as high as 60% in case of Osteosarcoma. The analyses is performed on two datasets, feature, and image dataset. Therefore, our work not only focusses on prediction using images, but it also focuses on prediction of malignant cells using features of the cells which include the count of red and blue bodies, and nuclei thickness. As compared to other studies done on particularly image dataset, we were able to ramp up the accuracy using Convolutional Neural Networks and take it as high as 94.7%. The concatenation of datasets gave good results too. The accuracy of Non-tumour versus Viable Tumour was the highest, i.e. 94.56% which means that it can be used for clinical settings.

References

1. Mirabello L, Troisi RJ, Savage SA (2009) Osteosarcoma incidence and survival rates from 1973 to 2004. Cancer 115:1531–1543
2. Arndt CA, Crist WM (1999) Common musculoskeletal tumours of childhood and adolescence. N Engl J Med 341:342–352
3. Im HJ, Kim TS, Park SY, Min HS (2012) Prediction of tumour necrosis fractions using metabolic and volumetric. Eur J Nucl Med Mol Imaging 39(1):39–49
4. Schulte M, Brecht-Krauss D, Werner M, Hartwig E (1999) Evaluation of neoadjuvant therapy response of osteogenic sarcoma using FDG PET. J Nucl Med 40(10):1637–1643
5. Mirabello L, Troisi RJ, Savage SA (2009) International osteosarcoma incidence patterns in children and adolescents, middle ages and elderly persons. Int J Cancer 125(1):229–234
6. Geller DS, Gorlick R (2010) Osteosarcoma: a review of diagnosis, management, and treatment strategies. Clin Adv Hematol Oncol 8:705–718
7. Shin H-C, Roth HR, Gao M, Lu L, Xu Z, Nogues I, Yao J, Mollura D, Summers RM (2016) Deep convolutional neural networks for computer-aided detection: CNN architectures, dataset characteristics and transfer learning. IEEE Trans Med Imaging 35(2016):1285–1298
8. Rawat W, Wang Z (2017) Deep convolutional neural networks for image classification: a comprehensive review. Neural Comput 29:2352–2449
9. Wang S, Yang DM, Rong R, Zhan X, Xiao G (2019) Pathology image analysis using segmentation deep learning algorithms. Am J Pathol 189:1686–1698
10. Serag A, Ion-Margineanu A, Qureshi H et al. (2019) Translational AI and deep learning in diagnostic pathology. Front Med, vol 6
11. Chang J, Yu J, Han T, Chang HJ, Park E (2017) A method for classifying medical images using transfer learning: a pilot study on histopathology of breast cancer. In: 2018 IEEE 19th international conference on e-health networking, applications and services (Healthcom). IEEE, pp 1–4
12. Russakovsky O, Deng J, Su H et al (2015) ImageNet large scale visual recognition challenge. Int J Comput Vis 115:211–252
13. Esteva A, Kuprel B, Novoa RA et al (2017) Darmatologist-level classification of skin cancer with deep neural networks. Nature 542:115–118
14. Sun W, Zheng B, Qian W (2017) Automatic feature learning using multichannel ROI based on deep structured algorithms for computerized lung cancer diagnosis. Comput Biol Med 289:530–539
15. Naqi SM, Sharif M, Jaffar A (2020) Lung nodule detection and classification based on geometric fit in parametric form and deep learning. Neural Comput Appl 32:4629–4647

16. Arunachalam HB, Mishra R, Daescu O, Cederberg K, Rakheja D, Sengupta A, Leonards D, Hallac R, Leavey P (2019) Viable and necrotic tumor assessment from whole slide images of osteosarcoma using machine learning and deep learning models. PLoS ONE 14:e0210706
17. Pan SJ, Yang Q (2009) a survey on transfer learning. IEEE Trans Knowl Data Eng 22:1345–1359

Detection of Age and Gender from Facial Images Using CNN

Devesh Kumar Srivastava, Esha Gupta, Snigdha Shrivastav, and Riya Sharma

Abstract This work focuses on the existing application of gender and age classification from an image. In this paper, convolutional neural network is used to accurately classify a person's gender and age through a single face image. Gender can be predicted to one of 'Male' or 'Female' and age predicted to be one of the following: (0–2), (4–6), (8–12), (15–20) (25–32), (38–43), (48–53), (60–100) from 8 nodes. Guessing the accurate age of an individual through image can be difficult due to variables such as make-up, lighting, and expressions. Therefore, rather than using regression, we consider it a classification model. Evaluations on the 'Audience' dataset, with a combination of simple preprocessing, obtained the performance in gender and age recognition. We experienced some patterns like face recognition, face detection, age recognition by implementing convolutional neural networks (CNNs) for age and gender predictions. The automatic age and gender detection has become increasingly important, particularly as social networks and social media which are rapidly grown. This demonstrates that the use of deep neural networks (CNNs) is increasing significantly to showcase the performance of observed accuracy for age and gender classification which are higher than another existing claim. We used the cloud services to compute and accessing the dataset from cloud storage.

Keywords CNN models · Performance metrics · Accuracy · Recall · Precision

1 Introduction

CNN models have various applications when it comes to pattern recognition. In this paper, we use it in detecting age and gender. The models are as follows: Deep convolutional neural convolutional neural networks—ResNet: Dense Net: Xception: AlexNet: VGG neural networks. Several concerns about age and gender classification, however, remain unresolved. Despite recent advancements made by the computer vision community in new techniques that improve age and gender

D. K. Srivastava (✉) · E. Gupta · S. Shrivastav · R. Sharma
SCIT Manipal University Jaipur, Jaipur, Rajasthan, India
e-mail: devesh988@yahoo.com

V. K. Gunjan and J. M. Zurada (eds.), *Proceedings of 3rd International Conference on Recent Trends in Machine Learning, IoT, Smart Cities and Applications*, Lecture Notes in Networks and Systems 540, https://doi.org/10.1007/978-981-19-6088-8_42

predictions of unfiltered real-life faces. Many strategies have been put forward in recent years to solve the classifications problem. These traditional hand-engineered approaches rely on differences in facial feature dimensions and face descriptors, which are incapable of handling the varied degrees of variation seen in these unconstrained imaging scenarios. Deep learning-based algorithms have recently demonstrated promising results in this sector, particularly in the age and gender categorization of poorly face images. All the computation are done through software as a service (SaaS) and platform as a service (PaaS). The dataset is stored on public cloud.

2 Literature Survey

In the field of computer vision, deep learning has found several applications. The domains that deal with facial data are some of the most important applications of computer vision. Face recognition and detection are extensively used in security-related applications. The authors of this study used CNN models to identify emotion detection, or facial expression recognition, and age as well as gender prediction from a facial image [1, 2]. The author of this article created a model to detect age and gender. They combined emotion detection and object detection into a single model, resulting in a fully functional application. In this study, we look at a black-box classifier that uses layer-wise relevance propagation (LRP) to predict age and gender using facial characteristics. On the challenging audience dataset, the authors compared different image preprocessing, model initialization, and architectural options and explored how these affect the performance. Authors achieved state-of-the-art gender categorization performance on the audience benchmark dataset using a combination of basic preprocessing procedures. The authors used the VGG architecture of the deep convolutional neural network (D-CNN) to improve the earlier result. Because of the proliferation of online social networking Web sites and social media, automatic gender detection is highly used in a wide variety of software and hardware. However, the performance of existing systems is not remarkably good, especially when compared to the results of tasks involving face recognition. The authors investigated a classification method, and with the help of the (Deep-CNN) technique, they were able to achieve a satisfactory increase in performance [3–5]. The author suggested an end-to-end CNN approach for robust age and gender categorization. They used a strong image preprocessing approach to prepare and process the unfiltered real-world faces before feeding them into the CNN model, which addressed the huge variability in the unfiltered real-world faces. Technically, their network was trained on an IMDb-WIKI dataset with noisy labels, then fine-tuned on MORPH-II, and finally on the OIU-Audience (original) dataset's training set. When the experimental findings were compared to the OIU-Audience standard for classification accuracy, it was discovered that their model outperformed the competition in both age group and gender categorization [6, 7]. In this paper, they applied a deep learning-based double-check layer validator. Also, they made a Web service to validate the process. The dataset for the same has been taken from the University of Palestine through an enterprise system and form validators were implemented to reduce data entry errors. The validation of the data is done by comparison between predicting and entered value by the user. In the end, good results in gender prediction were received, but the model suffered in age prediction [8]. In this paper, they presented a deep learning framework based on an ensemble of attentional and residual convolutional networks for accurately predicting the gender and age group of facial photographs. They used an attention model to focus on the important part of the face which further helps in accurate prediction. The study showed that models gave better accuracy when combined for both age and gender rather than when they are taken separately [9]. In this paper, the algorithms used are CNN and ResNet50 which works on UTKFace dataset. The algorithm gave an approximate prediction of 80% in gender prediction and 60% in age prediction though the gender of the person is correctly predicted but as we proceed for age prediction after the 20th iteration validation loss start to increase. Thus, stating that the validation loss is directly proportional to the age accuracy [10, 11].

3 Methodology

3.1 Dataset

The dataset 'Audience Benchmark Gender and Age classification' for this work is gathered from 'Kaggle repository' [12]. This dataset is a collection of face images and text file, consisting of 19,370 total face images. This dataset is the benchmark for face images and includes many image conditions in the real world, such as noise, light, position, and presence. The photographs are collected from Flickr albums. In this dataset, there are images of every age group; here are some samples as shown in Table 1 and Fig. 1.

Dataset consists of the text files those has the data features like—user_id, original_image, face_id, age, gender, x, y, dx, dy, tilt_ang, fiducial_yaw_angle, fiducial_score.

3.2 Data Preprocessing

In this step, from raw dataset, some data are removed that contain null or missing values. After mapping the age label in the dataset, irrelevant data are removed, and labeling of age and gender is done. Age is labeled into 8 classes/ranges, i.e., '0–2', '4–6', '8–13', '15–20', '25–32', '38–43', '48–53', '60+ '., whereas gender is labeled in two classes, i.e., female is 'f' classified as '0' and male is 'm' classified as '1'.

Table 1 Number of images for every age group in the dataset [8]

Age groups	No. of images labels	Age groups	No. of images labels
(25, 32)	5004	35	293
(0, 2)	2488	13	168
(38, 43)	2293	22	149
(4, 6)	2140	34	105
(8, 12)	2124	23	96
(15, 20)	1642	45	88
(60, 100)	872	(27, 32)	77
(38, 42)	46	55	76
36	56	3	18
57	24	29	11

Fig. 1 Sample images of audience dataset

All the face images or data have been collected from Internet and loaded of size or resize of 227 × 227 pixels.

3.3 *Split the Train and Test Dataset*

Before starting the modeling, the preprocessed dataset is split in the ratio of 70:30, respectively, for training and testing with the image size (227,227). For age classification and gender classification, dataset splits accordingly. Like—train images shape (12216, 227, 227, 3); test images shape (5236, 227, 227, 3) (Fig. 2).

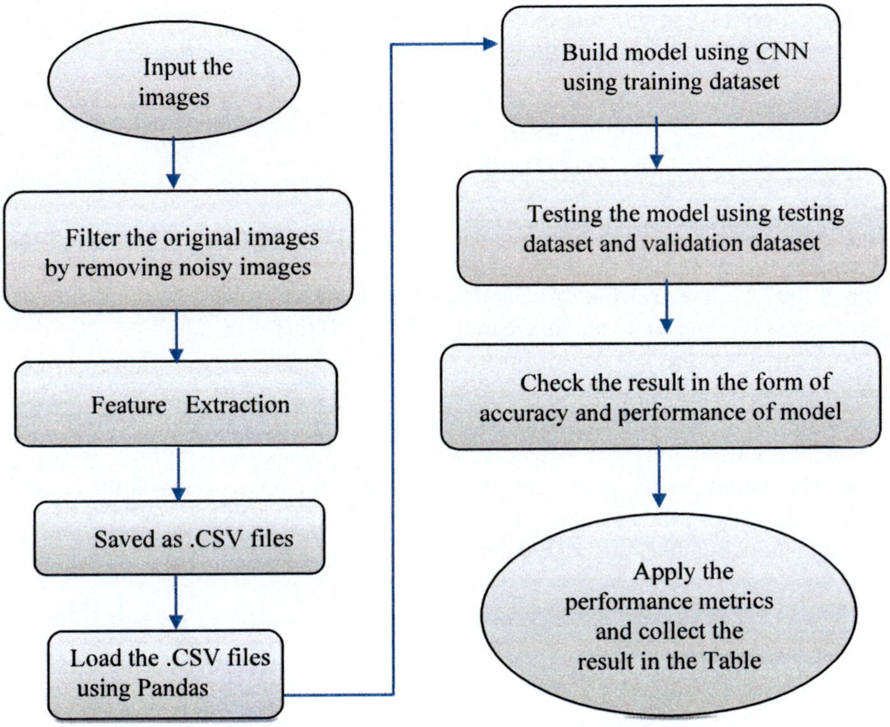

Fig. 2 Flowchart of problem formulation on detection of age and gender using feature extraction technique using CNN method

3.4 Building Up the Model

- First layer is the convolutional layer (Conv2D). In this layer, shape of the input image is of 227 × 227 × 3; filters are 96; kernel size is 7 × 7; strides are 2 × 2; padding is 'valid', and activation is 'ReLU'.
- Add a max pooling layer with pool size 2 × 2 and strides 2 × 2.
- Adding a normalization layer.
- The second convolutional layer (Conv2D) is there. In this layer, filters are 256; kernel size is 5 × 5; strides are 1; padding will be 'same', and activation will be 'ReLU'.
- Add the second max pooling layer with pool size 2 × 2 and strides 2 × 2.
- Add the second normalization layer. Again, add the third convolutional layer (Conv2D) is added. In this layer, filter size is 256; kernel size is 3 × 3; strides are 1; padding will be 'same', and activation will be 'ReLU'.
- Add third max pooling layer with pool size 2 × 2 and strides 2 × 2.
- Add the third normalization layer.
- Add a flatten layer.
- Add a dense layer with unit 512 and activation function is ReLU.

- Add a dropout layer with rate 0.25.
- Add the second dense layer with unit 512 and activation function is ReLU.
- Add the second dropout layer with rate 0.25.
- Add the dense layer with unit 2 and activation function is softmax.

3.5 Algorithm

- Input: the face images from training dataset.
- Preprocess the images to obtain cleared and filtered images.
- Resize the images to 227, 227. Make conversion to NumPy array and implement to the test and train dataset.
- Defining the CNN model, using a sequential model, i.e., data will travel sequentially from one layer to another without any jumps and splitting.
- Train the model (more epochs are likely to give better accuracy) with the set of images, to classify the age and gender, respectively.
- Test the model with testing image dataset.

3.6 Classification

Fit the model using the ADAM optimizer and batch size '32' with epochs '20'. In the final step of the proposed framework, first perform the gender classification/prediction on the test images. Also, we have to find the accuracy for gender classification. Then, perform the age classification/prediction on the test images. Also, we have to find the accuracy for age and gender classification (Figs. 3 and 4).

3.7 Performance Metrics

To measure the performance, we used the following equation:

$$\text{Accuracy} = (\text{TN} + \text{TP})/(\text{TN} + \text{TP} + \text{FN} + \text{FP})$$
$$\text{Recall} = \text{TP}/(\text{TP} + \text{FN})$$
$$[\text{Precision} = \text{TP}/(\text{TP} + \text{FP})]$$
$$[\text{F1 - Score} = 2 \times ((\text{Precision} \times \text{Recall})/(\text{Precision} + \text{Recall}))]$$

where TP $\rightarrow$ true positive
FP $\rightarrow$ false positive
TN $\rightarrow$ true negative and
FN $\rightarrow$ false negative, respectively.

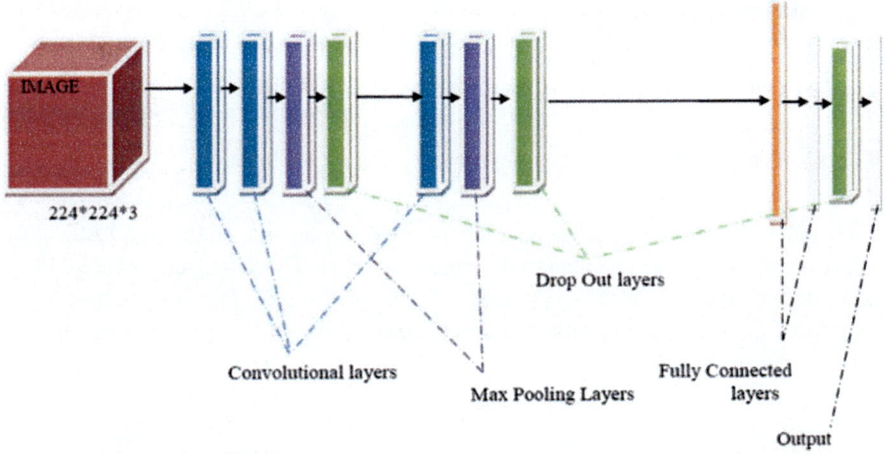

Fig. 3 CNN model illustration

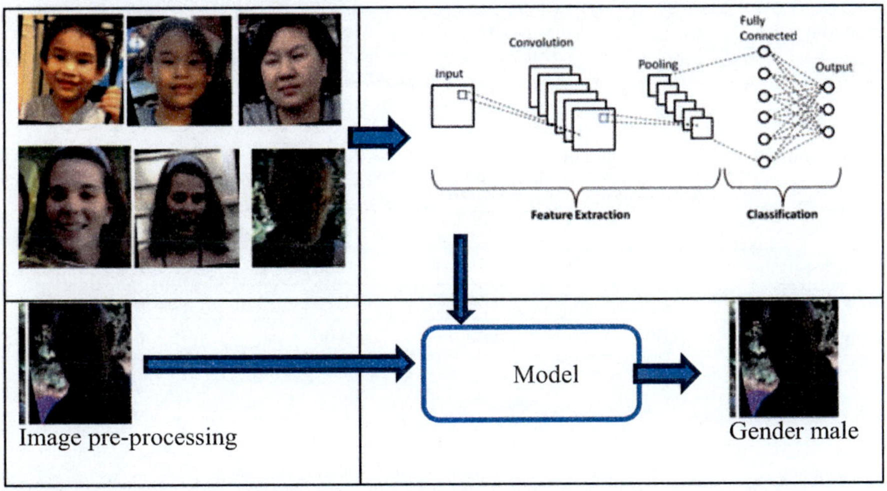

Fig. 4 Flow of framework for age group and gender

For both obtained age and gender classification, test accuracy is 36.65% and 52.36%, respectively.

4 Results and Analysis

- Confusion Matrix for Gender Classification/Detection (Table 2):
- Confusion Matrix for Age Classification/Detection (Table 3):

For gender detection, precision, recall, F1-score, support are 1.00, 1.00, 1.00, and 5236, respectively. For age detection, precision, recall, F1-score, support are 0.00, 0.00, 0.00, and 5236, respectively (Figs. 5 and 6).
Graphs of Training and Testing Accuracy/Loss:

Table 2 Confusion matrix for gender classification

Matrix	Precision	Recall	F1-score	Support
0	1.00	1.00	1.00	53.38
Accuracy			1.00	5236
Marco avg	1.00	1.00	1.00	5236
Weighted	1.00	1.00	1.00	5236

Table 3 Confusion matrix for age classification

Matrix	Precision	Recall	F1-score	Support
0			1.00	53.38
Accuracy			0.00	5236.0
Marco avg	0.00	0.00	1.00	5236.0
Weighted avg	0.00	0.00	0.00	5236.0

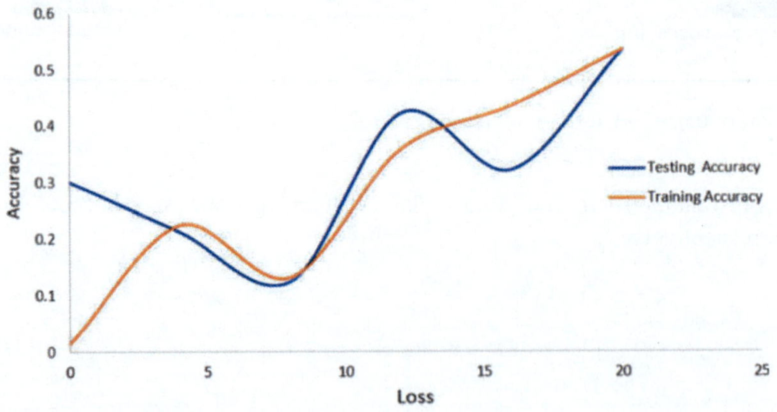

Fig. 5 Training accuracy and testing accuracy graph

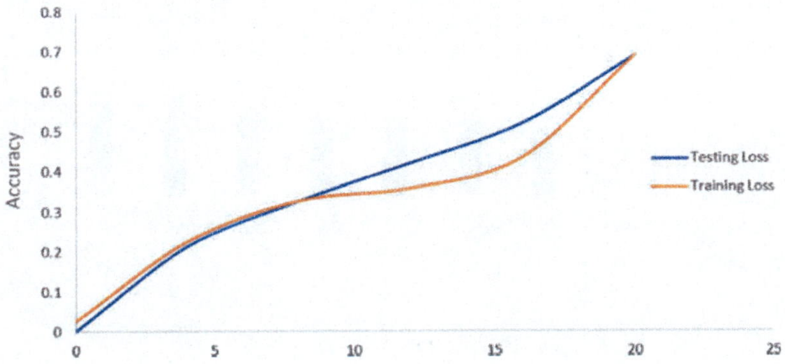

Fig. 6 Training loss and testing loss graph

Table 4 Observed accuracy of age and gender selection

Ref. no	Observed accuracy of age	Observed accuracy of gender
9	–	93.2
10	45.1	–
11	50.7	86.8
13	52.3	–
12	54.0	–
15	58.5	88.2
14	66.5	–
Experimental study	80.2	94.1

5 Conclusion

The problem of gender and age detection and classification is closely related to face detection. Current approaches use the neural network architecture for age as well as gender prediction. In this work, we detected and predicted the age and gender of images, using the convolutional neural network. Based on this CNN architecture results, the experimental study performs better accuracy compared to others as shown in Table 4 and in Fig. 7. Therefore, the model detected the age and gender using various images in an efficient and accurate way.

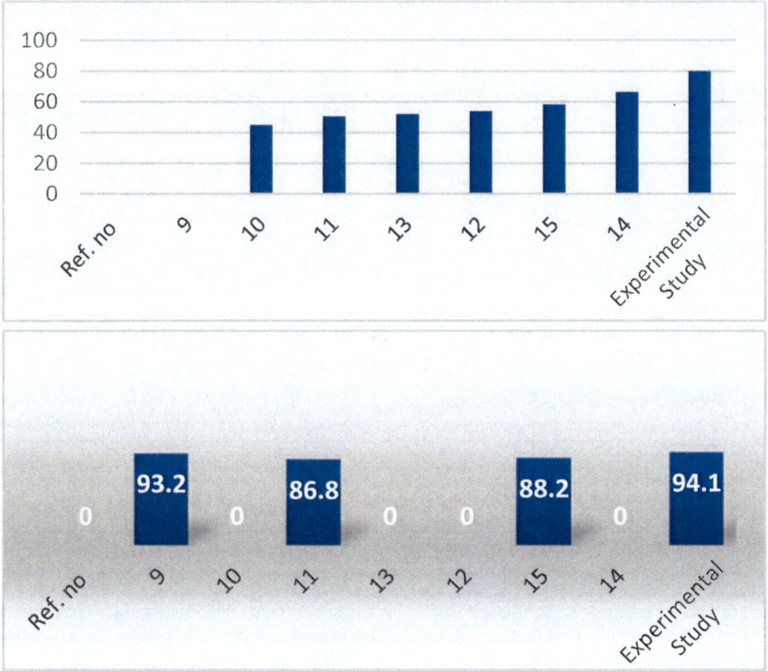

Fig. 7 Graph presentation of observed accuracy of age and gender selection

References

1. Agbo-Ajala O, Viriri S (2020) Deeply learned classifiers for age and gender predictions of unfiltered faces. Hindawi Sci World J 2020:12
2. Hussain SJ et al (2021) Age classification using convolution neural networks using a local dataset. Turk J Comput Math Educ 12(9):2038–2045
3. Zhang Y, Xu T (2018) Landmark-guided local deep neural networks for age and gender classification. J Sens 2018. https://doi.org/10.1155/2018/5034684
4. Lapuschkin S, Binder A, Muller KR, Samek W (2017) Understanding and comparing deep neural networks for age and gender classification. In: International conference on computer vision workshops ICCVW 2017, vol 2018-Janua, pp 1629–1638
5. Dhomne A, Kumar R, Bhan V (2018) Gender recognition through face using deep learning. Procedia Comput Sci vol 132. https://doi.org/10.1016/j.procs.2018.05.053
6. Rodríguez P, Cucurull G, Gonfaus JM, Roca FX, Gonzàlez J (2017) Age and gender recognition in the wild with deep attention. Pattern Recognit 72:563–571
7. Agbo-Ajala O, Viriri S (2020) Deeply learned classifiers for age and gender predictions of unfiltered faces. Sci World J vol 2020. https://doi.org/10.1155/2020/1289408
8. Abu Nada AM et al (2020) Age and gender prediction and validation through single user images using CNN. http://dstore.alazhar.edu.ps/xmlui/handle/123456789/632.
9. Abdolrashidi A, Minaei M, Azimi E, Minaee S (2020) Age and gender prediction from face images using attentional convolutional network. https://arxiv.org/abs/2010.03791. Published date: 7th Dec 2020

10. Meghana AS, Sudhakar S, Arumugam G, Srinivasan P, Prakash KB (2020) Age and gender prediction using convolution, ResNet50 and Inception ResNetV2. Int J Adv Trends Comput Sci Eng 9(2)
11. Agbo-Ajala O, Viriri S (2020) Deeply learned classifiers for age and gender predictions of unfiltered faces. Sci World J vol 2020
12. Dataset Used. https://www.kaggle.com/ttungl/adience-benchmark-gender-and-age-classification

A Survey on Detecting Location-Based Faults in Wireless Sensor Networks Using Machine Learning and Deep Learning Techniques

Neha Jagwani and G. Poornima

Abstract A Wireless Sensor Network is a group of multiple detection stations referred as nodes which performs multiple actions (sensing, communicating and computing) in a collaborative manner. Wireless Sensor networks are broadly used in many applications which includes healthcare, defense, smart security systems and many more. But they are facing many issues with Fault Tolerance and reliability. Various investigations are going on for increasing the fault tolerance capabilities of WSNs so that they can be used in critical applications effectively. Different hard and soft faults occur in WSN due to several factors which commonly includes energy depletion, failure of communication link, hardware failure, dislocation of sensor nodes, radio interference, congestion problem, malicious attack, network lifetime problem, etc. The use of Machine Learning technique is a very effective way to overcome these faults as it is important to make the system fault-free. Machine Learning are self-learning process which works without re-programing and human interference. This paper presents the survey of Machine Learning, Deep Learning and Time Series methods used to detect anomalies, data centric faults and system centric faults in WSN which are one of the most commonly occurring type of faults in sensor networks.

Keywords Anomaly detection · Data centric faults · Deep learning · Fault classification · Fault tolerance · Machine learning · Neural networks · System centric faults · Time series · Wireless sensor networks

---

N. Jagwani (✉) · G. Poornima
BMS College of Engineering, Bangalore, Karnataka, India
e-mail: jagwanineha@gmail.com

G. Poornima
e-mail: gpoornima.ece@bmsce.ac.in

V. K. Gunjan and J. M. Zurada (eds.), *Proceedings of 3rd International Conference on Recent Trends in Machine Learning, IoT, Smart Cities and Applications*, Lecture Notes in Networks and Systems 540, https://doi.org/10.1007/978-981-19-6088-8_43

1 Introduction

With the development in technology and recent advances in electronics and communication field, Wireless Sensor Network (WSN) has gained much attention by research community because of their advantages and uses in real time application. In brief, WSN is a set of thousands of autonomous, tiny, low power and low cost self-directed sensor nodes. The function of these nodes is to continuously monitor the environment where permanent presence of human is impossible. They sense and collect the data at regular interval, convert into electronic signals and transfer them to monitoring center (commonly known as sink node or base station) [1, 2]. Originally used for battlefield surveillance, WSNs have gradually spread its usage and scope in Industrial control and monitoring, health monitoring, disaster management, environmental monitoring, habitat monitoring, banks, airports and other places for protection, smart traffic control system, home automation, to mention a few [1].

Generally, the sensor nodes are densely located in adverse environments and left unattended for long time which makes them prone to failure. Also, Sensor Nodes (SNs) are normally deployed once in their life span. Thus it is very necessary to make the system error free. In-spite of best efforts to remove all faults and errors, they are bound to be faulty in most of the operational systems [3]. This can be better understand with an example. Let us take the case of health care monitoring system in which patients share their health information with their caregivers or doctors in real time. If fault occurs in communication link of health monitoring system, the patient's information will not reach to the doctor at proper time which can be dangerous in case of critical patients. Further, problem can exists with hardware components attached in the network or SNs can send faulty data to the caregiver which is also a matter of concern. Also, privacy of information can be disturbed due to malicious attack or failure of component. In this way, a WSN can fail due to various reasons. Thus, detection of faults in sensor network is very important in WSN.

Machine Learning (ML) and Deep Learning (DL) has come up with the solution of the problems in WSN. ML is a subset of Artificial Intelligence (AI) which automatically learns and improves from the study or experience and works without being explicitly programmed [4]. Over time, many application oriented problems have found the remarkable use of ML which increases accuracy, reliability, performance and speed of events being detected. ML comes up with a bundle of techniques to improve and optimize the ability of WSNs to survive in the ever changing behavior of the environment.

In this paper, we have presented a survey based on various fault detection techniques in WSN using ML and DL algorithms. Our survey emphasize on detecting location-based faults which consists of data centric and system centric related issues. The rest of the paper is organized as: Sect. 2 presents various types of faults present in WSN with their reason of origin. Section 3 is divided into 3 parts which presents the role of ML and DL in detecting anomalies, data centric faults and system centric faults in WSN. Finally Sect. 4 presents the conclusion.

Tables 1 and 2 show the list of Acronyms used in the paper.

Table 1 Acronyms of performance matrices used

Acronyms of performance matrices used	Full Form
DA	Detection Accuracy
FAR	False Alarm Rate
FPR	False Positive Rate
FCR	False Classification Rate
DL	Detection Latency
MCC	Mathews Correlation Coefficients
ROC	Receiver Operating Characteristics
MAE	Mean Absolute Error
MSE	Mean Squared Error
MAPE	Mean Absolute Percentage Error
MMSE	Minimum Mean squared Error
TRP	True Positive Rate

2 Types of Faults in WSN

The presence of anomalies and faults can affect the functionality of a SN or whole network of WSN. Thus, it is important to identify mostly occurring anomalies and faults in WSN.

Faults in WSN can be classified in various forms. In broader sense, faults can be categorized based on their origin as [5–7]:

1. Hardware level:—Defects in sensing unit, memory, battery or any other hardware components are referred to as hardware level faults. This can be due to quality of components used, limited energy and the hostile environment in which the WSN is placed.
2. Software level:—This layer consists of system software (such as operating system) and middle-ware like communication, routing and aggregation. The fault in this layer is due to bugs in software programming. The error can also be due to malicious attacks.
3. Network Communication layer:—Even if the state of the hardware and software is fine, the communication between SNs (or with sink nodes and gateways) is affected by various parameters such as signal strength, weather conditions, antenna angle obstacles. The effective radio range is reduced due to propagation in wireless medium which causes loss due to reflection, scattering and dispersions.
4. Application layer:—Another issue that arises with WSN is the coverage, connectivity and localization problem. In this the network should be designed in such a way that every targeted point and interested fields should be inside the range of

Table 2 Acronyms of ML and DL algorithms used

Acronyms of ML and DL Algorithms used	Full Form
RF	Random Forest
DT	Decision Tree
NN	Neural Network
ARIMA	Auto Regressive Integrated Moving Average
ARMA	Auto Regressive Moving Average
AR	Auto Regressive
MA	Moving Average
SVM	Support Vector Machine
KNN	K-Nearest Neighbor
PCA	Principal component Analysis
NB	Naïve Bayes
RNN	Recurrent Neural Network
FNN	FeedForward Neural Network
SGD	Stochastic Gradient Descent
MLP	Multilayer Perceptron
PNN	Probabilistic Neural Network
BRNN	Bidirectional Recurrent Neural Network
HMM	Hidden Markow Model
SODEN	Spatially organized distributed echo state networks

the sensor elements deployed in the network. These are application-dependent problems. Figure 1 shows the classification of faults in WSN.

Faults in WSN can also be classified on the basis of behavior, time, component and location. A brief description of the various types of faults are given below:

I. Based on their behavior, faults can be classified as: [8–10]

1. Hard Faults:—SNs are unable to communicate among themselves due to failure in certain modules, e.g., Black hole in case of energy depletion.
2. Soft Faults:—In soft faults, SN can continue to work even in case of failures but sense, process or transmit faulty data.

II. Time-Based faults can be classified as: [8, 9, 11–14]

1. Transient Faults: Faults occurs because of environment conditions such as temperature, humidity, cosmic rays, vibrations. These types of errors usually occur once and then disappears and thus are difficult to manage.

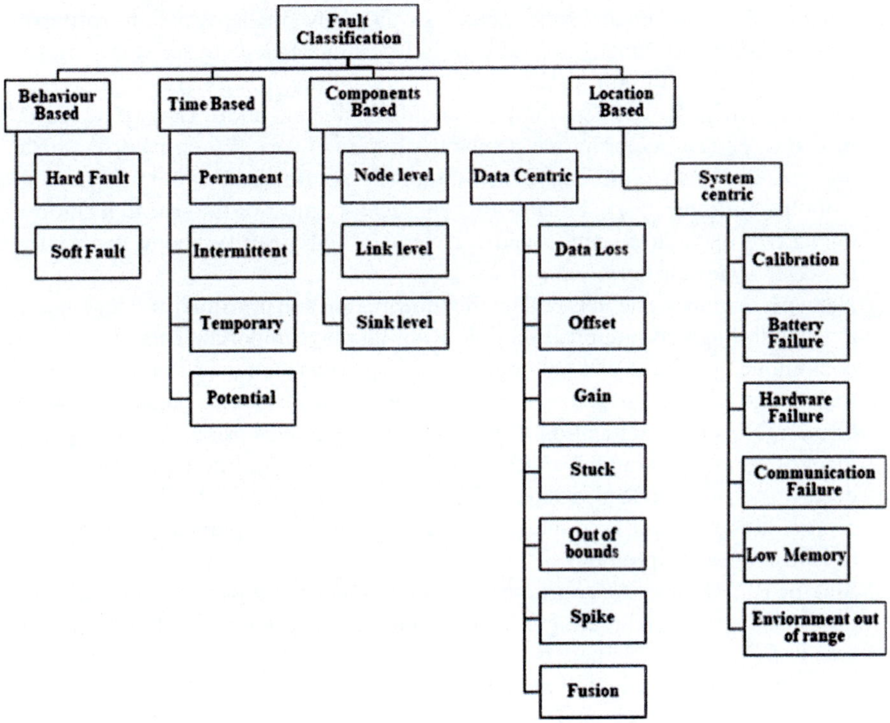

Fig. 1 Classification of Faults in WSN

2. Intermittent Fault: These types of errors do not occur continuously, as they appear and vanishes repeatedly. This type of error includes loose connection, obsolete/aged components, etc.
3. Permanent Faults: These faults include built-in defects such as fault in chip manufacturing, burned out of electronic components. The effects of permanent faults remain until the faulty components are totally removed from the circuit.
4. Potential Faults: This occurs due to depletion of hardware resources which ultimately reduces the network lifespan. The most common is the energy depletion of the nodes that impacts lifetime of the node. SNs require energy for various operations such as data sensing, data collection, communication and processing. Thus, it is necessary to charge or change the battery after they have been consumed. But it is not possible every time as WSNs is generally installed in hostile or hazardous environment. Moreover, a sensor node should be cost effective, which is a critical hardware constraint in the design of WSN.

III. Based on components of WSN, faults can be divided into: [7, 8]

1. Node faults:—At node level, faults can either be in hardware or software malfunction. Additionally, a hardware failure can also lead to software failure. For e.g., If sensing unit suffers with battery failure, it may collect incorrect readings which can affects the data acquisition application. Generally, faults in nodes lead to inappropriate decisions in the network. When incorrect data are collected at sink, the final Base Station (BS) collect inaccurate information from the entire network. For this reason, recently most of the research studies emphasize on fault detection and recovery in high priority nodes, especially manager nodes/cluster heads.
2. Network Faults:—The SNs collect the information and transmit the data toward the sink through communication link. Routing plays important role in this. So, communication links and routing layer are another cause of faults in WSN. Path faults, Radio interference, temporary/permanent blocks in paths may cause errors in SNs communication. Congestion problem also arises due to deployment of enormous amount of SNs simultaneously transmitting the data on the occurrence of interested event. This can lead to packet loss. Thus, software programming should be done in such a way that the applied algorithms can reduce congestion problem.
3. Sink or BS Faults:—A fault in the sink can hinder the process of transmitting tasks to sensors and directing information to final BS. Thus, failure in sink can lead to failure in whole network. Another issue in sink arises due to its limited energy and software malfunctioning like node.

IV. Location based faults can be classified as System centric and Data centric [5, 6, 12–15]

   1. System centric faults:—These faults occur due to calibration error, battery failure, hardware failure, communication failure, low memory, out of range from the environment.
   2. Data centric faults:—These faults are classified as:
      a. Offset—Output data is deviated from normal operation by a constant value
      b. Drift—Sensor output decreases or increases linearly at a constant rate
      c. Spike—Sensor output contains high amplitude values at constant periods in an intermittent manner
      d. Stuck—Sensor output undergoes nearly zero variations.
      e. Data loss—When sensed data is missing in output

Thus, a node or network may fail due to various reasons which results in the unexpected performance of sensor nodes and low reliability. Overall functionality of the wireless sensor network should not be affected due to failure of nodes within the network. A system build with fault tolerant capabilities will handle the situation and will continue operating probably at a degraded level, in the presence of these faults [3]. As a result, fault tolerance is prime necessity and critical research area in WSN. In this paper, we have focused on detecting anomalies and location-based faults as data centric and system centric faults are the maximum occurring faults in WSN.

3 ML and DL in Improving Fault Tolerance in WSN

In this section, we have explored the applications of ML, DL and time series algorithms in detecting data centric and system centric faults in WSN. Before this, it is advantageous to know about anomaly detection in WSN as anomalies are any unexpected events or observations that can occur due to any fault, outliers, deviations or noise. Anomalies are diagnosed as any deviation from the rest of the data pattern or any difference between the sensed and constructed value. This section is divided into 2 parts: In Sect. 3.1, a small survey on role of ML and DL in detecting anomalies has been done. Section 3.2 is dedicated for detecting location-based faults.

3.1 Role of ML and DL in Anomaly Detection

In [16], stacked LSTM networks have been used for anomaly/fault detection in time series. The network is trained on fault-free data for four different datasets and this trained data is used as input variable over a number of time steps. The prediction errors are modeled as a multivariate Gaussian distribution, which is used to estimate the probability of anomalous behavior. This LSTM-based Anomaly Detection gives good results on all four datasets which involve modeling small-term and long-term temporal dependencies. This is because stacking recurrent hidden layers captures the time series pattern more efficiently.

Jeong et al., in [17] focuses on anomaly detection and reconstruction of sensor data using BRNN. This algorithm is used to form a data-driven model which is based on the spatiotemporal correlation among the sensor data. Random noises were applied to dataset and anomaly is detected by comparing the reconstructed errors and thresholds. BRNN gives best results when compared with other existing techniques like PCA, MMSE, FNN and RNN as it considered both spatial and bidirectional temporal correlation into account.

Vamsi et al., [18] have also proposed a ML-based hybrid model for anomaly and fault detection in WSN. The dataset is divided into three timestamps based on hours of the day known as Time Of the Day (TOD). This TOD is used as window for detecting anomalies. The work mainly focuses on detecting faults in multivariate time series data using three unsupervised machine learning models namely Histogram-based Outlier Score (HBOS), Minimum Covariant Determination (MCD) and Isolation Forest (IF). Precision and AUC score found that IF performs best among three hybrid models.

The authors in [19] have detected anomaly in medical WSNs. The detection procedure is divided into two parts: first the classification algorithms are used to detect the abnormality in data and then regression algorithms are used to find which attribute exactly crossed the threshold. The attributes considered in the problem are heart rate, body temperature, pulse, SpO2 and respiration rate. For classification problem, RF showed the best results and for regression problem, Additive regression

Table 3 Role of ML and DL in anomaly detection

Algorithms used	Performance evaluation matrices used for comparison	Application area	Inferences drawn
LSTM, RNN [16]	Precision, Recall, F1 score	ECGs, Space Shuttle Marotta value time series, Power demand dataset, Multi-sensor engine data	LSTM is more robust than RNN
BRNN, PCA, FNN, RNN [17]	RMSE	Vibration sensor dataset for bridge monitoring	BRNN performed best as both spatial and bidirectional temporal correlation were considered
Histogram-based outlier score (HBO), Minimum Covariant determination (MCD) and Isolation Forest (IF) [18]	AUC, Precision	IBRL dataset	All three models are based on different concepts. IF performed best and was able to detect faulty sensor/least trusted sensor among the deployed sensors. This concludes that domain-based anomaly detection works best in case of time series data
For Classification: J48, RF, KNN For Regression: Linear Regression, Additive Regression [19]	ROC-curve, MEA	MIMIC dataset (Medical WSN where sensing device is attached to patient body)	Random forest performed best for classification job and Additive regression performed best for regression jobs

techniques performed better than other algorithms. Table 3 shows anomaly detection in WSN using ML and DL.

3.2 Role of ML and DL in Detecting Location-Based Faults

This section is divided into 2 sections—First section is for detecting data centric faults and second is for detecting system centric faults in WSN.

Data Centric Fault Detection using ML and DL. In [5], authors have done a comparative analysis among various ML classifiers that include SVM, SGD, RF, MLP, CNN and PNN. The above classifiers had been used to classify offset fault, out of bounds fault, gain fault, spike fault, stuck at fault, data loss fault and noise fault at the sensor level. The faults had been inserted in the dataset at various fault rates like 10%, 20%, up to 50%. The RF classifier performed best in terms of TPR, DA, F1-score and Matthews Correlation Coefficients (MCC).

In [20] and [6], the authors have used SVM to classify sensor faults like erratic fault, hard-over fault, spike, drift fault and stuck faults. In [20], SVM is trained by different kernel functions. The efficiency of the classifier has been increased by increasing the training data and the number of features but created overfitting problem on further increase. To overcome the issue of overfitting, cross validation techniques is used. Moreover, the accuracy of the classifier is increased by increasing the input sample size. The ROC-curve is used to prove the efficiency of SNM classifier over NN. In [6], the faults have been injected in the real data set at various fault rates. By using the kernel functions, SVM algorithm shows good capacity for the nonlinear classification cases. Two matrices (DA and FPR) have been used for the comparison among HMM, Bayes, SODSEN, Cloud with SVM method. SVM provides best result for both parameters. Moreover, the SVM technique does not require any overhead on sensors as it can be performed at sink level unlike the other techniques (SODSEN, Bayes and HMM) which are performed on cluster head.

An improved NN architecture based on LSTM has been suggested in [21]. The new approach, namely, Stateful Long Short-Term Memory Neural Networks (S-LSTM NN) can achieve the best performance to provide an automated method to detect the faulty sensor in WSN in terms of stability and accuracy in Structural Health Monitoring (SHM) system. The algorithm used the backpropagation algorithm to tune the parameters. The algorithm is used to detect constant, gain, bias and bottom noise faults assuming the whole structure is working correctly and only a single sensor is faulty in a period of time.

A lightweight Context Aware Fault Diagnostic (CAFD) scheme has been proposed in [12] to detect data centric faults and anomalies in WSN. This scheme utilizes Extra-Trees (ET) classifier which is an ensemble method based machine learning algorithm, advantageous in terms of bias and variance. The CAFD model uses two ML classifiers—first is based on multi-label classification in which faults are classified on the basis of context (causes) such as communication, hardware, calibration. This data is then trained by ET to find out the causes of each fault. Then, this output of first ET classifier is given to the second ET as an additional feature which detects the sensor faults such as hard, drift. The simulations shows that performance of the classifier improves when contexts were added to data as features in comparison to without context method. Moreover, this CAFD with ET classifier shows better results in terms of DA and training time when compared to SVM and NN. The same ET classifier has been used in [14] for fault detection and has been compared with RF, SVM, MLP and DT and found more efficient in terms of DA, precision and F-1 score. The proposed scheme also reduces bias and variance error.

A new nature inspired algorithm has been proposed by Gupta et al., in [22], known as Improved Fault Detection Crow Search Algorithm (IFDCSA). CSA is inspired by intelligent behavior of crows. The IFDCSA method is used to induce various data centric faults in three different datasets, collected by WSN. Then RF classifier is used to classify different faults. On grounds of DA, IFDCSA is more efficient than other ML classifiers (KNN, DT and SVM).

The authors in [23] have enhanced four famous ML classifiers using decision-based belief function fusion approach. These are used for decentralized approach as in centralized approach, the decisions regarding the reliability of data are taken by the central node. Thus, in decentralized approach, decision fusion rule is used to detect the reliability of collected data. The enhanced classifiers used are Enhanced KNN (EKNN), Enhanced Extreme Learning Machine (EELM), Enhanced SVM (ESVM) and Enhanced Recurrent Extreme Learning Machine (ERELM). These enhanced classifiers are used to detect the induced data centric faults. These classifiers are compared with each other along with their basic classifiers and found that ERELM performs best in terms of DA, TPR and ER. This improved belief function fusion approach also performed better when compared to NB and Weighted Majority Vote (WMV) approaches. Table 4 shows detection of data centric faults using ML and DL in WSN.

System Centric Fault Detection using ML and DL. The paper [24] focuses on failure of communication links in WSNs as decisions can be affected due to problems in communication link even if all the nodes are functioning correctly. Thus, the link faults can create disjoint sets of nodes which hampers overall functionality and purpose of WSN. This automated link failure detection is based on FeedForward Neural Network (FFNN) that can adapt and learn from gradient decent learning algorithm. The parameters used to consider the quality of links are Packet Delivery Ratio (PDR) and Latency. A link is considered as fail when latency is very high and PDR is very low. These parameters are considered as input (features) to neural networks. The proposed methodology is validated by indoor and outdoor testbed experiment as well as with software simulation. Various faults are introduced in network such as: for permanent fault, the reduction in communication power was done at every time instance so that the packets could not arrive at targeted place and hence lowering the PDR toward 0; for intermittent fault and behavior fault, interrupts were initiated. The FAR and DA values shows that it is quite easy to detect permanent faults as it is easy to detect a link that does not work at all.

The faults caused due to calibration issue and low battery in WSN have been considered in [25]. The work is focused on comparison and analysis of short-term fault, noise fault and fixed fault, occurs due to low battery and improper calibration. The machine learning methods used for the purpose are NB, SVM and Gradient Boosting Decision Tree (GBDT). The faults have been injected in the already available dataset and the performance of various techniques has been evaluated on criterion like TPR, FPR, DA and precision. The comparison shows that noise fault is better detected by SVM while GBDT has better detection ability for all the types of faults. GBDT is based on CART tree which provides it a natural advantage of finding a variety of features and feature combinations.

The comparison of different prediction/forecasting techniques for time series data has been done in the paper [26]. The techniques used are MA, AR and ARMA. These techniques provide great reduction in power consumption in WSNs. The efficiency of the algorithms is determined by Percentage of Insufficiently Correct Predictions

Table 4 Data centric faults detection in NWSN using ML and DL

Algorithms used	Performance evaluation matrices used for comparison	Application area	Inferences drawn
SVM, CNN, SGD, MLP, RF, PNN [5]	DA, MCC, FPR, F1-score	2 outdoor temperature and humidity sensors	RF algorithm performed best among all algorithms as it did not overfit the model
SVM, HMM, Bayes, SODSEN, cloud [6]	DA, FPR,	Humidity and temperature	The proposed SVM technique uses decision function to classify normal and faulty class and thus performed best in case of FPR. It is good technique for the multidimensional classification and can be used at cluster heads for fault detection
Extra Tree classifier [12]	DA, training time, ROC-AUC, F1-score	Humidity and temperature	CADF scheme based on extra tree (ensemble technique) is more effective than SVM and NN
Extra Tree classifier [14]	accuracy, precision and F-1 score	Temperature and humidity	The training time of this algorithm was found lower than the state-of-the-art algorithms like SVM, MLP, RF and DT
SVM with three different kernels, NN [20]	ROC-curve	Temperature sensor	SVM showed higher accuracy. Increase in number of training samples increased accuracy but further increase lead to overfitting problem which was overcome by k-fold cross validation technique
Stateful Long Short-Term Memory Neural Networks (S-LSTM NN) [21]	Residuals between true value and predicted values were used to determine threshold	Acceleration response of 3 span continuous bridge	S-LSTM is considered as very quick method to detect type of fault in sensors

(continued)

Table 4 (continued)

Algorithms used	Performance evaluation matrices used for comparison	Application area	Inferences drawn
IFDCSA, RF, KNN, DT [22]	DA	3 datasets: 1. Intel lab, 2. SensorScope data 3. Multihop labeled data	Nature inspired IFDCSA algorithm is better than rest three algorithms
EKNN, EELM, ESVM, ERELM [23]	DA, TPR, ER	Vehicle detection sensor	The improved belief function fusion approach performs better than NB, WMV and existing fusion techniques

(PICP) and MSE. The simulation result shows that performance of the algorithms is influenced by entropy of dataset, i.e., with lower entropy, ARMA performs best and with higher entropy MA shows the best results. Dynamic behavior of time series data using ARIMA model has been addressed by Herbert et al. in [27]. The authors have used concept of Low Energy Adaptive Clustering Hierarchy (LEACH) protocol to obtain parameters for ARIMA model. This model saves energy by decreasing the amount of transmitted data. The approach re-calculates the parameters according to the MSE technique and improves accuracy of ARIMA forecasting. It also checks the parameters before forwarding to the BS and thus avoids repeated transmissions. This improves the fault tolerance of the system. In [28], an automatic ARIMA-based data aggregation scheme has been developed for WSNs. This is a battery saving scheme which eliminates the inherent redundancy of raw data in WSNs. This is done by decreasing the number of transmitted data values between SNs and aggregators by using time series prediction model. The sensor node will build a model to forecast future data based on newly sensed data values and send the model parameters to the aggregator automatically. If the difference between the sensed and predicted value is within the threshold range, node will not send the sensed value to the aggregator and predicted value will be considered as the current value. If the difference between the sensed value and predicted value is not within the threshold range, the SN will again build the model and send the sensed value along with updated model parameters to the aggregator. The simulation result shows that the predicted value of the scheme suits the real time data very well and less number of messages are required to transmit. Table 5 shows detection of system centric fault using ML and DL in WSN.

4 Conclusion

Wireless Sensor Network has been considered as promising technology that can perform the function of sensing, processing and communicating to attain numerous

Table 5 Detection of system centric fault using ML and DL in WSN

Algorithms used	Performance evaluation matrices used for comparison	Application area	Type of fault taken into consideration	Inferences drawn
FFNN [24]	Packet Delivery Ratio (PDR) and Latency	Testbed experiment with 10 temperature and humidity sensors	Communication link failure	Delay, diagnosis time and energy efficiency are also considered for performance evaluation. The algorithm has high detection rate in link failure detection
NB, SVM and GBDT [25]	TPR, FPR, DA and precision	Temperature in IBRL	calibration issue and low battery	GBDT is better in detecting noise, short-term and fixed faults occurs due to calibration and low battery issue
MA, AR, ARMA [26]	PICP, MSE	Time series prediction	Battery Failure	Performance of algorithm depends on entropy of dataset: For lower entropy: ARMA For higher entropy: MA
ARIMA [27]	MSE	Temperature	Battery Failure	Coupling of LEACH protocol with ARIMA can be used to avoid retransmission which can save energy
ARIMA [28]	MSE, MAE, MAPE	Real Sense TAO	Battery Failure	Used in data aggregation scheme to reduce number of messages transferred which saves the precious battery energy in WSN

remote sensing applications in the field of automation, military, environment, security etc. The sensor nodes consist of a large number of hardware and software units. There are various issues and challenges in WSN which hinders the functionality of sensor nodes and network. We have given a detailed classification of various types of faults that can occur in WSN. The classification is based on behavior, time, location and components. This paper focuses on location-based faults which is categorized into data centric faults and system centric faults as these are the most frequently occurring faults in WSN. The presence of various faults has made fault detection, a prime and major area of research in the discipline of WSN technology. One of the most effective ways to detect these faults is the use of Machine Learning algorithm. Thus, in this paper, we have surveyed on the use of machine learning and deep learning algorithms for detecting anomalies and location-based faults in WSNs. Different ML algorithms have been used to detect and diagnose faults in WSN. We have summarized these ML-based algorithms for detecting data centric and system centric faults in WSN in tabular form. The study shows that ML algorithms are very effective for detecting system centric and data centric faults in WSN. In most of the cases, techniques like RF, SVM and NN has proved better than other algorithms. Many researchers have used modified and advanced versions of already existing algorithms for detecting faults in WSN. Still, the problems in WSN are not fully resolved and need attention from research community. Also, there is less research in models dealing with multivariate time series data. Thus, there is a scope of using some new, more advanced and modified ML and DL techniques to detect multiple faults in WSN such that sensor networks can be used more effectively in variety of applications.

References

1. Bangotra DK, Singh Y, Selwal A (2018) Machine learning in wireless sensor networks: challenges and opportunities. In: 5th IEEE international conference on parallel, distributed and grid computing (PDGC-2018), Solan, India, pp 534–539
2. Karimi H, Medhati O, Zabolzadeh H, Eftekhari A, Rezaei F, Dehno SB (2015) Implementing a reliable, fault tolerance and secure framework in the wireless sensor-actuator networks for event reporting. In: The international conference on advanced wireless, information and communication technologies (AWICT 2015), ELSEVEIR
3. Heimerdinger WL, Weinstock CB (1992) A conceptual framework for system fault tolerance. Software Engineering Institute, Carnegie Mellon University, Pittsburgh, Pennsylvania 15213, (October 1992)
4. Kumar DP, Amgoth T, Annavarapu CS (2018) Machine learning algorithms for wireless sensor networks: a survey. Int J Inf Fusion
5. Noshad Z, Javaid N, Saba T, Wadud Z, Saleem MQ, Alzahrani ME, Sheta OE (2019) Fault detection in wireless sensor networks through the random forest classifier. Sensors 2019:1568
6. Zidi S, Moulahi T, Alaya B (2018) Fault detection in wireless sensor networks through SVM classifier. IEEE Sens J 18(1):340–347
7. Paradis L, Han Q (2007) A survey of fault management in wireless sensor networks. J Netw Syst Manage
8. Moridi E, Haghparast M, Hosseinzadeh M, Jassbi SJ (2020) Fault management frameworks in wireless sensor networks: a survey. Comput Commun pp 205–226

9. Effah E (2018) Survey: faults, fault detection and fault tolerance techniques in wireless sensor networks. Int J Comput Sci Inf Secur
10. Manisha M, Nandal D (2015) Fault detection in wireless sensor networks. IPASJ Int J Comput Sci (IIJCS) 3(3)
11. Poornima G, Babu KS, Raja KB, Venugopal KR, Patnaik LM (2013) Fault diagnosis approach for WSN using normal bias technique. ACEEE Int J Commun 4(2):29–36
12. Saeed U, Lee YD, Jan SU, Koo I (2021) CAFD: context-aware fault diagnostic scheme towards sensor faults utilizing machine learning. Sensors
13. Muhammed T, Shaikh RA (2016) An analysis of fault detection strategies in wireless sensor networks. J Netw Comput Appl
14. Saeed U, Jan SU, Lee YD, Koo I (2021) Fault diagnosis based on extremely randomized trees in wireless sensor networks. Elsevier
15. Gupta D, Sundaram S, Rodrigues JJ, Khanna A (2019) An improved fault detection crow search algorithm for wireless sensor network. Int J Commun Syst
16. Malhotra P, Vig L, Shroff G, Agarwal P (2015) Long short term memory networks for anomaly detection in time series. In: ESANN 2015 proceedings, European symposium on artificial neural networks, computational intelligence and machine learning
17. Jeong S, Ferguson M, Law KH (2019) Sensor data reconstruction and anomaly detection using bidirectional recurrent neural network. In: Sensors and smart structures technologies for civil, mechanical, and aerospace systems
18. Vamsi PR, Chahuan A (2020) Machine learning based hybrid model for fault detection in wireless sensors data. EAI Endorsed Trans Scalable Inf Syst 7(24) (2019–2020)
19. Pachauri G, Sharma S (2015) Anomaly detection in medical wireless sensor networks using machine learning algorithms. In: 4th International conference on eco-friendly computing and communication systems. Elseveir
20. Jan SU, Lee YD, Shin J, Koo I (2017) Sensor fault classification based on support vector machine and statistical time-domain features. IEEE Access, vol 5
21. Li L, Liu G, Zhang L, Li Q (2018) Deep learning-based sensor fault detection using S-Long Short Term Memory Networks. Struct Monit Maintenance 5(1)
22. Gupta D, Sundaram S, Rodrigues JJ, Khanna A (2019) An improved fault detection crow search algorithm for wireless sensor network. Wiley
23. Javaid A, Javaid N, Wadud Z, Saba T, Sheta OE, Saleem MQ, Alzahrani ME (2019) Machine learning algorithms and fault detection for improved belief function based decision fusion in wireless sensor networks. MDPI—Sens
24. Swain RR, Khilar PM, Dash T (2019) Neural network based automated detection of link failures in wireless sensor networks and extension to a study on the detection of disjoint nodes. J Ambient Intell and Humanized Comput 10:593–610
25. Yuan Y, Li S, Zhang X, Sun J (2018) A comparative analysis of SVM, Naïve Bayes and GBDT for data faults detection in WSNs. In: IEEE international conference on software quality, reliability and security companion
26. Stojkoska BR, Mahoski K (2013) Comparison of different data prediction methods for wireless sensor networks. In: 10th Conference for informatics and information technology, (CIIT 2013)
27. Herbert D, Modelo-Howard G, Perez-Toro C, Bagchi S (2007) Fault tolerant ARIMA-based aggregation of data in sensor networks
28. Li G, Wang Y (2013) Automatic ARIMA modeling-based data aggregation scheme in wireless sensor networks. EURASIP J Wirel Commun Network. Springer

Gesture Recognition and Conversion to Speech for Specially Abled

Mukul Chaudhari, Chinmay Mukhedker, Jaspreet Singh Pannu, Varun Prasannan, and Sonali Patil

Abstract There are around 1 million people who suffer from hearing and speech impairment. They might find it difficult to express their thoughts because of their limited capabilities. However, using the latest technologies we can overcome this problem and help the specially abled with their inabilities. The latest trends in human-computer interaction, artificial intelligence and machine learning made it possible to build a system which can act as a mediator for specially abled people. In this paper, a comparative analysis is provided with respect to few existing gesture recognition methods. The intent of this paper is to compare and analyse the existing methods of gesture recognition and conversion of text to speech to help find out the most efficient algorithm among the algorithms compared to other fellow researchers.

Keywords Gesture recognition · Human–computer interaction · Machine learning · Deep learning · Artificial intelligence · Contour detection · Convolution neural network (CNN) · Double channel CNN

1 Introduction

The diversified technology and its continuous growth have made it possible to interact with the computer without any physical touch by using gestures, facial expressions and even intellectual thoughts. Hand gestures are primarily divided into two parts: static gestures and dynamic gestures. Static gestures are the constant and still positions of the hand to indicate some sign or some sort of activity. Dynamic hand gestures are the gestures with continuous motion of the hand in a particular time frame. The mentioned gestures can be recognized in multiple ways. The main two ways are by using software and by using hardware.

M. Chaudhari · C. Mukhedker · J. S. Pannu · V. Prasannan (✉) · S. Patil
Department of Information Technology, Pimpri Chinchwad College of Engineering, Pune, Maharashtra 411044, India
e-mail: vprasannan4@gmail.com

S. Patil
e-mail: sonali.patil@pccoepune.org

V. K. Gunjan and J. M. Zurada (eds.), *Proceedings of 3rd International Conference on Recent Trends in Machine Learning, IoT, Smart Cities and Applications*, Lecture Notes in Networks and Systems 540, https://doi.org/10.1007/978-981-19-6088-8_44

In the hardware, hand gloves wired with sensors are attached to the tip of the finger, and the motion is detected and parsed with the help of those sensors. In the software approach, there are multiple machine learning tools and algorithms that make it easy to detect the gestures through cameras.

2 Literature Survey

The survey provides an analysis of existing methods and technologies. For gesture detection, data gloves are the most used technology worldwide. But the latest trends in machine learning have made it possible to achieve better accuracy in gesture recognition using cameras. The approach by Aashni et al. [1] uses a camera to detect the gestures. The image frames are obtained in the form of video. In the later process, the frames are pre-processed to remove extra noises, useless background and to convert colours if required. Followed by this, contour detection is carried out to focus only on the required part of the image. Variable accuracy is seen in the approach with respect to the number of fingers used to show the gestures and type of background in the obtained frame.

In a convolutional neural network, the network is formed with multiple nodes, also called neurons. These nodes are connected to each other via links. Each link has its own weight, and as a whole, they produce the output. Generally for the visual classification of images, CNN is a better option than others. The pre-trained model MobileNet as proposed by Nishi et al. [2] states that the bottleneck (the second last layer, i.e. layer prior to last layer) does the main classification.

CNN is likewise a technique used for gesture reputation extra accurately. CNN usually acknowledges the entire neighbourhood functions after which it merges those neighbourhood functions at a better level. This technique is used to gather the all-inclusive traits of the photo and its structure, and then the properties of the pictures may be obtained. So in short, CNN has the upper hand in pixel price of the processing units.

In edge detection algorithm, by Wu et al. [3], an exceptionally speedy guided filter is followed to reconstruct the unique gesture image. Therefore, as a way to optimize the training method of gesture recognition, the double channel convolutional neural network is proposed. This shape is made from fairly impartial convolutional neural networks or CNNs. These two channels have their own two different inputs which usually have independent weights. These are linked to the relational layer, and a relational map is performed.

In orientation histogram approach by Deepali et al. [4], we find feature vector to classify the image. Firstly, the program reads the image database. Then, we resize all the images. The edges are being found by using two filters: one in x-direction and the other in y-direction. The divider method will give the gradient orientation. Then the image blocks are rearranged into columns using simulation software. The column values of the matrix are then converted to degree from radian.

Using optical character recognition and Festive Software, it consists of two models: image processing model and voice processing model. Image processing model will convert the image to text, and the voice processing model will convert the text to understandable speech.

The flow is as follows: Read the object as we take the picture with the camera and convert the image into grayscale. Then we determine the region of interest, OCR processing is initiated, and the text is achieved as output. The output of OCR is stored in .txt format, and Festive software is being used to convert text to speech in different languages [5].

The algorithm by Guo et al. [6] uses data gloves as its base. Initially, it uses the least square technique, and also the calculation of Hausdorff distance is done. The curve fitting method has finished before and so the shapes of the curves of the same gesture were similar, and the algorithm could transform a motion recognition method into a curve recognition method. The benefits of implementing this method are robustness, high accuracy, and reliability. The foremost downside of this method would be the speed and potency of the system. Once the recognition result is applied, the HCI is increased within the virtual scene.

In the study by Nascimento et al. [7], a method has been developed to control the playback function in Netflix using gesture recognition with the help of a smartwatch-like device. Here, the user performs an already defined gesture, and after it is processed by the smartwatch-like device, the command is sent to the communication platform and is executed on the Netflix application. A continuous gesture recognition algorithm is used and is instantaneously executed in Netflix.

The method proposed by Pinto et al. [8] detects the hand gestures using convolutional neural network (CNN). Later analysis of the results is done using cross-validation technique. In the proposed method, images are taken as input using a camera. The obtained images could have variable formats, scenarios, or backgrounds. So to bring the uniformity in the images, they are sent for pre-processing. During pre-processing, the images are passed through certain operations as follows:

- Colour segmentation: In this, the algorithm identifies uniformity in colours in the image and identifies the cluster of pixels in the image.
- Morphological operations: In this, digital images are processed on the basis of their shapes. The two specific operations in this case are erosion and closing. Erosion removes pixels on the object boundaries. Closing first adds the pixel on object boundaries and then removes it simultaneously.
- Contour generation: This is used to focus on the required area in the image.
- Polygon approximation: In this, the exact required area is detected and used. A logical AND operation is performed on the resultant pre-processed image so that it maintains the information accommodated on the fingers and surface of the hand. The images are then fed to CNN to train the model, and the results are stored. The stored results are then analysed using cross-validation techniques.

The approach by Bhagat et al. [9] has used image processing and deep learning to identify the gestures. The proposed method focuses mainly on Indian sign language translation. Different static images are trained using CNN. The area of focus is on

Indian sign language alphabets and numbers. A five-layer CNN model along with computer vision techniques is used in the method. The results showed the accuracy of 99.81%. With this, a method is proposed to identify the dynamic gestures too. For dynamic gestures, convolutional LSTM is used to train the video input of gestures. Accuracy in case of dynamic gestures is found to be 99.08%.

Varun et al. [10] The system proposed in this paper consists of acquiring the gesture specimen, processing the gesture specimen, gesture recognition at runtime and a control system. Then each and every image in the specimen data provided by the dataset will be worked on by the proposed system. The colour images are converted into black and white images which are called masks. Then input is converted into computer understandable images by our proposed system.

Shelke et al. [11] The points of defect are identified in gesture. With the help of these points, the number of fingers present in that particular gesture is evaluated. The obtained result of the gesture is supplied to a 3D CNN one by one to identify and recognize the current gesture. The system assists skeletal structure detection, skin colour detection, adjusting lighting and camera effects. Then we will implement hand localization using histogram clustering method. The resultant image is compared with the trained dataset using R CNN.

3 Comparative Analysis

Table 1 gives comparison of a few hand gesture recognition techniques and their conversion to text and speech based on parameters such as accuracy while testing accuracy in real time. With this, remarks have been added to each of the techniques.

Table 1 summarizes that the technique used in [9] is the most accurate one, but it lags behind in terms of speed. This CNN technique is only suitable and useful for static images. Technique used in [6] is the most accurate one for dynamic gestures. But data glove is not economical as compared to the other ideas.

4 Conclusion

In this paper, we have analysed some papers and the techniques used in them and shortlisted those in such a way that one would get a perfect idea of which algorithm will be most efficient for them.

Here we can conclude that a lot of advancements are needed in the field of hand gesture recognition systems, and more accurate and faster methods are needed to be developed for helping in communication using them. The recent updates in technologies have made it easier to achieve the same with just using softwares. The use of hardware devices like sensors and data gloves gives improvised accuracy, but it is costly to implement. The paper clearly identifies and analyses different existing approaches with their own pros and cons.

Table 1 Average accuracy of some hand gesture recognition techniques

Title	Average accuracy	Remark
1. Hand gesture recognition for human–computer interaction	92.28% with plain background and 64.85 with non-plain background	The analysis clearly shows that the accuracy is less in case of a non-plain background as compared to plain background
2. Indian sign language converter using convolutional neural networks	Accuracy of 96% for the testing phase with images and 87.69% for the images taken in real time	This analysis shows that for the images taken in real time, the accuracy might deteriorate as compared to just training image sets
3. Hand gesture recognition algorithm based on double channel CNN	98.02% with DC-CNN and 97.04% with SC-CNN	The analysis clearly shows that the accuracy is less in case of DC-CNN as compared to SC-CNN
4. Hand gesture recognition on Indian sign language using neural networks	Accuracy achieved is 93.32% with real-time images	The analysis shows that the accuracy of copy gestures is less than cut, open, close, and refresh gestures
5. A novel method for data glove-based dynamic gesture recognition	Rate of statistical recognition using several types of gesture has been achieved 98%	This study concludes that the accuracy of the system does not deviate with different shapes and sizes of hands
6. Static hand gesture recognition based on convolutional neural networks	For CNN with different layers, the average accuracy is found to be 96.21%	The method proves that performing segmentation and other techniques, CNN gives more accuracy with less computational cost
7. Indian sign language gesture recognition using image processing and deep learning	A five-layer CNN model used in this showed the accuracy of 99.81% for static images	The results show that using a multilayer CNN model has a positive impact on the accuracy

References

1. Haria A, Subramanian A, Asokkumar N, Poddar S, Nayak JS (2017) Hand gesture recognition for human computer interaction. Procedia Comput Sci 115(2):367–374. https://doi.org/10.1016/j.procs.2017.09.092
2. Intwala N, Banerjee A, Gala N (2019) Indian sign language converter using convolutional neural networks. In: IEEE 5th international conference for convergence in technology (I2CT), pp 1-5. 1109/I2CT45611.2019.9033667
3. Wu X (2020) A hand gesture recognition algorithm based on DC-CNN. Multimedia Tools Appl, vol 79. https://doi.org/10.1007/s11042-019-7193-4
4. Kaushik D, Bhardwaj A (2016) Hand gesture recognition on indian sign language using neural network
5. Venkateswarlu S, Duvvuri B, Kamesh K, Jammalamadaka, S, Rani R (2016) Text to speech conversion. Indian J Sci Technol 9. https://doi.org/10.17485/ijst/2016/v9i38/102967. PY—2016/10/01

6. Guo X et al. (2017) A novel method for data glove-based dynamic gesture recognition. In: 2017 International conference on virtual reality and visualization (ICVRV), pp 43–48. https://doi.org/10.1109/ICVRV.2017.00018
7. Nascimento TH, Soares FA, Nascimento HA, Vieira MA, Carvalho TP, de Miranda WF (2019) Netflix control method using smartwatches and continuous gesture recognition. In: 2019 IEEE Canadian conference of electrical and computer engineering (CCECE), pp 1–4. https://doi.org/10.1109/CCECE.2019.8861610
8. Pinto RF, Borges CD, Almeida A, Paula IC (2019) Static hand gesture recognition based on convolutional neural networks. J Electr Comput Eng. pp 1–12. https://doi.org/10.1155/2019/4167890
9. Bhagat NK, Vishnusai Y, Rathna GN (2019) Indian sign language gesture recognition using image processing and deep learning. In: 2019 Digital image computing: techniques and applications (DICTA). IEEE, pp 1–8
10. Varun KS, Puneeth I, Jacob TP (2019) Hand gesture recognition and implementation for disables using CNN'S. pp 0592–0595. https://doi.org/10.1109/ICCSP.2019.8697980
11. Shelke T, Nerkar V, Nandanwar T, Barapatre N, Umare D (2021) Development of real time hand gesture recognition for dumb and deaf using machine learning and deep learning. Int J Creative Res Thoughts (IJCRT) 9(3):2107–2114, ISSN: 2320-2882

Modified Mobility Aware MAC Algorithm to Improve the Lifetime of Dynamic Wireless Sensor Network

Rajiv Ramesh Bhandari, Khyati R. Nirmal, Mahesh R. Sanghavi, D. S. Rajnor, and Amit R. Bramhecha

Abstract Cluster head selection and data processing are two of the most significant approaches for extending the lifetime of a wireless network. The approach entails a clustering mechanism focused on the deployment of nodes and the sensor's coverage area. Cluster head gathers data from cluster nodes and compiles it before sending it to the base station. The election of cluster head and cluster members on a regular basis is a big challenge in a mobility conscious network. This paper represents new approach for appropriate cluster head selection and its member based on probability distribution function. The one-hop connection distance to reach all of the covered sensor nodes determines the starting probability of each node. The mobility aware clustering routing algorithm (MACRON) is used to elect cluster head based on LEACH algorithm, coverage and one-hop connectivity distance in terms of packet distribution, HOP distance, throughput, obtained signal power, and other factors. Furthermore, by evaluating the ratio of the current probability value to the cumulative probability value of all sensors, the MACRON algorithm dynamically calculates the shortest distance and normalizes the predicted probability. The simulation results show that the MACRON method is more successful than MEMAC and MMAC in extending the lifetime of mobility aware wireless sensor networks.

Keywords MAC protocols · MACRON · MEMAC · MMMAC and mobility aware wireless sensor network

---

R. R. Bhandari (✉)
Department of Computer Science and Engineering, K L Education Foundation, Vaddeswaram, Guntur Dist., AP 522502, India
e-mail: bhandari.rrcoe@snjb.org

K. R. Nirmal · M. R. Sanghavi · D. S. Rajnor · A. R. Bramhecha
Department of Computer Engineering, Late Sau. K. B. Jain College of Engineering, Chandwad, India

V. K. Gunjan and J. M. Zurada (eds.), *Proceedings of 3rd International Conference on Recent Trends in Machine Learning, IoT, Smart Cities and Applications*, Lecture Notes in Networks and Systems 540, https://doi.org/10.1007/978-981-19-6088-8_45

1 Introduction

The mobile wireless sensor network has to communicate various sensing parameters; therefore, network should have the transceivers device so that they can transfer the data to base station. Since communication through radio transceiver requires more power, an upgrade in the MAC protocol could dramatically prolong the existence of a mobile network. A network's lifetime, performance, node mobility, throughput, and other parameters are tightly regulated by a MAC protocol, which regulate and decide when packets are transferred and receiver is going to receive. Nowadays in most of the real-time applications, data transfer rate and mobility in WSN the total power usage are just as critical as network reliability and throughput. The MAC protocol [1] is built in such a way that in a dynamic environment, there must be a proper balance between packet transmission ratio, delay, received signal power, and hop count. The efficient handling of node mobility empowers the performance of wireless networks. Few strategies have been identified and implemented for mobility models. The early and simplest node mobility was proposed by Raja and Su with the concept of strong and weak mobility, in which the network reformation is required whenever the node changes the physical location or leaves the network due to power failure.

To regulate and minimize the energy consumptions for specific coverage area, the sink mobility model is designed. The different approaches which have adopted sink mobility to extend the lifetime of network are surveyed [2]. The sink mobility model demonstrated better results for on demand services and multi-hop environment.

In addition to this, some application-based schemes were also proposed like event, vehicular, dynamic medium mobility for wind and fluid medium, etc. The mentioned approaches are either static or designed for particular application. It is significant to prefer an algorithm that is independent of the application and also adopted for dynamic network. In light of the above scenarios, the MEMAC is the appropriate choice [3].

The organization of paper is done as follows: Sect. 2 discusses the modifications and variations added to the MEMAC algorithm. Section 3 describes the proposed modified MEMAC protocol. In Sect. 4, the system performance is measured through simulation experiments. Finally, the conclusion is derived and discussed in Sect. 5.

2 Related Work

The wireless sensor networks are significantly changing their state from stationary to mobile. Since mobile networks use the same architecture as stationary networks, they are limited by energy and processing constraints. Since one of the most critical aspects of a sensor network is energy usage, energy conservation has been a focus of current work on the medium access control protocol for sensor networks. The assumption that the majority of the nodes in the MAC protocol are stationary degrades efficiency when the nodes are mobile.

Zareei et al. have reviewed four main categories of MAC protocol like scheduling in MAC, TDMA-based MAC, hybrid MAC, and preamble sampling MAC. MEMAC and MHMAC are examples of hybrid MAC protocols that save more resources than the other strategies listed above [2, 4].

Bashir's MEMAC protocol [4] is built on a hybrid scheme of time division multiple access and code division multiple access that allows it to adapt to changing mobile and traffic conditions. To save energy, the MEMAC protocol informs sensor nodes when to wake up and when to sleep, but synchronization across all nodes necessitates additional overheads and computing effort.

The hybrid cluster-based MAC (CH-MAC) protocol is proposed by Cheng bin et al. in 2012 [5]. The cluster head selection is based on center of mass which is based on two parameters like mobility and priority of inactive nodes. In order to prevent detachment of mobile nodes from the network, the CH selection time interval reduces as mobility increases. Finally, contention is eliminated by assigning time slots to each node in a cluster, which prioritizes mobile node data transition demands and thereby decreases latency.

The architecture comprises two techniques like MAC, and routing was proposed by Gonga et al. [3]. The issue with Mobisense is that the cluster head starts synchronization with the rest of the participants, which uses more storage.

A few factors like quality of service model, cross-layer approach, and hybrid protocols have been identified for efficient management of mobility of wireless sensor networks and conserving the energy of nodes. These factors have a major effect on network performance while dealing with mobility. The MEMAC algorithm has been modified, and a novel algorithm has been proposed that focuses primarily on one-hop distance for routing, clustering of network for node mobility handling for the preparation of the volume.

3 Proposed Work

The clustering process is initiated by the proposed MACRON algorithm based on its geographical area and node deployment. Clustering is achieved by repeating the sequence in a systematic manner with probability identification. The one-hop distance to reach all covered sensor nodes is used to calculate all node starting probability. Calculate updated probability for each node by applying the LEACH algorithm with the energy parameter. Calculate the ratio of the current probability value to the cumulative probability value of all sensors to normalize the estimated probability. Of all the nodes in the list, choose the node with the highest probability as cluster head (CH). Members are nodes that are just a single hop away from CH. Delete CH and CM from the list, and then do the same with the rest of the network's nodes [2, 6] as shown in (Figs. 1 and 2).

The tentative cluster is selected by the cluster members by considering the ease of approachability. There may be possibility that the cluster member may associate with multiple clusters. In such cases, the respective cluster member associates with

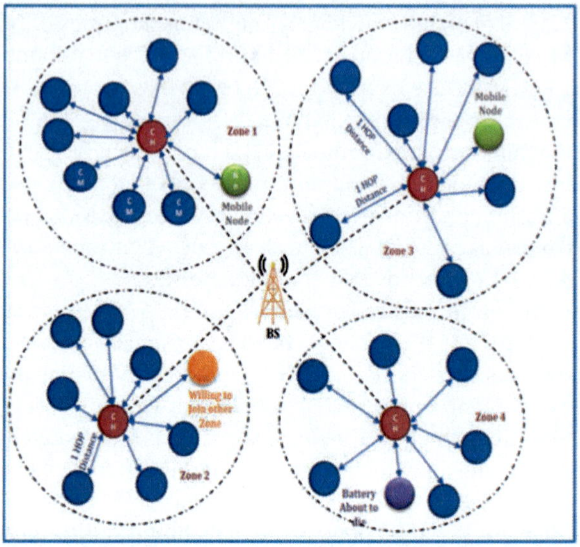

Fig. 1 Architecture of network creation using cluster approach

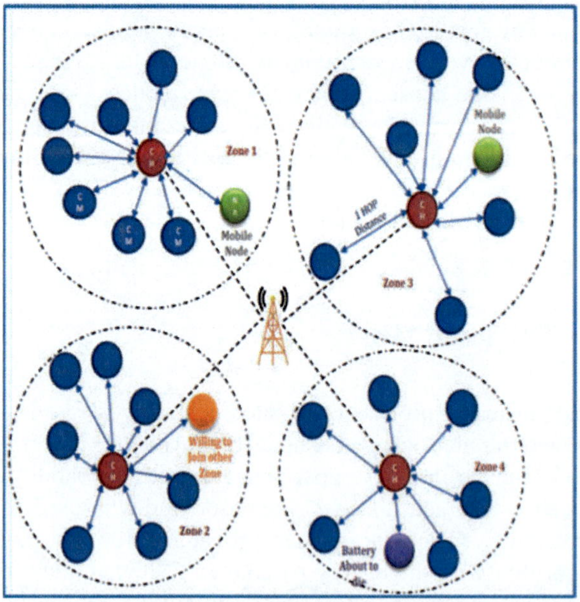

Fig. 2 Clustered network with leave and join operation

the new cluster head and sends the leave message to previously selected cluster. The newly selected cluster has minimum distance with cluster member with respect to previous one. Meanwhile if the cluster member receives the message from the cluster head, having minimum distance than the current one, the cluster allocation will be revised. In such overlapping cases, the proposed algorithm will introduce the probability criteria, to choose the appropriate cluster head.

The cluster head with uppermost probability will remain as cluster head, and the one with lowermost probability has to shift the cluster head responsibility to newly chosen cluster. This newly selected cluster head is having the minimum distance with cluster member. The cluster head sends the transfer message to all other cluster members and also to the old overlapping cluster. The scope of overlapping cluster is eaisly accessible; all the remaining members are joined with newly selected cluster head [7].

The proposed MACRON algorithm is divided into two sections, each of which is briefly explained below.

MACRON is a protocol that combines contention-based and scheduled-based protocols to save a lot of energy. MACRON modifies the frame size regularly based on sensor node mobility information and the number of nodes with data to broadcast; this avoids slots from being wasted by eliminating nodes that are intended to leave or join the cluster, as well as nodes that have no data to send, from the TDMA schedule and switching nodes to sleep mode, while they are not in the communication phase. We compared the performance of the MEMAC protocol to that of the MMAC protocol using simulations [8, 9].

Algorithm 1: Creation of Network with Clustering Approach

Step 1 The cluster creation is initiated by base node (BS) as per node's placement and its coverage.
Step 2 Clustering was executed by an iterative process with probability identification.
Step 3 Initially, the total number of nodes covered by each node was determined based on its location.
Step 4 The one-hop distance to reach all sensor nodes in specific area of network is determined by calculating initial probability of every node.
Step 5 Update probability for each node by applying the LEACH method with the energy parameter.
Step 6 Perform estimated probability normalization.
Step 7 Choose max probability node as cluster head among all other nodes in the list.
Step 8 Mark one-hop node as a member.
Step 9 Remove CH and CM from the list, repeat the same to all the nodes in the network.
Step 10 The cluster head is selected by cluster member based on the fastest reachability.

Algorithm 2: Network with Leave, Join, and Shift Algorithm

Step 1 If cluster member is willing to join to another shortest reachability cluster head, then it passes the leave message to existing cluster head and elects the shortest reachability cluster head.

Step 2 If cluster member is the part of other cluster area and got information from another cluster head which is in the nearby vicinity, then it shifts the cluster member to new cluster head.

Step 3 If the cluster head hears announcement from the other overlapping cluster head, then it measures the probability.

Step 4 The cluster member with highest probability keeps the cluster head position and cluster head with minimum probability to perform cluster shifting process.

Step 5 The transferred cluster members are joined to overlapped cluster head if cluster head is in reachable distance.

Step 6 Process continues until all cluster members are covered.

Step 7 A uniform cluster communication slot is created for all CM.

Step 8 CH sends slot message to CM to collect the sensed data

4 Performance Analysis

The NS2.34 framework is used to study and evaluate the performance of the MACRON protocol and compare it with MEMAC and MMAC protocol. The reason for using NS2.34 is that it allows the programmer to handle mobile networks and greatly aids the clustering operation. This section investigates the performance of MACRON in cluster mobile network topology and compares it against MEMAC and MMAC protocol. The hop count, throughput, packet delivery ratio, and average energy against interval, packet size, and simulation time are the efficiency metrics used to evaluate MACRON [11].

The practical setup includes 101 nodes in the area of approximately 500 m × 500 m. The traffic is generated in cluster network at variable rate and results are explained in figures ranging from Figs. 3, 4, 5, 6, 7, 8, 9, 10, 11, 12, 13 and 14.

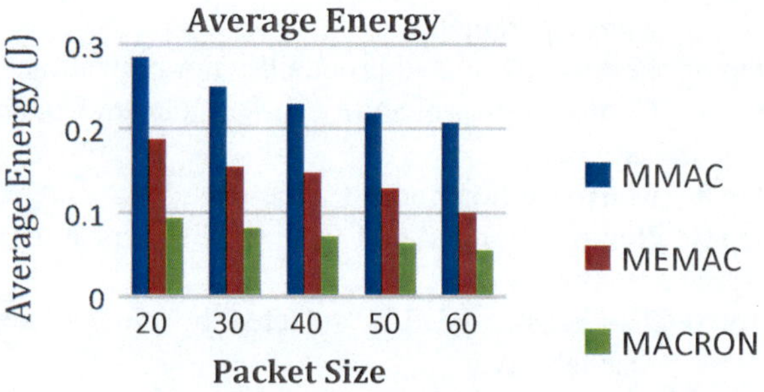

Fig. 3 Packet size versus average energy

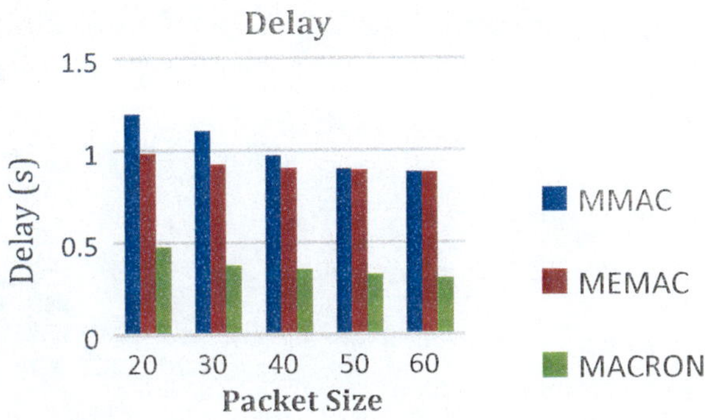

Fig. 4 Packet size versus delay

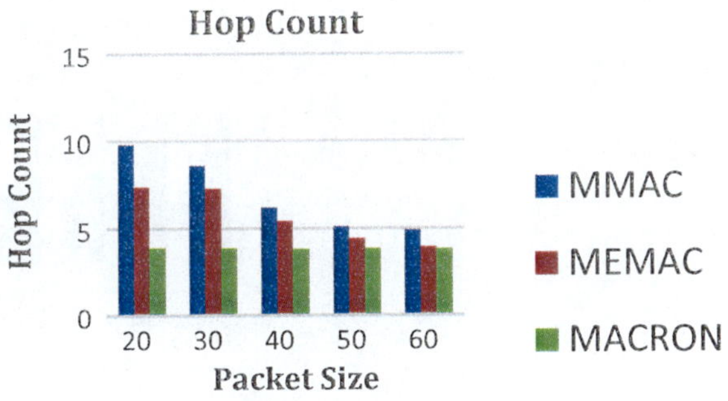

Fig. 5 Packet size versus hop count

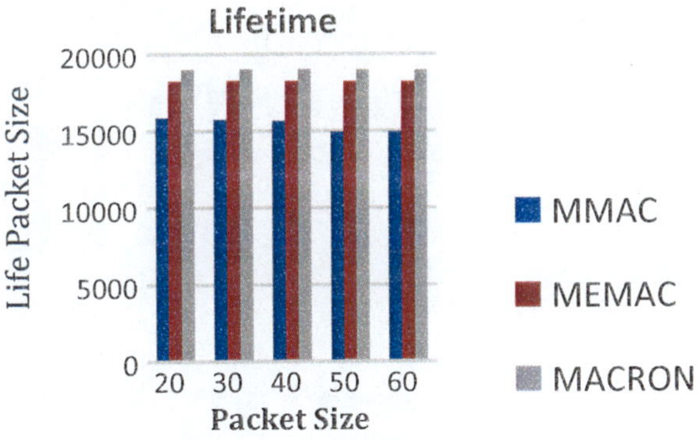

Fig. 6 Packet size versus lifetime

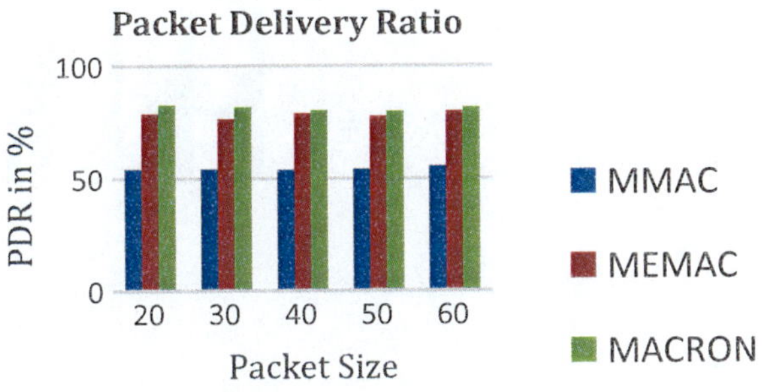

Fig. 7 Packet size versus packet delivery ratio

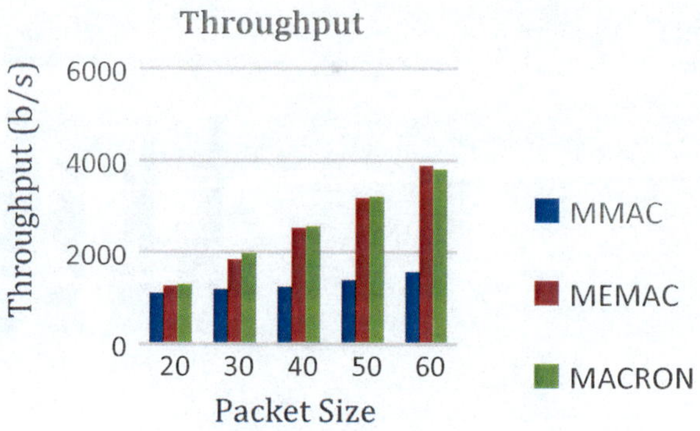

Fig. 8 Packet size versus throughput

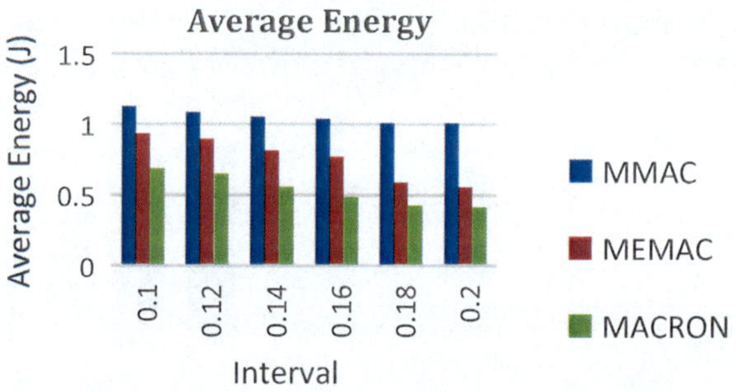

Fig. 9 Interval versus average energy

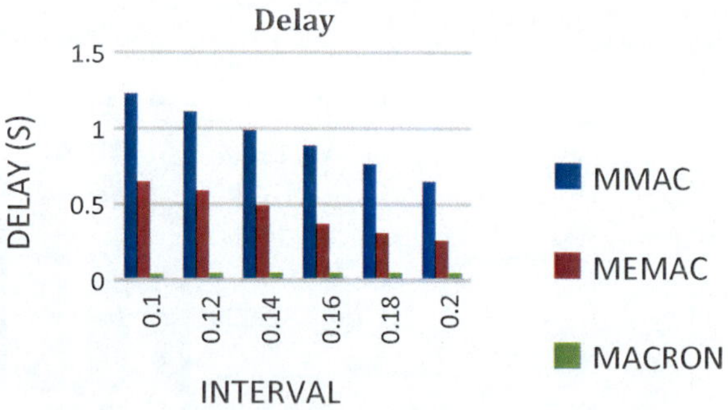

Fig. 10 Interval versus delay

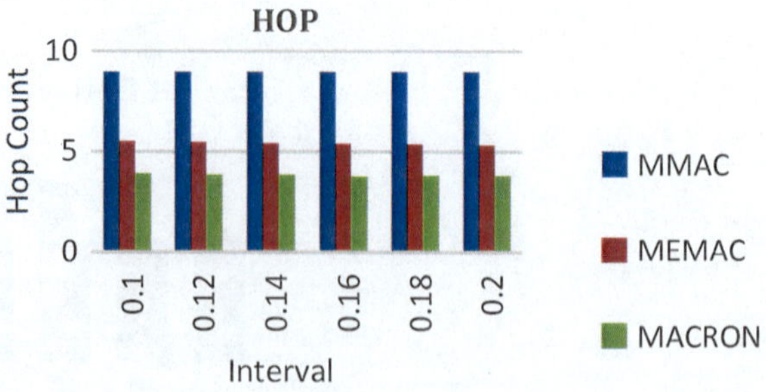

Fig. 11 Interval versus HOP

Fig. 12 Interval versus lifetime

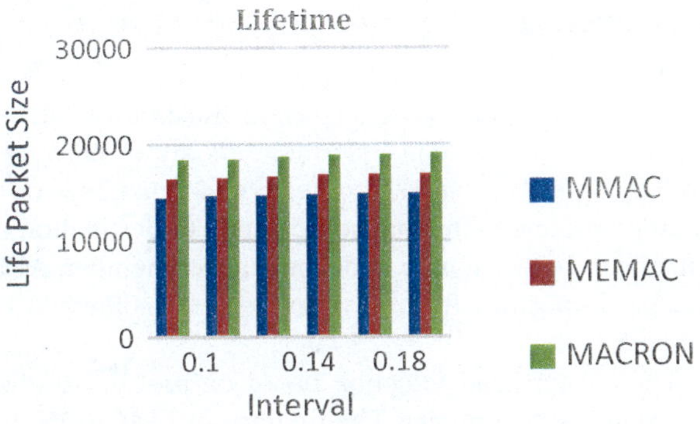

Fig. 13 Interval versus PDR

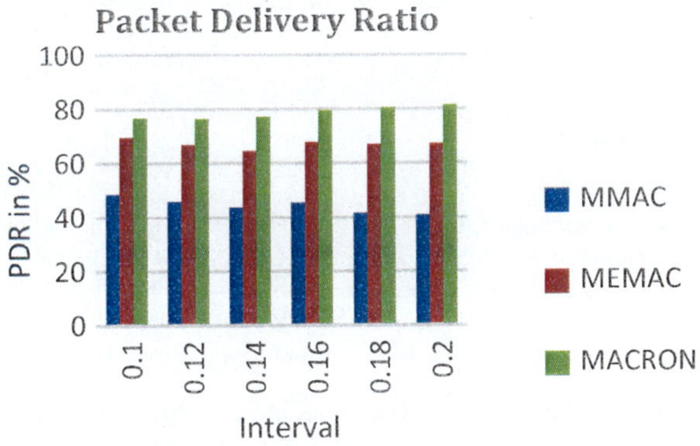

Fig. 14 Interval versus throughput

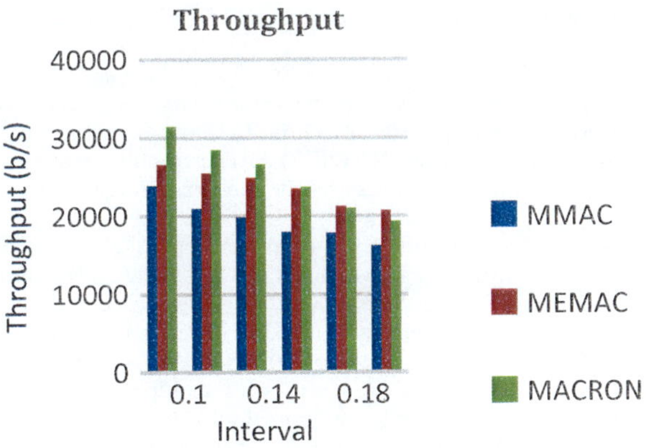

5 Conclusion

In wireless sensor network, all sensor nodes are awake whether it is in working state or idle state. It will consume more energy of network. Mobility aware clustering routing algorithm can easily create cluster and select cluster head which result in finding absolute path from source node to destination node. Cluster head works as a mediator between base node and cluster member. Also in Modified MAC, if node changes its position from current cluster to another cluster, it will not effect on whole network.

The cluster head selection based on probability distribution plays a great role in dynamic environment. The findings of MACRON over MEMAC and MMAC in terms of delay, hop count, lifetime, packet delivery ratio and throughput, and average energy are all much better in the performance analysis section.

References

1. Ali M, Suleman T, Uzmi ZA (2005) MMAC: a mobility-adaptive, collision-free MAC protocol for wireless sensor networks. In: PCCC 2005, 24th IEEE international performance, computing, and communications conference, pp 401–407. https://doi.org/10.1109/PCCC.2005.1460597
2. Gu Y, Ren F, Ji Y, Li J (2016) The evolution of sink mobility management in wireless sensor networks: a survey. IEEE Commun Surv Tutorials 18(1):507–524. https://doi.org/10.1109/COMST.2015.2388779
3. Srie Vidhya Janani E, Ganesh Kumar P (2015) Energy efficient cluster based scheduling scheme for wireless sensor networks. Sci World J 2015(185198):9. https://doi.org/10.1155/2015/185198
4. Ahmad A, Jabbar S, Paul A, Rho S (2014) Mobility aware energy efficient congestion control in mobile wireless sensor network. Int J Distrib Sens Netw 2014:1–13. https://doi.org/10.1155/2014/530416
5. Zareei M, Islam AM, Vargas-Rosales C, Mansoor N, Goudarzi S, Rehmani MH (2018) Mobility-aware medium access control protocols for wireless sensor networks: a survey. J Netw Comput Appl 104:21–37, ISSN 1084-8045. https://doi.org/10.1016/j.jnca.2017.12.009
6. Wan R, Xiong N, Loc NT (2018) An energy-efficient sleep scheduling mechanism with similarity measure for wireless sensor networks. Hum Cent Comput Inf Sci 8:18. https://doi.org/10.1186/s13673-018-0141-x
7. Ye D, Zhang M (2018) A self-adaptive sleep/wake-up scheduling approach for wireless sensor networks. IEEE Trans Cybern 48(3):979–992. https://doi.org/10.1109/TCYB.2017.2669996
8. Bhandari RR, Raja Sekhar K (2021) Priority-mobility aware clustering routing algorithm for lifetime improvement of dynamic wireless sensor network. Int J Adv Comput Sci Appl (IJACSA) 12(2). https://doi.org/10.14569/IJACSA.2021.01202100
9. Oh H, Ngo CT (2018) A slotted sense multiple access protocol for timely and reliable data transmission in dynamic wireless sensor networks. IEEE Sens J 18(5):2184–2194
10. Yahya B, Ben-Othman J (2009) An adaptive mobility aware and energy efficient MAC protocol for wireless sensor networks. IEEE Symp Comput Commun 2009:15–21. https://doi.org/10.1109/ISCC.2009.5202382
11. Ayushree SKA (2017) Comparative analysis of AODV and DSDV using machine learning approach in MANET. J Eng Sci Technol 12(12):3315–3328

An Analysis of Levenshtein Distance Using Dynamic Programming Method

J. Arockiya Jerson and N. Preethi

Abstract An edit distance (or Levenshtein distance) amongst dual verses refers to the slightest amount of replacements, additions and omissions of signs essential to turn one name addicted to the additional is referred to as the edit distance (or Levenshtein distance) amongst dual verses. The challenge of calculating the edit distance of a consistent verbal, that is the set of verses recognised by a fixed mechanism, is addressed in this research. The Levenshtein distance is a straightforward metric for calculating the distance amongst dual words using a string approximation. After witnessing its efficiency, this approach was refined by combining certain comparable letters and minimising the biased modification between associates of the similar set. The findings displayed a considerable enhancement over the old Levenshtein distance method.

Keywords Levenshtein distance · Similarity between words · String · Approximation · Similarity calculation · The longest common subsequence

1 Introduction

In the arena of computer science, figures repossession is a crucial problematic in the arena of computer science. String matching is very important in numerical information repossession. A series of letters or patterns is examined to check whether they match an input sequence. When a pattern and an input sequence are identical, inexact string matching shows that the pattern and the input sequence have been matched. On the other hand, approximate string matching uses a measure to determine how similar the input structure and pattern are. In the big-data age, approximation string matching may receive increasing attention due to advanced data analysis and a variety

J. Arockiya Jerson · N. Preethi (✉)
Christ University, Pune, Lavasa, India
e-mail: preethi.n@christuniversity.in

J. Arockiya Jerson
e-mail: arockiya.jerson@science.christuniversity.in

V. K. Gunjan and J. M. Zurada (eds.), *Proceedings of 3rd International Conference on Recent Trends in Machine Learning, IoT, Smart Cities and Applications*, Lecture Notes in Networks and Systems 540, https://doi.org/10.1007/978-981-19-6088-8_46

of applications [1–6]. The least amount of fundamental actions that make an input sequence equivalent to the goal pattern can be used to analyse the similarity between two strings. The addition, omission and swap of characters in a string are traditionally assumed to create the difference in approximation string matching [1, 7]. As fundamental operators, they are used to determine the distance amongst the input sequence and the goal pattern. String matching can be simplified in the Levenshtein distance computation since each fundamental operator has an uniform cost of one. After relating bitwise exclusive-OR, the Hamming distance [8] computation sums "1" bits to estimate the distance between two strings.

The semantics or connection amongst subsequences, on the other hand, makes approximation string matching complicated. Though the Levenshtein distance from the outline "catch" is one for both input arrangements, a humanoid can perceive that shape "catch" is additional related to input arrangement "cotch" than input arrangement "ctch." As a result, a more difficult edit distance metric, such as the generalised edit distance [9, 10] or the normalised edit distance [11, 12], can be used. When complicated roles generalise the edit distance, however, a lot of computing power is required. The edit distance amongst each input subsequence and sub-pattern, also called as step, is necessary to determine the edit distance between the input arrangement and shape. Furthermore, owing to the data reliance in manipulative steps, all steps in classic consecutive dynamic programming [13] must be computed in sequence, which is an extremely time-consuming task. Several mathematical techniques [14, 15] have been shown to improve computing complexity.

2 Related Works

In this study topic, a framework for checking and correcting misspelt words has been proposed. Many difficulties in NLP can be solved using the Levenshtein distance technique. String matching may be accomplished by calculating the edit distance amongst the sequences. The Levenshtein distance is the smallest quantity of modifications essential to transmute the sequence "a" to sequence "b" (by adding, removing, or replacing a character in string "a"). The closer the strings are, the smaller the Levenshtein distance. This is a relatively prevalent issue in dynamic programming applications.

3 Dynamic Programming

Use dynamic programming, an algorithmic approach, when you have issues that can be broken down into similar sub-problems and the solutions reused. These methods are mostly used for optimisation. The dynamic algorithm will try to evaluate the findings of previously solved sub-problems before tackling the in-hand sub-problems. Given two strings "CART" and "MARCH" find its edit distance. To change "CART"

Table 1 Edit distance of CART and MARCH

ED	ϕ	M	A	R	C	H
ϕ	0	1	2	3	4	5
C	1	1 D(n, m)	2	3	3	4
A	2	2	1	2	3	4
R	3	3	2	1	2	3
T	4	4	3	2	2	3

to "MARCH," conduct three operations on the string "CART" (1) Modify "*C*" to "*M*," (2) Modify "*T*" to "*C*," (3) Insert "*H*" (Table 1).

3.1 The Sub-problems

I. If the prior letterings of the two strings are identical, nothing has to be done. Calculate the number of strings left after ignoring the last characters. As a consequence, repeat for m-1 and n-1 distances. (For example, str1 and str2 may be "Sunday" and "Saturday," respectively.)
II. If the prior letterings vary, assess all actions on "str1," then all three actions on the last character of the original sequence, recursively, compute the cheapest price for all three actions and choose the lowest of the three results.

1. Addition: Reappear for m and n-1
2. Remove: Reappear for m-1 and n-1
3. Substitute: Reappear for m-1 and n (Fig. 1)

4 Methodology

The system was built using Levenshtein methods. The Levenshtein distance (LD) is a measure for computing the comparison of dual structures (the source string (s) and the destination string (s)) (t). The amount of omissions, additions, or replacements obligatory is the distance between s and t. As a sample,

- If s and t are both "test," LD(s, t) equals 0 since no alterations are required. Already, the sequences are similar.
- If s is "test" and t is "tent," then LD(s, t) = 1 since s can be transformed into t with just one substitution (changing "s" to "n"). Equation of Levenshtein distance

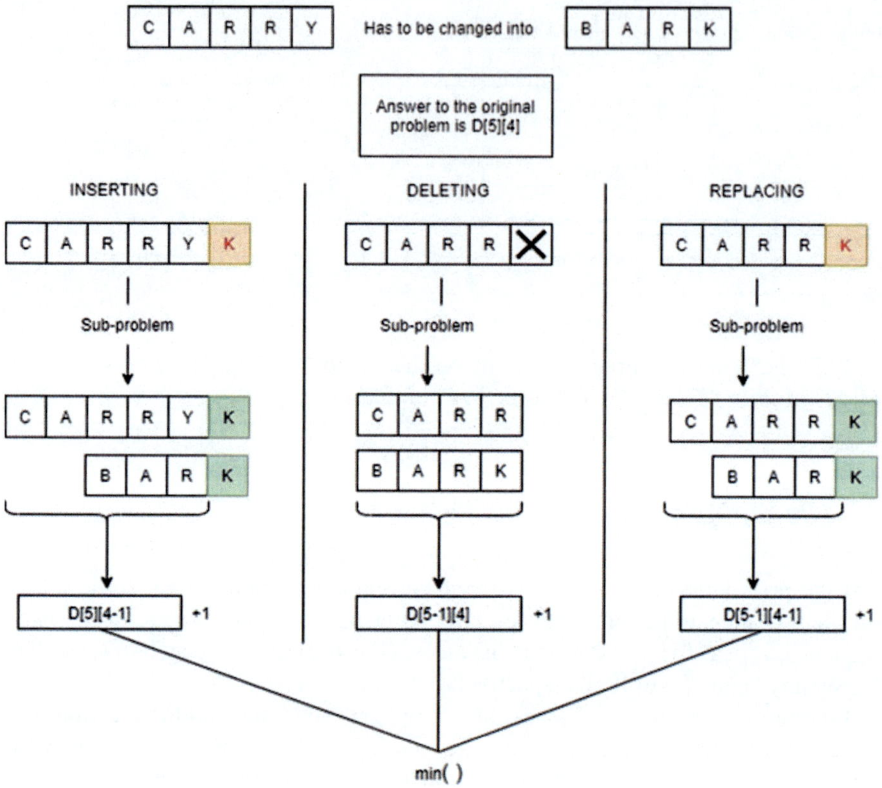

Fig. 1 Flowchart of sub-problems

$$lev_{a,b}(i,j) = \begin{cases} \max(i,j) & \text{if } \min(i,j) = 0, \\ \min \begin{cases} lev_{a,b}(i-1,j)+1 \\ lev_{a,b}(i,j-1)+1 \\ lev_{a,b}(i-1,j-1)+1_{a(i)\neq b(j)} \end{cases} & \text{otherwise.} \end{cases} \tag{1}$$

The similarity of the two strings may be represented by LD; obviously, the bigger the LD, the lower the similarity. If two strings S and T have lengths of m and n, respectively, and you use LD to represent their edit distance and Sim(S, T) to indicate their similarity, then there is:

$$\text{Sim}(S,T) = 1 - \frac{\text{LD}}{\max(m,n)} \tag{2}$$

The criterion edit distance is another name for Levenshtein distance. The Levenshtein distance approach has been used in spell testing, speech recognition, DNA analysis and plagiarism discovery.

4.1 The Algorithm

Stage 1: Initialisation

(a) Set the measurement of *s* to *n* and the measurement of *t* to *m*.
(b) Build a environment with 0...*m* columns and 0...*n* rows.
(c) Set the initial commotion's value to 0…*n*.
(d) Set the initial pillar's value to 0…*m*.

Stage 2: Handling

(a) Inspect, respectively, of *s*'s characters (*i* from 1 to *n*).
(b) Inspect, respectively, of *t*'s characters (*j* from 1 to *m*).
(c) The rate is 0 if *s*[*i*] matches *t*[*j*].
(d) The rate is 2 if *s*[*i*] does not match *t[j*].
(e) Set the minimum of: in cell *d*[*i*, *j*] of the environment

Stage 3: The distance is obtained in cell *d*[*n*, *m*] when iteration stage (2) is completed.

5 Experiment and Result

When the source string is "INTENTION" and the destination text is "EXECUTION," the Levenshtein distance is determined in this section (Table 2).

If there is a match, copy the diagonal value to LevDist[*i*][*j*]; if there is a mismatch, pick the lowest of the top, left, and diagonal values and add 2 to it.

The distance is obtained in cell *d*[*n*, *m*] when iteration stage (2) is completed.

Finally, the distance is in the matrix's upper right-hand corner, i.e. 8 (Table 3).

Table 2 Initialisation, iteration (i) = 1

N	9									
O	8									
I	7									
T	6									
N	5									
E	4									
T	3									
N	2									
I	1									
#	0	1	2	3	4	5	6	7	8	9
	#	E	X	E	C	U	T	I	O	N

Table 3 Result of LD example, iteration $(i) = n, j = m$

N	9	8	9	10	11	12	11	10	9	8
O	8	7	8	9	10	11	10	9	8	9
I	7	6	7	8	9	10	9	8	9	10
T	6	5	6	7	8	9	8	9	10	11
N	5	4	5	6	7	8	9	10	11	10
E	4	3	4	5	6	7	8	9	10	9
T	3	4	5	6	7	8	7	8	9	8
N	2	3	4	5	6	7	8	7	8	7
I	1	2	3	4	5	6	7	6	7	8
#	0	1	2	3	4	5	6	7	8	9
	#	E	X	E	C	U	T	I	O	N

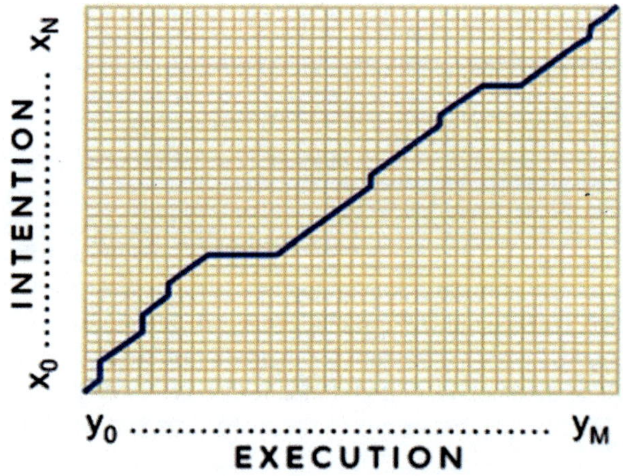

Fig. 2 Graph of distance matrix

5.1 *The Distance Matrix*

As a result, *D* is the distance between *X* and *Y* (*n*, *m*). Here, *X* identifies the source string, which is "INTENTION," and *Y* decides the destination string, which is "EXECUTION" (Fig. 2).

6 Conclusion

The fact that this modification reduces inaccuracy in a subtle way is most obvious from the outcome. Consider how this significant reduction might be effective for a larger number of test samples implementing this approach in a way that is relevant for all intents and purposes. It has also been demonstrated that, even though this technique does not minimise the fundamentally conceivable alternatives, the outcomes are frequently comparable to, if not better than, the totally normal LD method. Contrary to common assumption, this strategy has only been used on terms with lengths of three and five letters, and it has yet to be proved on arguments with fluctuating/extra dimensions. In order to make these Levenshtein procedures much more efficient as the amount of arguments in the wordlist rises, must preserve in mind, for the most part, the ideal balance between the wordlist's size and the relatively high computing cost, or so they believed. In terms of the semantic web, this feature may be utilised as a web service for uploading and downloading scanned manuscripts, which is significant for all intents and purposes. Furthermore, many websites currently create a random sequence of alphanumeric characters that the user must fundamentally enter as a particularly method of authentication in a huge way. Several periods, those letterings are indistinct, and in those cases, this method can create the procedure for all intents and purposes easier for the user, demonstrating how, in addition, several websites nowadays produce an accidental arrangement of alphanumeric letterings that, contrary to popular belief, must be arrived by the operator as a resources of confirmation.

References

1. Navarro G (2020) A guided tour to approximate string matching. ACM Comput Surv (CSUR) 33(1):31–88. https://doi.org/10.1145/375360.375365
2. Mateless R, Segal M (2019) Approximate string matching for DNS anomaly detection. In: International conference on security, privacy and anonymity in computation, communication and storage. Springer, pp 490–504
3. Hakak SI, Kamsin A, Shivakumara P, Gilkar GA, Khan WZ, Imran M (2019) Exact string matching algorithms: survey, issues, and future research directions. IEEE Access 7:69614–69637. https://doi.org/10.1109/ACCESS.2019.2914071
4. Ryu C, Lecoq T, Park K (2020) Fast string matching for DNA sequences. Theoret Comput Sci 812:137–148. https://doi.org/10.1016/j.tcs.2019.09.031
5. Al-Ssulami AM, Mathkour HI, Arafah MA (2020) Efficient string matching algorithm for searching large DNA and binary texts. In: Data analytics in medicine: concepts, methodologies, tools, and applications. IGI Global, pp 298–324
6. Kim T, Li W, Behm A, Cetindil I, Vernica R, Borkar V et al (2020) Similarity query support in big data management systems. Inf Syst 88:101455. https://doi.org/10.1016/j.is.2019.101455
7. Guo L, Du S, Ren M, Liu Y, Li J, He J et al (2013) Parallel algorithm for approximate string matching with K-differences. In: 2013 IEEE eighth international conference on Networking, architecture, and storage (NAS). IEEE, pp 257–261
8. Hamming RW (1950) Error detecting and error-correcting codes. Bell Syst Tech J 29(2):147–160. https://doi.org/10.1002/j.1538-7305.1950.tb00463.x
9. Kashyap RL, Oommen BJ (1981) An effective algorithm for string correction using generalized edit distances-I. Description of the algorithm and its optimality. Inf Sci 23(2):123–142. https://doi.org/10.1016/0020-0255(81)90052-9
10. Kashyap RL, Oommen BJ (1981) An effective algorithm for string correction using generalized edit distance? II. The computational complexity of the algorithm and some applications. Inf Sci 23(3):201–217. https://doi.org/10.1016/0020-0255(81)90056-6

11. Marzal A, Vidal E (1993) Computation of normalized edit distance and applications. IEEE Trans Pattern Anal Mach Intell 15(9):926–932. https://doi.org/10.1109/34.232078
12. Yujian L, Bo L (2007) A normalized Levenshtein distance metric. IEEE Trans Pattern Anal Mach Intell 29(6):1091–1095. https://doi.org/10.1109/TPAMI.2007.1078
13. Wagner RA, Fischer MJ (1974) The string-to-string correction problem. J ACM (JACM) 21(1):168–173. https://doi.org/10.1145/321796.321811
14. Allison L (1992) Lazy dynamic programming can be eager. Inf Process Lett 43(4):207–212. https://doi.org/10.1016/0020-0190(92)90202-7
15. Gusfield D (1997) Algorithms on strings, trees, and sequences: computer science and computational biology. ACM SIGACT News 28(4):41–60. https://doi.org/10.1145/270563.571472

Formula One Race Analysis Using Machine Learning

M. Keertish Kumar and N. Preethi

Abstract Formula One (also known as Formula 1 or F1) is the highest class of international auto-racing for single-seater formula racing cars sanctioned by the Fédération International de automobile (FIA). The World Drivers' Championship, which became the FIA Formula One World Championship in 1981, has been one of the premier forms of racing around the world since its inaugural season in 1950. This article looks at cost-effective alternatives for Formula 1 racing teams interested in data prediction software. In Formula 1 racing, research was undertaken on the current state of data gathering, data analysis or prediction, and data interpretation. It was discovered that a big portion of the league's racing firms require a cheap, effective, and automated data interpretation solution. As the need for faster and more powerful software grows in Formula 1, so does the need for faster and more powerful software. Racing teams benefit from brand exposure, and the more they win, the more publicity they get. The paper's purpose is to address the problem of data prediction. It starts with an overview of Formula 1's current situation and the billion-dollar industry's history. Racing organizations that want to save money might consider using Python into their data prediction to improve their chances of winning and climbing in the rankings.

Keywords Race · Simulation · Strategy · Motorsport · Machine learning · Decision-making · Predictions

M. Keertish Kumar · N. Preethi (✉)
Christ University, Pune, Lavasa, India
e-mail: preethi.n@christuniversity.in

M. Keertish Kumar
e-mail: m.kumar@science.christuniversity.in

V. K. Gunjan and J. M. Zurada (eds.), *Proceedings of 3rd International Conference on Recent Trends in Machine Learning, IoT, Smart Cities and Applications*, Lecture Notes in Networks and Systems 540, https://doi.org/10.1007/978-981-19-6088-8_47

1 Introduction

Formula 1 engineering necessitates a high level of precision since even little changes can have a significant impact on aerodynamics and, as a result, on performance. Teams of racers engineers utilize high-fidelity numerical simulations to evaluate the impact of design changes without having to build a physical prototype [1]. Many Formula One teams use statistical analysts to analyze race outcomes although these are often kept under wraps so that teams may maintain any tactical advantages that these studies provide [2]. This study uses cross-classified multilevel models to provide a more comprehensive picture of what drives F1 race performance. The models can partition variance to see the amount to which teams and drivers' matter [3], as well as produce rankings of F1 drivers that adjust for the influence of teams. One may argue that Formula One races are won at the factory rather than on the track. Formula One teams collect massive volumes of real-time data in order to forecast their race vehicles' performance [4]. An athletic championship is a collection of games or trials, the results of which are combined to determine the competition's final outcome [5]. Because minimal increases are made owing to engine and mechanical upgrades to the car, aerodynamic performance of an F1 car is currently one of the most important factors of performance gain. As a result, it has become the key to success in this sport, with clubs investing millions of dollars each year in research and development in this area [6]. A driver who wins a race receives a total of 25 points [7]. It is the ninth-largest sports industry in terms of revenue [8]. F1's international popularity and enormous success present a significant challenge for chassis and engine designers. The upshot of the engineering work is a steady rise in engine power, faster cars, and, as a result, shorter lap times. Such high-level competitions necessitate the utilization of the best and most up-to-date technology resources for the development of current and future engines, which can only be afforded by larger automobile companies [9].

2 Related Works

A system for automating race strategy decisions in circuit motorsport is presented in this study. The focus is on the Formula One racing series and, as a result, on the best pit stop timing to replace worn-out tires. A quadratic optimization problem was set up to obtain an initial estimate of the ideal race strategy, which reduces a driver's race time in the absence of opponents [10]. Aerodynamics is a crucial part of any vehicle's design. An attempt was made in this study to develop an F1 racing car model and assess the car's stability, including the drag created during high performance [11]. Information visualization, visual analytics, and data science are all becoming increasingly interested in sports data [12]. To anticipate the performance of F1 cars, artificial neural network prediction models were developed [13]. We believe that this lap segmentation is the correct method because the first architecture was miles ahead of the second [14]. The purpose of this study is to develop a method that employs machine learning techniques to forecast driver success in the FIA Formula One Championship based on a driver's career path, that is, his or her performance in other single-seater auto-racing championships [15]. We make three important points by tying the threads of this study together. First and foremost, we have broadened the sociological lens to include a technical elite at the forefront of a successful and vibrant business, but one that we contend requires historical understanding and is not easily represented by existing sociological themes. Can discern a marked crystallization of a technical habitus at the pinnacle of all British teams. Along with demonstrating strong passions and dedication in the field [16]. Throughout this article, it has been demonstrated that no rule can be regarded totally just [17] using a formal parallel between the Formula 1 World Championship and a multi-decision maker process. The purpose of the study was to create a simulation model that Formula One teams could use to identify the best racing strategy from a set of options [18]. A multilevel approach that allowed us to achieve two related goals: finding a rating of Formula One drivers while controlling for team impacts [19]. The fundamental goals of this paper are twofold. First, different statistical analysis measures in FIA Formula One racing are discussed [20]. Part is devoted to the topic's historical background. Because they are now considered tools, machine learning algorithms are no longer taken into consideration. According to the literature, decision-making in sports and sports analytics has gained a lot of attention during the last two decades. This is due to the increasing availability of data and the efficacy of machine learning algorithms that enable meaningful data analysis. Sports analytics may help with outcome prediction, performance evaluation, talent finding, and strategy evaluation. As result, a wide range of stakeholders are interested in such judgments, including bookmarks and betters, fans, commentators, and others.

3 Methodology

3.1 Dataset Description

Races, results, drivers, constructors, qualification, and circuits are all part of a large dataset. Dataset link—https://github.com/SampathHN/Formula-1-Prediction. Races—Race-id, name, circuit-id, round, Year Results—results-id, race-id, grid, total laps, fastest lap Driver—driver-id, number, driver ref, nationality constructors—Constructor-id, constructor ref, name, nationality Qualifying—qualifying-id, driver-id, constructor, q1, q2, q3 Circuits—circuit-id, circuits ref, name, location, country.

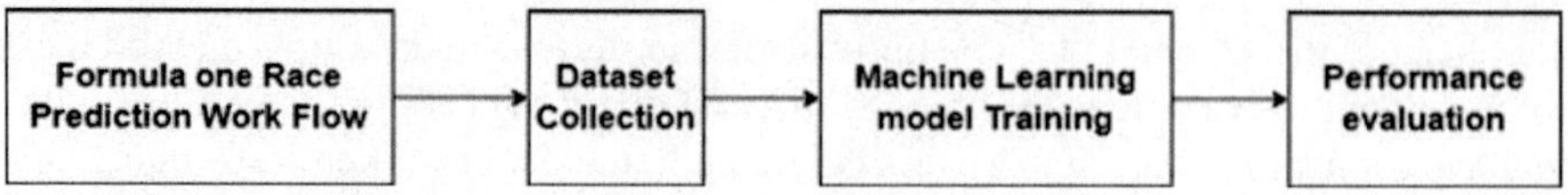

Fig. 1 Overview of the Formula One model progress

3.2 *Workflow of the Project*

See Fig. 1.

3.3 *Logistic Regression Algorithm*

Logistic regression is a statistical analysis method used to predict a data value based on prior observations of a dataset. A logistic regression model envisages a dependent data variable by analyzing the relationship between one or more existing independent variables. Consider the situation where we need to determine whether or not an email is spam. If we exploit linear regression to solve this problem, we will need to select a threshold by which we may classify the data. If the actual class is malignant, the projected continuous value is 0.4, and the threshold value is 0.5; the data point will be classed as non-malignant, potentially resulting in significant consequences in real time. Mathematically, (1) represents the logistic regression hypothesis function:

$$h_\theta(X) = \frac{1}{1 + \mathrm{e}^{-(\beta_0 + \beta_1 X)}} \tag{1}$$

The output is converted into a probability score using a sigmoid function to reduce the cost function represented by (2):

$$\mathrm{Cost}(h_\theta(x), y) = \begin{cases} \log(h_\theta(x)), & y = 1 \\ -\log(1 - h_\theta(x)), & y = 0 \end{cases} \tag{2}$$

3.4 *Decision Tree Algorithm*

A decision tree is a graphical representation of all possible solutions to a problem based on a set of criteria. Strive to construct a condition on the features at each step or node of a decision tree used for classification to segregate all the labels or classes included in the dataset to the fullest purity.

3.5 Random Forest Algorithm

Random forest is a supervised machine learning algorithm that is commonly used to solve classification and regression problems. It constructs decision trees from several samples and uses the majority vote for classification and the average for regression. It ranks augmenting for the current node split over considering how that split affects the whole tree. A greedy technique speeds up decision trees, but it also makes them more susceptible to overfitting. An overfit tree is a learning model with a large variance that is highly augmented at predicting values in the training dataset. The Gini index is applied as a cost function to calculate the split in the feature set for the classification task. Mathematically, if P_i is the probability of each class, (9) represents the Gini index as:

$$\text{Gini index} = 1 - \sum_{i=1}^{k} P_i^2 \tag{3}$$

3.6 Support Vector Machine Algorithm

SVC is a nonparametric clustering technique that makes no assumptions about the size or structure of the data clusters. Hyperplanes are decision boundaries that aid in data classification. Different classifications can be assigned to data points on each side of the hyperplane. The hyperplane's dimension is also determined by the number of features. If there are just two input appearances, the hyperplane is merely a line. The hyperplane becomes a two-dimensional plane when the number of input appearances reaches three. When the number of appearances exceeds three, it becomes intolerable to envision. Cost function for the SVM model is represented as:

$$J(\theta) = \frac{1}{2} \sum_{j=1}^{n} \theta_j^2 \tag{4}$$

such that

$$\theta^T x^{(i)} \geq 1, \quad y^{(i)} = 1, \tag{5}$$

$$\theta^T x^{(i)} \leq -1, \quad y^{(i)} = 0 \tag{6}$$

4 Experiment and Results

Table 1 shows the six models that were used to govern the accuracy of both drivers and constructors or some commonly thought.

Table 2 depicts the six models that for all intents and purposes were used to generally regulate the correctness of solely driver data, or so, it particularly is popularly believed, or so, they specifically thought.

Table 3 basically shows the six models that, for all intents and purposes, really were used to for all intents and purposes supervise the validity of only constructor's data, or so, it for all intents and purposes was widely accepted, or so, they thought, which definitely is quite significant.

The CV score of 6 different models of consequences for the mean results of the combined data, basically mean results of the drivers, and definitely mean results

Table 1 **Model** of driver's and constructor's performance

S. No.	Models used	Accuracy (%)
1	Logistic regression	90.7
2	Decision tree classifier	90.6
3	Random forest classifier	92.5
4	Support vector machine	92.1
5	Gaussian NB	83.7
6	K neighbor classifier	88.1

Table 2 Model of driver's performance

S. No.	Models used	Accuracy (%)
1	Logistic regression	92.8
2	Decision tree classifier	91.1
3	Random forest classifier	93.6
4	Support vector machine	93.09
5	Gaussian NB	86.5
6	K neighbors classifier	90.8

Table 3 Model of constructor's performance

S. No.	Models used	Accuracy (%)
1	Logistic regression	93.4
2	Decision tree classifier	91.9
3	Random forest classifier	93.5
4	Support vector machine	95.2
5	Gaussian NB	86.6
6	K neighbors classifier	93.02

of the constructors generally are illustrated in the graph above in Fig. 2, is fairly significant (Fig. 3 and Table 4).

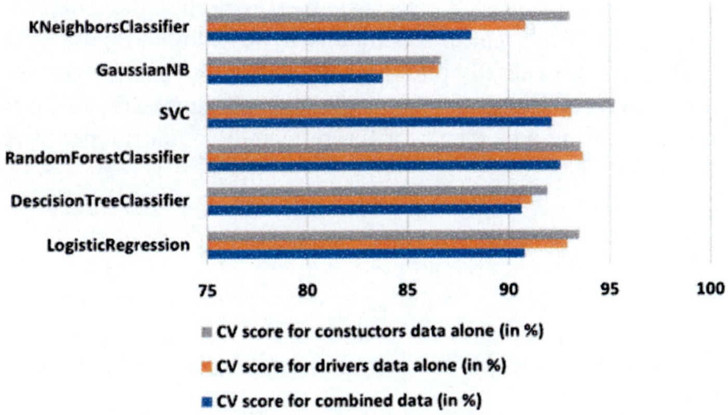

Fig. 2 Mean score of three assumptions

Fig. 3 Tuning model of random forest classifier

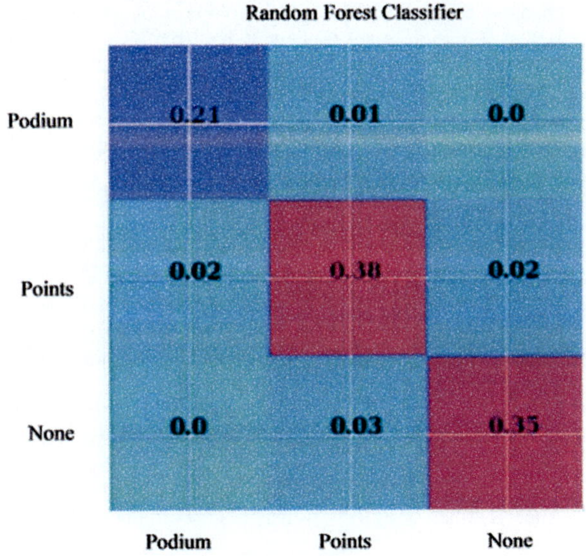

Table 4 Model performance score

S. No.	Score	Random forest classifier	SVC
1	Precision score	0.935338	0.901042
2	F1_score	0.934012	0.896870
3	Recall score	0.932923	0.893476

5 Conclusion

From an engineering standpoint, may deduce that the majority of classification issues in the literature definitely mostly are two-class (win, loss) and three-class (win, lose, draw), or so they kind of thought in a definitely major way. In this way, may also for all ideas and purposes generally refer to the Formula One race prediction, which streamlines the analysis of massive amounts of data regarding a driver's and a racing team's performance during a race and for future race in an understated way, which really is fairly significant.

References

1. Choo CLW (2015) Real-time decision making in motorsports: analytics for improving professional car race strategy
2. Heilmeier A, Thomaser A, Graf M, Betz J (2020) Virtual strategy engineer: using artificial neural networks for making race strategy decisions in circuit motorsport
3. Mourad A, Delzell P, McCabe P (2019) Automation of data analysis in Formula 1. Dec 2019
4. Marino A, Aversa P, Mesquita L, Anand J (2015) Driving performance via exploration in changing environments: evidence from Formula One racing. May 25
5. Bell A, Smith J, Sabel CE, Jones K (2016) Formula for success: multilevel modelling of Formula One driver and constructor performance, 1950–2014
6. Stoppels E (2017) Predicting race results using artificial neural networks
7. Abhisht NS (2021) Performance analysis of a Formula 1 car using CFD. Issue 03 | Mar 2021
8. Lampprecht T, Salb D, Mauser M, Van De Wetering H, Burch M, Kloos U (2019) Visual analysis of Formula One races. May 2019
9. Liu X, Fotouhi A (2020) Formula-E race strategy development using artificial neural networks and Monte Carlo tree search. 30 Mar 2020
10. Edinburg SJV, Luchsinger A (2019) Towards Formula One driver/vehicle optimization
11. Serapiglia A (2018) Formula One—a database project from start to finish. April 2018
12. Rosso G, Rosso AF (2016) Statistical analysis of F1 Monaco Grand Prix 2016. Relations between weather, tyre type and race stints. July 2016
13. Bhulai S, Hoogendoorn M (2018) A machine learning approach to predict the performance of a potential Formula 3 driver. Aug 30
14. Nichols G, Savage M (2017) A social analysis of an elite constellation: the case of Formula 1
15. de Mello JCCBS, Gomes LFAM, Gomes EG, de Mello MHCS (2019) Use of ordinal multi-criteria methods in the analysis of the Formula 1 World Championship
16. Budzinski O, Feddersen A (2016) Measuring competitive balance in Formula One racing
17. Alten H, Illien M (2017) Demands on Formula One engines and subsequent development strategies
18. Angulo Meza L (2015) Analysis of the 2014 Formula 1 constructors championship using a modified Condorcet method
19. Bekker R (2018) Simulating Formula One race strategies. 26 Feb 2018
20. Gutiérrez E, Lozano S (2018) Benchmarking Formula One auto racing circuits: a two stage DEA approach

IoT Enabled Water Conservation System Using LoRa LPWAN Networks

Arjuna Muduli, P. Kanakaraja, K. Raghavendra Reddy, M. Sai Santosh Anirudh, and B. Prasad

Abstract We currently have a large amount of water wastage in some major urban areas such as Chennai, Hyderabad, and numerous others. During the stormy season, people are also looking for water for their families. The groundwater ranges are rapidly dwindling. Water is becoming increasingly contaminated as a result of industrialisation, the use of composts and insecticides, city development, and other factors. As a result, we should have a dependable water top-notch checking device. We recommended this mission to address the problem in large urban regions. This duty is fully utilised in order to reduce water waste in cities. This project is being used to reduce water waste. The expense of a bound measure of water to a certain family is managed by this tool. We limit the amount of water phase in our own family of four people, just as we do in our own family of four people. It aims to make city administrations more efficient and to keep inhabitants informed. The present frameworks for water charming observation rely on remote innovation, which can be quick to reach and consume a lot of energy, and are not suitable for sensor hubs in WSN. The proposed framework uses a LoRa module based on the LoRaWAN convention, which is low-strength gigantic area organisation (LPWAN) innovation, to provide a low-cost, low-strength, large area, and adaptable procedure for water fantastic observing, which is a brilliant arrangement that is more noteworthy green than many existing frameworks. The proposed framework has a format that involves sensors being integrated into the microcontroller, a distant LoRa module for communicating and receiving sensor values, and a ThingSpeak IoT stage for assessing and visualising water noteworthy sensor values that have been transferred.

Keywords Internet of Things (IoT) · LoRa · LoRaWAN · The Things Network · Ubidots IoT platform

---

A. Muduli (✉) · P. Kanakaraja · K. Raghavendra Reddy · M. Sai Santosh Anirudh · B. Prasad
Department of Electronics and Communication Engineering, Koneru Lakshmaiah Education Foundation, Greenfields, Vaddeswaram, A.P. 522502, India
e-mail: arjunamuduli@gmail.com

V. K. Gunjan and J. M. Zurada (eds.), *Proceedings of 3rd International Conference on Recent Trends in Machine Learning, IoT, Smart Cities and Applications*, Lecture Notes in Networks and Systems 540, https://doi.org/10.1007/978-981-19-6088-8_48

1 Introduction

LoRa is a relatively new wireless technology. It denotes a long distance. LoRaWAN is a low-power wireless technology. It has the ability to handle a high number of hubs. It has a battery life capacity of more than ten years. It can collect data and focus on machine-to-machine and web-of-things applications. We use ESP32, Arduino UNO, and LoRa RX1276 in this project. Float sensor, and buzzer are the sensors. There is a tremendous quantity of water waste in some large urban areas. However, we use this effort to avoid wasting water. Prior to all of this, we attempted to rent condos in cities. Inside the loft, we provide a small amount of water to our own family. After the family has used all of the water allocated to them, they must purchase water. Following that, we give each household a bottle of water. This business could be quite beneficial in reducing water waste. The planned Smart City task is concerned with the design of water's nature. It enables more noteworthy unpractised and costly civil administrations to be made and keeps citizens informed. Unpolluted water in our bodies is a valuable resource in metropolitan areas. With the growing population of metropolitan areas, it is unavoidable that wastewater will expand and water resources will become contaminated. The Smart City project is engaged in continual water marvel observation. With the guidance of essential advancements, the contamination's unpleasant after-effects may be reduced. The Smart City idea wishes the incorporation of records technologies to render the usage of electricity and other assets extra effectively.

There are two hubs and data sources in Fig. 1 that are presented to Arduino. The result might be displayed in a flowing gem show. The acquired data is sent to the LoRaWAN door. It is possible that there is a web in the LoRa door. This information may be sent from the cloud, which is Ubidots. Low-power admittance and remote broad-spot local area is referred to as LPWAN. The majority of LPWAN advances make use of a technique known as cryptography. We can monitor air toxins, water level checking, astute stopping contraption, hearth recognising, and creature observation by utilising the LoRa technology. The freshest air is found indoors. We will contribute to the LoRaWAN entrance through an infinite number of hubs.

Fig. 1 Proposed conceptual layout diagram

2 Empirical Survey

By using infinite wireless technologies, GSM, GPS, and Mysignals, we monitored the water supply system. But we introduced the new technology called LoRa. It is used to monitor the water supply system from long range [1]. (Elakiya Bharathi Jaishnu; Padhmavathi; Premkumar & Sasikumar) The usage of LoRa in a water conservation framework helps to reduce water waste. These days, city private spots are filling up quickly around the world to avoid water shortages and meet the needs of buyers. Its goal is to provide enough water circulation networks that are naturally managed. Another issue within the water conveyance system is that the general public uses pull syphons to take water directly from the home street pipes. It is recommended that for an IoT-based completely remote water monitoring and burglary avoidance device to be built with the aid of recording, more substance is required [2]. It is extremely likely that sensors like as float sensors and pH sensors will be used.

3 System Design

Using this LoRa technology, we can check the water supply. At the same time, we can monitor an infinite number of different water supplies in a massive home.

The stream rate sensor can be seen in this image monitoring the water supply to the condos. Stream sensor is used to monitor the flow of water over a lengthy period of time. Figures 2 and 3 illustrate the system designs of the proposed transmitter and receiver sections, respectively. A solidnoid valve is a valve that outperforms the LoRa's ability and sends it to the LoRa's door in every slot. The buzzer and LCD are used to ring and display data on the screen [3]. These sensors are used to collect data from residences and transmit it via LoRa. The data is sent to the LoRaW A Gateway, while the network is being added to LoRa. Finally, the records have been sent to The Things Organisation's server. So, with the assistance of this LoRaW A Gateway and the issues network, all of the collected data could be updated in the "Ubidots Cloud". LoRaWAN Gateway is responsible for this because it has a web [4].

4 Methodology

In the proposed project, all the environmental sensors properties and specifications are explained in this section as follows.

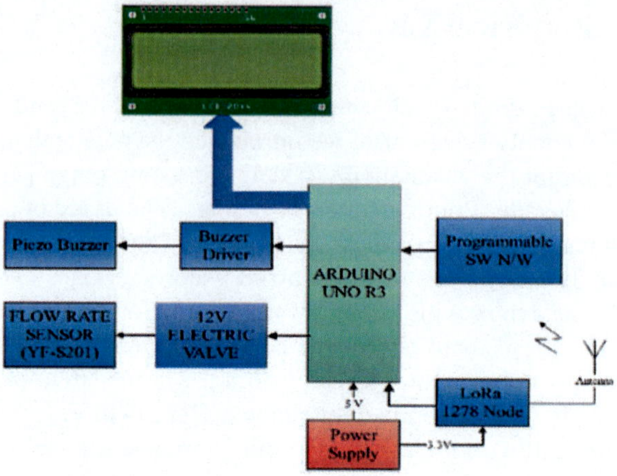

Fig. 2 Proposed transmitter section

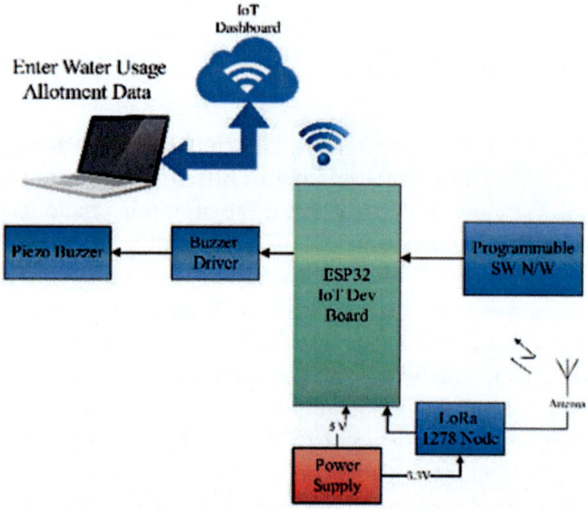

Fig. 3 Proposed receiver section

4.1 LoRa SX1276 Transceiver

A radio recurrence verbal trade convention is known as LoRa. It has layers of recurring bands. In all, an infinite number of LoRa modules could be linked. This could be a result of a long reach. There is no mandatory get section for connecting from one hub to another in LoRa [5]. As expected, LoRa has broadcasting hubs. When the LoRa module is connected to the Internet, the miles are expressed as LoRaWAN, and the measurements are sent to the cloud over Wi-Fi or Ethernet. The LoRa stop hubs come in a variety of shapes and sizes. They might be a radio module with an antenna, with the option of a microprocessor to organise the information. If the LoRa device contains a sensor, it can be used as a remote sensor [6]. It is a low-cost deviation

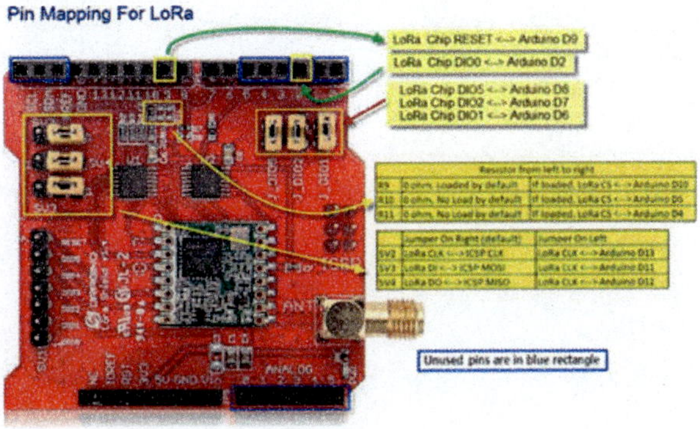

Fig. 4 Proposed LoRa transceiver module

with a low-strength use. It transmits data at a slower rate as shown in Fig. 4, the transmitter section of the proposed implementation.

4.2 *LPS8 LoRaWAN Gateway*

A LoRaWAN door is the LPS8. It is a gadget with an open stockpile. It creates a Wi-Fi scaffold between a LoRa remote organisation and a web convention local region. LoRa Wi-Fi will deliver data by letting the shopper to access it, and then the records will be accessed to lengthy levels with minimal information costs. They can be managed by the web GUI and SSH over Wi-Fi. It has ten equal demodulation techniques that can be programmed. LoRaWAN repetition groups have been favoured. The LPS8 is a demodulator that emulates 49 LoRa demodulators. LPS8 is capable of managing 1000 LoRaWAN stop hubs from multiple suppliers. The entryway is not accessible after the enactment. Customise the LoRaWAN neighbourhood borders with lts licences [7]. It is pre-configured to support the unique LoRaW settings in a rural environment.

There are two types of LoRaWAN passages. The first is the job passage, and the second is the channel gateway. There are two types of job gateway: indoor and outdoor position passages. The channel door is a type of door. There are two types of doors: single channel and multichannel [8] (Fig. 5). Within urban areas, LoRaWAN has a coverage area of 5 km, while in sub-city areas, it is 15 km. The less the power, the faster it burns. The low realities expenses are roughly the same as 27 Kbps.

4.3 *Flow Rate Sensor*

The sensor operates on the basis of evaluating the cooling impact of the fuel while ignoring the askinny, heated thread. The higher the float of gas, the better the cooling

Fig. 5 Multichannel LoRa WAN receiver

Fig. 6 Structure of water flow rate sensor

sway [9]. To be precise, the sensor needs to know the thickness of the fuel, which is sent to the estimation device through the gas assessment insights as shown in Fig. 6.

Additionally, the water stream sensor can be used with warm waters, bloodless waters, warm waters, clean water, and dirty water. Those sensors are available in a variety of sizes and with different skim charge levels [8]. The water stream sensor has a working range of 2–100 L/min and a water pressure of 2.0 Mpa.

4.4 Solinoid Valve Actuator

Uniqueness valves with actuators that can be controlled by temperature or stream sensors are known as programmed control valves. Electric-fueled, pressure-driven, or pneumatic indicators from sensors are used to limit the valves. The valves can be configured to open, close, or anywhere in between to manage coast uses such as water treatment plants, multi-story constructions, water storage towers, and recycled water systems [10]. A valve actuator is the device that opens and closes a valve. Valve stems that be operated by hand require someone to be present to change them using a temporary or equipped system connected to the valve stem. Energy-driven actuators, which use fuel pressure, water pressure, or energy, allow a valve to be altered from a

Fig. 7 Structure of solinoid valve sensor

distance or large valves to be activated quickly. Power operated valve actuators might be the extreme last components of a programmed supervise circle that consequently regulates some conformity, degree, or one-of-a-kind contraption [11]. Actuators can be fine for opening and closing the valve, or they can also allow halfway positioning; some valve actuators have switches or special means to remotely recommend the valve's location as shown in Fig. 7.

4.5 *Liquid Crystal Display (20 × 4)*

Here 20 × 4 LCD modules are regularly utilised in most implanted tasks, the explanation being their modest value, accessibility, software engineer cordial, and accessible instructive assets [12]. In the proposed project, we use 20 × 4 LCD to monitor all water allocation capacity slots in each and every individual flats, and at the same time, the allocation capacity levels visually displayed on the LCD screen as shown in Fig. 8 the project title and sensors data re displayed on the screen.

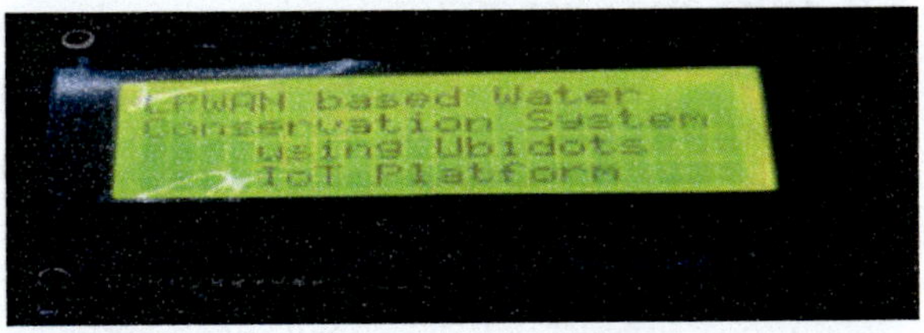

Fig. 8 Project title is displayed on LCD screen

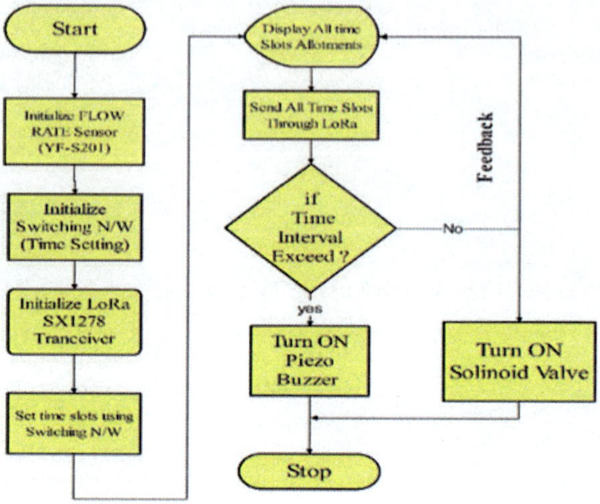

Fig. 9 Implimentation algorithm of transmitter section

5 Flowchart

The detailed process of the work contains two sections: One is transmitter section and the other is the receiver section. The working flow of water conservation system uses LoRa gateways and IoT platform. The flowchart shows the live data (Water Allocation Status in all the plots) monitoring using LoRa wireless technology. The sensor control mechanism uses SX1278 LoRa wireless transceiver modules and the 20 × 4 LCD for visualising real-time environments using LoRa Gateway. Transmitter section is used to measure all the sensors data using LoRa Node and 20 × 4 LCD. The interfacing algorithm is shown in the flowchart in Fig. 9. The receiver section contains LoRa SX1276 transceiver, Piezo buzzer, and programmable switching matrix. The interfacing algorithm is shown in Fig. 10.

6 Results and Discussions

The proposed project's real-time images and results are displayed in this section. A detailed description of the process and real-time implementation images of transmitter and receiver sections are shown in Figs. 11 and 12, respectively. The results of WAN data and the water allocation of each individual plots displayed on the Ubidots IoT platform are shown in Figs. 13 and 14a, b, respectively.

Fig. 10 Implimentation algorithm of receiver section

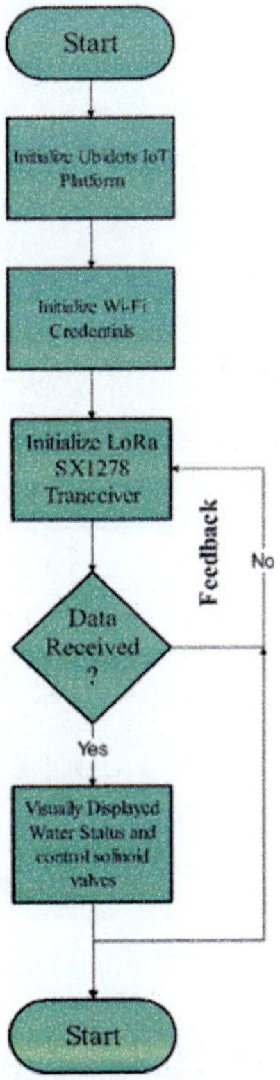

Fig. 11 Real-time implementation transmitter section

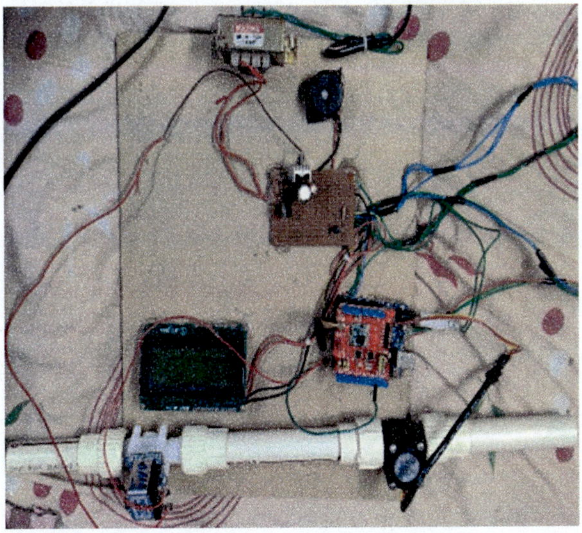

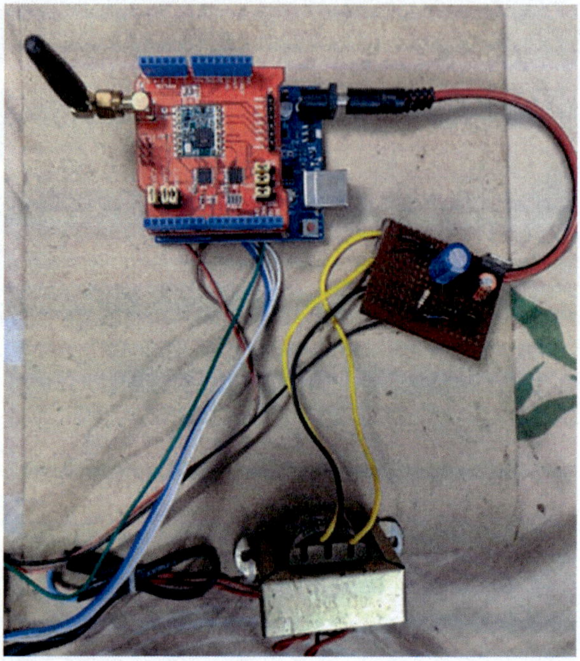

Fig. 12 Real-time implementation receiver section

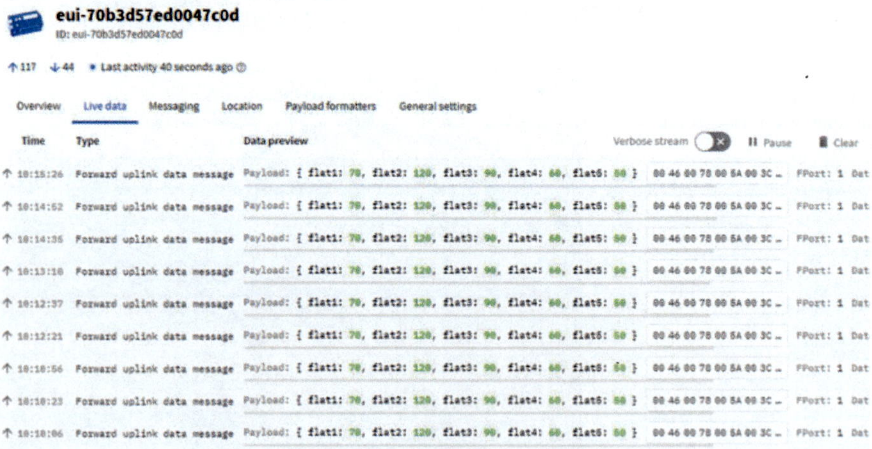

Fig. 13 LoRa WAN data displayed on The Things Network (TTN)

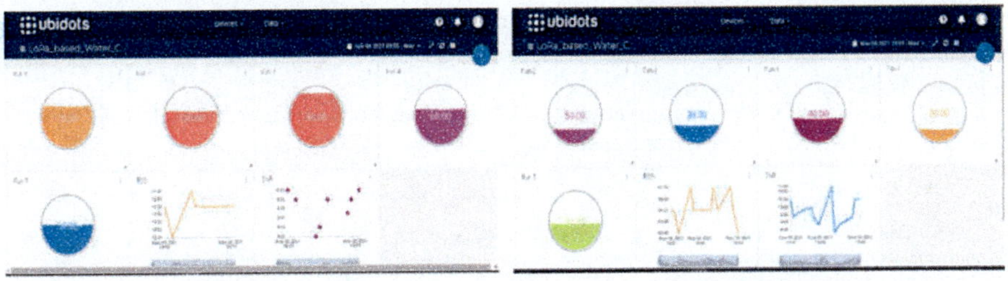

(a) (b)

Fig. 14 Water limits visually displayed on Ubidots IoT platform

7 Conclusion

In this paper, we discussed monitoring the water distribution to the residence by using LoRa technology. It is less power consuming wireless technology. In this, the readings of residence are seen, and the water level is monitored by using the flow sensor. Finally, we can conclude that we can monitor the water supply to the residence by using LoRa technology with less power consumption to long range.

References

1. Sidibé A (2005) Farm-level adoption of soil and water conservation techniques in northern Burkina Faso. Agric water Manag 71(3):211–224
2. Sawadogo H (2011) Using soil and water conservation techniques to rehabilitate degraded lands in northwestern Burkina Faso. Int J Agric Sustain 9(1):120–128
3. Windarto YE, Prasetijo AB, Damara GF (2018) A GIS-based waste water monitoring system using LoRa technology. In: 2018 5th International conference on information technology, computer, and electrical engineering (ICITACEE), pp 176–179
4. Wakindiki IIC, Ben-Hur M (2002) Indigenous soil and water conservation techniques: effects on runoff, erosion, and crop yields under semi-arid conditions. Soil Res 40(3):367–379
5. Joshitha C, Kanakaraja P, Kumar KS, Akanksha P, Satish G (2021) An eye on hydroponics: the IoT initiative. In: 2021 7th International conference on electrical energy systems (ICEES), pp 553–557
6. Kanakaraja P, Upadhyay S, Kotamraju SK, Suneela GVS, Neelesh R (2021) Design and implementation of smart energy meter using LoRa-WAN and IoT applications. J Phys: Conf Ser 1804(1):12207
7. Kumer SVA, Kanakaraja P, Mounika V, Abhishek D, Reddy BP (2021) Environment water quality monitoring system. Mater Today Proc
8. Kanakaraja P, Kumer SVA, Krishna BJ, Krishna VM (2021) Communication through black spot area using LoRa technology and IOT. Mater Today Proc
9. Raghuveera E, Kanakaraja P, Kishore KH, Sriya CT, Lalith BSKT (2021) An IoT enabled air quality monitoring system using LoRa and LPWAN. In: 2021 5th International conference on computing methodologies and communication (ICCMC), pp 453–459. https://doi.org/10.1109/ICCMC51019.2021.9418440
10. Wolka K, Mulder J, Biazin B (2018) Effects of soil and water conservation techniques on crop yield, runoff and soil loss in Sub-Saharan Africa: a review. Agric water Manag 207:67–79

11. Atnafe AD, Ahmed HM, Adane DM (2015) Determinants of adopting techniques of soil and water conservation in Goromti Watershed, Western Ethiopia. J. Soil Sci. Environ. Manag. 6(6):168–177
12. Ferguson BK (1987) Water conservation methods in urban landscape irrigation: an exploratory overview 1. JAWRA J Am Water Resour Assoc 23(1):147–152

Review on Impact of Attacks as a Malicious Node in MANET

Nitesh Ghodichor, Varsha Namdeo, V. Raj Thaneeghavl, and Gautam Borkar

Abstract The MANET is the collection of nodes which move from one place to another place and communicates with each other through wireless links. MANET is infrastructure less network having dynamic topology due to its dynamic topology network which is very susceptible to attacks. Preventing attack in MANET is a major issue. Malicious nodes in the network are the source of attacks. In MANET, there are various sorts of attacks that alter network behavior. It is the survey on impact of attacks in MANET. There are mainly two types of attacks occurring in the network data traffic attack and control traffic attacks. This paper explains different type of attacks and its impact on MANET and existing defense mechanism in MANET.

Keywords MANET · Malicious node · Attack · Security mechanism

1 Introduction

MANET is a network of wireless nodes that roam from one point to another to form a network structure without the use of infrastructure or a central administrator. MANET is the most popular network as it is easy to install. In MANET, information is transferred from one node to another node with the help of radio frequency, and it is peer-to-peer communication. Due its open nature, MANET is very susceptible for attack. Attacker may attack the network, while information or packet transfers from

N. Ghodichor (✉) · V. Namdeo · V. Raj Thaneeghavl
CSE, SRK University, Bhopal, India
e-mail: niteshgho@gmail.com

V. Namdeo
e-mail: Varsha_namdeo@yahoo.com

V. Raj Thaneeghavl
e-mail: rajthaneeghaivelv@gmail.com

G. Borkar
IT, RAIT, Mumbai University, Mumbai, India
e-mail: gautam.borkar2@gmail.com

V. K. Gunjan and J. M. Zurada (eds.), *Proceedings of 3rd International Conference on Recent Trends in Machine Learning, IoT, Smart Cities and Applications*, Lecture Notes in Networks and Systems 540, https://doi.org/10.1007/978-981-19-6088-8_49

one node to another and it also observes the traffic of network before attack occurs. Attack is divided into two types: data traffic attack and control traffic attacks.

In MANET, control information is transferred first to keep the network active and after that original information called data for communication is transferred, and this information is sent in the form of packet. Attackers may attack any one or both. Attack is the security threat which compromise the network or any action that compromise the security information. Security is the important concern in MANET. It is very difficult to establish end-to-end connection in open media. To provide security, different security mechanism and security protocols are implemented. Key-based protocols and IDS mechanism are used to attain authentication. It is important to provide security to both types of traffic: control traffic and data traffic [9]. Based on this, there are two types of attack: (a) control traffic attack and (b) data traffic attack [7].

2 Types of Attack

2.1 Control Traffic Attack

Traffic is the capability of network to send number of packets. In most cases, an attacker targets the network. As MANET is the infrastructure less network, the control information is sent continuously in the network to check the next hop availability as well as to keep the network active for information transfer. The attacker can easily attack on the control traffic and enter into the network by making the nodes malicious and eavesdropping the network, as a result of this there is loss of control packets [1]. Types of control traffic attacks are as follows:

Wormhole attack: In a wormhole attack, attackers grab messages from one point in the network, "tunnel" them to another spot, and then send them back into the network from that point. A single malicious node or two or more malicious nodes can be used in a wormhole attack. The wormhole attack is seriously damaging for many ad hoc network protocols, in which each node that detects a packet transmission directly from the source node considers itself to be in the range of that node [2].

Hello flood attack: The attacker node broadcasts a hello packet with very high power, other node thinks the active node is the parent node and forwards the packets toward this node considering the attacker node as next hope, and the route is the best route to the destination. In a flooding attack, the attacker floods the network by sending number of request, so it overflows the network resources. It consumes the more bandwidth computational energy. Battery power disrupts routing operation. This results in energy waste or more energy consumption to reach to the destination as well as delay in the network.

Bogus Registration attack: In bogus registration, attacker destination distinguishes itself like another node by sending stolen beacon or generating such false beacon registration to request him in the network. So other nodes consider him as a network part and send the packets through this malicious node.

Man-in-Middle attack: In man-in-middle attack, the attacker node becomes the part of network. It enters in between the valid path of network and catches the packet flowing through it. To attack the network, man-in-middle attacker needs to be a part of the network. In man in middle, the attacker comes in middle of the route through which the traffic is more and does not forward the packets that has come to him. Man-in-middle attack easily eavesdrops the network to become the part of route, and the destination does not receive the packets.

Rushing Attack: Rushing attack exploits duplicate suppression mechanism in MANET. Route discovery algorithm is used to find the shortest route or valid route for data transmission. AODV is the basic protocol used for the route discovery process. In route discovery protocol, RREQ message is sent to neighbor node to check the next hope and valid node sends the RREP to exchange the data. Some protocols limit the request and route reply in the network by using suppression mechanism. Rushing attack exploits this action by quickly forwarding malicious route reply on behalf of the valid node; due to this action, actual route reply message from the valid node is discarded and does not reach the source and the attacking node becomes part of a network. This process leads to delay in packet delivery.

Cache Poisoning Attack: In AODV protocol, route is active for some time until the timeout process occurs. In cache poisoning attack, attacker node remains in the route in between timeout schedule, where the route is active. The attacker nodes advertise him as a valid node which is out of reach for control message transmission. Thus, the malicious node acts as valid node and caches the traffic in active period.

Blackmail: In blackmail or cooperative attack, attacker node claims an innocent node as harmful node. Some protocol makes a list of node for packet transmission, as the blackmail attack shows valid node as harmful node; thus, the protocol may exclude these nodes from list and attacker node becomes part of MANET which affects security.

Sybil Attack: In Sybil attack, attacker node creates its multiple node identities. As it enters in the network, malicious behavior occurs in the network and single attacker node acts as multiple nodes, so it affects multiple valid nodes. Network assumes that this attacker node is the network part and forwards the traffic toward malicious node, as a result, traffic is not reached to destination and packets are lost. Malicious node alters the message which it receives and eavesdrops the network which appears large in size.

2.2 Data Traffic Attack

In MANET, control message and data message transmission is done by protocols. The meaningful information which transmits in the network is called data. The attack which specially attacks the data packets or data traffic is called data traffic attack.

Data traffic attacks are very harmful as it lost the meaningful message. Data traffic attacks are stated below [3]:

Black hole: Black hole is the active attack which leads to dropping of packet. In black hole attack, complete traffic disappears from the network like the energy disappears from universe. In a black hole, a malicious node claims itself of being the shortest path to destination but drops the complete routing packets and does not forward them to the neighborhood node [8].

Cooperative Black Hole: In a cooperative black hole attack, the cluster of black hole nodes attacks the network. Malicious nodes attract all the traffic in the network from different path and drop all the packets without forwarding to the receiver.

Gray Hole: It has two types of behavior as follows:

(1) Node-dependent link: Attacker node drops the packets which is reaching toward the destination node or coming from any source node.
(2) Time-dependent link: An attacker drops the packet as per pre-decided time.

Jellyfish: A jellyfish attack is a DoS type of attack in which end-to-end delay is increased by the attack node and in which the attacker node tries to increase packets end-to-end delay. The attacker node enters into the network, and instead of dropping the packets, it delays before delivering to destination. The attacker node may scramble the order of packet and send it in random order, so data may be lost as well as it leads to delay in receiving packets and order of packet is changed. Jellyfish attack disturbs the normal flow of traffic and results in end-to-end delay.

3 Existing Defense Mechanism in MANET

Defense mechanism is applied to nodes when they involve in routing the information. Different defense mechanism techniques are available, few of them discussed here are IDS mechanism, symmetric encryption, and asymmetric encryption methods.

(1) **Intrusion detection system (IDS)**: IDS detects malicious action in the system. Each mobile node acts as an IDS agent, when there is a need to discover some suspicious activity in the network. Node detects local intrusion, coordinates with its neighboring node, and then takes the decision of routing. Different types of IDS techniques are stand-alone IDS, cooperative IDS, and cluster-based IDS.

    Zing et al. (2007) presented principal of IDS techniques in which each node in ad hoc network is the part of IDS system, it recognized the processes and detected the local intrusion with anonymity, and collaboration process takes place while routing.
(2) **Symmetric key encryption**: Symmetric encryption is a form of cryptosystem in which encryption and decryption are performed using the same key. Data encryption standard (DES) and advance encryption standard (AES) both the algorithms based on symmetric key encryption differ on the basis of their data block size and key size used for the encryption and decryption.

(3) **Asymmetric key encryption**: It is the form of cryptosystem in which encryption and decryption are performed using different keys. One is the public key, another is private key, and it is also known as public key encryption. Asymmetric encryption transforms plain text to cipher text using one of the two keys and an encryption algorithm. Using the paired key and decryption algorithm, the plain text is recovered from cipher text. Asymmetric algorithm can be used for confidentiality and authentication. RSA algorithm is used as an asymmetric encryption method.

4 Impact of Control and Data Traffic Attack

See Table 1.

Table 1 Impact of control and data traffic attack on MANET

Types of attack in MANET	Attack name	Impact on MANET
Control traffic attack	Wormhole attack	It can significantly degrade networks performance and threaten network security Exposed wormhole attacks can be detected, but hidden wormhole attacks are difficult to detect. Form tunneling and drop the packets [6]
	Hello flood attack	Flooding of control messages
	Bogus registration attack	Stolen beacons become part of network
	Man-in-middle attack	It easily eavesdrops the network
	Rushing attack	Flooding of control messages and making bulky network
	Sybil attack	Increasing the size of network
	Cache poisoning attack	Fake route description to forward a data
	Blackmail	Affecting path by confusing networks
	Cooperative blackmail attack	The cluster of nodes affects multiple paths in the network
Data traffic attack	Black hole	Total drop of the packets, collapse the network [5]
	Gray hole	More drop of the packets, send fake RREP with respect to node and time [5]
	Cooperative black hole	Clustering of malicious node and affect the more number of path
	Jellyfish	End-to-end delay in networks

5 Proposed Scheme

Proposed scheme is a "secure routing protocol with wormhole attack prevention", and an implementation module provides security to network and protects the wormhole attack. Wormhole attack is a control traffic attack. Key-based mechanism and group signature are used for privacy protection of network, and it is compared with AODV protocol in terms of packet delivery ratio and energy consumption.

Wormhole attack: In wormhole, attacker receives the packet from source node and tunnels it to the other point in the network, whereas it also resents the packet into the network (Fig. 1).

"A secure routing protocol with wormhole attack prevention" [3] provides privacy protection using anonymous key establishment and secure route discovery process. In this protocol, data packet and control packet look random for outside world, and only authenticate node can identify them by employing anonymous key establishment and then can start route discovery. Each node establishes a session key with its neighbouring node during packet transmission following route discovery to the destination.

Anonymous key establishment: Each node in the network established a session key with its neighbor node for transmission of packet, when key gets verified, then only transmission occurs to achieve the secure transmission [4].

Secure route discovery: Under the anonymous key establishment, secure route discovery process started from source node to destination node. Secure node sends number of route request into the network, and only valid node sends route reply back to the source node [4].

Secure routing protocol provides following parameters

Privacy protection: Only valid node can receive the packet and is able to forward it to its neighbor, and by verifying key, it leads to privacy protection against attacker node.

Content un-observability: Anonymous key establishment established the secure route for data transmission, the data get protected from outside node, and the contents are unobservable to outer node which provides content un-observability [3].

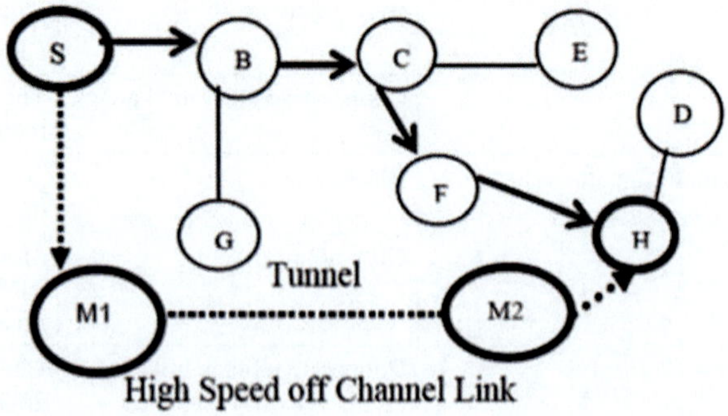

Fig. 1 Wormhole attack in MANET [10]

Wormhole attack prevention: Attacker node is unable to route the packet from one point to another point in the network, and as the traffic is not visible to outside world, only valid node can identify or observe the traffic result in wormhole attack prevention.

6 Result

The proposed scheme "secure routing protocol with wormhole attack prevention" results are compared with AODV routing protocol. The proposed protocol is a privacy protection routing protocol, and AODV is the only routing protocol which is unable to provide security, while routing drop of packet is more in AODV as compared to secure routing protocol. It affects the throughput of network. Thus, the expected throughput of secure routing protocol is more as compared to AODV routing protocol. The proposed methodology will be implementing on given configuration simulation tool: NS3, node placement: random, number of node: 50, and routing: multicast routing.

7 Conclusion

MANET contains mobile node which operates independently. It is a self-configuring mobile ad hoc network that contains a self-organizing network. In MANET, nodes communicate with each other directly without the aid of any centralized administration or fixed infrastructure. MANET operating and using the routing protocol within network due to registering and configuring in network start their exchange info to all neighbor node for updating routing table and other necessary information. On transmitting data and dynamically controlling itself within the network, the behaviour of any invalid node attempting to access data or a path. The attack on the control message and packet loss or network disruption.

We discuss here two types of attack control and data attack. Various attack possibilities are there. Every different behavior of attacker has defined different attack, and due to this, network and their valid node change their behavior and found that data and control of the message are dropping and executing all fake transmission and communication. These types of overhead network AODV protocol confuse and misbehave in communication; every type of attack impacts differently and eavesdrops the network in the form of invalid node, bulky network, tunneling between paths, and dropping packets and always changes network behavior due to attack. Secure routing protocol with wormhole attack prevention provides privacy protection and prevents the wormhole attack.

References

1. Bhattacharyya A, Banerjee A, Bose D, Saha HN, Bhattacharjee D (2011) Different types of attacks in Mobile ADHOC Network: prevention and mitigation techniques, Nov 2011
2. Abdelshafy MA, King PJ (2016) Resisting blackhole attacks on MANETs. In: 13th IEEE annual consumer communications & networking conference (CCNC)
3. Patil V, Fulare P, Ghodichor N (2014) An unobservable secure routing protocol with wormhole attack prevention for mobile Ad-Hoc network. Int J Current Eng Technol 4(3):1930–1932, E-ISSN 2277–4106
4. Patil V, Fulare P, Patil SS, Ghodichor N (2014) A review on an unobservable secure routing protocol with wormhole attack prevention in Manet. IOSR-JCE, E-ISSN 2278-0661:16-19
5. Usha U, Bose B (2012) Comparing the impact of black hole and gray hole attacks in mobile Adhoc networks. J Comput Sci 8(11):1788–1802
6. Kadam A, Patel N, Gaikwad V (2016) Detection and prevention of wormhole attack in MANET. IRJET 03(03):388–393
7. Arathy KS, Sminesh CNA (2016) A novel approach for detection of single and collaborative black hole attacks in MANET. Government Engineering College Thrissur, Kerala, 680009, India Global Colloquium in Recent Advancement and Effectual Researches in Engineering, Science and Technology (RAEREST 2016)
8. Dongarwar K, Ghodichor N (2017) Review on black hole attack detection in wireless MANET. IJERCSE 4(3):305–309
9. Younis OH, Essa SE, El-Sayed A (2017) Defense mechanisms in mobile Ad-hoc networks. Communications on Applied Electronics (CAE) 7(2):19–28, ISSN: 2394-4714, Foundation of Computer Science FCS, New York, USA
10. Niranjan P, Srivastava P, Soni RK, Pratap R (2012) Detection of worm-hole attack using Hop-count time delay analysis. IJSRP 2(4)

Automation in Project Management 4.0 with Artificial Intelligence

Somya Goyal, Harshit Jha, and Ayush Gupta

Abstract In today's world, digital transformation is common and innovative in each factor of life. Now, most of the human works are being done by robots and machines in project management. Artificial intelligence has the potential to change the way of project management (PM) tasks which are currently performed and can be changed and controlled in the recent future. Companies are searching out diverse approaches to beautify the performance with taking care of several options like reliability, accountability, etc., all simultaneously. Industries have initialized construction of AI embedded tools to assist the team to control their challenge(s). This study explores how our existing PM profession can lead to change the strategy by using AI inputs and how AI helps in enhancing project management and transforming it as Project Management 4.0. It enables to automate various time-taking venture of control that is crucial for success. AI also helps in evolving task automation to efficient analytics of the project, searching for resources to help, and building efficient strategies.

Keywords Artificial intelligence · Project management · Product management · Business intelligence

1 Introduction

Artificial intelligence (AI) includes the skills of making certain decisions and analysis usually present in humans which can be subsequent step of evolution. In project management, strategical decision-making is highly crucial toward successful completion of project [1]. The effort estimation to plan the resources and quality prediction is major decision areas in business intelligence [2–13]. Business planning, team coordination, and documentation are another range of tasks which requires automation.

S. Goyal (✉) · H. Jha · A. Gupta
Manipal University Jaipur, Jaipur, Rajasthan 303007, India
e-mail: somyagoyal1988@gmail.com

V. K. Gunjan and J. M. Zurada (eds.), *Proceedings of 3rd International Conference on Recent Trends in Machine Learning, IoT, Smart Cities and Applications*, Lecture Notes in Networks and Systems 540, https://doi.org/10.1007/978-981-19-6088-8_50

Before software tools, it just turned into a nightmare for the product managers to report documents and essentials and hold the project on track. Coordinating and doing alignment with several stockholders have been difficult. Traditional project management requires coordination, observation, and plenty of meetings and conferences to give the update the information to stockholders about status of the project.

Before this advancement, there was usually a single project manager allocated for a single module of the project. The whole project management team required lots of intervention to hold project on track. But with the help of AI, project management can evolve. Artificial intelligence enables project managers to use the tools for managing people and projects, helping them, and completing their project in the given timeline and on the budget. These tools are allowing managers to manage timelines, resources, dependencies, costs, and sending reminders to the required persons as needed to complete the project in effective manner. It helps mitigate project risk in a variety of ways.

With AI, project resources and task status can be updated, and questions can be posted as needed which significantly minimize coordination and complex tasks. According to some studies, project managers used to spend more of their time on tasks related to administration such as frequent check-in(s) processing and updating management [5–7].

Artificial intelligence reduces burden of project managers such as keeping records and sending reminders, notifying important information to the stake-owners, and updating project work. But on the other hand, there are many tasks that are currently into the limitations of AI. These tasks require influence, negotiation, and encouraging work to achieve the common goals that should not be ignored. Some of the important tasks like work experience, encouragement, and human influence play a key role in gaining success in the project [1]. Next, this study discusses the role of AI in Project Management 4.0.

2 AI in Project Management 4.0

AI supports the project managers in multiple ways to reduce their burden. Using AI in their application and concept, work of project manager can be easier in comparison with the traditional approach. Earlier, the assumptions are done for the project, and constraints provided to the individuals were considered to remain more intact throughout the life cycle of the project. But today, this is not the case, and we can keep a track on all the constraints as well as assumptions that can be performed throughout the life cycle [14]. Artificial intelligence allows the organizations to have crisis dashboards. On looking at these dashboards on the computer, officials can see the progress, challenges as well as completeness and perfectness of the projects [15, 16]. One can also see the level of criticalness in this case. During the project, AI practices may shed light on some of the immediate actions that project leaders should take. This reduces the response time for various conditions that are out of the defined acceptable range. Management is not aware that whether they can assign

Table 1 AI-based project management strategies

Task	Strategy based on AI
Predictive analytics	AI tools collect all the data to help make the tools smarter and make long-term decisions [2, 8]
Reporting and the status tracking	AI automation tools have created the ability to integrate many platform tools [21]
Documentation	AI tools to design the document templates
Strategic business intelligence	AI tools for resource forecasting, budgets, risks, and allocations [3, 4, 11, 13]
Insights and accuracy	AI automaton [5, 6, 9, 10]
Budgeting and resource allocation	AI-based automation [4–7]
Risk mitigation	Risk analysis and prediction tools [2, 8–13]
No bias	AI has no emotions like human, hence eliminates bias [22]
Research for growth	Quantitative and qualitative [18]

some extra work without overloading the workforce. Additions in the projects are typically done within the queue, with little consideration given to the required technology, required unit skill levels, and availability of the resources [4]. AI practices enable the development of a project portfolio that provides the best opportunities for an organization to increase the business value it can receive while simultaneously identifying effective practices of management of the resources [15].

According to [17], **successful artificial intelligence (AI) practices make optimization** by identifying and matching the past and current projects done in the organization [18]. Project managers are often asked to rely on for quick, intuitive decisions rather than step-by-step computer inference. Taking the references from the previous projects, decisions are made for the current project to achieve high rate of success [2, 11–13]. In business intelligence (BI), AI plays vital role by performing accurate **regression and classification** [3–10, 19, 20]. Table 1 gives a summary of the AI-enabled project management (PM) strategies.

This section discussed the strategies provided by AI for automation of PM tasks. It highlighted the AI-enabled tools and techniques for PM tasks. Next, we draw the inferences from the study conducted. Later, the work is concluded with remarks on the future scope of this work.

3 Results and Discussion

Based on the above study, it can be inferred that AI-enabled tools play an important role in the current environment for running projects. A project manager acts as a driver in the team and can help the system or project to run smoothly. With the help of AI tools, PM gets help to complete the job faster, as it is less prone to errors. Also, with the help of AI and machine learning, they can see and learn from the past projects and collect data which helps them in building a more fluent and effective strategy. Coordination and meetings to run the project can be significantly minimized. PM can be able to manage multiple projects as it broadens their bandwidth. Setting up AI tools at the beginning is quite complicated, and technical help is required to set it properly. In some cases, the initial cost is also very high but, we need not to purchase the service every year. In this way, we get the advantage as it solves the long-term problem and will also be cost-effective for a longer run. From the survey, it is found that AI tools provide multi-domain benefits to industries as plotted in Fig. 1.

The description is detailed under Table 2 where the first column shows the heads of domains, and second column shows the percentage benefit obtained from AI strategies in Project Management 4.0.

Next, the pace up an industry gets in this competitive era by using AI tools has been analyzed. Here, Fig. 2 displays the benefits to a company by AI in comparison

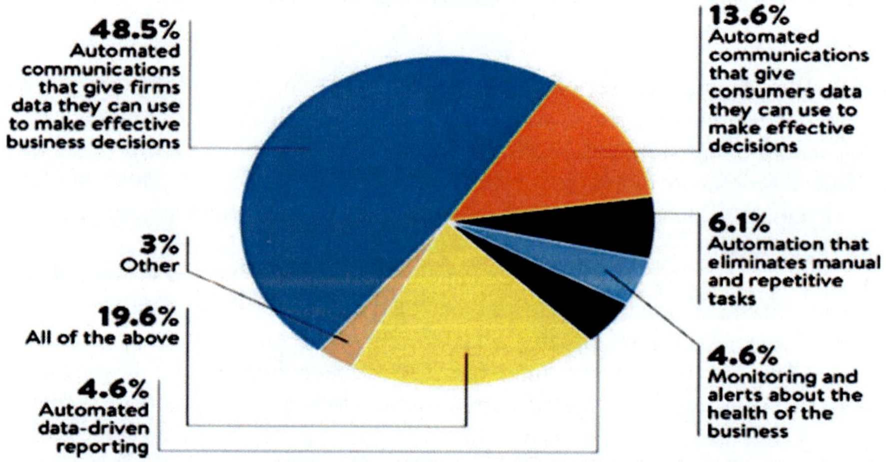

Fig. 1 Benefits of AI in business

Table 2 Percentage benefits from AI-based project management strategies in different domains

Domain	Benefit from AI-based project management strategies (%)
Automation in decision-making	48.5
Automated communication	13.6
Automation of repetitive tasks	6.1
Productivity by constant monitoring	4.6
Automated reporting	4.6
Financial benefits	19.6 (for all above reasons)

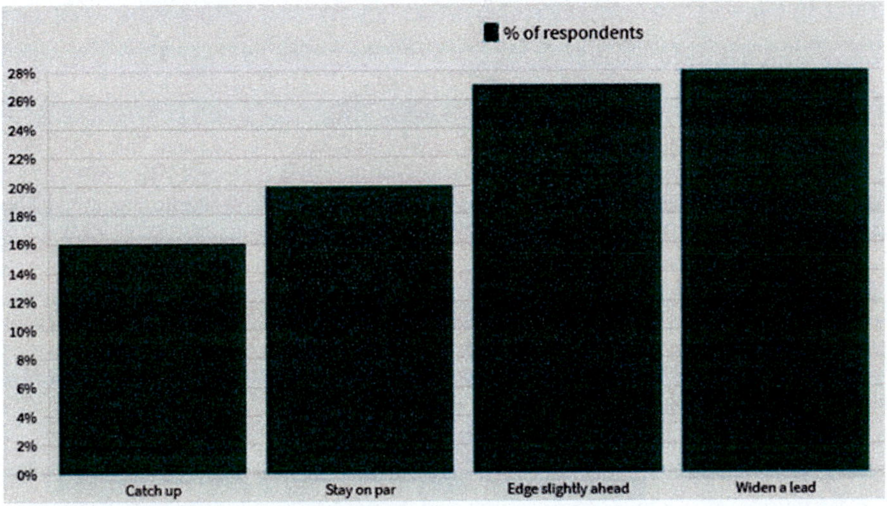

Fig. 2 Benefits of AI in competitive market

with its competitors. It raises 16% catch up rate and 20% stay on par rate. It allows a company to edge slightly ahead by 26% and widens the lead by 28%.

After analyzing the contribution of AI to an industry in both the aspects—as an individual and as a competitor in market, we conclude the work with the future expectations from AI which is also the future scope of the work.

4 Conclusion

AI usually supports rather than just competing the project manager. As with any technology currently, AI alone does not give any guarantee about the success of a project. However, when used in a targeted and efficient manner, artificial intelligence can be used as a catalyst and can be a game-changer for project managers as it helps in increasing the success rate of the projects. Technology can sometimes imagine beyond the "human" imagination can help the project managers by bringing positive and creative changes to a project. This will help in creating a better strategy for the project [17]. In future, project managers expect the contribution from AI in their business logic in very effective manner which is as shown in Fig. 3. The details are shown as Table 3. In future, AI technology is being proposed to contribute to PM tasks and transform it into Project Management 5.0.

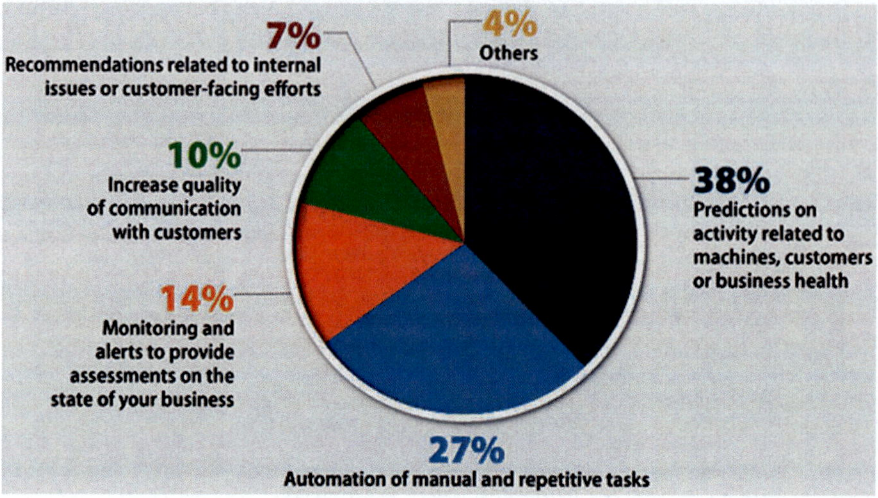

Fig. 3 Future expectations from AI in project management

Table 3 Future expectations from AI-based project management strategies in different domains

Domain	Future expectations from AI-based project management strategies (%)
Automated communication	10 + 7
Automation of repetitive tasks	27
Productivity by constant monitoring and reporting	14
Financial benefits from predictions	38

References

1. https://www.forbes.com/sites/cognitiveworld/2019/07/30/ai-in-project-management/?sh=4e18978fb4a0. Accessed on 01 Jan 2022
2. www.pwc.ch/ta ("AI will transform project management. Are you ready?"). Accessed on 01 Jan 2022
3. Mackey A, Gass SM (2015) Second language research: methodology and design. Routledge
4. Goyal S (2022) FOFS: firefly optimization for feature selection to predict fault-prone software modules. In: Data engineering for smart systems. Lecture notes in networks and systems, vol 238. Springer, Singapore. https://doi.org/10.1007/978-981-16-2641-8_46
5. Goyal S (2020) Heterogeneous stacked ensemble classifier for software defect prediction. In: 2020 Sixth international conference on parallel, distributed and grid computing (PDGC). Waknaghat, Solan, India, pp 126–130. https://doi.org/10.1109/PDGC50313.2020.9315754
6. Goyal S, Bhatia PK (2020) Software quality prediction using machine learning techniques. In: Innovations in computational intelligence and computer vision. Advances in intelligent systems and computing, vol 1189. Springer, Singapore, pp 551–560. https://doi.org/10.1007/978-981-15-6067-5_62
7. Goyal S, Bhatia PK (2020) Feature selection technique for effective software effort estimation using multi-layer perceptrons. In: Proceedings of ICETIT 2019. lecture notes in electrical

engineering, vol 605. Springer, Cham, pp 183–194. https://doi.org/10.1007/978-3-030-30577-2_15
8. Goyal S, Bhatia PK (2019) GA based dimensionality reduction for effective software effort estimation using ANN. Adv Appl Math Sci 18(8):637–649 © 2019 Mili Publications
9. Goyal S, Bhatia PK (2019) A non-linear technique for effective software effort estimation using multi-layer perceptrons. In: 2019 International conference on machine learning, big data, cloud and parallel computing (COMITCon), Faridabad, India pp 1–4. https://doi.org/10.1109/COMITCon.2019.8862256
10. Goyal S, Bhatia PK (2021) Heterogeneous stacked ensemble classifier for software defect prediction. Multimed Tools Appl. https://doi.org/10.1007/s11042-021-11488-6
11. Goyal S (2021) Effective software defect prediction using support vector machines (SVMs). Int J Syst Assur Eng Manag. https://doi.org/10.1007/s13198-021-01326-1
12. Goyal S, Bhatia PK (2021) Software fault prediction using lion optimization algorithm. Int J Inf Tecnol. https://doi.org/10.1007/s41870-021-00804-w
13. Goyal S (2021) Handling class-imbalance with KNN (neighbourhood) under-sampling for software defect prediction. Artif Intell Rev. https://doi.org/10.1007/s10462-021-10044-w
14. https://www.cio.com/article/228200/how-ai-could-revolutionize-project-management.html. Accessed on 01 Jan 2022
15. https://pmworldjournal.com/article/artificial-intelligence-ai-influence-in-project-management. Accessed on 01 Jan 2022
16. Munir M (2019) How artificial intelligence can help project managers. Global J Manage Business Res. Retrieved from https://journalofbusiness.org/index.php/GJMBR/article/view/2728. Accessed on 01 Jan 2022
17. Kerzner H (2018) Project management best practices: achieving global excellence. John Wiley & Sons
18. Willems LL, Vanhoucke M (2015) Classification of articles and journals on project control and earned value management. Int J Project Manage 33(7):1610–1634
19. Goyal S (2021) Predicting the defects using stacked ensemble learner with filtered dataset. Autom Softw Eng 28:14. https://doi.org/10.1007/s10515-021-00285-y
20. Goyal S, Bhatia PK (2020) Comparison of machine learning techniques for software quality prediction. Int J Knowl Syst Sci (IJKSS) 11(2):21–40. IGI Global. https://doi.org/10.4018/IJKSS.2020040102
21. Bond AH, Gasser L (eds) (2014) Readings in distributed artificial intelligence. Morgan Kaufmann
22. Kłosowski G, Gola A (2016) Risk based estimation of manufacturing order costs with artificial intelligence. Annals of computer science and information systems. In: Proceedings of the 2016 federated conference on computer science and information systems, vol 8. IEEE, pp 729–732. https://doi.org/10.15439/2016F323

Prediction in Industry 4.0 with Lean Six Sigma

Somya Goyal, Harshit Jha, and Ayush Gupta

Abstract Over the past few decades, the Lean Six Sigma methods and processes have improved performance and productivity in both product and service-based organizations. Artificial intelligence (AI) is used to expand that number and promote greater and faster development across the organization in Industry 4.0. To realize the full value of their AI investment, it is necessary for the organization to first develop key organizational skills, such as setting up the process, with the help of programs such as Lean Six Sigma. Hence, AI does not replace Lean Six Sigma, but rather develops it while benefiting from its contextual practices. In this paper, the impact of AI in lean Six Sigma strategies is highlighted. AI enhances lean six sigma methodologies and lean six sigma benefits the industries and transforming it into the culture Industry 4.0.

Keywords Artificial intelligence (AI) · Six Sigma · Lean Six Sigma · Industry 4.0

1 Introduction

Artificial intelligence (AI) is helping to work at a faster and efficient pace with maintaining accuracy [1, 2]. Artificial intelligence is the power of a computer system to replicate human cognitive functions such as learning and solving problems. With AI, a computer system uses mathematics and logic to replicate the thinking of people which use it to learn new information and make decisions [3]. The goal of AI is to make computers and computer programs smart enough to imitate the human mind behavior [4, 5]. For organizations to remain active and retain their top spots, they will need to free up routine tasks and adopt AI practices [6, 7]. The future is much brighter for those who update and can change according to the trend. Another major change is making the product and offers in such a way that it looks like a product even if it is a service. Consequently, it can be marketed and sold and also delivered frequently. In todays' world, flexibility is an essential thing in order to meet customer

S. Goyal (✉) · H. Jha · A. Gupta
Manipal University Jaipur, Jaipur-303007, Rajasthan, India
e-mail: somyagoyal1988@gmail.com

V. K. Gunjan and J. M. Zurada (eds.), *Proceedings of 3rd International Conference on Recent Trends in Machine Learning, IoT, Smart Cities and Applications*, Lecture Notes in Networks and Systems 540, https://doi.org/10.1007/978-981-19-6088-8_51

demands. AI can work in the background to direct the entire process and make the most current information easily accessible [4].

It also focuses on changing and increasing pace. The new invention is data-driven and has the ability to extract ideas from data to make decisions. The traditional engineering method of a new product may require testing to be done on the existing customers and reviewing their responses [8–10]. It usually takes time. Artificial intelligence here can help us. AI can filter data to identify patterns that we may miss in some way and combine that knowledge with our research results to allow us to make better decisions about the new product we want to create. AI can also direct many aspects of stakeholder engagement such as continuous email communication and feedback. We can build confidence by delivering an accurate, content-friendly, and personalized experience using AI. It is becoming increasingly important for us to use AI to do our job better than to stay comfortable and be distracted by it [4, 11].

Next section of this paper discusses the role of AI in six sigma strategy to lead Industry 4.0 revolution. Later, the inferences have been drawn in Sect. 3 from the study about how lean six sigma resulting into digital transformation of Industry. Then, the work is concluded in Sect. 4 with remarks on the future scope of the work.

2 AI and Six Sigma in Industry 4.0

Six Sigma is a platform that provides organizations with development tools that improve the capability of their business processes. This performance increase and reduction in variation of the process help improvement in profits, employee morale, and quality of products or services. Therefore, it focuses on process and its improvement as shown in Fig. 1 [12].

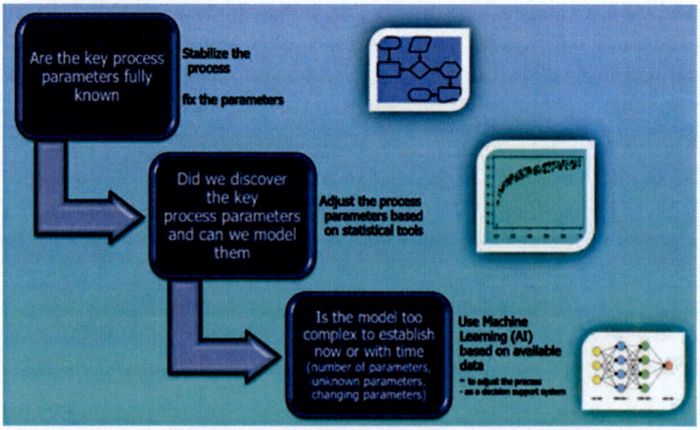

Fig. 1 AI in Six Sigma strategy

We can understand below those cases on how it creates an impact in lean six sigma methodologies. Here, we can get a walkthrough by understanding the cases [12].

1. If the process parameters are known: In this case, stabilization of the process is needed which can be done using traditional Lean and Lean Six Sigma tools.
2. If process parameters are discovered: In this case, we just need to adjust the process parameters based on tools such as regression [2, 7, 8].
3. If the model is complex due to the complexity in parameters like unknown parameters, number of parameters or changing parameters with time: In this case, machine learning has a vital role to play. Based on available data, it adjusts the processes and assists in making the decision. This reduces the complexity of the problem and provides a better vision of the decision [9].

Artificial intelligence and machine learning can help in this area too. As we have discussed earlier, artificial intelligence (AI) can do tasks which without use requires personal involvements. Businesses can issue a schedule function in AI to keep it fast. Artificial intelligence can direct processes and make most of the information accessible [13, 14]. Machine learning teaches the computer to predict and analyzes changes by monitoring the data. Insight and benefits provided by machine learning can lead to productivity and also reduce cost and wastage [15, 16]. One of the key benefits is that it increases the proficiency of the processes. It is the field of study that deals with the ability to learn without being explicitly programmed [13].

The main goal of the Six Sigma strategy is to increase efficiency and reducing errors. The methods used in this strategy have the aim to improve the quality of the process by identifying the errors and also removing it. It relies heavily on a combination of statistical analysis and operational evidence. There are two main methods used to implement Six Sigma—define—measure—analyze—improve—control (DMAIC) and define—measure—analyze—design—verify (DMADV). DMAIC strategy is used to focus on developing existing processes while DMADV strategy used to focus on creating new processes. The DMAIC approach consists of the following five categories [4, 17]:

- Explain and Define
- Measure
- Analyze
- Improve
- Control

The DMADV approach replaces the "development" and "control" categories with "design" and "verification." Producers who want to improve existing processes often use DMAIC methods, which begin with a description of the system and project objectives. After defining a project, technology is used to collect and measure relevant data. Machine learning helps in third phase, i.e., in analyzing. Machine learning software analyzes data to help identify causes of disability. Engineers use the effects of machine learning software to develop and control future production processes [4, 17, 18]. This leads to lean six sigma in Industry 4.0 as shown in Fig. 2.

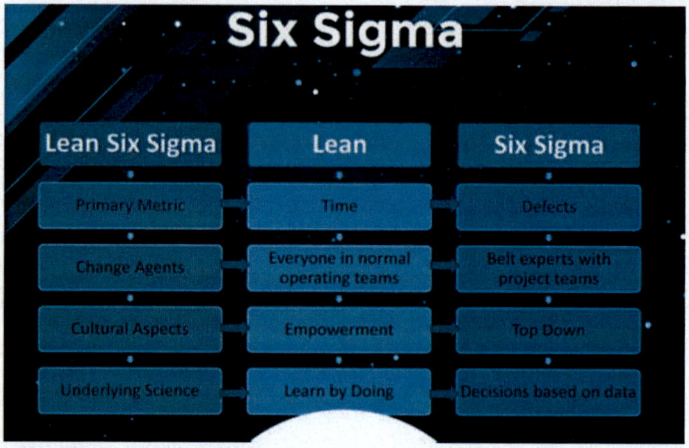

Fig. 2 Lean Six Sigma in Industry 4.0

Next, we discuss the inferences drawn from the study of market trends and impact of AI in Industry 4.0 with lean six sigma.

3 Results and Discussion

Lean Six Sigma is a time-tested solution and helps in increasing productivity. Every possible industry, we can think of, is well suited to use Six Sigma methods to improve the process. Be it transport, manufacturing, health care, or education, Six Sigma can be everywhere in reality [19, 20].

From the study, it can be inferred that artificial intelligence (AI) helps in enhancing various strategies and planning. It quickly turns into something new, has a great opportunity, and has shown a great impact across all businesses and applications. Lean Six Sigma is very benefitted with artificial intelligence. Machine learning (ML) is used to quickly separate designs into complex frameworks and data sets that cultural test will simply fail [19, 21, 22].

Also, native language process or simply NLP is a useful tool of AI that provides great value and information on common DMAIC techniques. NLP can be used to get another view from informal information like voice of customer and voice of process data sources and use media based on Web to provide the piece of information that can be simply missed [21, 22].

AI combining with Lean Six Sigma can be very helpful for productivity. Artificial intelligence can serve as a catalyst in Lean Six Sigma's projects [17, 21, 22].

AI and machine learning can be helpful in working in complex environments. In an example published on LinkedIn, machine learning is required if there are a large number of important process parameters, unknown parameters, or the parameters that

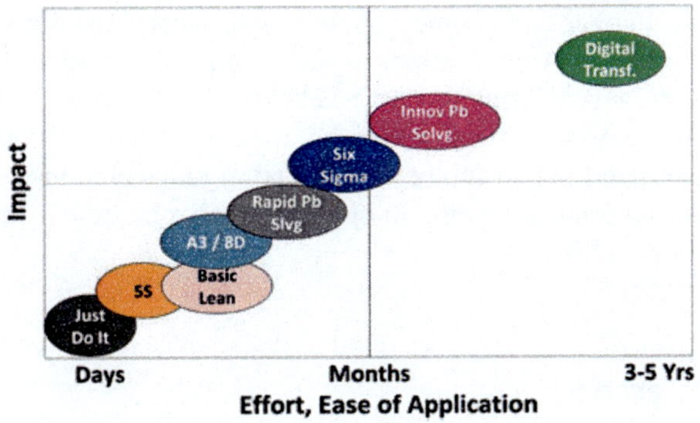

Fig. 3 Lean Six Sigma leading digital transformation

can change over time. Also, AI has the ability to quickly understand large amount of complex data which is useful for providing solutions [19, 21, 22].

In another example from LinkedIn, a pharmaceutical company that distributes medicines for chronic diseases wanted to predict which patients are to be treated with fewer medications. Through AI, the company has been able to successfully detect warning signs in patients' data sets and improve patient health by changing the timetable of the prescribed medicines [21, 22].

AI can see and analyze common examples in non-assembling applications, for example, task management and development of the project. AI helps in achieving greater reduction of cost and cycle time which can be possible with only Lean Six Sigma to attain competitive advantage. Therefore, taking help of artificial intelligence in lean six sigma methods are very helpful for the product [21, 22].

The performance of the learning machine depends on the data. Lean Six Sigma supports the accuracy of the collection of the data as it reduces variability or lack of consistency and also provides a way to review the entire process [13]. The lean six sigma with AI in Industry 4.0 has led the digital transformation as seen from Fig. 3.

Results obtained by using Lean Six Sigma techniques, it is possible to identify the various causes and put forward the best way to get rid of it and combining it with the machine learning, and AI increases the effectiveness of the processes involved in analyzing stage of Lean Six Sigma methodology [13]. It increases the efficiency and reduces time that usually takes in the process.

4 Conclusion and Future Scope

A significant change from a simple digital usage to newly advanced technologies, powered by artificial intelligence (AI), has forced organizations to think and re-evaluate some key industrial habits. One such practice is to improve the process involved in the business. Few decades ago, Lean Six Sigma's process of improving the process has brought proven benefits to efficiency and productivity in both productive and service-based organizations. Artificial intelligence (AI) can be used to expand that number and promote greater and faster development across the organization. However, for organizations to see the full value in their AI investment, it is necessary for the organization to first develop the necessary skills for the process.

Therefore, AI is very helpful after combining with Lean Six Sigma Methods in its projects [4]. In future, the lean six sigma techniques are proposed to be extended with futuristic AI methodologies to benefit our industry and transforming the era into Industry 5.0.

References

1. Goyal S (2020) Heterogeneous stacked ensemble classifier for software defect prediction. In: 2020 Sixth international conference on parallel, distributed and grid computing (PDGC). Waknaghat, Solan, India, pp 126–130. https://doi.org/10.1109/PDGC50313.2020.9315754
2. Goyal S, Bhatia PK (2019) GA based dimensionality reduction for effective software effort estimation using ANN. Adv Appl Math Sci 18(8):637–649 © 2019 Mili Publications
3. https://builtin.com/artificial-intelligence. Accessed on 01 Jan 2022
4. https://leanandsixsigma.org/Artificial-Intelligence-Transforming. Accessed on 01 Jan 2022
5. Goyal S (2022) FOFS: firefly optimization for feature selection to predict fault-prone software modules. In: Data engineering for smart systems. Lecture notes in networks and systems, vol 238. Springer, Singapore. https://doi.org/10.1007/978-981-16-2641-8_46
6. Goyal S, Bhatia PK Software quality prediction using machine learning techniques. In: Innovations in computational intelligence and computer vision. Advances in intelligent systems and computing, vol 1189. Springer, Singapore, pp 551–560. https://doi.org/10.1007/978-981-15-6067-5_62
7. Goyal S, Bhatia PK (2020) Feature selection technique for effective software effort estimation using multi-layer perceptrons. In: Proceedings of ICETIT 2019. Lecture notes in electrical engineering, vol 605. Springer, Cham, pp 183–194. https://doi.org/10.1007/978-3-030-30577-2_15
8. Goyal S, Bhatia PK (2019) A non-linear technique for effective software effort estimation using multi-layer perceptrons. In: 2019 International conference on machine learning, big data, cloud and parallel computing (COMITCon). Faridabad, India, pp 1–4. https://doi.org/10.1109/COMITCon.2019.8862256
9. Goyal S, Bhatia PK (2021) Heterogeneous stacked ensemble classifier for software defect prediction. Multimed Tools Appl. https://doi.org/10.1007/s11042-021-11488-6
10. Goyal S (2021) Effective software defect prediction using support vector machines (SVMs). Int J Syst Assur Eng Manag. https://doi.org/10.1007/s13198-021-01326-1
11. Goyal S, Bhatia PK (2021) Software fault prediction using lion optimization algorithm. Int J Inf Tecnol. https://doi.org/10.1007/s41870-021-00804-w
12. https://www.linkedin.com/pulse/when-does-process-improvement-lean-six-sigma-need-ai-alex-boussetta. Accessed on 01 Jan 2022
13. https://www.datasciencecentral.com/how-data-science-and-ai-complement-lean-six-sigma/. Accessed on 01 Jan 2020
14. Goyal S (2021) Handling Class-Imbalance with KNN (neighbourhood) under-sampling for software defect prediction. Artif Intell Rev. https://doi.org/10.1007/s10462-021-10044-w
15. Goyal S (2021) Predicting the defects using stacked ensemble learner with filtered dataset. Autom Softw Eng 28:14. https://doi.org/10.1007/s10515-021-00285-y
16. Goyal S, Bhatia PK Comparison of machine learning techniques for software quality prediction. Int J Knowledge Syst Sci (IJKSS) 11(2):21–40. IGI Global. https://doi.org/10.4018/IJKSS.2020040102

17. The Application of Machine Learning to Consolidate Critical Success Factors of Lean Six Sigma. IEEE. Accessed on 01 Jan 2022
18. https://corporatefinanceinstitute.com/resources/knowledge/strategy/six-sigma/. Accessed on 01 Jan 2022
19. https://vector-international.com/lean-six-sigma-in-the-era-of-artificial-intelligence-ai/. Accessed on 01 Jan 2022
20. https://www.novelvista.com/blogs/quality-management/top-5-sixsigma-certification-benefits. Accessed on 01 Jan 2022
21. https://www.novelvista.com/blogs/quality-management/ai-ml-in-six-sigma. Accessed on 01 Jan 2022
22. https://www.sixsigmadaily.com/can-artificial-intelligence-improve-lean-six-sigma/. Accessed on 01 Jan 2022
23. Johansson C (2020) How Lean Six Sigma can improve predictability in an Industry 4.0 perspective. Accessed on 01 Jan 2022

Optimized Buffer Management System in Delay Tolerant Network

Pratibha Kamal and Nanhay Singh

Abstract DTN is a kind of network that has an absence of instantaneous end-to-end path between two nodes and intermittent connectivity. When constructing a DTN network, there must be effective and efficient aspects that need to be taken under consideration for an optimized network. The prominent parameter in the DTN network is the buffer management system which determines the packets that need to be entered or dropped if the buffer is full. This paper will focus on various buffer management strategies for different conditions. All the various strategies have been taken into consideration and evaluated according to when the packet is dropped. Analysis has been done on the classification of different buffer management strategies based on various features.

Keywords Delay tolerant network · Buffer management system

1 Introduction

DTN works on store and the forward mechanism that come across two main problems. The first problem is communication between nodes which is not feasible due to transmitting a huge number of packets that burden the limit of nodes resources like energy, buffer, bandwidth, etc. This problem leads to congestion in the networks due to flooding in the network. The second problem is the buffer management strategies as there is a limit for every buffer in the network. In such a situation when new

P. Kamal (✉)
Guru Gobind Singh Indraprastha University, Delhi, India
e-mail: kamalpratibha1@gmail.com

N. Singh
Netaji Subhas University of Technology, Delhi, India

V. K. Gunjan and J. M. Zurada (eds.), *Proceedings of 3rd International Conference on Recent Trends in Machine Learning, IoT, Smart Cities and Applications*, Lecture Notes in Networks and Systems 540, https://doi.org/10.1007/978-981-19-6088-8_52

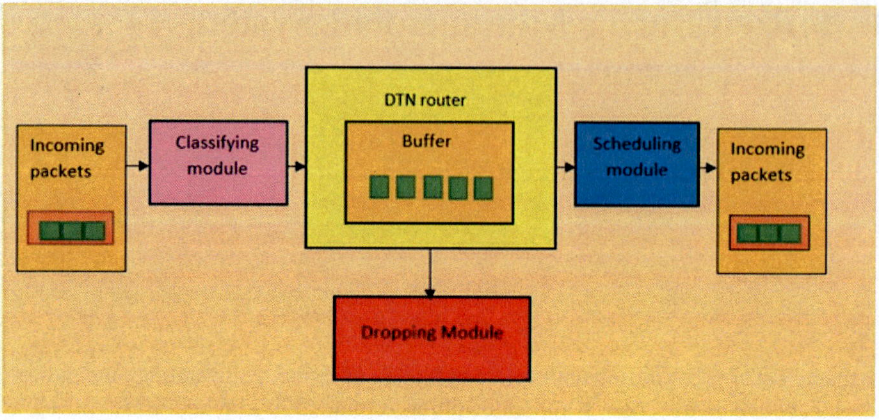

Fig. 1 Architecture of buffer management system

packets arise, either the node has to drop the packet or alert the next node to send the packet. These problems establish buffer management policies [1].

Buffer Management Architecture

Buffer management plays a very important role in resource allocation in space occupied by buffer in the memory as a mechanism of the buffer is responsible for all the scheduling and dropping of packets. An effective buffer is necessary to decide for dropping the packets by executing the right buffer management strategies [2].

In Fig. 1, buffer architecture is explained with incoming packets entered in the classifying module where it uses to classify packets according to their size, and in the DTN router, buffer space is present where all the packets encounter, and if the buffer is overflow, then dropping decision will take place in dropping module, and if the buffer does not overflow, it will enter the scheduling module where it will transfer the outgoing packets. This mechanism goes for every packet that enters the buffer.

In this section, two independent specifications are divided over existing policies into different types [3]:

- **Protocol-specific**. Buffer policies that plummet under protocol-specific are strategies that are made explicit for a certain routing protocol.
- **Non-protocol-specific**. Buffer policies that come under non-protocol-specific determine for any routing protocol.

Protocol-specific can be customized according to properties of routing protocol policies which results in a high delivery ratio and reduced delay in the network.

Now protocols are classified into static and dynamic dropping policies as in buffer management system some packets should be dropped to receive some packets coming in [4].

- **Static dropping policies**: Static dropping strategies do not need any information from the network to specify which packet should drop. Hence, it takes less time on processing.
- **Dynamic dropping policies**: In dynamic dropping, the strategy needs information from the network to specify which packet should drop to evaluate better policy to drop packets. It takes more time to acquire information from the network to obtain accurate processing.

In this section, we have divided the strategies and shown the relationship between different categories.

1.1 Static and Protocol-Specific Strategies

- **N-drop congestion control**: In N-drop congestion control, strategy epidemic routing is used, and the node will check for congestion when a packet is received in the buffer. If there is congestion, the node will check and compare the number of packets (fq) to n, where $n = f(\text{buffer size})$. If the size of fq is bigger than the buffer size, then it will be dropped. If not, the last packet in the buffer will be dropped [5].

1.2 Static and Non-protocol-Specific Strategies

- **Drophead (FIFO)**: It has a mechanism to drop a message from the front every time the buffer is overflowing. If the buffer is full and a packet needs to be dropped, then the drophead will drop the first packet by using the first in first out mechanism [1].
- **Drop tail (LIFO)**: The mechanism used in drop tail is last in first out that is if the buffer is full, the dropping mechanism takes place by dropping the last packet entered in the buffer.
- **Equal drop (E-drop)**: In equal drop, searching of message takes place and finds if the message is equal or greater to the size of the waiting message. The message is dropped if it is found equal when the buffer is overflowed [3].
- **Threshold drop (T-drop)**: In threshold drop, a message within the calculated threshold will be dropped mechanism is used to keep the place for the next packet when the buffer is overflow.
- **Flood-Based drop (F-drop)**: Knowledge about the message during a flood is used to decide for dropping policy mechanism is used flood-based drop [6].

1.3 Dynamic and Protocol-Specific Strategies

- **Message and scheduling and drop strategy on spray and wait to route (SDSRF)**: In this priority, the mechanism is used when the buffer is overflow. According to the priority, the message is dropped [7, 8].
- **Evict least probable first (LEPR)**: The probability of message that found the least to be delivered to destination is dropped first when the buffer is overflow.
- **Message drop control (MDC)**: In message drop control, the largest message found is dropped the first mechanism is used.

1.4 Dynamic and Non-Protocol-Specific Strategies

- **Least recently forward (LRF)**: The packet in the buffer live longest will be dropped first when the buffer is overflowed. This mechanism is used when it is found difficult to drop the packet.
- **Most forwarded (MOFO)**: In most forward mechanisms, the packet that is been forwarded the most is dropped first. The local node keeps track of how many times a packet is entered in the buffer when the buffer is overflowed.
- **Source Prioritized-Based Drop**: In source prioritized-based to drop the oldest relay packet is dropped first when the buffer is overflow. A relay packet comes from another node like a broadcast message. If the incoming packet is a source node and the already packet in the buffer is packet in the buffer, then the already source packet will be dropped.
- **Global Knowledge-Based Scheduling and Drop (GBSD)**: In global knowledge scheduling, global information is used to calculate the utility of the node which helps to decide to drop packet when the buffer is overflow [9–12].
- **History-Based Scheduling and Drop (HBSD)**: History-based scheduling is an extension of global knowledge scheduling, as a substitute user local information with history is used to decide to drop a message to make place for another node when the buffer is overflowing [13, 14].

2 Related Work

In et al. [14], Qaisar has showcased various city scenarios based on buffer management policy where they encounter different problems based on various scenarios like a message not arrived on the destination if TTL expires, node unable to find a suitable carrier, the node replicates copies of message based on these cases they have explained and evaluated data structure and proposed a priority queue-based reactive buffer management strategy. It is based on the weight, size, and age of the message, if the weight is less and the size of the message is small with small age, then that message is given the highest priority to enter the buffer system.

In et al. [1], Obaid has proposed a SS drop scenario to enhance buffer management in a delay tolerant network; wherein in this mechanism, they have found the inception size of the message analyzed the size of the message compared performance and showed results of dropping concerning various metrics.

In et al. [15], G. R Sreekanth has proposed a message vector in the algorithm and explained message age in a packet scheduling scenario. In et al. [9], Rachana has showcased comparisons between various buffer management schemes using different parameters and explained various cases. In et al. [3], Fabie has showcased buffer management strategies based on static and dynamic based on the network that has knowledge about the network and that does not have any knowledge of the network and discussed the best and mostly used classification.

3 Discussion

From the above classification, we can observe that dynamic strategies are more efficient as in static, the size of the buffer is also static, and we cannot use complex scenarios with static. Also, the non-specific protocol is much better as compared to specific as in specific, the system is more customized and optimized. For generic and dynamic optimization, dynamic and non-specific dropping are required.

4 Conclusion

In this paper, we have focused on the existing research strategies of the buffer management system and classified two different types of dropping policies. The future scope of this classification is that the researchers may take this classification to another level by implementing dynamic strategies to optimize the buffer management system of the delay tolerant network.

References

1. Obaid, Irshad A, Hythem H (2021) SS-drop: a novel message drop policy to enhance buffer management in delay tolerant networks. Hindawi, Wireless Commun Mobile Comput Wiley
2. Gupta S, Poonia RC, Gao XZ (2019) Performance evaluation of flooding based routing protocol for delay tolerant networks. Int J Recent Technol Eng (IJRTE) 7(6S4)
3. Ezife F, Li W, Yang S (2017) A survey of buffer management strategies in delay tolerant networks. In: IEEE 14th international conference on mobile Ad Hoc and sensor system
4. Du Z, Wu C, Yoshinaga T, Ji Y (2018) A prophet-based DTN protocol for VANETs. In: SmartWorld, ubiquitous intelligent & computing, advanced & trusted computing scalable computing & communication

5. Kumar S, Kumar A (2018) Efficient prophet with buffer management for multicasting in DTN. In: Proceeding of the international conference on inventive research in computing application (ICIRCA)
6. Wang JS, Zhang B (2016) Cache management based on node status in delay tolerant networks. In: Advance in computer science research, 3rd international conference on wireless communication and sensor network (WCSN), vol 44
7. Sobin CC (2016) An efficient buffer management policy for DTN. In: 6th International conference on advance in computing & communication (ICACC), Elsevier, Procedia Computer Science, Cochin, India
8. Bjurefors F, Gunningberg P, Rohner C, Tavakoli S (2014) Congestion avoidance in a data-centric opportunistic network, 15 Dec 2014
9. Mhatre RR, Deshmukh M (2016) Buffer management in delay tolerant network. Int J Modern Trends Eng Res (IJMTER) 3(2)
10. Liu Y, Wang J, Zhang S, Zhou H (2011) A buffer management scheme based on message transmission status in delay tolerant networks. In: Proceeding IEEE Globecom 2011
11. Rashid S, Ayub Q, Zahid MS, Abdullah AH (2011) E: drop: an effective drop buffer management policy for DTN routing protocols. Int J Comput Appl 13(7):0975–8887
12. Li Y, Qian M, Jin D, Su L, Zeng L (2009) Adaptive optimal buffer management policies for realistic DTN. In: Proceeding IEEE Globecom
13. Krifa A, Barakat C, Spyropoulos T (2008) Optimal buffer management policies for delay tolerant network. In: Proceeding IEEE SECON
14. Ayub Q, Ngadi A, Rashid S, Habib HA (2018) Priority queue based reactive buffer management policy for delay tolerant network under city based environments. PLOS One Open Access
15. Sreekanth GR (2019) Performance improvement of DTN routing protocols with enhanced buffer management policy. Int J Recent Technol Eng (IJRTE) 8(4)
16. Wang H, Feng G, Wang H, Lv H, Zhou R (2018) RABP: delay/disruption tolerant network routing and buffer management algorithm based on weight. Int J Distrib Sensor Netw
17. Matzakos P, Spyropoulos T, Bonnet C (2016) Buffer management policies for DTN application with different QoS requirements. In: Mobile computing department, EURECOM
18. Jain S, Chawla M (2014) Survey of buffer management policies for delay tolerant networks. J Eng
19. Li Y, Zhao L, Liu Z, Liu Q (2009) N-Drop: congestion control strategy under epidemic routing in DTN. In: Research center for wireless information networks

A State of the Art Approaches to Question Generation Techniques

Khushbu Khandait and Sohel A. Bhura

Abstract Automatic question creation for educational purposes has recently piqued the interest of scholars from several fields. In this study, we look at a few different techniques to create instructional question generating software. Although there are several approaches for automatic question generating, only a few educational systems that use question generation have been created and used in real classrooms. We also provide research recommendations for using question technology in computer-assisted educational systems.

Keywords Automatic question generation (AQG) · Multiple-choice question (MCQ) · Natural language processing (NLP) · Seq2seq

1 Introduction

The principle objective of any instructive field is to further develop understudy quality. Beforehand, question papers were created physically, which took additional time and exertion with respect to instructors. This is on the grounds that making an assessment paper depends on the essayist's information, experience, and style, so we really want new cutting-edge innovation, for example, a programmed question paper age framework. Utilizing this procedure to consequently deal with unmistakable uncommon gatherings of inquiries requires less exertion (Fig. 1).

Exams are held on a regular basis in this competitive environment to assess and enhance students' knowledge and performance. The researcher examined the test procedure and determined if questions might be created automatically using a computerized application, therefore reducing the educator's workload. Natural language processing (NLP) is a branch of study and application that examines how

K. Khandait (✉)
Babasaheb Naik College of Engineering, Pusad 445215, India
e-mail: khushbukhandait1991@gmail.com

S. A. Bhura
Jhulelal Institute of Technology, Nagpur 441111, India
e-mail: sabhura@rediffmail.com

V. K. Gunjan and J. M. Zurada (eds.), *Proceedings of 3rd International Conference on Recent Trends in Machine Learning, IoT, Smart Cities and Applications*, Lecture Notes in Networks and Systems 540, https://doi.org/10.1007/978-981-19-6088-8_53

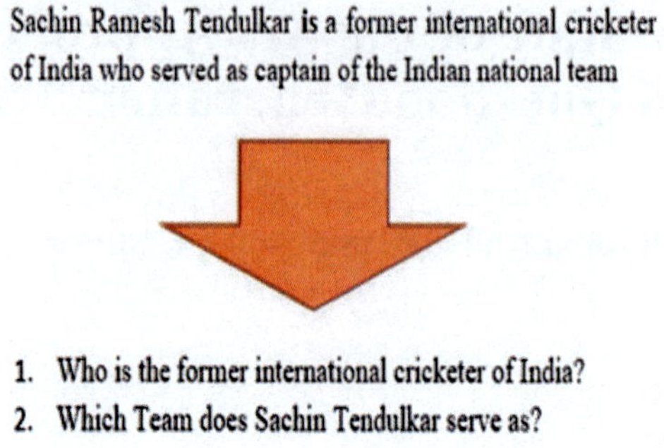

Fig. 1 Simple example of generated question from an input text file

computers can understand and change natural language text or speech to accomplish useful tasks. Humans will be able to communicate to the computer in their own language rather than having to learn a specialist language of computer commands, which offers enormous potential for designing computer interfaces that are easier to use for people. The goal of NLP researchers is to learn how humans understand automatic question generation (AQG), multiple-choice questions (MCQ), and language so that proper instruments and methods can be created to help PCs comprehend and control regular dialects to play out the assignments they are customized to do [1]. From the days of punch cards and batch processing to the age of Google, NLP research has progressed. Automatic question creation is a subset of NLP, or processing a natural language, which is a research area in which many academics have published their findings and which is continuously being researched to improve accuracy. Many scholars have worked in the subject of automatic question production using NLP, and multiple strategies and models have been created to automatically produce various sorts of questions in a variety of languages, including English, Punjabi, China, Spain, and others. Many advancements have been made in numerous industries that are moving away from manual systems and toward automated solutions. These automated methods assist us in saving money [4].

2 Literature Review

Lelkes et al. [1] created NewsQuizQA, the first dataset including 20 K human-written question–answer pairs derived from 5 K news item summaries, with the goal of generating quiz-style question answers. They presented a number of unique ways for using this dataset, apply large pre-trained transformer encoder-decoder models like PEGASUS and T5 to the tasks of question–answer generation and distractor generation. They divided the problem into two jobs as generating question–answer (QAG) and erroneous answers called distractor creation (DG). Multiple-choice questions (MCQs) creation for CBTE utilizing natural language processing (NLP) techniques is a system that assists teachers in creating multi-choice questions from text files and then providing solutions to the questions created. As a result, it may be used in any field of education to assess knowledge or skill levels. Nwafor and Onyenwe [3] described an NLP-based approach for automating MCQG in computer-based testing exams (CBTE). They did this by using a natural language processing (NLP) approach to extract keywords that are essential terms in a specific course content. Five instructional materials were utilized to test the system's efficacy and efficiency to ensure that it was not perverse. The results demonstrated that the system was capable of extracting keywords from class materials in establishing examinable questions when compared to the manually derived keywords by the teacher. Kai [4] devised a straightforward way for self-learning visual suggestions without the need for extra human annotations.

To capture these complex interactions, the authors created a novel double-clues guided graph-to-sequence learning framework that models them as a dynamic graph and learns the implicit topology end-to-end before using a graph-to-sequence model to produce questions with double hints. Experiments on the VQA2.0 and COCO-QA datasets show that this model outperforms existing state-of-the-art baselines by a substantial margin, according to the authors. The questions with answer-awareness and region-reference are generated using the DH-Graph2Seq model, which can examine the relationships among the objects in the image. Ma et al. [2] suggested a neural question generating model that included two concrete modules: sentence-level semantic matching and response location inference. They also used the answer-aware gated fusion approach to improve the decoder's starting state. They also noted that on the SQuAD and MARCO datasets, experimental findings show that the model outperforms the SOTA models. Prasetyanto et al. [5] established the AQG system, which was created to provide parallel questions for each school level's subject.

The previous method of producing questions, which could only recognize text and numeric variables, will be improved to detect picture variables and geometric types based on number variables. When provided a script, numeric, graphic, or mathematical notation, the system is meant to act like a human who can ask inquiries. The system may produce recommendations of questions by combining the stem, multipart parser algorithm, and custom media type parser algorithm in the Python-Django framework based on mathematical geometry-based input by merging the stem, multipart parser algorithm, and custom media type parser algorithm in the Python-Django framework. In this study, the notion of manipulating keywords represented by variables will be used to generate objects. Text, numeric, and picture values are stored in this variable. The designed AQG system has the capacity to alter text, numeric, and visual data in order to produce questions on the material to be examined. AQG is predicted to have a higher degree of accuracy in detecting text, numeric, documentation, and picture input which are identified, and the form of geometry-based mathematical problems is the output based on text, numeric, notation, and image input.

Neural question generation (QG) seeks to create questions from a text and a response, according to Chen et al. [6]. The previous work on generating questions either ignores the rich structural data concealed behind the text and relies simply on

CEL, cross-entropy loss, resulting in difficulties such as exposure bias and measurement inconsistencies between train and test, or fails to correctly utilize the information provided in the response. To tackle these restrictions, we present a reinforcement learning (RL)-based graph-to-sequence (Graph2Seq) model for QG in this paper. This model contains a Graph2Seq generator with a novel bidirectional gated graph neural network-based encoder to incorporate the passage and a hybrid evaluator with a mixed aim including both cross-entropy and RL losses to ensure the development of syntactically and semantically valid text. The authors also devised a powerful deep alignment network for embedding response data into the passage on a word-by-word and context-by-context basis. Liu et al. [7] presented the clue copy guided network for query production (CGC-QG), a copying mechanism in a sequence-to-sequence generative model that incorporates a number of innovative components and strategies to improve question generation presentation.

They used a multi-task labeling technique in CGC-QG to determine if a question word from the input material should be duplicated or created instead, allowing the model must learn the exact distinctions between copying and generating. Moreover, the encoder input passage accepts as input, among the variety of other variables, a clue word predictor's prediction, which aids in determining if each word in the input section might be used as a clue in the target question. The sign word indicator depends on a clever utilization of graph convolutional networks to a syntactic reliance tree portrayal of every section, permitting it to figure hint words dependent just upon their setting in the entry and relative situations to the tree's answer. They use multi-task learning and a variety of practical ways to lessen the difficulty of the clue prediction and question production. Benmalek et al. [8] demonstrated the usefulness of the mechanism of a scratchpad, a novel neural network design based on sequence-to-sequence (seq2seq), in boosting the overall confidence of seq2seq models for natural language production jobs. Scratchpad may use the encoder as a "scratchpad" memory allowing the decoder to write to all of the encoder output layers at each time step allows the decoder to maintain track of what has been generated thus far and hence guide future creation.

Scratchpad was tried with regards to three all around concentrated on natural language creation jobs: machine interpretation, question generation, and text outline, as indicated by the creators.

To solve the issue of generating questions from sentences and paragraphs, [9] proposed an adaptive copying recurrent neural network model that is unique. The suggested model incorporates a coping mechanism into an LSTM bidirectional architecture to adaptively create more appropriate queries from the data inputted to the system. To conduct autonomous question creation, the system combined a bidirectional long short-term memory network (LSTM) with a global attention mechanism. It generates questions for you automatically and copying technique onto this neural model in order to include original input vocabulary information in the decoding phase and create appropriate queries. Du et al. [10] researched the impacts of encoding sentence-level versus section-level data. Unlike earlier work, the model does not rely on rules that have been handcrafted or a complex NLP pipeline; instead, it can be trained from the beginning to end using sequence-to-sequence learning.

The approach greatly outperforms the current cutting-edge rule-based system, according to the authors. In human evaluations, machine-generated questions are also seen to be more natural (i.e., grammaticality, fluency) and more difficult to answer in terms of syntactic and lexical divergence from the original text and reasoning needed to answer. Quan et al. [11] created a graphical model for extracting fundamental concepts from various corpora and indicating significant relationships using word embedding technology. This method employs semantic match between several domain-specific terminologies to produce more perplexing distractors. This activity will be combined with employee performance assessment to give effective guidance for individual self-development as well as the company's overall employee evaluation.

The authors use word embedding technology to describe the semantic relationship between distinct domain terminologies, as well as an unsupervised extraction of core idea words from knowledge sources using a graphical model. Kim et al. [12] introduced seq2seq with answer separation, which makes greater use of information from both the passage and the target answer. Our program learns the interrogative word to employ by substituting the intended response in the original paragraph with a unique token. They also suggested keyword-net, a novel module that aids the model in capturing the relevant information in the target answer and generating a suitable inquiry. The encoder is used to convert a variable-length input sequence into a fixed-length vector that incorporates contextual characteristics and reflects interdependence among input tokens. Following that, the decoder creates an output sequence based on the encoder's output. Elsahar et al. [13] proposed a neural model for generating questions from triples in a knowledge base, in "Zero-Shot" scenario, that is, for triples comprising predicates, topic types, or object kinds that were not observed during training. To create questions, this model uses triple occurrences in the natural language corpus, as well as an innovative part-of-speech copy action mechanism, in an encoder-decoder architecture.

This is a novel neural model for generating questions from knowledge bases, focusing on predicates, topic kinds, and object types that were not observed during the training phase (Zero-Shot question generation). This approach is built on an encoder-decoder architecture that makes use of triples' textual contexts, two attention layers for triples and textual contexts, and a part-of-speech copy action. According to the authors, this strategy produces much better results for Zero-Shot QG than a collection of strong baselines, which includes KB's state-of-the-art question generation. Wang et al. [14] was a Chinese method for generating questions from an organized knowledge source. There are two pieces to this system. First, a neural creation method based on long short-term memory was used (LSTM).

Second, the system was given a new input sequence format, which improves the model's presentation. With CopyNet and attention, the seq2seq model is a full-fledged end-to-end model that can be trained with backpropagation. Batches of the model were trained. The purpose of training is to decrease cross-entropy loss (CE). Raynaud et al. [15] created a whole system to address the problem of automated question generating.

The themes and their borders were determined by the authors using the structure and content of Wikipedia. They created a template-based method to question creation that allows complicated inquiries to be generated from binary and *n*-ary expressions. To build templates automatically, the authors devised a method that returns patterns used for answering the questions, letting us to the introduction over 2000 patterns. They use Wikipedia's structural and direction-finding data to generate a list of consistent themes and rank its member units. They have proposed a method for converting answering question templates into generation question templates, permitting us to reuse a huge number of existing formats and their different summarizes, and they have changed the point/question issue into a topic/entity issue, resolving this issue by joining Wiki-data types, latent semantic analysis, and our positioned subject information base.

Liang et al. [16] suggested a model for learning to identify distractors that are comparable to those in actual test questions, which differs from other previous unsupervised ontology- and similarity-based techniques. They use trials on the dataset SciQ and their dataset MCQL to investigate ranking based on features and neural networks (NN) algorithms. *G* is a generative model that attempts to represent the conditional likelihood of creating distractors given stems and replies. *D* is a discriminative model that calculates the likelihood that a distractor sample is drawn from genuine training data rather than *G*.

3 Conclusion

We propose a review of a method for automatically generating questions from a given text in this work. As the previously stated, the automatic question generating system is constructed using a variety of algorithms and methodologies. The text is processed using NLP, and the semantic relationship is identified using NER and SRL. The majority of the work is done in English and in the form of multiple-choice questions (MCQs). Automatic question generating system is an open domain where further study is needed to provide approaches by identifying difficulties and the types of questions that need to be generated, such as one-word answers and true or false.

References

1. Lelkes AD, Tran VQ, Yu C (2021) Quiz-style question generation for news stories. In: The web conference 2021—proceedings of the world wide web conference, WWW 2021, Apr 2021, pp 2501–2511. https://doi.org/10.1145/3442381.3449892
2. Ma X, Zhu Q, Zhou Y, Li X, Wu D (2022) Improving question generation with sentence-level semantic matching and answer position inferring. Accessed 23 Feb 2022. [Online]. Available: arxiv:1912.00879v3
3. Nwafor CA, Onyenwe IE (2019) An automated multiple-choice question generation using natural language processing techniques

4. Kai S (2021) Ask question with double hints: visual question generation with answer-awareness and region-reference. In: ICLR conference, pp 1–17
5. Prasetyanto AAB, Adji TB, Hidayah I (2020) Automatic question generator system conceptual model for mathematic and geometry parallel question replication. J Phys: Conf Series 1577(1). https://doi.org/10.1088/1742-6596/1577/1/012023
6. Chen Y, Wu L, Zaki MJ (2019) Reinforcement learning based graph-to-sequence model for natural question generation, Aug 2019, [Online]. Available: http://arxiv.org/abs/1908.04942
7. Liu B et al (2019) Learning to generate questions by learning what not to generate. In: The web conference 2019—proceedings of the world wide web conference, WWW 2019, May 2019, pp 1106–1118. https://doi.org/10.1145/3308558.3313737
8. Benmalek RY, Khabsa M, Desu S, Cardie C, Banko M (2019) Keeping notes: conditional natural language generation with a scratchpad mechanism. In: Proceedings of the 57th annual meeting of the association for computational linguistics, pp 4157–4167
9. Lu X, Guo Y (2019) Learning to generate questions with adaptive copying neural networks
10. Du X, Shao J, Cardie C (2017) Learning to ask: neural question generation for reading comprehension. In: ACL 2017–55th annual meeting of the association for computational linguistics, proceedings of the conference (Long Papers), vol 1, pp 1342–1352. https://doi.org/10.18653/v1/P17-1123
11. Quan P, Shi Y, Niu L, Liu Y, Zhang T (2018) Automatic Chinese multiple-choice question generation for human resource performance appraisal. Procedia Comput Sci 139:165–172. https://doi.org/10.1016/j.procs.2018.10.235
12. Kim Y, Lee H, Shin J, Jung K (2018) Improving neural question generation using answer separation. Assoc Adv Artif 1–9. [Online]. Available: http://arxiv.org/abs/1809.02393
13. Elsahar H, Gravier C, Laforest F (2018) Zero-shot question generation from knowledge graphs for unseen predicates and entity types. In: Proceedings of NAACL-HLT, pp 218–228, [Online]. Available: http://bit.ly/2GvVHae
14. Wang H, Zhang X, Wang H (2018) A neural question generation system based on knowledge base. In: Lecture notes in computer science (including subseries lecture notes in artificial intelligence and lecture notes in bioinformatics), vol 11108 LNAI, pp 133–142. https://doi.org/10.1007/978-3-319-99495-6_12
15. Raynaud T, Subercaze J, Laforest F (2018) Thematic question generation over knowledge bases. In: Proceedings—2018 IEEE/WIC/ACM international conference on web intelligence, WI 2018, Jan 2019, pp 1–8. https://doi.org/10.1109/WI.2018.0-114
16. Liang C, Yang X, Dave N, Wham D, Pursel B, Giles CL (2018) Distractor generation for multiple choice questions using learning to rank. In: Proceedings of the thirteenth workshop on innovative use of NLP for building educational applications, pp 284–290

Solid Waste Management Using Machine Learning Algorithm for Smart Cities

Meera Khandekar, Sudhir D. Agashe, and Raunak Pawar

Abstract In big cities, waste management is becoming challenging because of variety of wastes and quantity of the waste generated. If waste management is not proper, it is directly affecting environment and human life. For systematic and effective waste management, it is important to know the quantity of waste generated in a particular area. To know the quantity of waste and types of waste generated per day in a particular area, we can use prediction algorithm. Linear regression model is developed using past available data. For varying number of household, we can predict the quantity of solid waste generated per day for a particular residential area.

Keywords Smart cities · Solid waste · Linear regression · Data analytic

1 Introduction

In big cities, waste management is becoming challenging because of variety of wastes and quantity of the waste generated. If waste management is not proper, it is directly affecting environment and human life. Solid waste management is important to save planet earth!!!

As per the available information, "the quantity and nutritional quality of food support human health, and 95% of food production depends on soils [1]. Only healthy soils can provide the needed ecosystem services and secure supplies of more food and fiber. Soil pollution reduces food security both by reducing crop yields due to toxic levels of contaminants and by causing the produced crops to be unsafe for

M. Khandekar (✉) · S. D. Agashe · R. Pawar
Department of Instrumentation and Control Engineering, College of Engineering, Pune 411005, India
e-mail: mgj.instru@coep.ac.in

S. D. Agashe
e-mail: sda.instru@coep.ac.in

R. Pawar
e-mail: pawarrr20.instru@coep.ac.in

V. K. Gunjan and J. M. Zurada (eds.), *Proceedings of 3rd International Conference on Recent Trends in Machine Learning, IoT, Smart Cities and Applications*, Lecture Notes in Networks and Systems 540, https://doi.org/10.1007/978-981-19-6088-8_54

consumption [2]. Unsafe water kills more people each year than war and all other forms of violence combined. Our drinkable water sources are finite: Less than 1% of the earth freshwater is actually accessible to us. Nearly, half of our rivers, streams, and more than one-third of our lakes are polluted and unfit for swimming, fishing, and drinking. Nutrient pollution, which includes nitrates and phosphates, is the leading type of contamination in these freshwater sources."

Urbanization can also inevitably bring about a variety of negative effects, resulting in more challenges and problems faced by cities: The ecological environment has been devastated; natural resources are gradually being depleted; pollution (air, water, and sound) is increasing; and infectious diseases and cancer cases are growing; criminal activities remain rampant; and so on [3]. It is reported according to statistics that, in the 1950s, only about 30% of the world's population lived in cities. By 2014, the level of global urbanization has reached 54%. The United Nations (UN) has analyzed and predicted the urbanization problem of the world's population, predicting that by 2050, the proportion of urbanization around the world will reach 66% [4].

Pinar Oguz-Ekim implemented three different machine learning algorithms, namely backpropagation neural network (BPNN), support vector regression (SVR), and general regression neural network for prediction of waste [5]. Drinking water quality parameters like pH and turbidity are very important. These parameters are measured in real time using the sensors that send the data to monitoring room/control station (base station). For continuous monitoring, sensor nodes with networking ability are installed using wireless sensor network for water quality monitoring [6]. Water treatment plant operation can be improved with support of machine learning techniques. Water filter clogging may affect water quality, water filtration capacity, and caused unplanned down time. Based on real-time data, data-driven model is designed to predict the filter bed backwash time. Decision tree regression algorithm has been used to predict remaining backwash time of water filter bed [7] (Fig. 1).

Municipal and industrial waste discharges contribute their fair share of toxins. Industry and individuals dump all the random junk directly into waterways. So it is time to think what can be done with waste management in big cities? It is important

Fig. 1 Municipal dumping landfill

to know quantity of solid waste generated per day in the big city. It is important to know types of wastes generated per day. To know the quantity of waste and types of waste generated per day in a particular area, we can use prediction algorithm. Linear regression model is developed using past available data. For varying number of household, we can predict the quantity of solid waste generated per day. Section two discusses about possible types of waste in a city and its processing techniques. Section three discusses about implementation of algorithm for prediction of solid waste for a particular residential area.

2 Types of Solid Waste

We can classify waste as biomedical waste, biodegradable waste, sanitary waste, e-waste, plastic waste, industry waste, building, and structure waste.

2.1 Biomedical Waste

Every big city has around 25–30 hospitals which are in service of the society. Significant amount of biomedical waste is generated per day; we need to handle biomedical waste separately and safely. Separate transport system can be used to collect biomedical waste from the hospitals. Biomedical waste management rule was defined in 1998 with an amendment in 2003. The organizations generating biomedical waste are accountable of ensuring that all biomedical waste is segregated, transported, processes, and disposed off without any adverse effect to human health and the environment.

2.2 Biodegradable Waste

Any municipal corporation area has nearly 20 vegetable markets. Vegetable market waste is biodegradable, and we can generate compost form the waste. Around 40 big temples are in the big city area, where for worship flowers garlands are used. Waste from the temples is also biodegradable. Around 50 hotels are giving service to the society. Food waste is generated at the hotels every day. The quantity of biodegradable waste generating per day is significant. A dedicated transport system can be used for effective collection of waste from temples, vegetable markets, and hotels, so that waste processing can be done. Composting and vermicompost process can be successfully used for biodegradable waste management. Composting is a process where the organic solid waste is rotten under controlled conditions. Microbes metabolize the organic solid waste, and its volume is reduced by almost 50%. The final product is called compost. It looks like soil in consistency and odor and may be used as a soil conditioner. Vermicompost is the outcome of the decomposition operation using a variety of species of worms, usually red wigglers, white worms, and other earthworms. Vermicompost comprises water-soluble nutrients and is nutrient-rich organic manure and soil conditioner.

2.3 E-waste

Computers, televisions, printers, stereos, copiers, and fax machines are everyday electronic products. Every household has many electronic appliances like laptops, mobiles, and charging batteries. Electronic waste contains toxic components such as mercury, lead, cadmium, polybrominated flame retardants, barium, and lithium. Improperly disposed electronics in landfills release toxic chemicals, affecting the earth's air, soil, water and ultimately, human health. Unused laptops, mobiles, chargers, batteries, and electronic appliances produce substantial amount of e-waste. There should be e-waste collection drive for housing societies. So that e-waste processing will be more systematic and effective.

2.4 Plastic Waste

Plastic has become integral part of our daily life. Plastic waste is the accumulation of plastic objects into the environment that adversely affects wildlife, wildlife habitat, and humans. Improper management of landfills will make way for harmful chemicals in plastic wastes to leach into the environment, polluting the soil, air, and underground water. Plastic waste collection, treatment, and disposal process must be defined by local authorities. Authorities must make sure all the defined process must be practiced in everyday life. There should be plastic waste collection drive for housing societies. So that plastic waste processing will be more systematic and efficient. Plastic waste is transported to pyrolysis plant where it is converted into diesel and gasoline. Through pyrolysis, we can handle wide variety of plastic forms. Shredded plastic is stored in reactor. After that catalyst is applied, and plastic is heated upto 150 °F. The released gases, such as methane and propane, are stored in separate gas tank and used as source of heat for the system to operate. The final oil product is refined and processed.

It is not possible to recycle all types of solid waste, and there will always be remains from treatment processes that will eventually require disposal underground. It is becoming challenging to find sites that offer required capacity, accessibility, and environmental conditions.

As per CPCB and NEERI reports, solid waste collected by corporation follows distribution percentage of waste as shown in graph [8] of Fig. 2.

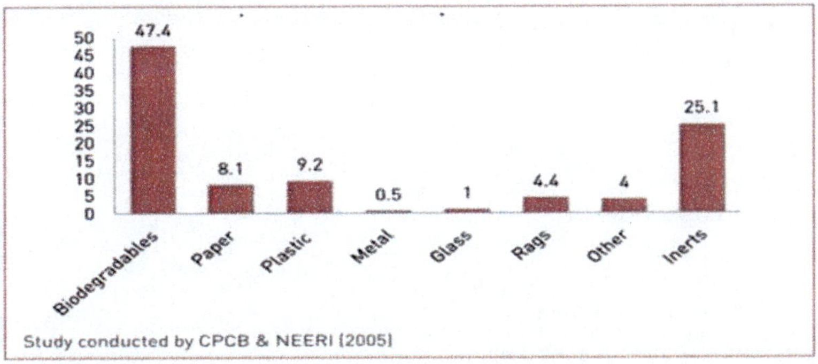

Fig. 2 Percentage of waste

3 Case Study

Let us do a case study to understand how much solid waste is generated per day for Pimpri Chinchwad Municipal Corporation (PCMC). Table 1 shows sample data available for solid waste generated per day for PCMC. Data like total zone A to H, different wards, number of households in a ward and door to door (D2D) collection status of garbage from households are available on PCMC portal. For solid waste collection, different types of community bins, compactor bins, and dhalaos are provided for residential areas. For effective collection of waste, it is important to understand what types of wastes are generated in a city area. Data analytic can be used to predict total solid waste generated per day for given number of households for particular residential area. In addition to that we can have segregation for types of waste like biodegradable, metal, plastic, paper waste, etc. Once we understand the total generated waste per day, we can use proper technology for processing the waste.

Table 1 Sample data available from PCMC

Zone name	Ward number	Total no. of household established	D2D collection status	No. of household with waste segregation	Waste quantity (TPD)	No. of street bins
A	10	18,521	17,924	14,339	33	43
A	14	18,910	16,106	12,884	29	27
A	15	20,567	17,500	14,000	32	95
A	19	22,331	22,000	17,600	40	69

Table 2 Regression model performance evaluation

R^2 score	0.807
Adjusted R^2 score	0.71
Sum of squared error (SSE)	68.78
Mean squared error (MSE)	6.878
Root mean squared error (RMSE)	2.622

Table 3 Solid waste segregation for predicted waste quantity

No. of HH	Predicted solid waste quantity (TPD)	Solid waste segregation							
		Biodegradable	Paper	Plastic	Metal	Glass	Rags	Other	Inert
35,000	53.23	25.23	4.31	4.9	0.27	0.53	2.34	2.13	13.36
40,000	59.91	28.4	4.85	5.51	0.3	0.6	2.64	2.4	15.04
45,000	66.59	31.56	5.39	6.13	0.33	0.67	2.93	2.66	16.71
50,000	73.26	34.73	5.93	6.74	0.37	0.73	3.22	2.93	18.39

4 Results

Data analytic is carried out for available data set, which identify trends and patterns in the data. Linear regression machine learning algorithms are used for prediction of quantity of solid waste generated per day. To develop linear regression model, Jupyter notebook platform has been used. *R* square and adjusted *R* values are observed to know how good regression model is. After data analytic, the following table is prepared. If a specific PCMC ward included, number of households are 35,000, 40,000, 45,000, and 50,000 predictions for quantity of solid waste generated per day can be available using developed regression model. Knowing the exact waste generation per day helps to manage solid waste effectively. To evaluate model performance, mean squared error (MSE) and root mean squared error (RMSE) have been observed as depicted in Table 2.

As per CPCB and NEERI reports, distribution percentage of solid waste for predicted quantity of waste generated per day is given in Table 3.

5 Conclusion

For systematic and effective waste management, it is important to know the quantity of waste generated in a particular area. If a specific PCMC ward included, number of households is 35,000, 40,000, 45,000, and 50,000; predictions for quantity of solid waste generated per day can be available using developed regression model.

In addition to that we can have segregation for types of wastes like biodegradable, metal, plastic, paper waste, etc. Once we understand the total generated waste, we can use proper technology for processing the waste.

Acknowledgements We are thankful to Pimpri Chinchwad Municipal Corporation to make available the data for solid waste management.

References

1. Oliver MA, Gregory PJ (2015) Soil, food security and human health: a review. Eur J Soil Sci
2. Food and agriculture Organization of the United Nations and Intergovernmental Technical Panel on soils, Rome, Italy. FAO ITPS (2015) Status of the world's soil resources (SWSR). http://www.fao.org/3/a-i5199e.pdf
3. Song Y, Ding CR (2009) Smart urban growth for China. Lincoln Institute of Land Policy Cambridge. Washington, DC, USA
4. Zvolska L, Lehner M, Voytenko PY, Mont O, Plepys A (2019) Urban sharing in smart cities: the cases of Berlin and London. Local Environ 24(7):628–645
5. Oguz-Ekim P (2021) Machine learning approaches for municipal solid waste generation forecasting. Environ Eng Sci 38
6. Suryawanshi V, Khandekar M (2018) Design and development of wireless sensor network (WSN) for water quality monitoring using Zigbee. In: Proceedings of the second international conference on intelligent computing and control systems (ICICCS 2018)
7. Kale SJ, Khandekar MA, Agahse SD (2021) Prediction of water filter bed backwash time for water treatment plant using machine learning algorithm. In: 2021 IEEE 6th international conference on computing, communication and automation (ICCCA). Aurel Vlaicu University of Arad, Romania, Dec 17–19, 2021
8. Management of waste in India, CAG environment report

Estimation of the Variance of Parameters in a Model Proposed to Study Prognosis of Lung Cancer

Manjula S. Dalabanjan and K. R. Nataraj

Abstract Cancer is one of the major health problems persisting worldwide. The data for the prognosis of cancer is taken from the National Cancer Registry Program (www.ncrpindia.org.in) [1]. We have analyzed the underlying pattern of distribution of incidence rates of lung cancer in males for the two regions such as Bengaluru and Mumbai and fitted model A by observing the pattern of the incidence rates of lung cancer in males. By intuition; we divided the data into 2 groups. For Group 1, the second-degree equation fitted well. For Group 2, the cubic spline model fitted well. The estimation of parameters involved in both Group 1 and Group 2 was estimated by using least squares method. Expressions for the variance of parameters of second-degree curves were derived.

Keywords Incidence rates · Risk · Fit · Parameter · Residues · Cancer · Regions

1 Introduction

The underlying pattern of distribution of incidence of lung cancer in males was studied. There were no cases of lung cancer observed for persons aged below 20 years. Starting from 20 to 60 years of age, the secondary data available were classified according to age, in 5 years age groups [1].

Second-degree equation was observed to be a best fit for male young adults of age ranging from 20 to 60 years. In the old adult's age groups 60–64, 65–69, 70–74, and 75+, there was a downturn in the incidence of lung cancer [2].

M. S. Dalabanjan (✉) · K. R. Nataraj
Department of Mathematics, Don Bosco Institute of Technology, Bengaluru, India
e-mail: manjulasd@dbit.co.in

K. R. Nataraj
e-mail: director.research@dbit.co.in

M. S. Dalabanjan
Department of Research and Development, Don Bosco Institute of Technology, Bengaluru, India

V. K. Gunjan and J. M. Zurada (eds.), *Proceedings of 3rd International Conference on Recent Trends in Machine Learning, IoT, Smart Cities and Applications*, Lecture Notes in Networks and Systems 540, https://doi.org/10.1007/978-981-19-6088-8_55

Kenneth et al. [3] analyzed the population and mortality count for male and female cohorts. The data collected were evaluated according to the distribution of age and hazard rate. Twelve models were fitted. The likelihood ratio χ^2—test statistics for all the models was calculated. Among these 12 models, 8 models were rejected because of heterogeneity.

Tetsuji et al. analyzed the data collected from 47 prefectures yearly from 1975 to 2002 on large bowel cancer mortality in Japanese males. They introduced a non-parametric model with time-varying mixed effects for cancer and constructed parameter estimators based on a local linear approximation. The proposed non-parametric regression model was applied to the data collected [3].

Christopher et al. [4] used the data from Australian Institute of Health Welfare (AIHW) published in AIHW (2006) a study of major causes of death in Australia during twentieth century. Generally, five year age intervals were used, but they obtained a selection of numbers of deaths owing to heart or circulatory disease, cancer, and other causes by single years for selected charts. The data used are of 31,147 women aged 74 in 1968 and 105 in 1999 and are classified by cause of death in 3 groups—Group 1 being those alive in 1968 but ultimately died of heart or circulatory disease; Group 2 being those alive in 1968 but untimely died of cancer and Group 3 comprising of the remainder of 31,147 women dying of other causes. They observed that in all three cases, Gompertz model is reasonable up to age 100.

Balgobin et al. [2, 5] used the geographical units, Health Survey Areas (HSAs), as in the Atlas (Pickle et al. 1996) which included the deaths of residents in the contiguous 48 states during 1988–1992. Here, they started investigating alternative models for inference about age-specific and age-adjusted mortality rates for chronic obstructive pulmonary disease (COPD).

Next, we observed the pattern of incidence of cancer for old adults (persons above 60 years). In the older ages, the rate of incidence of cancer was observed to be deteriorating. This may be due to an increase in the rate of mortality in older age groups in general.

2 Model A and Notations

Thus, let us classify the data into two groups. Let Group 1 consists of young male adults, age ranging from 20 to 60 years, and Group 2 consists of old male adults, aged above 60 year.

Let l_{1ij}, n_{1ij}, and λ_{1ij} denote, respectively, the number of males suffering from cancer, population at risk, and the observed rate of incidence for Group 1.

$$\lambda_{1ij} = \frac{l_{1ij}}{n_{1ij}} \text{for age class } j \text{ and in the region } i.$$

Let l_{2ij}, n_{2ij}, and λ_{2ij} denote, respectively, the number of males suffering from cancer, population at risk, and the observed rate of incidence for Group 2.

$$\lambda_{2ij} = \frac{l_{2ij}}{n_{2ij}} \text{for age class } j \text{ and in region } i.$$

We fitted a cubic spline model for old adults' age group 60–64, 65–69, 70–74, and 75+, which is a good fit. The estimated values obtained were almost coinciding with observed values. Thus if x_{ij} denotes age, then the estimate of the incidence of lung cancer for males in the region *i* and age group *j* is given by the following model A.

In the age group 0–20, the incidence rates of lung cancer in males for the 2 regions such as Bengaluru and Mumbai are negligible. The parameters a_{1i}, b_{1i}, c_{1i} and $a_{2i}, b_{2i}, c_{2i}, d_{2i}$ can be estimated by using the least squares method for region *i* and for Group 1 and Group 2, respectively. Instead of using the least squares method, we can also use the least absolute sum of squared deviations method which gives better estimates of a_{1i}, b_{1i}, c_{1i} and $a_{2i}, b_{2i}, c_{2i}, d_{2i}$. However, the estimates obtained by using the least squares method are unbiased (Tables 1, 2, 3, and 4; Figs. 1 and 2).

$$\lambda_{0ij} = 0 \text{ if } x_{ij} < 20 \quad \text{(Group 0, which is excluded fromour study)}$$

$$\lambda_{1ij} = a_{1i}x_{ij}^2 + b_{1i}x_{ij} + c_{1i} + \alpha_{1ij} \text{ if } 20 \le x_{ij} < 60 \text{ for } i = 1, 2, 3, 4; \\ j = 1, 2, 3, 4, 5, 6, 7, 8$$

$$\lambda_{2ij} = a_{2i}x_{ij}^3 + b_{2i}x_{ij}^2 + c_{2i}x_{ij} + d_{2i} + \beta_{2ij} \text{ if } x_{ij} \ge 60 \text{ for } i = 1, 2, 3, 4; \\ j = 9, 10, 11, 12$$

The standardized cross-validation residuals for Group1 and Group 2 are given by the following test statistic.

$$r_{1ij} = \frac{\lambda_{1ij} - E[\lambda_{1ij}]}{\sqrt{\text{var}(\lambda_{1ij})}} \text{ and } r_{2ij} = \frac{\lambda_{2ij} - E[\lambda_{2ij}]}{\sqrt{\text{var}(\lambda_{2ij})}}$$

Table 1 Bengaluru region, Group 1

j	Age	λ_{11j}	$E[\lambda_{11j}]$	Residues	r_{11j}
1	27.5	8.250261871	25.60265	−17.35238813	0.09062
2	32.5	15.34074087	−1.95675	33.85291187	0.17679
3	37.5	32.08494743	16.35225	−17.35238813	0.09062
4	42.5	84.37869157	80.52965	17.29749087	0.09033
5	47.5	172.1532547	190.57545	15.73269743	0.08216
6	52.5	326.5121594	346.48965	3.849041568	0.02010
7	57.5	567.151784	548.27225	−18.42219535	0.09620

Table 2 Bengaluru region, Group 2

j	Age	λ_{21j}	$E[\lambda_{21j}]$	Residues	r_{21j}
8	62.5	658.8827486	657.5684813	1.3142673	0.01737
9	67.5	828.9381394	864.4535438	−35.5154044	0.4696
10	72.5	759.850481	724.3348563	35.51562478	0.4696
11	77.5	719.6248665	720.9399188	−1.315052229	0.01737

Table 3 Mumbai region and Group 1

j	Age	λ_{11j}	$E[\lambda_{11j}]$	Residue	r_{11j}
1	27.5	4.316	15.86116	−11.54521	0.14043
2	32.5	6.588392	−2.26344	4.324993	0.05261
3	37.5	9.357483	1.46225	7.89533	0.09603
4	42.5	32.746406	27.03773	5.70876	0.06944
5	47.5	80.583533	71.00152	9.58215	0.11655
6	52.5	105.401240	143.73874	−38.33742	0.46631
7	27.5	255.151297	234.86425	20.28714	0.24676

Table 4 Mumbai region and Group 2

j	Age	λ_{21j}	$E[\lambda_{21j}]$	Residue	r_{21j}
8	62.5	409.648698	409.59125	0.05735	0.00038
9	67.5	596.8260789	596.65125	0.17482	0.00117
10	72.5	762.448708	762.28125	0.16746	0.00111
11	77.5	780.3826662	780.48125	0.09858	0.00066

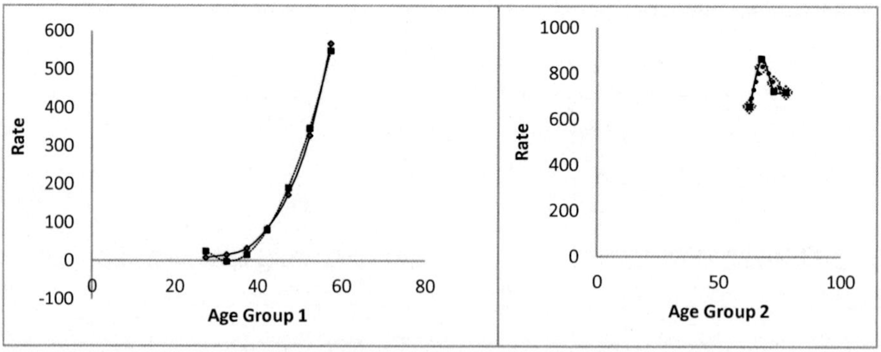

Fig. 1 Pattern of observed and estimated incident rates in Bengaluru

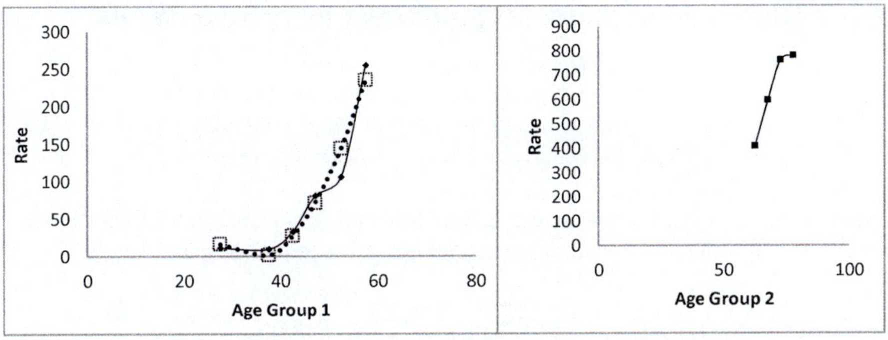

………. Estimated rates ________Observed rates

Fig. 2 Pattern of observed and estimated rates in Mumbai

Both residuals are assumed to be normally distributed. Thus, we consider $\alpha_{1ij} \approx N\left(0, \sigma_{1i}^2\right)$ and $\beta_{2ij} \approx N\left(0, \sigma_{2i}^2\right)$ (Tables 1, 2, 3 and 4; Figs. 1 and 2).

3 Expressions for the Variance of Parameters

The simple linear fit for data is given by $y_i = ax_i + b + \varepsilon_i$. Here, y_i is called as the response variable or dependent variable or uncontrolled variable. x_i is a deterministic variable or independent variable or controlled variable. ε_i is a random error component. Now, we can make certain basic assumptions that for the model

$$y_i = ax_i + b + \varepsilon_i \quad i = 1, 2, 3, \ldots n$$

(i) $E(\varepsilon_i) = 0,\ \mathrm{Var}(\varepsilon_i) = \sigma^2,\ \mathrm{Cov}\left(\varepsilon_i, \varepsilon_j\right) = 0,$ for $i \neq j$
(ii) $\varepsilon_i \approx N\left(0, \sigma^2\right)$
(iii) x_i is a not a r.v

$$E(y_i) = E(ax_i + b + \varepsilon_i) = ax_i + b$$

$$\mathrm{Var}(y_i) = \mathrm{Var}(ax_i + b + \varepsilon_i) = \mathrm{Var}(\varepsilon_i) = \sigma^2 \tag{1}$$

Therefore, $y_i \approx N\left(ax_i + b,\ \sigma^2\right)$

3.1 Derivation of Variance of Parameters for a Parabolic Fit

For the parabola, $y_i = ax_i^2 + bx_i + c$ e a, the variance of parameters *a, b, and c* is estimated by constructing normal equations. Shift the origin of variable x_i to $(x_i - \overline{x})$. Let $(x_i - \overline{x}) = X_i$. Thus, the value of c changes to C (say). Here, $C = c + K_1(x)$. Therefore, $\mathrm{Var}(C) = \mathrm{Var}(c)$.

Since the residuals of fitted quadratic equation and cubic equation follow normal distribution with mean zero and variance. Thus, normal equations are

$$\sum_{i=1}^{n} y_i = a\sum_{i=1}^{n} X_i^2 + b\sum_{i=1}^{n} X_i + nC \tag{2}$$

$$\sum_{i=1}^{n} X_i y_i = a\sum_{i=1}^{n} X_i^3 + b\sum_{i=1}^{n} X_i^2 + C\sum_{i=1}^{n} X_i \tag{3}$$

$$\sum_{i=1}^{n} X_i^2 y_i = a\sum_{i=1}^{n} X_i^4 + b\sum_{i=1}^{n} X_i^3 + C\sum_{i=1}^{n} X_i^4 \tag{4}$$

Using the fact that

$$\sum_{i=1}^{n} X_i = 0 \quad \text{and} \quad \sum_{i=1}^{n} X_i^3 = 0$$

and solving the above normal equations, we get

$$b = \frac{\sum_{i=1}^{n} X_i y_i}{\sum_{i=1}^{n} X_i^2}$$

$$\mathrm{Var}(b) = \frac{\sigma^2}{\sum_{i=1}^{n} X_i^2} \tag{5}$$

$$a = \frac{\sum X_i^2 \sum y_i - n\sum X_i^2 y_i}{\left(\sum X_i^2\right)^2 - n\sum X_i^4}$$

$$\mathrm{Var}(a) = \frac{\sigma^2}{\sum_{i=1}^{n}\left(X_i^2 - \overline{X^2}\right)^2} \tag{6}$$

$$nC = \sum y_i - a\sum X_i^2$$

$$\mathrm{Var}(c) = \frac{\sigma^2}{n}\left[n - \frac{\left(\sum_{i=1}^{n} X_i^2\right)}{\sum_{i=1}^{n}\left(X_i^2 - \overline{X}^2\right)^2}\right]$$

4 Results and Conclusion

We find that the values of all standardized cross-validation residual are less than 1.96 for all the four regions and both groups. Thus, we can conclude that the second-degree model fits well for Group 1, and cubic spline model fits well for Group 2.

For Bengaluru region, the probability that a person having lung cancer belongs to Group 1 is around (4/10) times the probability that he belongs to Group 2. For Group 1, the odds that a person is susceptible to lung cancer to that he is not susceptible to lung cancer is 0.00120. For Group 2, the odds that a person is susceptible to lung cancer to that he is not susceptible to lung cancer is 0.0299.

For Mumbai region, the probability that a person having lung cancer belongs to Group 1 is around 0.1938 (1/5) times the probability that he belongs to Group 2. For Group 1, the odds that a person is susceptible to lung cancer to that of he is not susceptible to lung cancer is 0.000494. For Group 2, the odds that a person is susceptible to lung cancer to that of he is not susceptible to lung cancer is 0.002555.

References

1. www.ncrpindia.org.in
2. Nandram B, Sedransk J, Pickle LW (2000) Bayesian analysis and mapping of mortality rates for chronic obstructive pulmonary disease. J Am Stat Assoc 95(452), Application and Case studies (Dec 2000)
3. Tonda T, Satoh K, Nakayama T, Katanoda K, Sobue T, Ohtaki M (2011) A non-papRametric mixed-effects model for cancer mortality. Aust NZ J Stat 247–256
4. Heathcote CR, Puza BD, Roberts SP (2009) The use of aggregate data to estimate Gompertz—type old-age mortality in heterogeneous populations. Aust NZ J Stat 51(4):481–497
5. Nandram B, Sedransk J, Pickle L (1999) Bayesian analysis of mortality rates for U.S. health service areas. Indian J Stat 61(Series B)

Mono Versus Multilingual BERT: A Case Study in Hindi and Marathi Named Entity Recognition

Onkar Litake, Maithili Sabane, Parth Patil, Aparna Ranade, and Raviraj Joshi

Abstract Named entity recognition (NER) is the process of recognizing and classifying important information (entities) in text. Proper nouns, such as a person's name, an organization's name, or a location's name, are examples of entities. The NER is one of the important modules in applications like human resources, customer support, search engines, content classification, and academia. In this work, we consider NER for low-resource Indian languages like Hindi and Marathi. The transformer-based models have been widely used for NER tasks. We consider different variations of BERT like base-BERT, RoBERTa, and AlBERT and benchmark them on publicly available Hindi and Marathi NER datasets. We provide an exhaustive comparison of different monolingual and multilingual transformer-based models and establish simple baselines currently missing in the literature. We show that the monolingual MahaRoBERTa model performs the best for Marathi NER whereas the multilingual XLM-RoBERTa performs the best for Hindi NER. We also perform cross-language evaluation and present mixed observations.

1 Introduction

Named Entity Recognition [1], a term coined in 1995, refers to a popular technique of the information extraction process in natural language processing. It is a two-step process that involves (a) detection of a named entity and (b) categorization of the entity. These categories include a myriad of entities like names of persons, locations, organizations, numerical expressions like percentages, monetary values, and temporal values like date, time. The applications of these entity recognitions include text summarization [2], customer support [3], machine translation [4], efficient search algorithms [5], etc.

O. Litake (✉) · M. Sabane · P. Patil · A. Ranade
SCTR's Pune Institute of Computer Technology, L3Cube, Pune, Pune, India
e-mail: onkarlitake@ieee.org

R. Joshi
Indian Institute of Technology Madras, L3Cube Pune, Pune, India

V. K. Gunjan and J. M. Zurada (eds.), *Proceedings of 3rd International Conference on Recent Trends in Machine Learning, IoT, Smart Cities and Applications*, Lecture Notes in Networks and Systems 540, https://doi.org/10.1007/978-981-19-6088-8_56

The process of NER can be performed in numerous ways. The English language has a large body of NER literature; however, very few efforts have been recorded in the Hindi and Marathi language. It is because of the deficit of well-annotated corpus and available tools. Additionally, these languages have various intricacies like the lack of capitalization and the equivocation between proper nouns and common nouns, i.e., the Indian languages contain a myriad of common words which can be used as proper nouns. Furthermore, Hindi lacks a rigid grammar pattern. These shortcomings make it difficult to use the existing deep learning approaches that have successfully been used in English processing, to be used for these languages. Along with that, standard tools like Stanford NER [6] and other POS and NER taggers do not have support for Hindi and Marathi. Recently, Hindi and Marathi text classification has received some attention [7–11].

Named Entity Recognition can mainly be performed using three major approaches. These include NER using Machine Learning [12], Rule-based NER [13], and Hybrid NER [14]. NER using Machine learning involves building a model using tagged text. Some examples include Conditional Random Fields (CRF) [15], Support Vector Machine (SVM) [16], Hidden Markov Model [17] (HMM), and tools like Spacy [18] and Stanford NER. Rule-based NER uses rules defined typically by linguists. It includes Lexicalized grammar, Gazetteer list, list of triggered words, and so on. Hybrid NER uses an amalgamation of machine learning and rule-based approaches. It could be a combination of HMM model with CRF or the Gazetteer method with HMM, etc.

Deep Learning [19] is becoming increasingly popular as a result of its superior accuracy when trained with large amounts of data. Its architecture is also adaptable to new challenges. Deep Learning approaches outperform others when domain awareness is lacking for feature introspection since feature engineering is less of a concern and these techniques tend to solve problems end to end. Transformer-based systems have become increasingly popular in recent years due to their highly efficient architectures [20, 21]. In this paper, we try to establish baseline numbers on various publicly available datasets for Marathi and Hindi languages by training various transformer architectures. We make use of pre-trained BERT-based masked language models. The multilingual variants of these language models have been very popular recently for low-resource languages. We also try to provide a comparative study of multilingual and monolingual variants of these language models. The monolingual variants are only pre-trained on Hindi or Marathi data. We are focusing on transformer architectures like BERT[22], RoBERTa [23], and their variants such as mBERT, RoBERTa-Hindi, Indic Bert [24], mahaBERT [7] etc. We also perform a cross-language evaluation of these BERT models since both Hindi and Marathi share the Devanagri script. The main contribution of this work is as follows:

- We show that monolingual Marathi models based on mahaBERT perform better than their multilingual counterpart thus showing the importance of language-specific pre-training.
- For the Hindi language the multilingual models perform better hence there is a need to develop better monolingual language models.

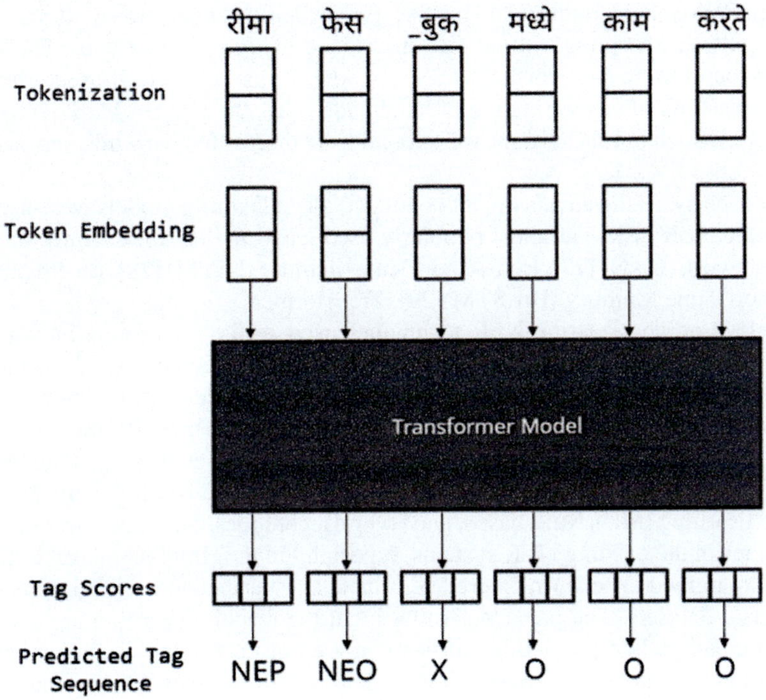

Fig. 1 Model architecture

- During cross-language evaluation the results favor Marathi models. The Marathi monolingual models based on the mahaBERT scale well on Hindi NER datasets but the same is not true for publicly available Hindi BERT models as they perform poorly on Marathi NER datasets. Again highlights the need to have better resources for Hindi.

The rest of the paper is structured as follows. Section 2 surveys the advancement in Named Entity Recognition and focuses on Indian languages. Section 3 explains how we set up our experiments to test various models. Section 4 summarizes the findings from all of the experiments. Our paper's conclusion is presented in Sect. 5.

2 Related Work

The concept of Named Entity Recognition originated in 1995 at the Message Understanding Conferences (MUC) [1] in the US. However, it was not until 2008 that the study on Indian languages received widespread recognition. Krishnarao et al. [25] presented a comparative study based on two algorithms viz. Support Vector

Machine (SVM) and Conditional Random Field(CRF). The paper provides a comparison between these algorithms that are trained using the same data. The CRF model is found to be superior to the SVM model. Srihari et al. [26] used a hybrid system consisting of a combination of handmade rules, the Hidden Markov Model, and MaxEnt. Such hybrid systems were found to be more effective while performing NER.

Subsequently, as the advancement continued, deep learning models were used to perform the NER task. The most popularly used NER models were Convolutional Neural Network (CNN) [27], Long-Short Time Memory (LSTM) [28], Bi-directional Long-Short Time Memory (BiLSTM)[29], Transformers.

Shah [30] emphasizes the NER techniques used so far for various Indian languages. The paper provides a comparative analysis of the methods used for identifying named entities. It compares the Hidden Markov Model (HMM) method and the Conditional Random Field (CRF) method and finds that the CRF method is the most effective approach for Indian languages. Bhattacharjee et al. [31] examine various techniques for NER in Indian languages, with a focus on Hindi. It compares the Machine Learning (ML), Rule-based, and Hybrid techniques. The paper aims to identify the gaps in the existing NER systems, especially in the Hindi language, as these systems are trained to perform on predetermined datasets and do not produce results on universal datasets. The paper determines that the machine learning approach is more systematic while predicting entities that are not known, but it showcases an accuracy lower than the rule-based system. Patil [32] discusses the importance of NE recognition for Marathi along with the concerns and obstacles that come with NE recognition in the Marathi language. It also looks at various methodologies and techniques for creating learning resources that are necessary for extracting NEs from unstructured natural language material.

Further, in the deep learning domain, a variety of models have been proposed and tested to perform the NER tasks on Indic languages including SVM by Singh et al. [33], Conditional Random Field by Shishtla et al. [34], and Hidden Markov model. But with the advancement of deep learning architectures, methods were proposed to identify entities from text without adhering to language-specific rules. However, Shah et al. [30] and Shelke et al. [35] have illustrated encouraging results by utilizing BiLSTM networks to simplify the NER complexities. Our work builds upon theirs and adds other models that are trained to obtain the required accuracies. Additionally, we have performed the NER task on the Marathi dataset.

Murthy et al. [36] showcases the influence of the differences in tag distributions of common named entities between primary and helping languages on the efficacy of multilingual learning. The paper proposes a measure based on symmetric KL divergence using neural networks like CNNs and Bi-LSTMs to filter out the highly divergent training examples in the helping language to solve this challenge.

3 Experimental Setup

3.1 Dataset

We are limiting our work of performing Named Entity Relation(NER) tasks on Hindi and Marathi which are among the top 3 languages spoken in India. We are carrying out NER tasks on all publicly available datasets for these languages [37].

For Hindi, we are using datasets released in IJCNLP [38] in 2008 and WikiAnn NER Corpus released by Pan et al. [39] in 2017. The IJCNLP dataset contains a total of 11,400 sentences. It has a total of 12 categories named as a person, organization, location, abbreviation, brand, title-person, title-object, time, number, measure, designation, terms. No split for the data was provided, hence we split the data into 70-15-15 train, test, tune respectively. We have corrected a few tags which were improperly annotated. For example, The 'Term' entity is ununiformly tagged throughout the dataset. It is tagged using 'B-NETE', 'I-NETE', 'B- nete', 'B-N ETE', 'I-NETE/=', and many more such tags. We have replaced all of them with either 'B-NETE', or 'I-NETE' to keep the tagging uniform. We have also discarded ambiguous tags and replaced them with 'O' tags. Following is the list of such tags: 'B- NET', 'B-NEB', 'B-NET/=', 'B- NET', 'B-NEB', 'B-NET/=', 'B-NTA', 'B-Terms', 'B-k1', 'I- NET', 'I-NEB', 'I-NET/=', 'I-k1'. The dataset released by Pan et al. for Hindi contains a total of 11,833 sentences and has been divided into 3 categories namely Organization, Person and Location. It is a "silver-standard" dataset.

For Marathi, we are using a dataset released by Murthy et al. [36] named IIT Bombay Marathi NER Corpus in 2018 and WikiAnn NER Corpus released by Pan et al. [39] in 2017. The dataset contains a total of 5,591 sentences. It has a total of 3 categories named Location, Person, Organization. Train-Test-Tune split of the dataset was provided beforehand. Both the datasets used were in IOB format. The dataset released by Pan et al. for Marathi contains a total of 14,978 sentences and has been divided into 3 categories namely Organization, Person and Location. It is a "silver-standard" dataset.

We removed the IOB formatting from the IJCNLP 200 NER Corpus in accordance with the previous work carried out on it. Challenges faced while working with these datasets were:

- The IJCNLP dataset and IIT Bombay Marathi NER Corpus included English words.
- Tagging was non-uniform in the IJCNLP dataset and also included some ambiguous tags.
- More than 68% of the sentences in the IJCNLP dataset and 39% of sentences in the IIT Bombay Marathi NER Corpus included sentences that had only O tags

There are more datasets for Hindi and Marathi on NER but they aren't publicly available. Following are the names of such datasets. (a) FIRE-2013- Named Entity Recognition Indian Languages. (b) FIRE 2014 - Named Entity Recognition Indian Languages. (c) FIRE 2015- Entity Extraction from Social Media Text Indian

Languages (ESM-IL) (d) FIRE 2016 - Shared Task on Code Mix Entity Extraction in Indian Languages (CMEE-IL) (e) TDIL- Named Entity Annotated Corpora for Marathi. (f) TDIL- Named Entity Corpora for Hindi, Marathi, Punjabi (Tables 1, 2, 3, 4 and 5).

Table 1 Count of sentences and tags in the datasets

Dataset	Count of sentences			Count of tags		
	Train	Test	Tune	Train	Test	Tune
IJCNLP 200 NER Corpus	7979	1711	710	208,750	44,692	44,146
IIT Bombay Marathi NER Corpus	3588	1533	470	67,775	32,214	8370
WikiAnn NER Corpus(Marathi)	10,674	4304	–	76,006	32,572	–
WikiAnn NER Corpus(Hindi)	8356	3477	–	48,601	20,829	–

Table 2 Count of individual tags of IJCNLP 200 NER Corpus

Tags	Train	Test	Tune
O	18,172	38,692	38,049
NETE	6468	1315	1379
NEN	4529	1062	1009
NEP	3893	886	806
NEL	3257	692	880
NEO	2119	459	392
NETI	2210	482	524
NEM	2017	463	550
NETO	1630	397	348
NED	796	141	145
NEA	459	103	64

Table 3 Count of individual tags of WikiAnn NER Corpus (Hindi)

Tags	Train	Test
O	20,015	9243
I-ORG	8195	4638
B-ORG	2907	1512
I-PER	7445	1705
B-PER	5570	973
I-LOC	2314	1224
B-LOC	2155	1534

Table 4 Count of individual tags of IIT Bombay Marathi NER Corpus

Tags	Train	Test	Tune
O	61235	28,215	7349
B-LOCATION	3372	1871	598
I-LOCATION	1449	1277	198
B-PERSON	974	432	131
I-PERSON	572	328	63
I-ORGANIZATION	98	56	16
B-ORGANIZATION	75	35	15

Table 5 Count of individual tags of WikiAnn NER Corpus (Marathi)

Tags	Train	Test
O	46,011	19,633
I-ORG	7076	3047
B-ORG	3053	1263
I-PER	6686	3020
B-PER	4469	1651
I-LOC	3019	1330
B-LOC	5692	2628

Example:
Sentence (Hindi): रीमा फेसबुक में काम कर ती है
Tag: NEP NEO O O O O

Sentence (Marathi): रीमा फेसबुक मध्ये काम करते
Tag: B-PERSON B-ORGANIZATION O O O

3.2 Model Architecture

In natural language processing, the Transformer seeks to solve sequence-to-sequence tasks while also resolving long-range dependencies. The Transformer NLP model includes an "attention" mechanism that analyzes the association between all the words in a sentence. It generates differential weightings to suggest which components in the sentence are most important for determining how a word should be interpreted. This accounts for the quick and efficient resolution of ambiguous elements. For example, the input given is a sentence, the transformer recognizes the context that grants the meaning of each word in the sentence. As the feature improves parallelization, the training time is reduced. The general model setup is shown in Fig. 1.

BERT: Originating from the pre-training contextual representations, BERT is a transformer-based technique for NLP pre-training developed by Google. It is a deep bidirectional model, meaning that it grasps the details from both sides of a token's context while training. The most important characteristic of BERT is that it can be fine-tuned by adding a few output layers.

mBERT: The next stage in developing models that grasp the meaning of words in context is MBERT, which stands for multilingual BERT. By simultaneously encoding all of their information on MBERT a deep learning model was trained on 104 languages

ALBERT: Google AI open-sourced ALBERT, a transformer architecture based on BERT which uses much fewer parameters than the state-of-the-art model BERT model. As compared to BERT models, these models have higher data throughput and can train about 1.7 times faster than the BERT model. IndicBERT, a multilingual ALBERT model, trained on large-scale datasets covers 12 major Indian languages. Many public models like mBERT and XLM-R contain more parameters as compared to IndicBERT, yet the latter performs very well on a variety of tasks.

RoBERTa: RoBERTa is a self-supervised transformers model, trained on a large corpus of English data. This implies that it was pre-trained on raw texts solely, with no human labeling, and then used an automatic method to build labels and inputs from those texts. XLM-RoBERTa is a multilingual model that has been trained in 100 languages. It does not require lang tensors to recognize which language is used, unlike some XLM multilingual models. It is also capable of determining the proper language from the input ids.

4 Result

In this section, we report and discuss the F1 score obtained by training various models [1] on respective datasets. Table 6 represents the results of transformer models trained on the IIT Bombay Marathi NER Corpus and WikiAnn NER Corpus (Marathi). Table 7 represents the results of IJCNLP 200 NER Hindi Corpus and WikiAnn NER Corpus(Hindi). The number of sentences in the training set for Hindi is double that of Marathi, due to which the models trained on the Hindi dataset have a better F1 score

[1] Multicase BERT: https://huggingface.co/bert-base-multilingual-cased
Indic BERT: https://huggingface.co/ai4bharat/indic-bert
Xlm-roberta: https://huggingface.co/xlm-roberta-base
Roberta-Marathi: https://huggingface.co/flax-community/roberta-base-mr
Roberta-Hindi: https://huggingface.co/flax-community/roberta-hindi
indic-transformers-hi-roberta: https://huggingface.co/neuralspace-reverie/indic-transformers-hi-roberta
MahaBERT: https://huggingface.co/l3cube-pune/marathi-bert
MahaRoBERTa: https://huggingface.co/l3cube-pune/marathi-roberta
MahaAlBERT: https://huggingface.co/l3cube-pune/marathi-albert-v2.

Table 6 F1 score(macro), precision and recall of various transformer models using the Marathi datasets

Dataset	IIT Bombay				WikiAnn			
	F1	Precision	Recall	Accuracy	F1	Precision	Recall	Accuracy
Multicase BERT	58.35	63.67	54.58	92.42	86.49	86.25	86.73	95.18
Indic BERT	60.79	66.05	53.76	92.57	87.03	87.06	87.00	95.13
Xlm-Roberta	62.32	64.14	60.60	93.00	87.38	86.92	87.85	95.48
Roberta-Marathi	43.81	42.64	45.03	91.34	82.00	80.26	83.82	93.73
MahaBERT	62.57	64.67	60.61	92.97	88.18	88.22	88.14	95.77
MahaRoBERTa	**64.34**	65.64	63.08	92.90	**88.90**	88.59	89.20	96.06
MahaAlBERT	60.00	63.77	56.52	92.52	87.15	87.19	87.11	95.14
RoBERTa Hindi	42.19	41.52	42.88	91.10	82.50	81.69	83.33	95.29
Indic-transformers-hi-roberta	36.80	36.81	36.7	90.49	80.00	78.73	81.32	94.27

Table 7 F1 score(macro), precision and recall of various transformer models using the Hindi datasets

Dataset	IJCNLP 200				WikiAnn			
	F1	Precision	Recall	Accuracy	F1	Precision	Recall	Accuracy
Multicase BERT	72.74	70.64	74.97	95.09	81.21	79.64	82.85	91.50
Indic BERT	71.63	70.50	72.87	95.12	82.65	81.44	83.90	92.01
Xlm-Roberta	**75.90**	74.19	77.70	95.62	**83.04**	82.80	83.27	91.68
Roberta-Hindi	69.06	67.27	70.95	95.29	80.52	78.52	82.63	90.85
indic-transformers-hi-roberta	64.36	62.44	66.41	94.27	73.79	70.71	77.15	87.75
Roberta-Marathi	61.69	59.39	64.17	94.48	79.46	77.20	81.85	90.82
MahaBERT	72.91	70.71	75.21	95.09	81.95	80.99	82.93	91.71
MahaRoBERTa	75.30	73.64	77.04	95.61	80.66	79.71	81.63	91.37
MahaAlBERT	69.83	69.62	70.03	94.82	81.68	80.87	82.50	92.23

compared to that of Marathi. MahaRoBERTa model which has RoBERTa as its base architecture performs the best on both the Marathi datasets. Roberta-Marathi in spite of being trained in Marathi has the least F1 score. The monolingual Marathi models based on MahaBERT perform better than the multilingual models. Whereas for Hindi NER datasets the multilingual models perform better than the Hindi monolingual counterparts. During the cross-language evaluation, we test the Marathi models on Hindi datasets and vice versa. This is desirable as both Marathi and Hindi share the same Devanagari script. We observed that models based on mahaBERT perform competitively on Hindi datasets. However, this is not true of Hindi models as they perform poorly on Marathi datasets. Both the models released by flax-community do not perform well on either of the languages. We, therefore, highlight the need for developing better resources for the Hindi language. In general, we observe that language-specific fine-tuning does not necessarily guarantee better performance and the factors underlying this disparity need to be investigated.

One hypothesis for this disparity could be attributed to the fact that the Marathi models have been trained on top of multilingual models whereas Hindi models have been trained from scratch.

5 Conclusion

Many NER systems have been deployed in English and other major languages. However, there hasn't been much work done on Hindi and Marathi languages. This study seeks to examine transformer-based deep learning-based NER solutions to Hindi and Marathi NER tasks. We benchmark for a host of monolingual and multilingual transformer-based models for Named Entity Recognition that includes multilingual BERT, Indic BERT, Xlm-Roberta, mahaBERT, and others. We show that monolingual training doesn't necessarily ensure superior performance. Although Marathi monolingual models perform the best same is not true for Hindi. Moreover, we observe that the mahaBERT models even generalize well on Hindi NER datasets. It is worthwhile to investigate the poor performance of monolingual models and is left to future scope.

Acknowledgements This work was done under the L3Cube Pune mentorship program. We would like to express our gratitude towards our mentors at L3Cube for their continuous support and encouragement.

References

1. Grishman R, Sundheim BM (1996) Message understanding conference-6: a brief history (1996)
2. Maybury M (1999) Advances in automatic text summarization. MIT Press
3. Davenport TH, Klahr P (1998) Managing customer support knowledge. California Manage Rev 40(3):195–208
4. Wu Y, Schuster M, Chen Z, Le QV, Norouzi M, Macherey W, Krikun M, Cao Y, Gao Q, Macherey K et al (2016) Google's neural machine translation system: bridging the gap between human and machine translation. arXiv preprint arXiv:1609.08144
5. Savelsbergh MW (1990) An efficient implementation of local search algorithms for constrained routing problems. Eur J Operat Res 47(1):75–85
6. Finkel JR, Grenager T, Manning CD (2005) Incorporating non-local information into information extraction systems by gibbs sampling, pp 363–370
7. Joshi R (2022) L3cube-mahacorpus and mahabert: marathi monolingual corpus, marathi bert language models, and resources. arXiv preprint arXiv:2202.01159
8. Joshi R, Goel P, Joshi R (2019) Deep learning for hindi text classification: a comparison. In: International conference on intelligent human computer interaction. Springer, pp 94–101
9. Kulkarni A, Mandhane M, Likhitkar M, Kshirsagar G, Jagdale J, Joshi R (2022) Experimental evaluation of deep learning models for marathi text classification. In: Proceedings of the 2nd

international conference on recent trends in machine learning, IoT, smart cities and applications. Springer, pp 605–613
10. Kulkarni A, Mandhane M, Likhitkar M, Kshirsagar G, Joshi R (2021) L3cubemahasent: a marathi tweet-based sentiment analysis dataset. In: Proceedings of the eleventh workshop on computational approaches to subjectivity, sentiment and social media analysis, pp 213–220
11. Velankar A, Patil H, Gore A, Salunke S, Joshi R (2021) Hate and offensive speech detection in hindi and marathi. arXiv preprint arXiv:2110.12200
12. Seon CN, Ko Y, Kim JS, Seo J (2001) Named entity recognition using machine learning methods and pattern-selection rules. In: NLPRS. Citeseer, pp 229–236
13. Alfred R, Leong LC, On CK, Anthony P (2014) Malay named entity recognition based on rule-based approach
14. Shao Y, Hardmeier C, Nivre J (2016) Multilingual named entity recognition using hybrid neural networks
15. **Xu K, Zhou Z, Hao T, Liu W (2017) A bidirectional lstm and conditional random fields approach** to medical named entity recognition, pp 355–365
16. Ekbal A, Bandyopadhyay S (2010) Named entity recognition using support vector machine: a language independent approach. Int J Electr Comput Syst Eng 4(2):155–170
17. **Patil NV, Patil AS, Pawar BV (2017) Hmm based named entity recognition for inflectional** language, pp 565–572. https://doi.org/10.1109/COMPTELIX.2017.8004034
18. Matthew Honnibal Ines Montani SVL, Boyd A (2020) spacy: industrial-strength natural language processing in python. https://doi.org/10.5281/zenodo.1212303
19. Goodfellow I, Bengio Y, Courville A (2016) Deep learning. MIT Press
20. Lothritz C, Allix K, Veiber L, Bissyand T, Klein J (2020) Evaluating pretrained transformer-**based models on the task of fine-grained named entity recognition, pp 3750–3760.** https://doi.org/10.18653/v1/2020.coling-main.334
21. Vaswani A, Shazeer N, Parmar N, Uszkoreit J, Jones L, Gomez AN, Kaiser L, Polosukhin I (2017) Attention is all you need
22. Devlin J, Chang MW, Lee K, Toutanova K (2019) Bert: pre-training of deep bidirectional transformers for language understanding
23. Liu Y, Ott M, Goyal N, Du J, Joshi M, Chen D, Levy O, Lewis M, Zettlemoyer L, Stoyanov V (2019) Roberta: a robustly optimized Bert pretraining approach
24. Kakwani D, Kunchukuttan A, Golla S, Gokul N, Bhattacharyya A, Khapra MM, Kumar P (2020) inlpsuite: monolingual corpora, evaluation benchmarks and pre-trained multilingual language models for Indian languages, pp 4948–4961
25. Krishnarao AA, Gahlot H, Srinet A, Kushwaha D (2009) A comparative study of named entity recognition for Hindi using sequential learning algorithms, pp 1164–1169
26. Srihari RK (2000) A hybrid approach for named entity and sub-type tagging. In: Sixth applied natural language processing conference, pp 247–254
27. Albawi S, Mohammed TA, Al-Zawi S (2017) Understanding of a convolutional neural network. In: 2017 international conference on engineering and technology (ICET), pp 1–6. IEEE
28. Schmidhuber J, Hochreiter S et al (1997) Long short-term memory. Neural Comput 9(8):1735–1780
29. Yang G, Xu H (2020) A residual Bilstm model for named entity recognition. IEEE Access 8:227,710–227,718. https://doi.org/10.1109/ACCESS.2020.3046253
30. Shah H, Bhandari P, Mistry K, Thakor S, Patel M, Ahir K (2016) Study of named entity recognition for Indian languages. Int J Inf 6(1):11–25
31. Bhattacharjee K, Mehta S, Kumar A, Mehta R, Pandya D, Chaudhari P, Verma D et al (2019) Named entity recognition: a survey for Indian languages 1:217–220
32. Patil N, Patil AS, Pawar B (2016) Issues and challenges in Marathi named entity recognition. Int J Nat Lang Comput (IJNLC) 5(1):15–30
33. Singh TD, Ekbal A, Bandyopadhyay S (2008) Manipuri POS tagging using CRF and SVM: a language independent approach, pp 240–245 (2008)
34. Shishtla PM, Gali K, Pingali P, Varma V (2008) Experiments in telugu ner: a conditional **random field approach**

35. Shelke R, Thakore DS (2020) A novel approach for named entity recognition on Hindi language using residual Bilstm network
36. Murthy R, Kunchukuttan A, Bhattacharyya P (2018) Judicious selection of training data in assisting language for multilingual neural NER. In: Proceedings of the 56th annual meeting of the association for computational linguistics (Volume 2: Short Papers). Association for Computational Linguistics, Melbourne, Australia, pp 401–406. https://doi.org/10.18653/v1/P18-2064. https://aclanthology.org/P18-2064
37. Murthy R, Kunchukuttan A, Bhattacharyya P (2018) Judicious selection of training data in assisting language for multilingual neural NER, pp 401–406. https://doi.org/10.18653/v1/P18-2064
38. Ijcnlp-08 workshop on NER for south and south east Asian languages. http://ltrc.iiit.ac.in/ner-ssea-08/
39. Pan X, Zhang B, May J, Nothman J, Knight K, Ji H (2017) Cross-lingual name tagging and linking for 282 languages, pp 1946–1958. https://doi.org/10.18653/v1/P17-1178. https://aclanthology.org/P17-1178

Role of Remote Sensing in Precision Agriculture

Bhuvika Sharma, Upanshu Aggarwal, Komalpreet Kaur, Hitika Saini, Sujata Rani, Parteek Kumar, Karun Verma, and Yosi Shacham

Abstract Over the period of time, weather conditions have changed drastically. These changes have led to many serious issues such as soil nutrition depletion, reduction in soil water holding capacity, and increase in pesticide resistant weeds. Precision Agriculture (PA) is the pioneer that is changing the way people farm, offering benefits in profitability, productivity, sustainability, crop quality, environmental protection, on-farm quality of life, food safety, and rural economic development. In addition, with the aid of the recent advancements in the sensor technology, it can develop into an intelligent crop production system. Specifically, remote sensing technology that allows non-destructive acquisition of information about the earth's surface can facilitate the implementation of PA. In this paper, a specific area of Punjab is analyzed for crop production using remote sensing data. To increase crop production and make management decisions, PA entails the application of remote sensing technology.

Keywords Precision Agriculture · Remote sensing · NDVI · Land surface temperature

1 Introduction

Agriculture was the first occupation of man, as it embraces the whole earth. It is the foundation of all other industries. Precision Agriculture (PA) is a key component of sustainable agricultural systems in today's world. Over the years, it has provided various benefits in crop production, quality, and profit. Information-based management, site-specific crop management, target farming, variable rate technology, and grid farming are some other names used synonymously for PA. It consists of a

B. Sharma · U. Aggarwal · K. Kaur · H. Saini · S. Rani (✉) · P. Kumar · K. Verma
CSED, Thapar Institute of Engineering and Technology, Patiala 147001, India
e-mail: sujata.singla@thapar.edu

Y. Shacham
Electrical Engineering, Tel Aviv University, Tel Aviv, Israel

V. K. Gunjan and J. M. Zurada (eds.), *Proceedings of 3rd International Conference on Recent Trends in Machine Learning, IoT, Smart Cities and Applications*, Lecture Notes in Networks and Systems 540, https://doi.org/10.1007/978-981-19-6088-8_57

management strategy that uses a variety of advanced information, communication, and data analysis techniques in the decision-making process (e.g., application of water, fertilizer, pesticide, seed, fuel, labor, etc.), which helps in enhancing crop production and reducing water and nutrient losses and negative environmental impacts.

All aspects of the environment—soil, weather, vegetation, and water—vary from place to place. All these factors determine crop growth and farming success. Farmers have always been aware of this, but they lacked the tools to measure, map, and manage these variations precisely. Thus, precision farming can make a difference to food production facing the challenge of a rising world population. It can help farmers to achieve an information and technology-based farm management system that identifies, analyzes, and manages variability in fields by conducting crop production practices at the right place and time and in the right way, for optimum profitability, sustainability, and protection of the land resource [1].

The rest of the paper is structured as follows. Section 2 discusses the introduction of remote sensing and its importance of precision agriculture. The state-of-the-art work in this area is covered in Sect. 3 followed by dataset acquisition in Sect. 4. Section 5 includes the different experiments, and Sect. 6 covers the conclusions and future indirections.

2 Remote Sensing in Precision Agriculture

Remote sensing process involves acquiring information about the earth's surface measured by its reflected and emitted radiation without coming into direct contact with the object. The primary source of the energy observed by satellites is the Sun. This energy is observed by the remote sensors that could be of any form such as drones and satellite. The most useful electromagnetic radiation in remote sensing includes visible light (VIS), near infrared (NIR), shortwave infrared, thermal infrared, and microwave bands. Remote sensing can be used variably for PA that is influenced by many factors such as type of platforms, i.e., satellite, air or ground used for data collection; number and width of spectral bands captured by the sensor (multi-versus hyperspectral); and spatial (high, medium, and low), temporal (hourly, daily, and weekly) and radiometric (8-, 12-, and 16-bit) resolutions at which sensors collect quality of spectral information represented in acquired images data. Many sensors acquire data at different spectral wavelengths.

Over the years, remote sensing for PA has observed rapid advancements. Ground-based sensors have been developed for on the go monitoring of crop and soil characteristics. Different land cover types can be discriminated more readily, by using image classification algorithms. Overall, there is an ample variety of remote sensors and platforms available to generate high-resolution (spatial, spectral, radiometric, and temporal) images critical to develop and implement site-specific management [2].

3 Previous Research

Remote sensing has garnered a fair amount of interest over the years be it for its easy availability or accessibility. Thus, a comprehensive research has been performed to know more about the role it will play in the future, some of which are summarized as follows. Navalgund et al. [3] discussed different types of remote sensing such as microwave remote sensing and hyperspectral remote sensing and also discussed different techniques that can be used to sharpen the output images in order to extract useful information from them. Rekha et al. [4] analyzed about the satellite remote sensing with going in detail about the spectral wavelength and properties that are needed to get the desired output along with applications in the field of agriculture. Patil et al. [5] provided an insight on use of remote sensing data for PA. The data provided includes useful indices such as NDVI. Nemecek [6] discussed classification of irrigated areas using time–series NDVI data. Ge et al. [7] provided knowledge about the quantitative analysis of soil using the detailed spectral information obtained from the spectroscopic reflectance measurement of soil samples. Weiss [8] talked about the recent research development that has led to strengthen applicative capabilities in remote sensing. Over the years, many technologies have been applied in the agriculture sector so that the products that are made are safer and have less adverse impact on the environment among which remote sensing sure has upheld its name which is supported by Liaghat and Balasundram [9] by illustrating the remote sensing as a valuable component of the agriculture framework.

From the literature review, it can be observed that the researchers have experimented with different types of remote sensing techniques along with their various applications mainly in the agriculture sector [10]. Majority research work implying satellite remote sensing plays the key role for agriculture. The researchers indicated the vast range of factors that can be evaluated using remote sensing such as soil moisture, surface temperature, and irrigation facilities thus giving further indication of what the future of remote sensing might hold. In this work, the analysis of various factors such as deforestation, NDVI, and land surface temperature (LST) extracted for the target area has been performed using remote sensing. The various MODIS products enable to access various datasets [11]. Further drawing time series analysis for these factors is done to get an insight of the trends over the years which helped in filtering of the images for many factors such as the irrigation sights and surface temperature. This analysis helps in drawing inferences for the target area and takes further measures accordingly. The description about the processes involved is discussed in the next section.

1	Time IST	SM1	SM2	SM3	TEMP	HUMD	PRSR	LMNS
2	3/28/2020	2650.19	2982.13	2835.613	22.89	100	98909.47	3372
3	3/28/2020	2673.859	2963.604	2834.021	24.13143	91.12571	98909.51	3419.143
4	3/28/2020	2717.523	2961.507	2835.582	25.36	85.42667	98907.45	3400.167
5	3/28/2020	2739.772	2970.317	2843.607	24.99833	81.75167	98906.01	3494.167
6	3/28/2020	2784.721	2978.736	2854.871	27.07143	74.15143	98901.56	4713
7	3/28/2020	2865.442	2991.04	2873.623	28.73667	67.025	98895.59	5195.167
8	3/28/2020	2938.44	2997.015	2895.77	28.755	66.30333	98890.67	5365.167
9	3/28/2020	2996.649	2999.161	2906.97	29.74286	61.98143	98888.62	4681.571
10	3/28/2020	3036.607	3012.04	2923.97	30.485	59.45833	98880.25	4938.5
11	3/28/2020	3081.07	2999.161	2941.17	30.99	56.16857	98867.47	5145.857
12	3/28/2020	3112.06	3000.02	2949.87	31.31333	54.33333	98856.45	5737.333
13	3/28/2020	3141.458	2994.01	2973.32	32.05	49.04167	98841.26	5686.167
14	3/28/2020	3161.744	2976.19	2976.247	31.84	51.19857	98830.76	4451.286
15	3/28/2020	3198.403	2982.13	3003.025	33.22333	46.095	98812.84	3214

Fig. 1 Snippet of the LIBELIUM dataset

4 Dataset Acquisition

Satellite products such as MOD11A2 V6 are used to monitor soil and vegetation health, hydrologic and climatic parameters which are necessary for precision agriculture, e.g., (soil moisture, NDVI, rainfall, and groundwater). Landsat 8's band one acquires data at 0.433–0.453 micrometers, and MODIS's band one acquires data at 0.620–0.670 micrometers. Landsat 8 has 11 bands, whereas MODIS has 36 bands, all measuring different regions of the electromagnetic spectrum. For a true color (RGB) image from Landsat, bands 4, 3, and 2 are combined, respectively. In this work, experimentation is done on a field in Patiala, Punjab, India, using Google Earth Engine API. The field has the coordinates [30.46366, 76.31545; 30.46327, 76.31542; 30.46369, 76.31459]. For data collection, LIBELIUM hardware is installed in the target field. The precision agriculture-based Waspmote device is composed of six sensors. These sensors are connected at different nodes of the device. These sensors give the readings of soil moisture at different depths, temperature, humidity, pressure, and luminosity. By this process, data has been collected in the form of a csv file for various time stamps as shown in Fig. 1. In this study, the major factors that are considered with respect to the Punjab region are land surface temperature (LST) and soil moisture. The data about LST using remote sensing is extracted using MOD11A2 product of MODIS which is shown in Fig. 2.

5 Experimentation Analysis

For the selection of area of vegetation, different factors play an important role such as the determination of the elevation above sea level, type of terrain, water bodies nearby, and LST. To analyze this, different experiments have been performed on

Fig. 2 Snippet of the LST dataset extracted through remote sensing

1	system:time_sta	LST_Day_1km
2	Jan 1, 2020	
3	Jan 9, 2020	15.13
4	Jan 17, 2020	17.11
5	Jan 25, 2020	
6	Feb 2, 2020	17.43
7	Feb 10, 2020	18.13
8	Feb 18, 2020	20.93
9	Feb 26, 2020	23.59
10	Mar 5, 2020	21.79
11	Mar 13, 2020	23.41
12	Mar 21, 2020	24.55
13	Mar 29, 2020	28.11
14	Apr 6, 2020	33.67
15	Apr 14, 2020	34.43

a particular area. The necessary analysis of data and its experiments is provided at the given URL www.cropportal.com. Figures 3 and 4 show the homepage and data catalog pages of the portal which tell about data collection through various experiments using remote sensing technology.

Fig. 3 Homepage

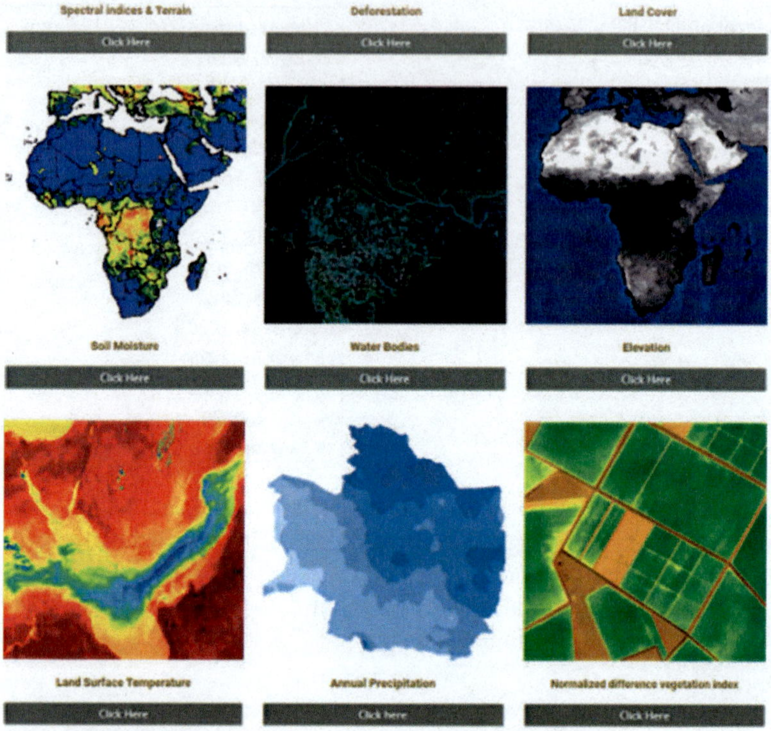

Fig. 4 Data catalogue

5.1 Determining Elevation Above Sea Level

Determining elevation above sea level plays an important role for selection of area of vegetation as it may affect sunlight, water, nutrients, and other factors affecting plants. Because certain plants can grow very well in higher elevations, others can only be grown in middle or lower elevation. For this experiment SRTM Digital Elevation Data, 30m dataset is used. After adding image to the interactive map using Map.addLayer() method, layer transparency options are used to create draped images for colorized hillsides. As shown in Fig. 5, the lowest points are shown in blue, and the highest points are shown in red. And it has been observed that target area or Punjab lies in red region, which means a region with higher percentage vegetation given the elevation. The region around Punjab is yellow and blue indicating less vegetation.

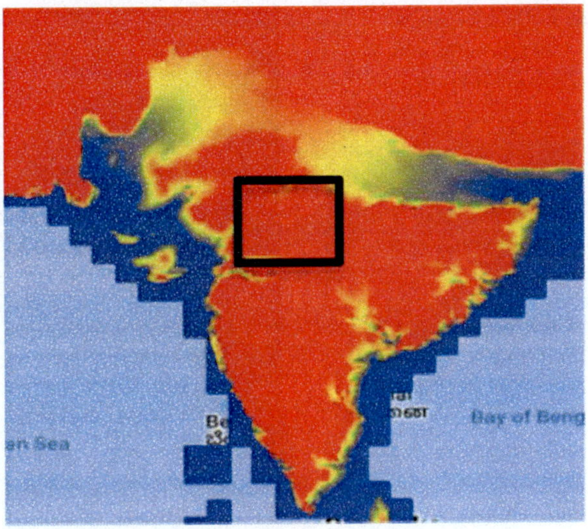

Fig. 5 Determining elevation

5.2 *Distinguishing Between Water and Terrain Bodies*

To distinguish the water and terrain bodies around the target field, a multispectral image collected by European Space agency's Sentinet-2 Satellite is used. A false color composite image as shown in Fig. 6 is generated by putting bands 8, 4, and 3 into red, green, and blue channels, respectively. The figure shows the photosynthetically active vegetation in vibrant red and water bodies in blue. Further terrains ranging from mountain ranges, grasslands, etc., can be classified to determine whether the conditions are feasible for the vegetation to be grown in the area. Machine learning classifiers such as SVM and cart can be used for this task. For experimentation, a custom dataset is created by collecting representative samples of reflectance spectra from each band for each land cover class of interest and then overlaying these training points on the image. The training points are classified into different land cover classes using the cart classifier as shown in Fig. 7. The terrains included here are the plains, and the mountain range (blue region) which is further clarified as the points near the target area are classified in the plains (red region).

5.3 *Determining Spectral Indices*

Spectral indices are combinations of spectral reflectance from two or more wavelengths. Out of these indices, the ones that measure the greenness, i.e., the relative density and health of the vegetation, for each pixel in a satellite image is NDVI. Its values range from + 1.0 to − 1.0. Areas of barren rock, sand, or snow usually show very low NDVI values (e.g., 0.1 or less). Sparse vegetation such as shrubs

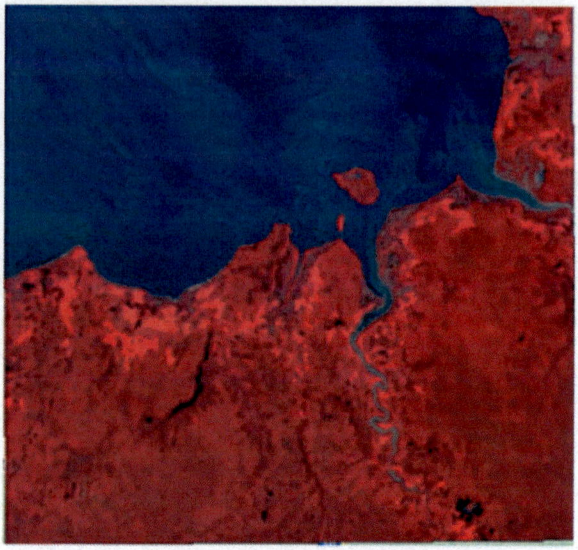

Fig. 6 Determining water and terrain bodies

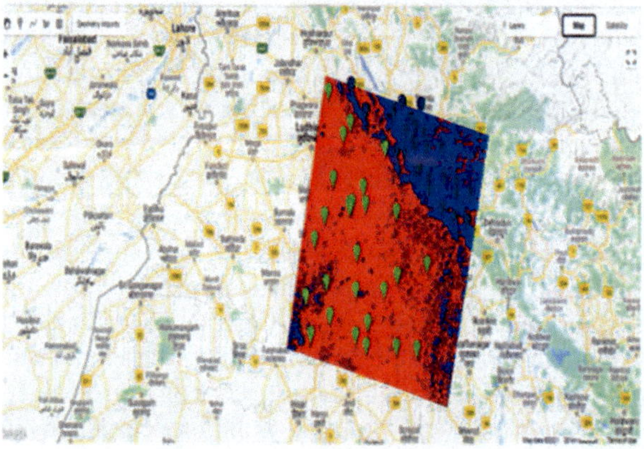

Fig. 7 Training points classifiers using cart classifier

and grasslands or senescing crops may result in moderate NDVI values (approximately 0.2–0.5). High NDVI values (approximately 0.6–0.9) correspond to dense vegetation such as that found in temperate and tropical forests or crops at their peak growth stage. NDVI is calculated in accordance with the formula (1), where NIR is the reflection in the near-infrared spectrum and RED is the reflection in the red range of the spectrum.

$$\mathrm{NDVI} = \frac{\mathrm{NIR} - \mathrm{RED}}{\mathrm{NIR} + \mathrm{RED}} \tag{1}$$

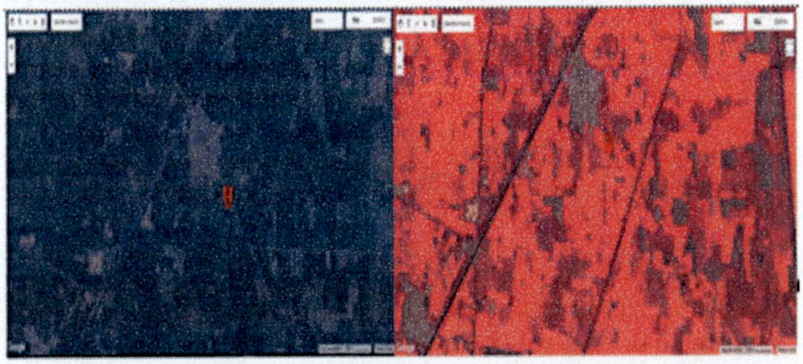

Fig. 8 True color image and false color composite image around the target area

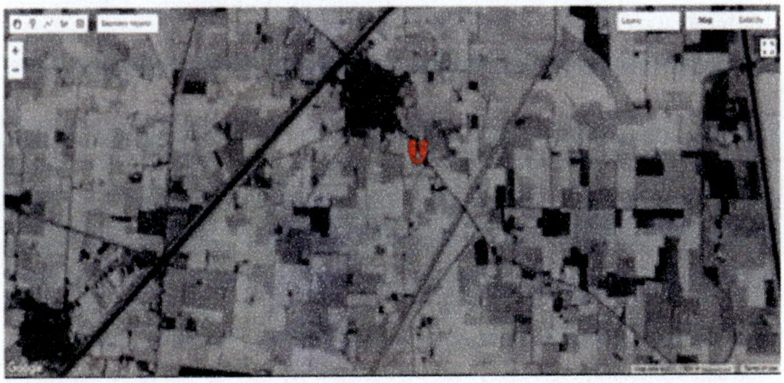

Fig. 9 Image showing NDVI

For this experiment, images around the target area are extracted from the Sentinel-2 archive and sorted by the most recent cloud free image. Then these images are viewed using different spectral wavelength to obtain and to find the best relevant image especially for NDVI. The spectral combination of bands 4, 3, and 2 gives a true color image as shown in Fig. 8 (left). The spectral band 8 that is the infrared band provides the response to the chlorophyll content in the green leaves as shown in Fig. 8 (right). In Fig. 8 (left region), the vegetation that appears dark green in color appears bright red in Fig. 8 (right region) indicating high chlorophyll content in the plants.

Figure 9 shows the NDVI calculated using the values from the RED and near-infrared (NIR) bands according to the formula (1).

5.4 *LST Analysis*

LST is described as the exchange of energy and water between the land surface and atmosphere which thus influences the rate and timing of plant growth. It is a mixture

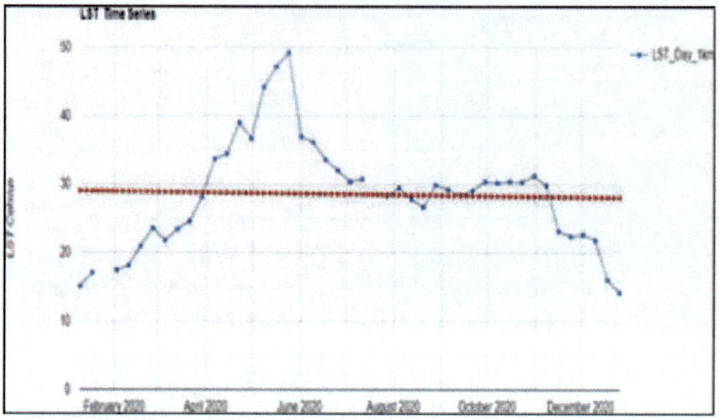

Fig. 10 Time series analysis of LST dataset

of vegetation and bare soil temperatures. For this experiment, both the day- and nighttime surface temperature bands are extracted from MOD11A2 product of AQUA MODIS. Then from these bands using the daytime and nighttime temperatures, the daily mean temperature is estimated. Daytime LST is more tightly coupled with the radiative and thermodynamic characteristics of the earth's surface than standard air temperature measurements. Figure 10 shows the time series graph of average 8-day land surface temperature (LST).

It can be observed from the graph that Punjab region has varying temperature pattern, i.e., November–February are the months with low temperatures, April–August are the months with high temperatures, May–June being the highest peak, and December is the month with the lowest peak. This will give an idea of the cultivation of kharif and rabi crops in Punjab. Although there is variation in the values of both datasets, however, the mapping of these parameters can help in getting an understanding of the trend/pattern as shown in Fig. 11. It has been observed from the figure that both the datasets follow similar pattern which could help us in getting accurate LST predictions. This pattern can further help us in finding the accuracy in both the datasets and take the one that gives more accurate predictions.

Figure 12 shows the RNN analysis performed on soil moisture at depths 10 and 80 cm. It can be observed from the figure that predicted results match with the actual results depicting the good accuracy of the system.

6 Conclusions and Future Work

Farming is becoming more scientific, with introduction of various technologies like remote sensing and GPS being added to farming techniques. Precision farming is about managing variations in the field accurately to grow more food using fewer

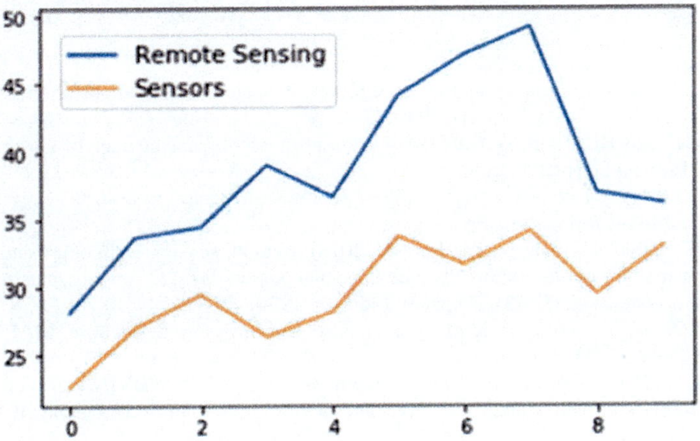

Fig. 11 Mapping of LST

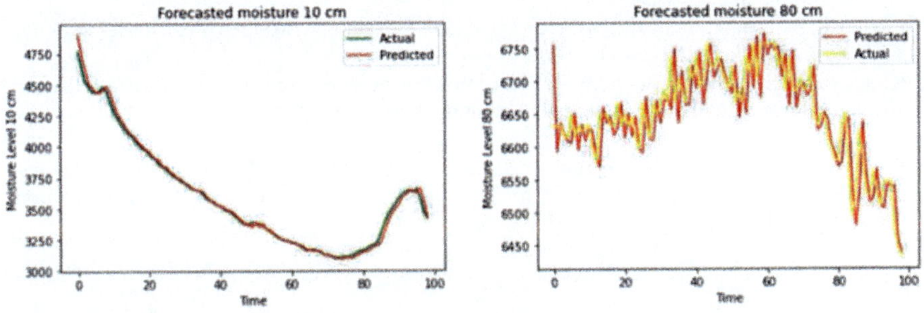

Fig. 12 RNN analysis soil moisture at different depths

resources that results in better production and fewer losses. With the help of vegetation, indices such as NDVI, the density of vegetation of the target area can be obtained. Another parameter that can be calculated through remote sensing is LST and normalized difference water index (NDWI) that can be used by farmers to recognize plant water stress and prepare for consequences like droughts. Thus, remote sensing can become the pioneer to bring the necessary changes in the old reforms.

Acknowledgements This work was supported by Thapar-TAU Center for Excellence in Food Security (T2CEFS), under research project "A Data-Driven Approach to Precision Agriculture in Small Farms Project".

References

1. Berry JK, Detgado JA, Khosla R, Pierce FJ (2003) Precision conservation for environmental sustainability. J Soil Water Conserv 58(6):332–339
2. Sishodia RP, Ray RL, Singh SK (2020) Applications of remote sensing in precision agriculture: a review. Remote Sens 12(19):1–31
3. Navalgund RR, Jayaraman V, Roy PS (2007). Remote sensing applications: an overview. Curr Sci 93(12):1747–1766
4. Rekha BU, Desai VV, Ajawan PS, Jha SK (2018). Remote sensing technology and applications in agriculture. In: International conference on computational techniques, electronics and mechanical systems (CTEMS), Belgaum, India, pp 193–197
5. Patil B, Chetan HT (2017) Role of remote sensing in precision agriculture. MARUMEGH Kisaan E-Patrika 2:34–37
6. Nemecek A (2019). Utilization of various methods and a landsat NDVI/google earth engine product for classifying irrigated land cover. Thesis, Master of Science, University of Montana, Missoula
7. Ge Y, Thomasson JA, Sui R (2011) Remote sensing of soil properties in precision agriculture: a review. Front Earth Sci 5(3):229–238
8. Weiss M, Jacob F, Duveiller G (2020). Remote sensing for agricultural applications: a meta-review. Remote Sens Environ:1–19
9. Liaghat S, Balasundram SK (2010) A review: the role of remote sensing in precision agriculture. Am J Agric Biol Sci 5(1):50–55
10. Kingra PK, Majumder D, Singh SP (2016) Application of remote sensing and GIS in agriculture and natural resource management under changing climatic conditions. Agric Res J 53(3):295–302
11. Doraiswamy PC, Sinclair TR, Hollinger S, Akhmedov B, Stern A, Prueger J (2005) Application of MODIS derived parameters for regional crop yield assessment. Remote Sens Environ 97(2):192–202

A Novel 18-Convolutional Layered Deep U-Net Architecture for COVID-19 Infection Diagnosis Through Object Detection on Lung CT Scan Segmentation

R. Aruna, M. Shyamala Devi, J. Arun Pandian, K. Kanchanadevi, P. Keshav Kumar, B. Sai Tarun, and N. Charan Sai

Abstract With the advancement of digital technology, a large number of medical images are created in the field of digital medicine. Object detection is a critical field of study in medical image analysis, detection and processing. Object detection is the categorization of the pixels in an image that determines where the objects in the image are located. Image classification is the global classification of an image's pixels. The objects are identified but not located from all of the pixels in the image. The intelligent identification process for the adjuvant diagnosis to assist medical doctors of various disciplines is in high demand. The need for the high accurate object detection method for the medical images remains a major challenge toward disease detection in the health care. With this summary overview, a novel model is proposed for lung CT scan image detection as 18-Convolutional Layered Deep U-Net architecture for diagnosis of COVID-19 detection through object detection. Model fitting is done with 18-Convolutional Layered Deep U-Net architecture, and the model design is formed with four contraction path and four expansive path layers. The customized model design is fine-tuned with the parameter optimization. The object

R. Aruna (✉) · M. S. Devi · K. Kanchanadevi · P. K. Kumar · B. S. Tarun · N. C. Sai
Computer Science and Engineering, Vel Tech Rangarajan Dr. Sagunthala R&D Institute of Science and Technology, Chennai, Tamil Nadu, India
e-mail: drraruna@veltech.edu.in

M. S. Devi
e-mail: shyamaladevim@veltech.edu.in

K. Kanchanadevi
e-mail: kanchanadevi95@gmail.com

P. K. Kumar
e-mail: vtu14879@veltech.edu.in

B. S. Tarun
e-mail: vtu11866@veltech.edu.in

J. A. Pandian
School of Information Technology & Engineering, Vellore Institute of Technology, Vellore, India
e-mail: aparunpandian@gmail.com

V. K. Gunjan and J. M. Zurada (eds.), *Proceedings of 3rd International Conference on Recent Trends in Machine Learning, IoT, Smart Cities and Applications*, Lecture Notes in Networks and Systems 540, https://doi.org/10.1007/978-981-19-6088-8_58

detection is done, and the performance is analyzed. The dataset is also fitted with existing deep convolution neural network models like region-based, threshold, edge-based and clustering-based method, and the performance is analyzed and compared with proposed model with metrics like pixel accuracy, intersection over union and dice coefficient. Implementation results show that the proposed models have pixel accuracy of 98.32%, intersection over union of 48.7% and dice coefficient of 97.56% compared to existing object detection models.

Keywords CNN · Pixel accuracy · Object detection · Dropout · Dice coefficient

1 Introduction

Coronavirus is an inescapable illness causing a large number of passing every day. Early analysis of this illness ended up being one of the best techniques for disease tree pruning. It has caused a staggering impact on day-to-day routines, general well-being and the worldwide economy. It is basic to distinguish the positive cases as soon as could be expected to forestall the further spread of this scourge and to rapidly treat affected patients. The requirement for helper demonstrative devices has expanded, as there are no exact mechanized tool compartments accessible. Late discoveries got utilizing radiology imaging procedures proposing that such pictures contain remarkable data about the COVID-19 infection. Artificial intelligence (AI) procedures combined with radiological imaging can be useful for the precise location of this illness and can likewise be assistive to defeat the issue of an absence of specific doctors in distant towns. The huge number of COVID-19 patients is delivering medical care frameworks in numerous nations overpowered.

Deep learning has turned into a traditional technique for building networks prepared to do effectively demonstrating higher-request frameworks to accomplish human-like execution. Growths have been immediate focuses for DL-helped division of clinical pictures. Until the present time, many exploration projects have been led for COVID-19 recognition utilizing DL examination of clinical pictures, for example, X-ray and CT filters and uncovered huge outcomes. Notwithstanding, semantically dividing those pictures has been less engaging.

2 Literature Review

The clinical pictures investigation procedure has been proposed for naturally separating the lung parenchyma locale in CT pictures. The founded arrangement of these systems is clear and convincing for average lung division, yet they altogether bomb when we extend the articulation "lung" to address the common lung tissues as well as peculiar tissues and veins [1]. In this paper, [2] proposed a stepwise division procedure. Starting, an iterative thresholding is used to get a hidden divided locale.

Another division methodology is to use wavelet change and an ideal thresholding to get the essential division and then refine the gained divided region using mathematical morphology exercises [3]. This paper [4] manages cell breakdown in the lungs screening instrument which was executed using DL structures significance to cut down sham positive rate in cell breakdown in lungs screening with low-segment computerized tomography channels. In this study [5], numerous deep learning configurations were measured by experts to perceive coronavirus affected people using clinical pictures. This paper [6] utilized ResNet50V2 networks and Xception for coronavirus identification from computerized tomography checks, bringing about a precision of 0.97 for the objective class. This work [7] uses a significant CNN as a twofold portion or and stood out it from various plans. The makers showed up at a Sorensen Dice of 0.73, a familiarity with 0.74 and an exactness of 0.73 [8].

This paper performs [9] investigation on point by point arrangement of a unique deep neural network arrangements named Semi InfNet and InfNet to meaning fully segment polluted regions and association [10]. Chest radiological imaging like registered tomography and recognized by inspecting cathode rays play essential parts in early conclusion and treatment of this illness. This paper [11] fostered the profound learning model utilizing 224 affirmed coronavirus pictures. In this review, [12] it utilized CNN models to extricate picture highlights and afterward characterized them utilizing the SVM classifier. A 95.38% precision was accomplished involving ResNet50 and SVM in blend with 50 pictures. This exploration survey [13] utilizes significant model proposed for the customized assurance of coronavirus. This paper uncovers that recovered patients may continue to spread the disease. The primary preventions of chest radiography assessments are a weakness to distinguish the starting periods of coronavirus, as it will not have satisfactory responsiveness in GGO ID [10]. Radiologists accept a basic part in view of their colossal association with this area, and the artificial intelligence progresses in computed tomography can be dependable to get unmistakable finding [14].

3 Proposed 18-Convolution Layered Deep U-Net Architecture

The architecture of the proposed 18-Convolutional Layered Deep U-Net architecture is shown in Fig. 1. The following contributions are carried out in this paper.

- The Kaggle dataset consists of 11,191 lung CT scan images from the link https://www.kaggle.com/andrewmvd/covid-19-ct-scans-getting-started/.
- The lung CT scan images are first normalized along with data preprocessing.
- The dataset is split with training and testing dataset with 90:10. Model fitting is done with 18-Convolutional Layered Deep U-Net architecture.
- The design of customized 18-Convolutional Layered Deep U-Net architecture is formed with four contraction path and expansive path layers as shown in Fig. 2.

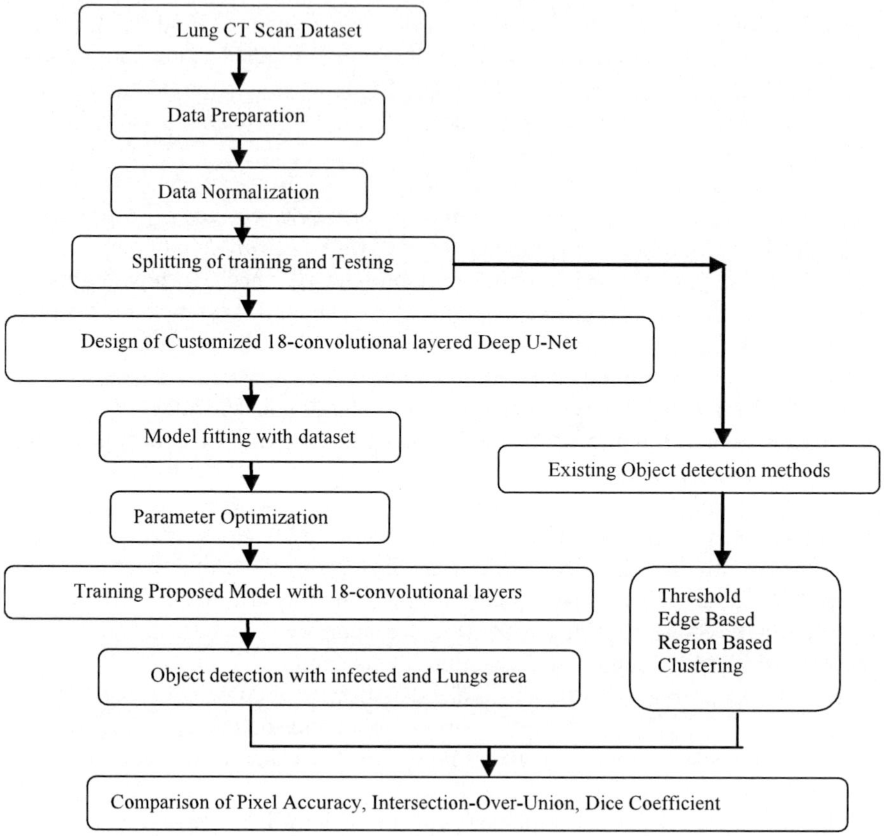

Fig. 1 18-Convolutional Layered Deep U-Net architecture system workflow

- The 18-Convolutional Layered Deep U-Net architecture is formed with the first contraction path layer forming 2 * Conv(16) and fourth expansive path layer forming 3 * Conv(16).
- The 18-Convolutional Layered Deep U-Net architecture is formed with second contraction path layer forming 2 * Conv(32) and third expansive path layer forming 3 * Conv(32).
- The 18-Convolutional Layered Deep U-Net architecture is formed with third contraction path layer forming 2 * Conv(64) and second expansive path layer forming 3 * Conv(64).
- The 18-Convolutional Layered Deep U-Net architecture is formed with third contraction path layer forming 2 * Conv(128) and second expansive path layer forming 3 * Conv(128).
- The contraction path layer of 18-Convolutional Layered Deep U-Net architecture is formed with four layers having first layer with conv2D, having second layer

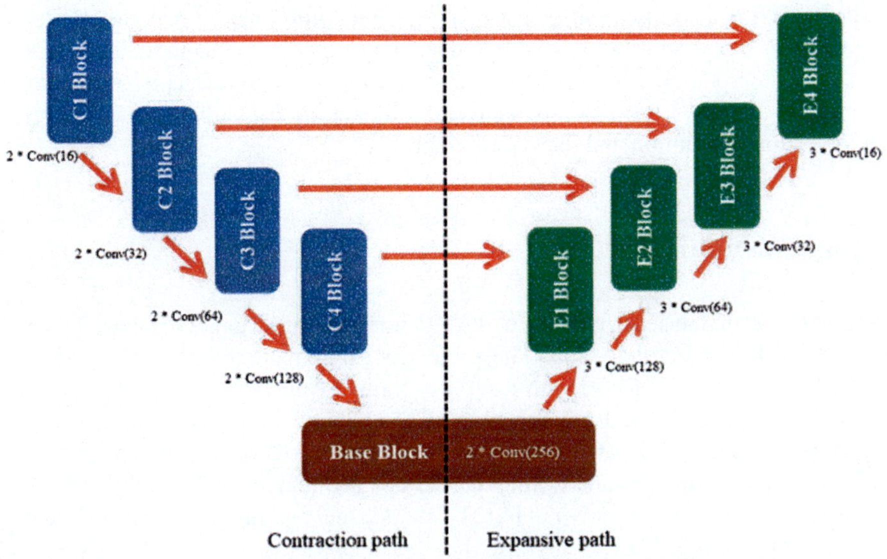

Fig. 2 Proposed 18-Convolutional Layered Deep U-Net architecture design

with dropout of insignificant features, having third layer with conv2D followed by fourth layer with Maxpooling2D layer and is shown in Fig. 3.

- The expansive path layer of 18-Convolutional Layered Deep U-Net architecture is formed with five layers having first layer with conv2DTranspose, having concatenation as second layer, having third layer with conv2D, having fourth layer dropout of insignificant features followed by fourth layer with conv2D layer. The customized model design is fine-tuned with the parameter optimization. The object detection is done, and the performance is analyzed.

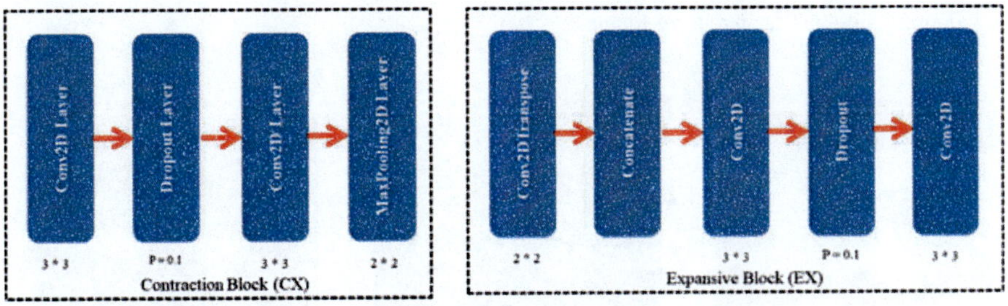

Fig. 3 Contraction and expansive path of 18-Convolutional Layered Deep U-Net

- The dataset is also fitted with existing deep convolution neural network models like region-based, threshold, edge-based and clustering-based method, the performance is compared with proposed 18-Convolutional Layered Deep U-Net architecture, and the performance is analyzed with the metrics like pixel accuracy, intersection over union and dice coefficient.

4 Implementation Setup

The lung CT scan dataset consisting of 11,191 images from Kaggle database repository is used for object detection. The project is implemented with Python under NVidia Tesla V100 GPU server with 1,940,817 trainable parameters, training epochs of 300, batch size of 32 and training time of 1531 min. The dataset lung CT scan images are normalized along with data preprocessing. The dataset is split into training dataset with 10,071 images and testing dataset of 1120 images in the ratio of 90:10, and the sample images are shown in Fig. 4. The original and predicted lung CT scan image is shown in Fig. 5.

The dataset is also fitted with existing deep convolution neural network models like region-based, threshold, edge-based and clustering-based method, the performance is compared with proposed 18-Convolutional Layered Deep U-Net architecture, and the performance is analyzed with the metrics like pixel accuracy, intersection over union, and dice coefficient. The training and validation loss, training and validation

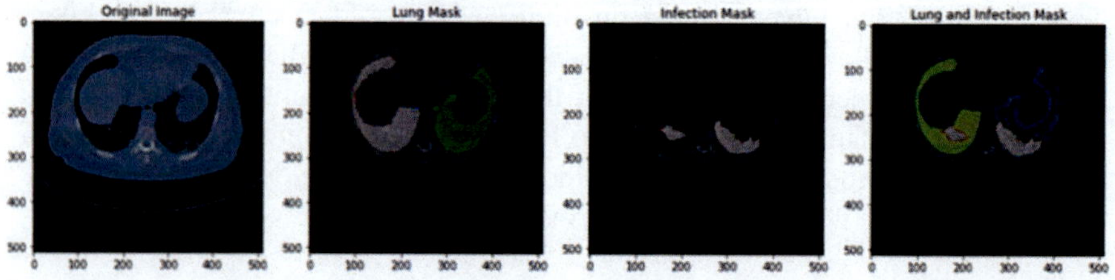

Fig. 4 Sample lung CT scan dataset images

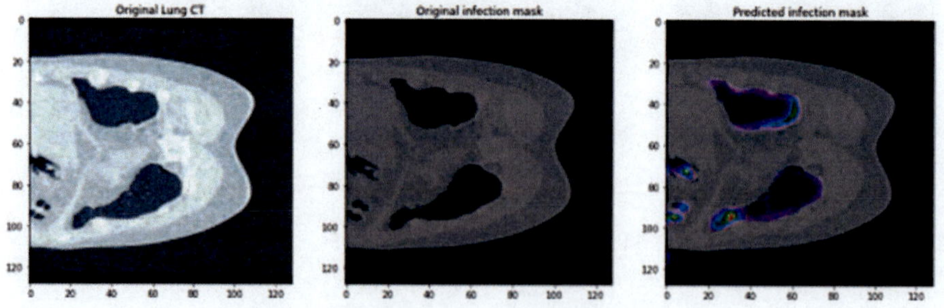

Fig. 5 Pixel distribution of original and predicted lung CT scan image

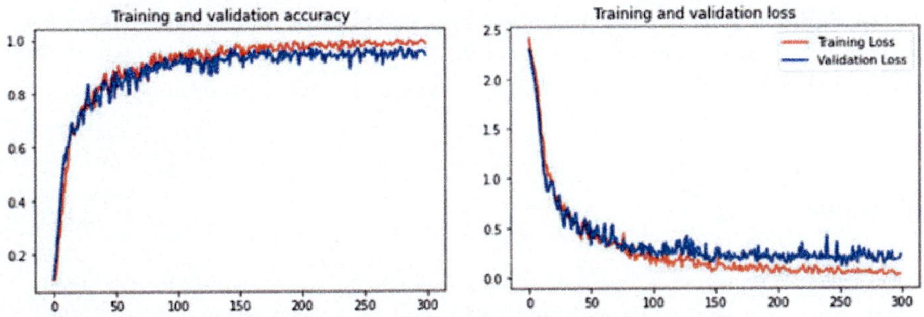

Fig. 6 Training, validation loss and accuracy of proposed model

Table 1 Performance metrics of 18-Convolutional Layered Deep U-Net architecture

Class type	Pixel accuracy (%)	Intersection over union (0–50)	Dice coefficient (%)
Threshold technique	85.49	35.28	83.74
Edge technique	87.64	38.94	86.23
Region-based technique	91.74	42.31	90.85
Clustering technique	92.42	42.67	92.17
Proposed deep U-Net technique	98.32	48.7	97.56

accuracy of 18-Convolutional Layered Deep U-Net architecture are shown in Fig. 6 and Table. 1.

5 Conclusion

This paper attempts to provide the object detection of lung area and infected area in the lung CT scan images using the proposed 18-Convolutional Layered Deep U-Net architecture. The project is implemented with Python under NVidia Tesla V100 GPU server with 1,940,817 trainable parameters, training epochs of 300, batch size of 32 and training time of 1531 min. This paper attempts to contribute in proving the proposed 18-Convolutional Layered Deep U-Net architecture performs more superior to object detection with lung CT scan images when compared to traditional object detection techniques like edge-based, region-based, clustering and Threshold methods. The object detection of lung CT scan images was done, and the pixel accuracy, intersection over union and dice coefficient are compared with our proposed

18-Convolutional Layered Deep U-Net architecture. Implementation results show that proposed model has pixel accuracy of 98.32%, intersection over union of 48.7% and dice coefficient of 97.56% compared to existing object detection models.

References

1. Nakagomi K, Shimizu A, Kobatake H, Yakami M, Fujimoto K, Togashi K (2013) Multi-shape graph cuts with neighbor prior constraints and its application to lung segmentation from a chest CT volume. Med Image Anal 17(1):62–77
2. Hu S, Hoffman EA, Reinhardt JM (2001) Automatic lung segmentation for accurate quantitation of volumetric X-ray CT images. IEEE Trans Med Imaging 20(6):490–498
3. Korfiatis P, Skiadopoulos S, Sakellaropoulos P, Kalogeropoulou C, Costaridou L (2007) Combining 2D wavelet edge highlighting and 3D thresholding for lung segmentation in thin-slice CT. Br J Radiol 80(960):996–1004
4. Sluimer I, Schilham A, Prokop M, van Ginneken B (2006) Computer analysis of computed tomography scans of the lung: a survey. IEEE Trans Med Imaging 25(4):385–405
5. Ozturk T, Talo M, Yildirim EA, Baloglu UB, Yildirim O, Acharya UR (2020) Automated detection of COVID-19 cases using deep neural networks with X-ray images, Comput Biol Med 6(121)
6. Rahimzadeh M, Attar A (2020) A modified deep convolutional neural network for detecting COVID-19 and pneumonia from chest X-ray images based on the concatenation of Xception and ResNet50V2. Inform Med Unlocked 19
7. Yan Q, Wang B, Gong D, Luo C, Zhao W, Shen J (2020) COVID-19 chest CT image segmentation: a deep convolutional neural network solution
8. Voulodimos A, Protopapadakis E, Katsamenis I, Doulamis A, Doulamis N (2020) Deep learning models for COVID-19 infected area segmentation in CT images, Cold Spring Harbor Laboratory
9. Fan DP, Zhou T, Ji GP, Zhou Y, Chen G, Fu H (2020) Inf-Net: automatic COVID-19 lung infection segmentation from CT images
10. Zu ZY, Jiang MD, Xu PP, Chen W, Ni QQ, Lu GM, Zhang LJ (2020) Coronavirus disease 2019 (COVID-19): a perspective from China. Radiology
11. Apostolopoulos ID, Bessiana T (2019) COVID-19: automatic detection from x-ray images utilizing transfer learning with convolutional neural networks. arXiv:2003.11617
12. Song Y, Zheng S, Li L, Zhang X, Zhang X, Huang Z, Chong Y (2020) Deep learning enables accurate diagnosis of novel coronavirus (COVID-19) with CT images. medRxiv
13. Lan L, Xu D, Ye G, Xia C, Wang S, Li Y, Xu H (2020) Positive RT-PCR test results in patients recovered from COVID-19. Jama 323(15):1502–1503
14. Caobelli F (2020) Artificial intelligence in medical imaging: game over for radiologists. Eur J Radiol 126:108940

An Innovative Cloud Security and Data Sharing Across Multi-clouds Through PACM Techniques

E. V. N. Jyothi, Kaja Masthan, and Maddala Kranthi

Abstract Cloud is providing virtual environment in order to share the data across multiple owners and users across pervasive computing environments. The security among virtual environment is the key challenge of the research in providing role-based access control policies and providing security to the confidential data of the cloud owners. Cloud generally provides low-cost services and provides the end-to-end solutions through managing the documents and various files of the data owner, and data of the owners is available in various formats. Key management and cryptographic techniques are widely used to provide security to the data transmitted across insecure networks. The secret key management is providing the valid keys to authenticate through the cloud-based services and access with valid keys and providing security without any loopholes in the transmission and sharing the data through owners and cloud users. A symmetric key management algorithm is used to provide a unique to each and every cloud user and a group key to access the permissions and allows particular used to gain access to the cloud services using key pairs for valid authentication process. Providing security across multi-cloud environment through PACM techniques and implementing most advanced secured key management algorithms to overcome the existing data leakages across cloud computing environment.

Keywords Advanced cloud · S3-buckets · Access controls · Privacy preserving-based key distribution · Sharing techniques

E. V. N. Jyothi (✉)
Department of Information Technology, Keshav Memorial Institute of Technology, Hyderabad, India
e-mail: jyothiendluri@gmail.com

K. Masthan
Department of Computer Science and Engineering, Sphoorthy Engineering College, Hyderabad, India

M. Kranthi
Department of Computer Science and Engineering, PACE Institute of Technology and Sciences, Ongole, India

V. K. Gunjan and J. M. Zurada (eds.), *Proceedings of 3rd International Conference on Recent Trends in Machine Learning, IoT, Smart Cities and Applications*, Lecture Notes in Networks and Systems 540, https://doi.org/10.1007/978-981-19-6088-8_59

1 Introduction

The next and advanced generation of virtual computing is providing pay as you go by cloud computing, in which we provide both centralized computing-based services and resources (hardware and software) through a network of centralized resources. Cloud computing is a huge saving, the application and processing, mobile-based applications, operating system, network computing using VPNs, and several services of AWS cloud are integrated with global infrastructures, and all the specified functions are centralized in a large server called cloud server. The following features come in a variety of forms required by the transmitter and can be accessed through systems, mobile, tabs, and other required media.

A brief discussion of common cloud applications is a sign of abstraction in complex infrastructure in a central location. Cloud computing provides reliability over long distances access to any media with user data, software, software, security, and computing.

The advanced cloud-based services like SaaS and PaaS are centralized and consist of resources of hardware, mobile-based and web-based software, and application resources. The party and other third-party users get the opportunity to use the resource in the way they want. These services often provide access to high-end networks of advanced software applications and server computers. The upcoming generation of web-based and network-based accessing resource through Internet and distributed computing will be cloud computing. With cloud computing, we can reduce infrastructure size, support large systems and green computing with a single centralized system that provides resource services to a wide range of users. The cloud architecture is used to overcome the investment and service gaps and get rid of the proposed attacker architecture. In a local network, we deploy servers and use the majority of the server and proxies integrated facilities through the connected network. Local network server performs HPC centrally using power and other underlying resources, typically performs the functionalities and accessing the services across user-centric applications such as a banking and financial information portfolio, and provides a database or power for large-scale, computer game delivery. The above technology is similar to that implemented in cloud computing to expand consumer networks of cloud-based servers, usually connected to various Internet environments and wireless communication. These servers operate at a very low cost for the user's computer or mobile phone. It contains a shared cloud server and information technology infrastructure connected to core systems and resources. Virtualization and sharing techniques on cloud servers consume resources and empower cloud computing for a wide range of users or researchers.

2 Literature Survey

Cloud computing with internal data exchange and low maintenance allows for better resource utilization. For cloud computing, the cloud service provider offers customers unlimited storage space to store data [1]. It can help customers reduce the financial costs of information management by moving local management systems to cloud servers. However, security is a major concern as we make the information garage more secured and confidential across the cloud-based vendors.

To keep statistics private, it is not uncommon for clients to encrypt data documents before uploading them to the cloud. Unfortunately, for cloud dynamic companies, for example, it is quite difficult to design a convenient and efficient data sharing scheme. Initially, the system has been provided a cryptographic storage system that allowed reliable statistics to be shared on trusted servers based on a file sharing strategy. Encrypt each file groups and file group with a document lock key. However, the device has overstated the distribution of keys because it wants the registry keys to be updated, and the user wants to pay for the cancelation. Other record sharing schemes provided with the reliable servers are recommended in [2, 3].

However, the single owner method may interfere with the implementation of any program. It was stated that the scheme would be available to users who had changed or revoked it, and that they would not be able to access the statistics to be shared again as soon as it was blocked. However, the scheme will suffer from a conspiracy attack through the blocked user and the cloud. The blocked user can use it, and its private key can be retrieved after decrypting the encrypted data record and deleting the game statistics in a cloud conspiracy. During the document access phase, the blocked user sends the request to the cloud and then the cloud examines and responds to the person who returned the relevant encrypted information document and revocation list.

The paper has described the main objectives of the proposed scheme design in terms of key distribution, information security, access control, and efficiency. Key distribution requirements: Users can obtain private management keys from the group manager without any certification. In other existing schemes, the communication channel achieves this goal if it is considered reliable, but according to our scheme, we can achieve it without this strong imagination. Data confidentiality: Benefits: A member of a group can save and share data files with other members of the group. User cancelation can be done without any other intervention, which means that the remaining users do not need private key update Network-based architecture provides different policies between the data owner, user, and cloud administrator. Cloud admin provides different policies regarding application sharing, uploading, downloading, and storing data based on policy key holders.

3 Key Exchange Mechanism

The group key integration cryptosystem includes the Group-Key Aggregate-based Cryptosystem algorithms. The data owner installs most of the people using Key-Setup and generates a public and non-public key and combines the use of Key_Gen. Confidential documents are encrypted using the encryption-based algorithm. The statistician will generate a combination using the master secret key to decrypt a group of documents.

The generated keys can be safely handed over to the delegates (via a calm email or convenient tools). Finally, any user with a mixed key can decode the statistical report and load it down. Figure 1 shows the structure of the gadget. The state of the users in this structure for example, 1 person, wants to upload a document to the cloud, while user 2 wants to download information from the cloud.

When user 1 imports data and files, the information is first encrypted using the DES rule set, and after the record, we received clouds. Separately, it generated private key. Sharing keys are across multiple parties of cloud. Only authorized user can access the key for one-time download.

The green building key-aggregation method is used to combine non-public keys and create a fixed size key called a mixed key. For the boot module, the receiver can load the recording using the mixing button sent by mail with the help of the sender. If the mixed key is legitimate, you will be allowed to download the document. When the recipient downloads the file, the mixed key is approved or the extraction is done with non-public keys.

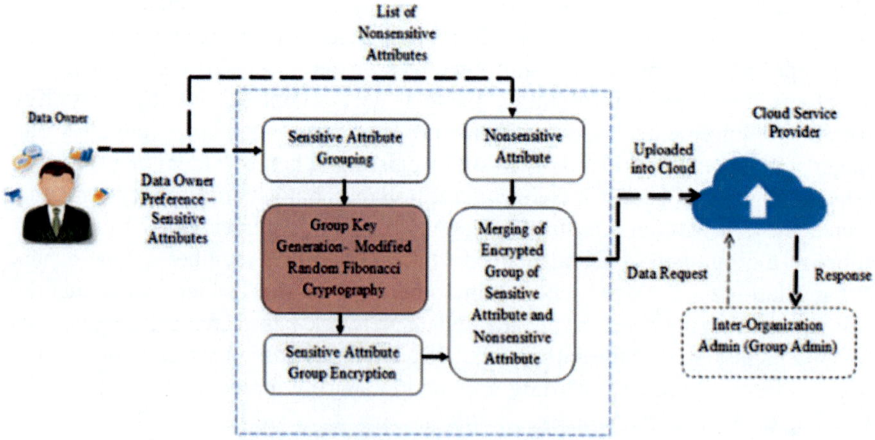

Fig. 1 Cloud-based key exchange architecture

4 Methodology of Key Exchange

In this program, the installation is performed by placing and extracting the various rar file formats on a cloud server. The data owner performs the configuration of the account on the server. The most effective configuration package requires covert protection option. This segment is made by the registry owner to create a public or master key pair. This segment is performed by someone who wants to send encrypted information. Encryption and encryption algorithms come in the form of public parameters such as pk and messages m, and I express the magnificence of the encrypted text. Different methods and techniques are applied in order to encrypt the given messages and retrieve the encrypted text C, so that the message can be used by the authorized user in the most efficient way. Such a flow of key exchange between client and server based on policies associated in cloud has been shown in Fig. 2.

4.1 Algorithm for Encryption

The key-based aggregation provides advanced cloud-based solution using encryption algorithms. In the key aggregate cryptosystem applying the hash function enc = A_HASH(m), the return value of the hash function is used as a signature. To encrypt the given message m1 using a key k1 to generate the cypher text c1,

Encryption C = E_(K)(P).
Decryption P = E_(K)^(-1)(C).
E_(K) is chosen from a family of transformations known as a cryptographic system.
The parameter that selects the individual transformation is called the key K, selected from a key space K.
More formally, a cryptographic system is a single parameter family of invertible transformations.
E_(K): K in K:P- > C.

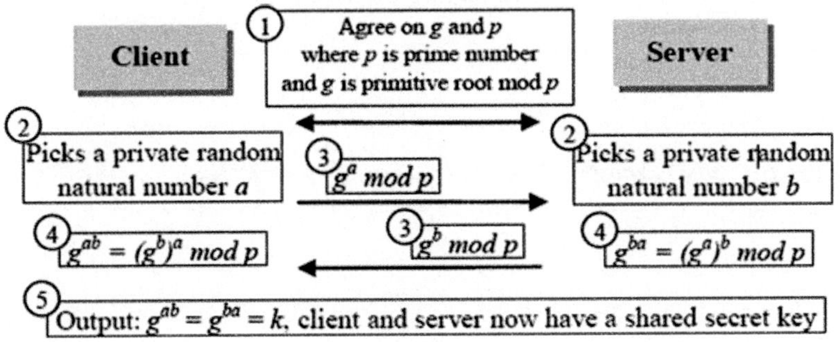

Fig. 2 Flow of key exchange between client and server

With the inverse algorithm E_(K))^(-1): K:C- > P such that the inverse is unique. Usually assume the cryptographic system is public, and only the key is secret information.

4.2 Extract My_Aggregate_Key1 (D1, S1)

aggr_key1 = D1.

s1 < - S1.size ().

i1 < - 1.

While (i1 < = s1).

aggr_key1 < - aggr_ key1 * S1[i1].

Return aggr_key1.

4.3 Algorithm for Decryption

The public key is (n = 3233, e = 17). For a padded Plain—Text message m, the encryption function is as follows:

c = m17 mod 3233.

Suppose m = 65, then cipher text c.

C = (65)17 mod 3233 = 2790.

The private key is (n = 3233, d = 2753). For an encrypted Cipher—Text c. The decryption function is as follows:

m = c2753 mod 3233.

In the above example c = 2790, if we get message m again then:

m = (2790)2753 mod 3233 = 65.

4.4 Extract_My_Decryption (C1, Aggr1, Key1, S1)

S1 < - S1.size().
i1 < - 1.
while(S1! = empty).
temp1 = temp1 * S1[i1].

dec_data_1 = aggr_ key1/temp1 i < - 1.
while S1! = empty1.
derypt1(key1,document1).

We compared the performance modeling with Mona using NS2 [4] and showed the first dynamic radiation encryption (ODBE) scheme [5]. Without losing the unity, the data representation size is 16 bits, which gives the group capacity of the data files. Similarly, user and group recognition is set to 16 bits. The process for both team members and group manager takes place on various virtual machines with different capacities that are enclosed.

5 Results and Discussion

We have chosen an elliptic curve to order a group of 160 bits. Computational costs of members have been figured by comparing with many algorithms in file uploading. As shown in Fig. 3, we listed a comparative cost of members for file uploads between ODBE-RBAC-Mona and our schema. The proposed scheme does not depend on the number of invalid users.

The reason for this is that in our scheme, the user authentication process is delegated to the group manager, so that legitimate clients encrypt the data file without the involvement of other clients, including both legitimate and non-legitimate ones, blocked, and deleted or blocked customers. Conversely, the computational price will be increased with the count of the users who are disabled based on ODBE. This is because the client must perform a number of operations, including point multiplication and exponential, to calculate the parameters.

6 Conclusion

In this paper, we developed anti-conspiracy documents. The sharing scheme of data in the virtual instance indicates cloud-based services offered by the multinational and software companies across the cloud environment. In advanced schemes, the customer's data is protected with advanced and secured methods. Clients obtain their non-public keys from the group manager's certification office communication channels. Our scheme is able to effectively assist dynamic organizations when and where a new customer registers and authenticates and join the organization or becomes a user will be blocked, other person's private keys users do not want to recalculate or update. In addition, our scheme allows finding a comfortable person if blocked or deleted users will not be able to get it as soon as the initial data documents appear, although they conspired with the cloud of unbelief.

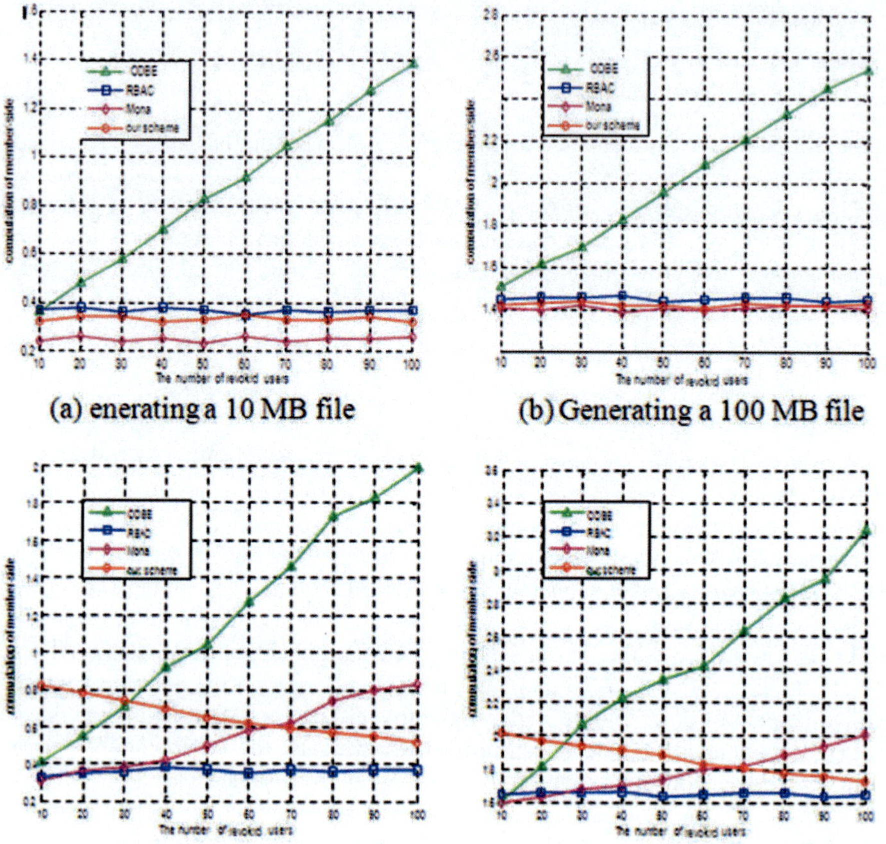

(a) enerating a 10 MB file (b) Generating a 100 MB file

(a) Downloading a 10 MB file (b) Downloading a 100 MB file

Fig. 3 Generating and downloading 10–100 MB files

References

1. Garg N, Yadav P (2014) Comparison of asymmetric algorithms in cryptography. J Comp Sci Mob Comput (IJCSMC) 3(4):1190–1196
2. Timothy DP, Santra AK (2017, Aug) A hybrid cryptography algorithm for cloud computing security. In: 2017 International conference on microelectronic devices, circuits and systems (ICMDCS). IEEE, pp 1–5
3. Maqsood F, Ahmed M, Ali MM, Shah MA (2017) Cryptography: a comparative analysis for modern techniques. Int J Adv Comput Sci Appl 8(6):442–448
4. Jimbo K, Iriyama S, Regoli M (2020) Implementation of a new strongly-asymmetric algorithm and its optimization. Cryptography 4(3):21
5. Sann Z, Soe T, Knin K, Win Z (2019) Performance comparison of asymmetric cryptography (case study-mail message). APTIKOM J Comp Sci Inf Technol 4(3):105–111

Performance Analysis of Classification Algorithms

A. Prakash and Vijender Kumar Solanki

Abstract Classification issues are crucial in machine learning and data mining. Classification difficulties are utilised in medical diagnostics, bank customer estimation, medicinal investigations and emotion analysis. Many categorisation methods have been created with many different parameter inputs. Using hyperparameter optimisation algorithms, this work aims to improve classification success. The 'heart and iris' data sets have been classified using K-nearest neighbour, SVM, decision tree and gradient boost techniques. The hyperparameter optimisation algorithms grid search and random search are applied to these selected classification algorithms. Experiments have shown that using hyperparameter optimisation algorithms improves the performance of all classification algorithms. The best parameter values are shown.

Keywords Optimisation · Hyperparameter optimisation · Machine learning · Supervised learning · Classification algorithms

1 Introduction

Machine learning is a branch of artificial intelligence that uses data to generate predictions. Machine learning uses training data to build predictive models. Many parameters in supervised and unsupervised machine learning algorithms must be exported before training. These are hyperparameters. It is critical to adjust these parameters correctly to properly train the model and receive accurate results. Because each algorithm's parameters have a large range of values, restricting the optimisation process helps. Machine learning is classified as supervised, unsupervised and reinforcement learning [1].

Supervised learning works by labelling each data set after adding the training and test data sets to the system to understand and estimate new data based on existing data [2]. In supervised learning, classification and regression;

A. Prakash (✉) · V. K. Solanki
CMR Institute of Technology, Hyderabad, India
e-mail: prakash15aug@gmail.com

V. K. Gunjan and J. M. Zurada (eds.), *Proceedings of 3rd International Conference on Recent Trends in Machine Learning, IoT, Smart Cities and Applications*, Lecture Notes in Networks and Systems 540, https://doi.org/10.1007/978-981-19-6088-8_60

Unsupervised learning is when data are loaded unlabelled without user input. The technique detects data relationships and is used for clustering [3].

Unlike supervised and unsupervised learning, reinforcement learning does not require labels or data. The reinforcement learning method is analogous to biological learning. The application interacts with a dynamic environment to teach. The environment rewards or punishes him for his activities. It then evaluates the feedback and takes action. This cycle continues until learning occurs [4].

Classification is a critical issue in machine learning and data mining. The approaches' performance demonstrates the model's efficacy. Hyperparameter optimisation approaches try to improve system performance by making predictions based on training data. The classification algorithms in this study are optimised using hyperparameters. The experiments improved the classification success of all classification algorithms.

2 Literature Review

A typical machine learning model was optimised by François-Lavet et al. [5]. The use of developed optimisation approaches to machine learning algorithms is reviewed.

Shaik et al. [6] explored algorithm selection and hyperparameter optimisation. They chose a learning algorithm while tuning hyperparameters.

Yang et al. [7] offer methods to fully solve a single-level optimisation problem using parametric programming and a two-level optimisation problem using K-fold cross-validation. Their methods are called (HY-POP) two-level parametric optimisation. Mishra et al. [8] suggest an initialisation technique for hyperparameter optimisation. Unlike current initialisation procedures, theirs do not rely on prior hyperparameter setups or meta-properties. Their proposed initialisation strategies, classical initialisation strategies, and other data transfer methods have been improved.

Thornton et al. [9] present hyperparameter optimisation results for neural networks and deep belief networks. They tuned the hyperparameters utilising RA and two new greedy sequential approaches. RA has been proved to efficiently learn neural networks for several data sets. Large hierarchical models will place an increasing load on hyperparameter optimisation strategies, according to Sahu et al. [10]. To evaluate success in developing adaptive hyperparameter optimisation methods, the researchers used RA.

Tso et al. [11] investigated the performance of machine learning models and their hyperparameters. Thus, the hyperparameter tuning problem is an optimisation problem solved by Bayesian optimisation. The suggested technique discovers the appropriate hyperparameters for popular machine learning models like the random forest algorithm.

The voting-based ensemble classification weights were optimised by Wistuba et al. [12].

Using hyperparameter optimisation to obtain the best values for group categorisation based on weighted voting has been proven useful in experiments.

3 Analysis and Design

As illustrated in Fig. 1, the data set is initially trained with K-fold cross-validation ($k = 10$), followed by classification methods such as K-nearest neighbour (KYK), support vector machines (SVMs), decision tree (Eng. Decision Tree) and gradient boost (GA). This method divides the data set into k subsets, each of which is treated as test data and the rest as training data. After k repetitions, the results are averaged. As a result, the standard deviation of the obtained data is not presented, only the average results from the literature. The classification success rates of each of these classification algorithms were then compared using the hyperparameter optimisation methods grid search (IA) and random search (RA). Truck. K-fold cross-validation using RA or IA for hyperparameter optimisation can be generalised.

A. Classification Algorithms

Classification algorithms learn from labelled data to classify fresh data. The study's classification algorithms are listed below.

(1) K-Nearest Neighbour (KYK): KYK is a simple supervised machine learning technique that uses distance between data points to identify them. In KYK, each test sample's projected class is the class in which most of its k-nearest neighbours are. The KYK algorithm parameters are as follows:

- N-neighbours: The nearest neighbour's number. This number of neighbours is used to classify.
- Weights: uniform or distance. Equally weights each neighbour. Distance, on the other hand, magnifies the impact of close neighbours.
- P: Manhattan distance is used when $P = 1$; Euclidean distance is used when $P = 2$.

(2) SVMs are supervised learning algorithms that can be utilised for both classification and regression issues. SVM techniques work by converting data points from low-dimensional to high-dimensional space and then using a hyperplane as a classification boundary.

The DVM algorithm parameters are as follows:

- A: Regularity parameter. Inversely proportional to C, regulatory power it should be positive.
- Kernel: Kernel parameters select the type of hyperplane used to separate data. 'linear' uses a linear hyperplane, while 'rbf' and 'poly' use a nonlinear hyperplane.
- These three terms use gamma as their core coefficient. More fitting to the training data means higher gamma. Excessive gamma might induce overfitting.

(3) Decision Tree (KA): A decision tree (KA) is a typical categorisation method that uses a tree structure to model decisions and alternative outcomes. All data are represented by the root node; numerous decision nodes display decision

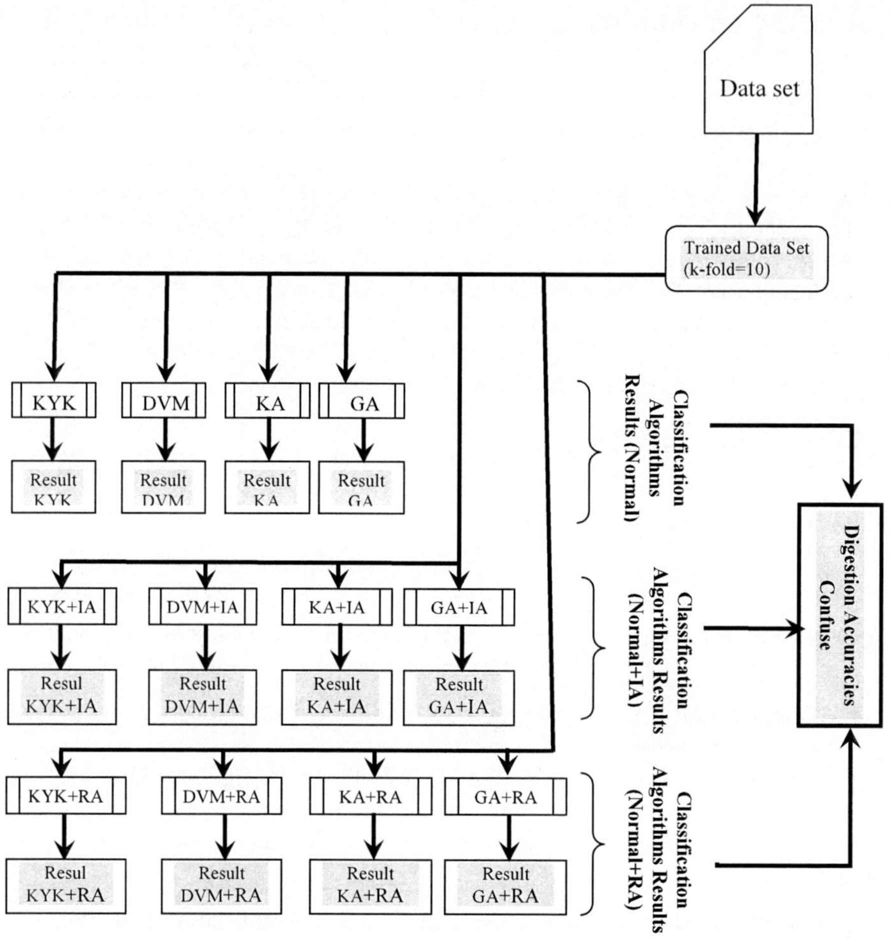

Fig. 1 System model

tests, and subnode splits on each feature; and several leaf nodes show outcome classes.

The KA algorithm parameters are as follows:

- Criterion: Selects the quality of the split. Takes the value 'gini' or 'entropy'.
- Max-depth: The tree's maximum depth. It is assumed that the value is limitless.
- Min-sample-split: Minimum sample count before splitting.
- Min-samples-leaf: The minimum number of samples a leaf should have.

(4) A machine learning strategy for regression and classification issues, gradient boosting (GA). This develops a decision tree model by combining weak predictive models. Its goal is to identify and reduce a loss function.

The GA algorithm's parameters are as follows.

- Loss: Loss function to be optimised.
- Learning rate: The rate at which each tree reduces its contribution.
- N-estimator: Represents the number of trees in the forest. In general, the higher the number of trees, the easier it is to learn the data. However, adding a large number of trees can greatly slow down the training time.
- Criterion: The function of measuring the quality of a split.
- Max-depth: Shows the maximum depth of the tree. The deeper the tree, the more it splits and the more information it has about the data.
- Min-samples-split: It expresses the minimum number of samples to split a node.
- Min-samples-leaf: Indicates the minimum number of samples that must be in a leaf node.
- Max-features: Refers to the number of features to consider when searching for the best bin.

B. Hyperparameter Optimisation Algorithms

Many hyperparameter optimisation approaches exist, including decision-theory, metaheuristic and Bayesian. Using training data, machine learning creates predictive and decision-making models. Machine learning methods require some parameters to be set up front. These characteristics are termed hyperparameters and must be established beforehand by the model designer. This is dependent on the problem and data set. The model's success depends on the hyperparameter setting. In this study, decision theory-based IA and RA optimisation approaches are applied.

This method explores extensively for optimal configuration in a fixed hyperparameter space [9]. Some hyperparameters accept infinity. As a result, when optimising hyperparameters, the value ranges are constrained by the problem. The ranges are stored as a list. IA selects the best parameter group by combining values within the defined ranges.

Given a limited execution time and resources, random search (RA) picks random hyperparameter combinations from the search space. Each hyperparameter configuration is processed in IA and RA. As illustrated in Fig. 2, IA tests only three places while RA tests all nine tracks for different values. IA searches for all potential parameter values in the search space, whereas RA searches them randomly.

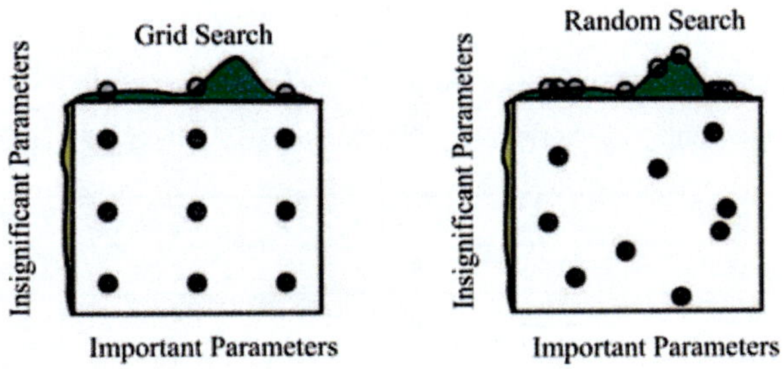

Fig. 2 Example demonstration of IA and RA optimisation methods [12]

4 Experimental Studies

A. **Data sets**

The study employed UCI and Iris data. The heart data collection has 14 characteristics and 303 samples. Age, gender, blood pressure, blood sugar shows the symptom of chest discomfort. This data is used to assess heart health. Features like sepal length, sepal breadth, petal length, petal width and variety are used to detect iris.

B. **Experimental Setup**

The scikit-learn library in Python (3.7.6) implements classification and optimisation methods. The computer used has an Intel(R) Core(TM) i7-9750H CPU, 2.60 GHz, 32 GB RAM and Nvidia GeForce RTX 2060.

C. **Experimental Results**

Classification accuracy is the ratio of successfully identified samples to total samples, as illustrated in (1). The performance metric employed is accuracy.

True Positives (TP): The observation is positive and has been predicted to be positive.
True Negatives (TN): The observation is negative and has been negatively estimated.
False Positives (FP): The observation is negative but predicted positively.
False Negatives (FN): The observation is positive but predicted negatively.

$$\text{truth} = \frac{\text{TP} + \text{TN}}{\text{TP} + \text{FP} + \text{TN} + \text{FN}}$$

Heart Data set Results: This section illustrates the classification algorithms' success rates in the heart data set. Table 1 compares the classification algorithms' accuracy percentages with the IA and RA optimisation methods' accuracy percentages. All classification methods improved between 2.08 and 9.24% when IA hyperparameter tuning was used.

Table 1 also compares the classification algorithms' accuracy percentages with the RA optimisation methods' accuracy percentages. All classification methods improved between 2.03 and 7.77% while using RA hyperparameter tuning.

Table 1 Classification accuracy in heart data set

Algorithms	Normal (%)	Normal + IA (%)	Increase (%)	Normal + RA (%)	Increase (%)
KYK	81.82	84.65	3.46	83.48	2.03
DVM	82.21	83.92	2.08	83.88	2.03
KA	76.22	83.26	**9.24**	82.14	**7.77**
GA	80.52	83.88	4.17	83.81	4.09

Iris Data Set Result: In this section, the classification algorithms' success rates are shown in the iris data set. Table 2 compares the classification algorithms' accuracy percentages with the IA and RA optimisation methods' accuracy percentages. All classification methods improved between 4.20 and 7.24% with IA hyperparameter adjustment. All classification methods improved between 3.86 and 6.75% when using RA hyperparameter tuning.

Figure 3 visualises the performance evaluation of the cardiac data set approaches. For this data set, the IA optimisation method outperforms the RA optimisation method. Optimisation approaches for SVM and GA algorithms produce fairly similar results. Applied optimisation strategies boost success in all data sets.

Figure 4 displays the performance evaluation of the iris data set approaches. In this data set, the IA optimisation method outperforms the RA optimisation method. Also, when applied to the SVM and KA algorithms, the RA optimisation method outperforms the IA optimisation method.

Table 3 displays the optimum parameter values for all classification methods in the heart and iris data sets after performing each optimisation strategy (IA and RA).

Table 2 Classification accuracy in iris data set

Algorithms	Normal (%)	Normal + IA (%)	Increase (%)	Normal + RA (%)	Increase (%)
KYK	92.01	98.67	**7.24**	98.22	**6.75**
DVM	94.61	98.58	4.20	99.34	5.00
KA	92.02	96.66	5.04	97.37	5.81
GA	93.33	98.55	5.59	96.93	3.86

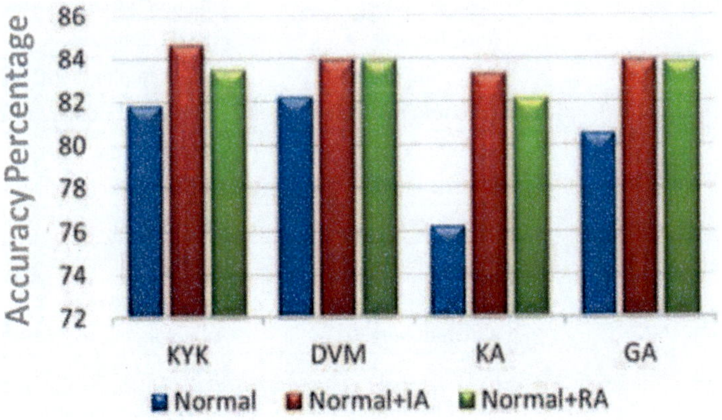

Fig. 3 Performance evaluation of algorithms (heart data set)

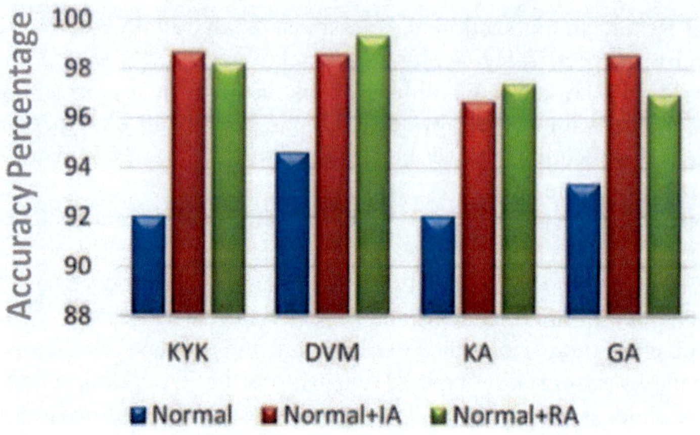

Fig. 4 Performance evaluation of algorithms (iris data set)

Table 3 Best parameter values in heart and iris data sets

Data set	Algorithms	Best parameters (IA)	Best parameters (RA)
HEART	KYK	n_neighbours: 5	n_neighbours: 6
		p: 1	p: 1
		weights: uniform	weights:distance
	DVM	A: 10	A: 100
		gamma: 0.1	gamma: 0.1
		kernel: rbf	kernel: poly
	KA	criterion: entropy	criterion: gini
		max_depth: 3	max_depth: 4
		max_features: 7	max_features: 5
		min_samples_split: 3	min_samples_split: 8
	GA	criterion: mae	criterion:friedman_mse
		learning_rate: 0.5	learning_rate: 0.1
		loss: deviance	loss: deviance
		max_depth: 1	max_depth: 1
		max_features: sqrt	max_features: sqrt
		n_estimators: 64	n_estimators: 32
IRIS	KYK	n_neighbours: 9	n_neighbours: 6
		p: 1	p: 2
		weights: uniform	weights: distance
	DVM	A: 1000	A: 1000
		gamma′: 0.001	gamma: 0.001
		kernel: sigmoid	kernel: rbf′

(continued)

Table 3 (continued)

Data set	Algorithms	Best parameters (IA)	Best parameters (RA)
	KA	criterion: gini	criterion: gini
		max_depth: 3	max_depth: 6
		max_features: 2	max_features: 3
		min_samples_split: 6	min_samples_split: 4
	GA	criterion: mae	criterion: friedman_mse
		learning_rate: 0.5	learning_rate: 0.5
		loss: deviance	loss: deviance
		max_depth: 2	max_depth: 7
		max_features: log2	max_features: sqrt
		n_estimators: 4	n_estimators: 2

5 Conclusions and Future Work

The performance of KYK, SVM, KA and GA algorithms was evaluated using the hyperparameter optimisation methods IA and RA. The algorithms' categorisation accuracy is compared using heart and iris data sets. The goal is to improve the success rate of machine learning algorithms in these data sets. The success of all classification methods using IA hyperparameter optimisation increases from 2.08 to 9.24% in the cardiac data set when using RA hyperparameter optimisation. It rose between 2.03 and 7.77%. Success of all IA hyperparameter optimisation techniques on the iris data set. It increased between 4.20 and 7.24% when RA hyperparameter optimisation was used, but only 3.86–6.75% when it was not. The IA optimisation method outperformed the RA method in terms of classification accuracy because it calculates all potential states in the solution space. However, when data sets grow in size and number of characteristics, the IA technique becomes less efficient, making the RA method more efficient.

Future work planned; (a) Evaluation of performance utilising different optimisation approaches. (b) Compare the performance of hyperparameter optimisation techniques on various study subjects using published data sets. (c) Vote weighted on optimisation methods, new hybrid approaches using the lama method.

References

1. Kızılkaya YM, Oğuzlar A (2018) Comparison of some supervised learning algorithms with R programming language. Karadeniz Int Scien J 37(37):90–98
2. Aydın S, Özkul AE (2015) Data mining and an application in Anadolu University open education system. J Res Educ Teaching 4(3):36–44
3. Alpaydın E (2010) Introduction to machine learning, 2nd edn. The MIT Press, Cambridge, S 4

4. Gunjan VK, Prasad PS, Mukherjee S (2020) Biometric template protection scheme-cancelable biometrics. In: ICCCE 2019. Springer, Singapore, pp 405–411
5. François-Lavet V, Henderson P, Islam R, Bellemere MG, Pineau J (2018) An introduction to deep reinforcement learning
6. Shaik F, Sharma AK, Ahmed SM, Gunjan VK, Naik C (2016) An improved model for analysis of diabetic retinopathy related imagery. Indian J Sci Technol 9:44
7. Yang L, Shami A (2020) On hyperparameter optimization of machine learning algorithms: theory and practice. Neurocomputing 415:295–316
8. Mishra B, Singh N, Singh R (2014, Dec) Master-slave group based model for co-ordinator selection, an improvement of bully algorithm. In: 2014 International conference on parallel, distributed and grid computing. IEEE, pp 457–460
9. Thornton C, Hutter F, Hoos HH, Leyton-Brown K (2013) Auto-WEKA: combined selection and hyperparameter optimization of classification algorithms. In: Proceedings of the 19th ACM SIGKDD international conference on knowledge discovery and data mining, pp 847–855
10. Sahu H, Singh N (2018) Software-defined storage. In: Innovations in software-defined networking and network functions virtualization. IGI Global, pp 268–290
11. Tso WW, Burnak B, Pistikopoulos EN (2020) HY-POP: hyperparameter optimization of machine learning models through parametric programming. Comp Chem Eng 139:106902
12. Wistuba M, Schilling N, Schmidt-Thieme L (2015) Learning hyperparameter optimization initializations. In: 2015 IEEE international conference on data science and advanced analytics (DSAA), pp 1–10

Intelligent Information System for Detection of COVID-19 Based on AI

Roshani Raut, Anuja Jadhav, Swati Jaiswal, Atul Kathole, and Sonali Patil

Abstract The pandemic is changing the clinical needs and potential for AI-driven computer-assisted diagnoses (CDS). Since the beginning, rapid identification of COVID-19 patients has been a significant difficulty, especially in areas with limited diagnostic testing capacity. Intelligent Information System (IIS) represents the knowledge progression of available data. It has been directed by recent technological integration, data processing, and distribution in multiple computational environments. Intelligent Information Systems are aimed to work like an advanced human brain, where, as per the requirement of changing circumstances, the optimal decision can be evolved. IIS tools are expected to be adaptive, which may vary according to their processing data. As a result, the goal of this study was to provide a complete analysis of various technologies for combating COVID-19, with a focus on their features, problems, and domiciliation nation. Our findings demonstrate the performance of developing technologies.

Keywords Information system · Machine learning · KDD · Natural language processing · Medical sciences

1 Introduction

An information system is a set of procedures, tools, and principles that permit us to cope with information intentionally, work competently, maintenance of business procedures, interconnect efficiently, and the most significant is to mark superior judgments. Information system with artificial intelligence is the most famous embodiment of Intelligent Information System. More specifically, it is an integration of artificial intelligence technologies, database management, and information system methodologies to create imminent information systems, known as Intelligent Information systems [1]. This research emphasizes AI-assisted methodologies and models that

R. Raut (✉) · A. Jadhav · S. Jaiswal · A. Kathole · S. Patil
Department of Information Technology, Pimpri Chinchwad College of Engineering, Savitribai Phule Pune University, Pune, India
e-mail: rosh513@gmail.com

V. K. Gunjan and J. M. Zurada (eds.), *Proceedings of 3rd International Conference on Recent Trends in Machine Learning, IoT, Smart Cities and Applications*, Lecture Notes in Networks and Systems 540, https://doi.org/10.1007/978-981-19-6088-8_61

can aid in the fight against COVID-19. The evolution of these two emerging fields, AI and IIS, for Intelligent Information Systems is explained. AI and IS have been in the study for a long time but gained recognition [2]. The primary goal of this work is to recognize and understand some conceptual order in the occurrence of IIS and AI categories proliferation.

Section 2 describes the use of IIS for detecting COVID-19. Proposed model is introduced in Sect. 3. Result and experiments are explained in Section 4, and finally, the research is concluded in Sect. 5.

1.1 History and Growth of Information System

The evolution of information systems has made substantial growth over the half-century. It was started with simple punch cards and electronic calculators. It has evolved to electronic business, electronic commercialisms, and the revolutions in mobile computing, cloud computing, and artificial intelligence. As this field has developed, new areas and research populations have appeared, and the quality of research has improved radically [3, 4]. The hierarchical levels of organizations and the way to organize information system types are shown in Fig. 1, called as pyramid model.

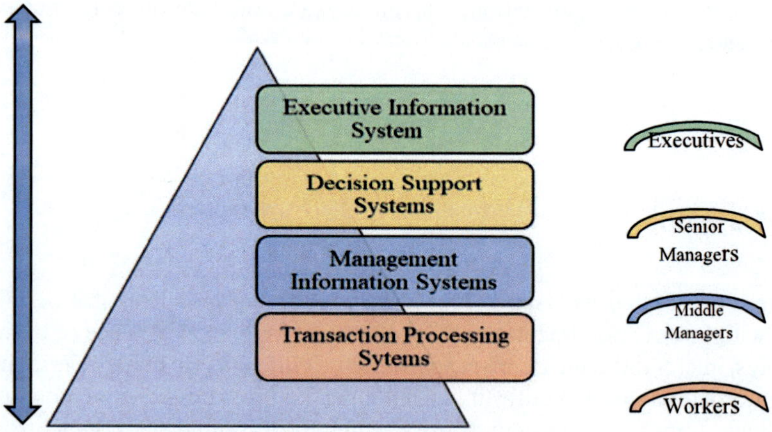

Fig. 1 Pyramid model of hierarchical level

1.2 DIKW (Data, Information, Knowledge, and Wisdom) Chain

The DIKW is a hierarchy related to data, information, knowledge, and wisdom. This hierarchy is helpful to describe the concepts related to these terms. The DIKW pyramid was presented by Ackoff [5]. These terms in the view of knowledge management can be defined as follows:

- Data: It is raw data or symbols which characterize the objects.
- Information: It is for understanding relations between raw data.
- Knowledge: It gives the patterns and helps to understanding patterns.
- Wisdom: It is for thoughtful philosophy or values. It is the capacity to improve efficiency.

The DIKW chain is shown in Fig. 2. Data always needs to be converted into information, knowledge, and wisdom to add more returns to the organization. Hence, DIKW hierarchy plays a significant role. There are various options, or alternative models of DIKW are available. DIKW model has been applied in various fields like information technology service management [6], manufacturing [7], security protection in IoT, etc.

There are various advanced machine learning techniques used in literature to solve the various real-life problems including health care [8, 9].

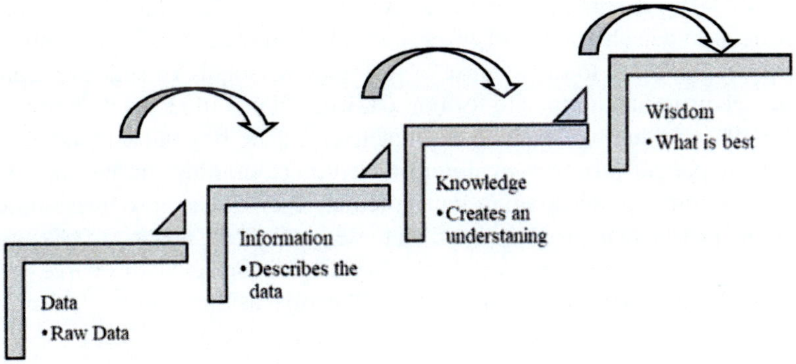

Fig. 2 DIKW chain

2 Use of IIS for Detecting COVID-19

2.1 Decision Support System

A decision-making tool is used for effectively detecting COVID-19 distribution. The reverse transcription-polymerase chain reaction (RT-PCR) is the standard gold test for COVID-19 diagnosis (RT-PCR). Despite this, RT-PCR is rarely used exclusively for diagnosis and treatment decisions due to its high false-negative rates. The clinical presentation and radiological findings of the patient are equally critical. COVID-19 has no distinguishing clinical features or computed tomography (CT) findings. Disease detection and community health protection are getting more complex due to these challenges. Even though AI-based clinical decision support appears to have matured, the use of AI-based tools in COVID-19 has been limited thus far. The prospects and requirements for AI-based clinical decision support systems are examined in this viewpoint essay, as well as the limitations that affect "AI readiness" for quickly expanding healthcare concerns. The pandemic is altering clinical requirements and the possibility for AI-driven computer-assisted diagnoses (CDS). New indications and symptoms of the disease emerge and guidelines for who should be considered persons under investigation (PUIs) for COVID-19 or tested for SARSCoV2 infection change all the time. Since the outset, identifying COVID-19 patients has been a serious challenge, especially in locations with inadequate diagnostic testing capabilities (such as most of the USA). Failure to quickly identify and isolate COVID-19 patients in hospitals allows the nosocomial infection to spread, putting patients and healthcare workers at risk. A lack of specificity in screening and isolating processes leads to the overuse of restricted personal protective equipment (PPE) and delays in obtaining care for non-COVID-related illnesses. Our approach to COVID-19 screening has been chiefly reactive to date. Beyond screening, identifying specific patient variables, symptoms, and clinical findings or measures (e.g., vital signs, laboratory, and imaging data) predictive of disease progression could lead to a more proactive approach. Based on these predictions, AI-driven CDS could be used to target earlier intervention or guide disposition (e.g., discharge versus inpatient) decision-making. Clinical criteria will evolve as the epidemic progresses.

3 Proposed Model

The introduction of artificial intelligence signaled a significant shift in information technology. Deep learning technology has been used in medical image processing in recent years. The rapid growth of deep learning has altered the field of image processing. Examining and identifying the object is beneficial in the area of medical data. COVID-19's outbreak surprised the world since it spread swiftly and became a pandemic. Its spread can be slowed if it is caught early [10, 11]. The suggested ID2S-COVID-19-DL model uses deep learning to provide a prognosis for early identification of the pandemic. The data gathering, preprocessing, and application layers are the three layers that make up this system. The data acquisition layer collects data from diverse sources such as cameras, X-rays, and CT scan devices (IoMT) through the Internet of Medical Things. Noise and blurriness appear with the gathering of this vast amount of data. The information acquired is now referred to as raw data. The preprocessing layer processes the raw data in two ways. It uses moving average, mean, and mode techniques to minimize the noise in the initial step. The red green blue (RGB) image is resized in Step 2 to preserve the desired information quality. The following layer is the performance layer, and the prediction layer is two elements of the application layer [12]. The performance layer evaluates the root mean square, mean absolute, and fundamental percentage error. The model must be retrained if the learning requirements are not met; otherwise, data is retained in the cloud, as shown in Fig. 3. If the intelligent decision support system identifies COVID-19 positivity, see a doctor; if COVID-19 is negative, no COVID-19 treatment is required [13].

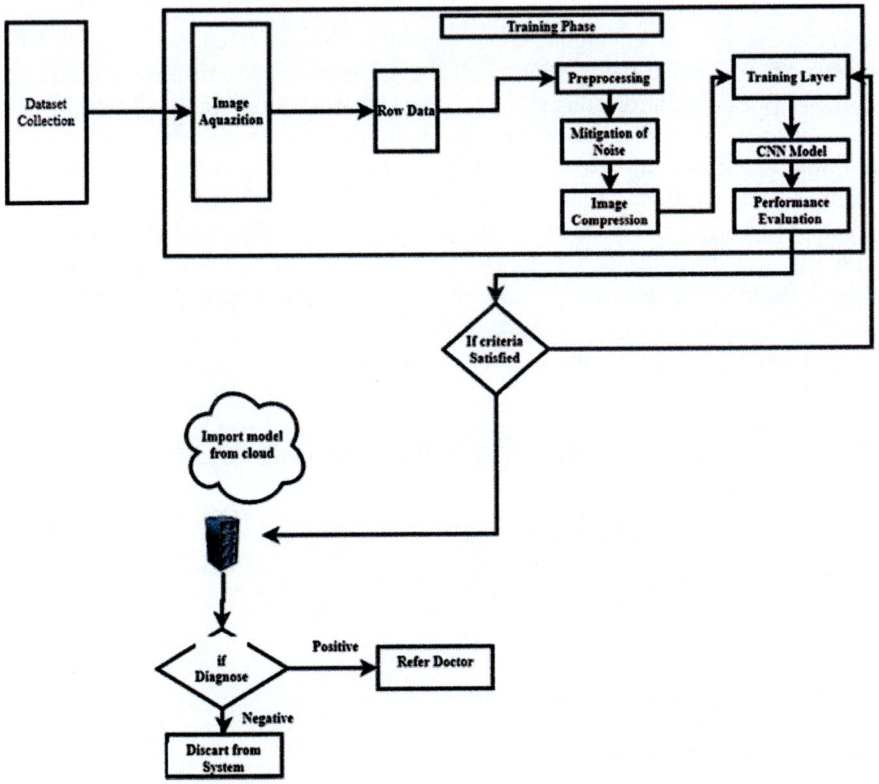

Fig. 3 Flow of proposed work for detecting the COVID-19

3.1 Performance Parameter

Each image conceals essential factual data that is helpful for AI models. Thought of the set number of qualities rather than images diminishes the computational time while accomplishing sensible outcomes. In this exploration, essential measurable data and the pre-prepared attributes were gotten from the images. Edge esteem was resolved as half of the most extreme pixel esteem inside the image, and the number of pixels more prominent and more modest than this worth was tallied. Then, at that point, the image was isolated into three portions upward, and the middle one was the largest to separate the district of revenue. The mean upsides of each section were determined independently. This interaction was performed to wipe out the corners and boundaries inside the image. The mean upsides of the Laplacian channel, honed images, and histogram leveling applied graphics were determined independently to give distinctive data to the AI models for a similar image simultaneously.

VGG1913 is similar to SVM, but it includes 19 charged layers, which provides the organization around 143 million boundaries. KNN has 50 layers left to address problems, e.g., when the organization uses time. The quality of the machine is based on the skip connections among character capability levels, which enhances model accuracy and reduces the preparation time. It has teachable borders of more than 23 million. The V315 initiation consists of 42 layers and 24 million limits. It factorizes convolutions to reduce the number of borders without reducing organizational skills. In addition, at the beginning of V3, new reverse scaling was suggested to lessen highlights. DT consists of 53 layers of teachable limits and more than 3.4 million. It includes other linkages, including extension, depth, and projection. The evolution convolutions transform the info tensor to an overhead tensor, depth convolutions are used on changing channels over tensors, and the projections convolution are finally projected to fewer tensors by the higher media.

The precision, affectability, and explicitness are determined as given in Eqs. (1–3), separately:

$$\text{Accuracy} = (\text{TP} + \text{TN})/(\text{TP} + \text{TN} + \text{FP} + \text{FN}) \tag{1}$$

$$\text{Sensitivity} = \text{TP}/(\text{TP} + \text{FN}) \tag{2}$$

$$\text{Specificity} = \text{TN}/(\text{TN} + \text{FP}) \tag{3}$$

4 Results and Discussions

The COVID-19/Normal arrangement conducted five analyses, including 14 highlights obtained from the images and five AI classificatory. For CNN and DT, conflicting results have been achieved. Furthermore, it formed the lowest mean affectability and most reduced average DT (63.10 and 81.33%), but it provided the least mean accuracy and mean peculiarity. CNN achieved the remarkable average explicitness rate (99.55%). It also produced the least medium affectable and mean DT (82.95 and 92%); the chart provides the results. The layout of COVID-19/Pneumonia included comparable AI classifiers and highlights. In the assessment, similar results were achieved, and DT provided for the COVID-19/Pneumonia accurate estimate assessments the most remarkable mean recommended approach to hybrid, mean affectability, and average accuracy scores (88.92, 80.00, and 96.96%). The hybrid method and SVM were the most notable explicit meaning (97.85% each).

Transfer Learning Experiments

Since the initial photographs achieved their most remarkable results in the hybrid technique tests presented, no pretreatment procedures have been used for the images. All investigating groups have been investigated. Moving training assessments were performed in three groups: COVID-19/Normal, COVID-19/Pneumonia, and Tests examined the two models, which would produce frequent findings for the COVID-19/Normal and COVID-19/Pneumonia groupings. SVM and DT produced the most horrible consequences during the COVID-19/Normal meeting. They are just prepared to learn one lesson and cannot group X-ray images of COVID-19. SVM and DT have provided almost favored results over MobileNet-V2 and VGG19. The average SVM and DT values were calculated independently as 65.78 and 72.64%. Beginning V3 has produced more significant results than other previously developed institutions; despite this, DenseNet 121 has obtained the most significant mean score in moving learning tests (96.48%). The result of transferring information for the COVID-19/Normal collection is shown in the Fig. 4. Comparable results were achieved in the COVID-19/Pneumonia bunch. Although their scores were extended by the SVM, DT, and CNN, they never reached them. A hybrid strategy has been used to complete the most remarkable mean score of COVID-19/Pneumonia in the transfer learning assessment.

In COVID-19/Normal arrangement, the most noteworthy mean explicitness (when the 100.0% scores of pre-prepared organizations are not considered due to not learning another class) and the most elevated mean precision results were obtained in Exp (99.78 and 99.11%, individually), which comprised the most profound design in

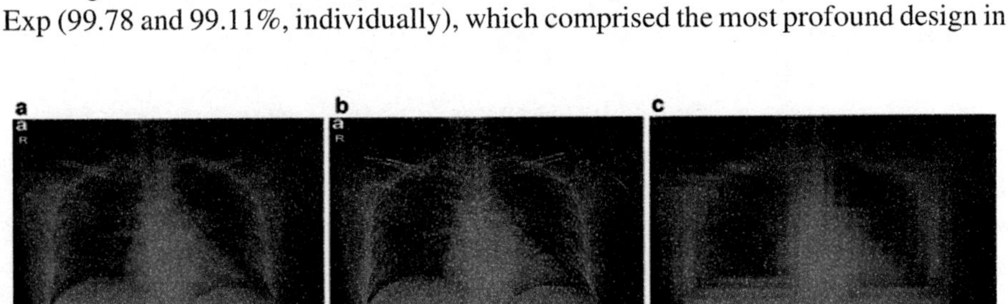

Fig. 4 Preprocess of X-ray images

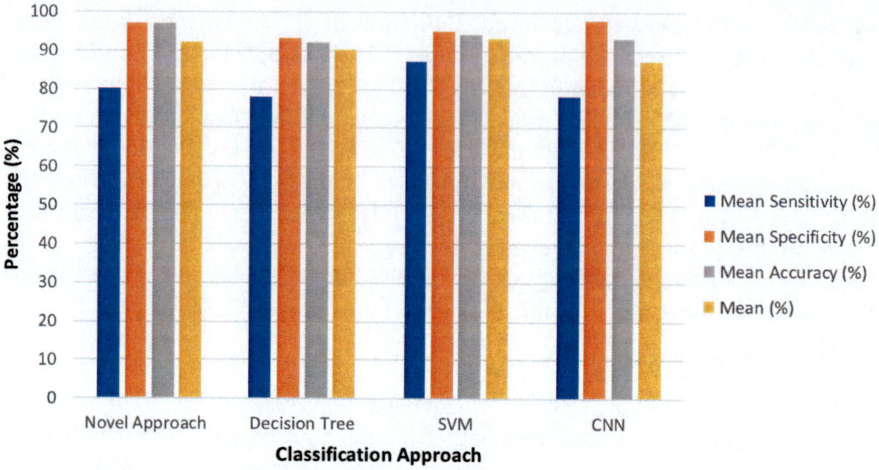

Fig. 5 Overall performance of the proposed system while detection

hybrid proposed approach tests. This fizzled, in any case, to create higher outcomes as far as to mean affectability, and this decreased the exhibition of the considered hybrid proposed approach in the critical execution marker for the two classes, mean score. As shown in Fig. 5, the highest mean affectability was accomplished by SVM (93.92%); however, other scores were not sufficiently high to beat different models in different measurements. AI classifiers could not create palatable outcomes utilizing the removed measurable information to characterize COVID-19 in this trial bunch. In COVID-19/Pneumonia arrangement, likewise to the past tests, the most elevated mean score was gotten in Exp (96.33%) with the hybrid proposed approach. In addition, the most elevated mean affectability and mean exactness results were likewise gotten in Exp (92.88 and 99.44%, separately). The most noteworthy mean explicitness was accomplished in move learning tests by the hybrid proposed approach (100%).

5 Conclusion

This project aims to look at the fundamentals of an Intelligent Information System. The most critical methodology utilized in enterprises is the information system with intelligence. A vital aspect of the information system is the knowledge discovery process, which analyzes data from multiple perceptions. Artificial intelligence, machine learning, natural language processing, and evolutionary algorithms are just a few of the tools that can help an Intelligent Information System succeed. IIS has a wide range of applications, including health care, education, marketing, finance, and insurance; with our method, we can detect COVID-19 more accurately in X-ray images. Researchers may use these tools in the future to compare them to other existing methodologies to handle the issues of these applications more effectively and efficiently.

References

1. Swanson EB (2011) Information systems. In: Understanding information retrieval systems: management, types, and standards
2. Boell SK, Cecez-Kecmanovic D (2015) What is an information system? https://doi.org/10.1109/HICSS.2015.587
3. W. Is (2011) @@Information systems: definitions and components. What is an information system? Management Information System Lecture
4. Bryant A, Black A, Land F, Porra J (2013) Information systems history: what is history what is IS history what IS history and why even bother with history. J Inf Technol. https://doi.org/10.1057/jit.2013.3
5. Ackoff R (1989) Ackoff's best. J Appl Syst Anal. CiteULike-article-id:6930744
6. Conger S, Probst J (2014) Knowledge management in ITSM: applying the DIKW model. Intell Syst Ref Libr. https://doi.org/10.1007/978-3-642-39928-2_1
7. Duan Y, Sun X, Che H, Cao C, Li Z, Yang X (2019) Modeling data, information and knowledge for security protection of hybrid IoT and edge resources. IEEE Access. https://doi.org/10.1109/access.2019.2931365
8. Ade R, Deshmukh PR (2015) Efficient knowledge transformation system using pair of classifiers for prediction of students career choice. Procedia Comp Sci 46:176–183. ISSN 1877-0509. https://doi.org/10.1016/j.procs.2015.02.009
9. Ade R, Deshmukh P (2015) Efficient knowledge transformation for incremental learning and detection of new concept class in students classification system. In: Mandal J, Satapathy S, Kumar Sanyal M, Sarkar P, Mukhopadhyay A (eds) Information systems design and intelligent applications. advances in intelligent systems and computing, vol 340. Springer, New Delhi. https://doi.org/10.1007/978-81-322-2247-7_77
10. Joshi AV (2020) Machine learning and artificial intelligence
11. Rana S, Li C, Gupta S, Nguyen V, Venkatesh S (2017) High dimensional Bayesian optimization with elastic Gaussian process
12. Purohit J, Bagwe A, Mehta R, Mangaonkar O, George E (2019) Natural language processing based Jaro-the interviewing chatbot. https://doi.org/10.1109/ICCMC.2019.8819708
13. Fang Y, Zhang C, Huang C, Liu L, Yang Y (2019) Phishing email detection using improved RCNN model with multilevel vectors and attention mechanism. IEEE Access. https://doi.org/10.1109/ACCESS.2019.2913705

Plant Quality Assessment and Disease Identification System Using AI

Divya Rajput, Hiral Rane, Janhavi Wagh, Devika Nikam, Roshani Raut, and Anuja Jadhav

Abstract Agriculture is the pillar of the economy and salient to accomplish comprehensive development. Presently, plant disease detection and quality assessment has encountered a rising scrutiny as a large dataset of plants is produced by monitoring its features. The naked eye inspection is the conservative approach endorsed in practice for plant leaves disease identification and inspection. Besides being subjective, it is becoming impracticable. This approach is followed by traditional machine learning methods for classification. In this paper, all the AI techniques from traditional to the most recent ones are concerned. Among all image classification techniques, we studied which method and algorithm surmounted every other and deployed it. Crop quality appraisal is an extremely fundamental assignment as it can assume an indispensable part in the understanding of the nature of the harvests and its sicknesses. Computer-based intelligence and ML is helping make that objective conceivable. Nowadays, deep learning and GAN-based techniques have been used for the quality appraisal of harvests and disease detection. Preliminary knowledge on plant health and disease identification will result in advancements in AI and management strategies. It has paved a way for constructing trained models with higher accuracy and effectiveness. This paper comprises various phases like image acquisition, dataset preparation, training, validation, and evaluation. Overall, all these play a pivotal role in shaping the agriculture infrastructure on a gigantic global scale. Iterating the training model by tuning different parameters has resulted in optimal training and validation accuracy and least training and validation loss. The final experimental results indicate that training accuracy achieved is approximately 98.5% followed by a training loss equivalent to zero and remains constant with respect to the epochs.

Keywords Artificial intelligence · Machine learning · Deep learning · Artificial neural network · Convolutional neural network · Plant disease detection

D. Rajput · H. Rane · J. Wagh · D. Nikam · R. Raut · A. Jadhav (✉)
Department IT, PCCOE, Savitribai Phule University, Pune, India
e-mail: anuja.jadhav@pccoepune.org

V. K. Gunjan and J. M. Zurada (eds.), *Proceedings of 3rd International Conference on Recent Trends in Machine Learning, IoT, Smart Cities and Applications*, Lecture Notes in Networks and Systems 540, https://doi.org/10.1007/978-981-19-6088-8_62

1 Introduction

Agriculture is a significant element of the global economy. India is known to be the world's largest producer of vegetables, grains, and spice products. Symptoms resembling plant damage caused by pathogens can be induced by abiotic (non-living) factors, such as nutrient deficiency and soil or air pollution. Frequently, diseases are a major source of plant damage that can be caused by a number of plant pathogenic (disease-causing) organisms [1]. Plant diseases and quality assessment is a less addressed problem which causes considerable menace to food security. Along with-it, plant weed identification is also a research problem in the agriculture field which relies on computational technology and sciences to detect unwanted plants growing wildly on the plants and crops. Manual diagnosis of plant illness is a tedious task, and the main purpose of farmers is to yield plants on a large scale. Plant disease identification and classification at millimeter level is important for better health of plants which in turn can increase the amount of yield produced [2]. So, plant quality should be accessed so as to provide fundamental information for comprehending quality and diseases suffered by plants.

The motivation for taking this topic is that it is one of the most eminent and promising topics and addressing it effectively can make a ton of difference. It could bring up optimistic changes globally. Disease detection in plants could bring out the best yields and provide us with surplus necessities, and we could uncover the problem of food scarcity and security. Food is a universal necessity; there is a never-ending opportunity for every individual to contribute and inculcate distinct techniques to unleash a perfect machine model with utmost efficiency.

The objectives of this paper are to study all the possible approaches and algorithms to diagnose the plant disease and implement the CNN architecture to obtain optimal results. To iterate the model by tuning various parameters. To deploy efficient algorithms to assess the plant disease. To enhance the recognition rate and training and validation accuracy of the trained model. To plot visualization of training and validation accuracy. Finally, to perform overall result analysis on the model by identification of distinct diseases.

The topics presented in this paper are organized into distinct sections as follows: After the introductory section, Sect. 2 describes the literature survey in which related and similar work is presented; Sect. 3 describes the proposed methodology; Sect. 4 presents experiments and discussion; Sect. 5 discusses the results, and finally, Sect. 6 holds the conclusion and future scope.

2 Literature Survey

Before, identifying the plant diseases was done visually which was very time-consuming, subjective, and also expensive. Many traditional ML algorithms were later proposed for cost reduction and subjectivity and also high accuracy [3]. Traditional machine learning algorithms have commonly been used to detect diseases like SVM, KNN, and NB. Classification between these different kinds of ML algorithms has been done in [4]. In this paper, through tests, it was proven that among these classifiers SVM was the best. But, classical ML models are only effective under specific setups. If changes are made in the conditions, the performance of these algorithms significantly decreases. Therefore, ML algorithms can only provide a foundation for further research for more advanced algorithms.

In this section, perspective of distinct methods for detecting the illness in plant leaves using different image processing techniques is discussed:

This paper [5] concentrates on numerous plant diseases in their growing phases. Early identification of it will be much more beneficial as it will positively affect the total yield. This paper introduces an advanced model that is amalgamation of convolutional neural network and convolutional autoencoder network. This approach is applied to detect bacterial spot disease using leaf images. While overall performance evaluation, we can observe that it accomplishes maximal training accuracy and testing accuracy using least training parameters. It exhibits better performance than the remaining methods till date.

The paper [6] aims to detect disease in plants using artificial intelligence. Different fields of AI in the agriculture sector like IoT-driven development, image-based insight, expert system, field management, and robotics in agriculture are discussed. AI techniques used for disease detection comprise image dataset collection and transformation followed by dataset vectorization, model training, and inference. CNN model is utilized from deep learning methodologies to score high precision and accuracy. Further, CNN and ANN are applied to computer vision applications. It is centric toward increasing the productivity of the primary sector.

In the following paper, different techniques for plant quality assessment are discussed.

In this paper [7], a machine vision system will be discussed to assess crop quality. It is reasonable to specify the main objectives of the vision plan. In crop quality testing, the main purpose is to measure the area of the plant with high accuracy and to assess the quality of the plant depending on its growth. Plant segregation is an important process of finding a plant and its boundary in photographs so that we can further analyze it. The leaves of the plants are intertwined, which is one of the main challenges of leaf division. In this way, different types of ideas serve the purpose of determining which types of ideas are useful for measuring plant area and crop quality assessments three different ideas: from top to bottom, front, and side.

In this paper [8, 9], developing CNN-based quality testing model CNN popular models that deliver CNN features such as AlexNet, GoogleNet, and Inceptionv3 will be used in this project. Images will be included in the CNN model developed as input. First, the model will be trained with training data and then tested with test data. The result of the model will give the result, based on where the level of the plant can be divided.

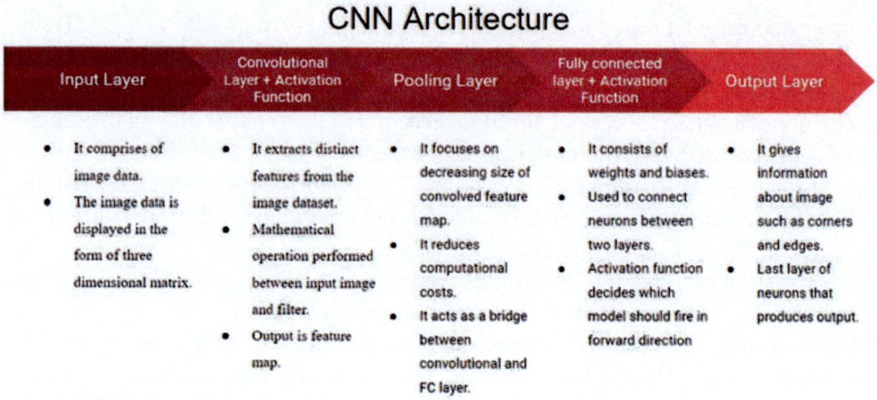

Fig. 1 Architecture of CNN [10]

3 Proposed Methodology

This paper on the whole fixates on effective plant disease identification. With the aid of traditional algorithms like SVM, KNN, and RF, the results obtained are examined and surveyed. Then, the modern approach based on CNN using TensorFlow and Keras is executed to evaluate the distinction.

3.1 CNN Architecture

Figure 1 illustrates the architecture of CNN which comprises an input layer, convolutional layers after that the pooling layer (to lessen feature maps and reduce parameters), fully connected layer, and output layer. The activation layer is in combination with both convolutional and fully connected layers.

3.2 Flowchart

See Fig. 2.

3.3 Data Collection and Inspection

For this report, we have utilized the public PlantVillage dataset curated by spMohanty's GitHub Repo [11]. This dataset consists of 15 directories and 20.6 k files of plant leaves images for implementation of the algorithm. The frequency of some

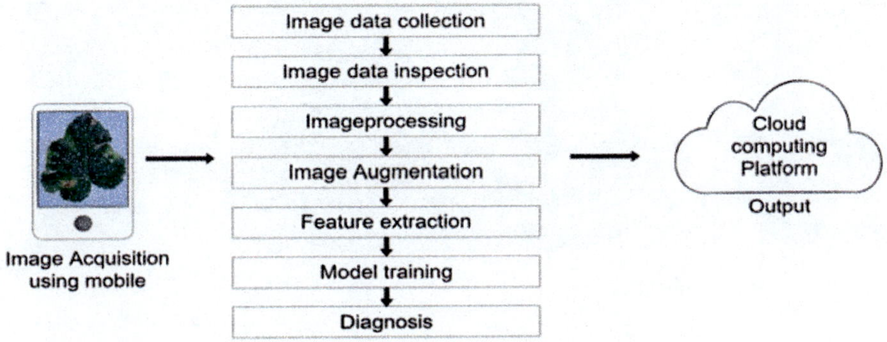

Fig. 2 Flowchart for proposed system

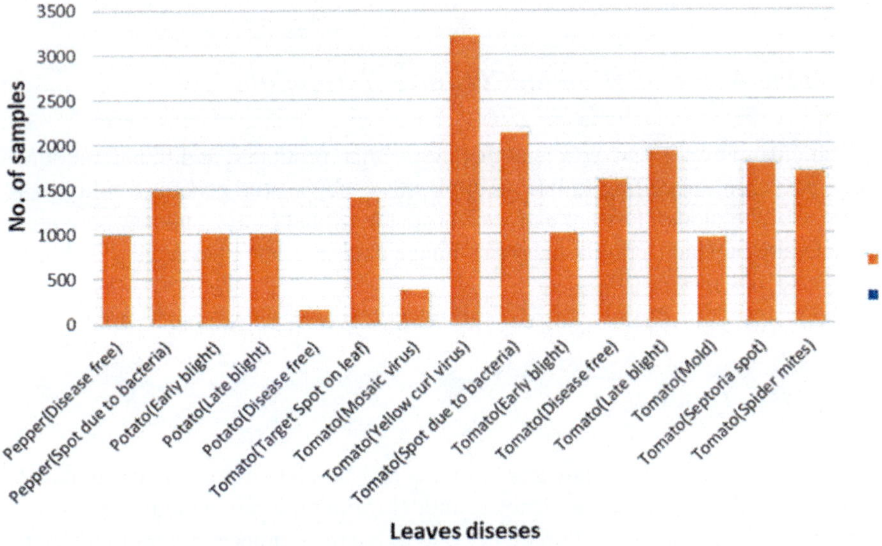

Fig. 3 Frequency of common diseases in the dataset

of the common diseases as well as that of healthy samples in the dataset has been illustrated in the graph of Fig. 3.

Some sample images from the dataset are shown in Fig. 4.

3.4 *Image Processing*

Image dataset processing is the necessary stride for disease detection in leaves which helps every individual working in the field achieve better precision and accuracy to produce disease healthy crops. The objective of preprocessing is to strengthen the

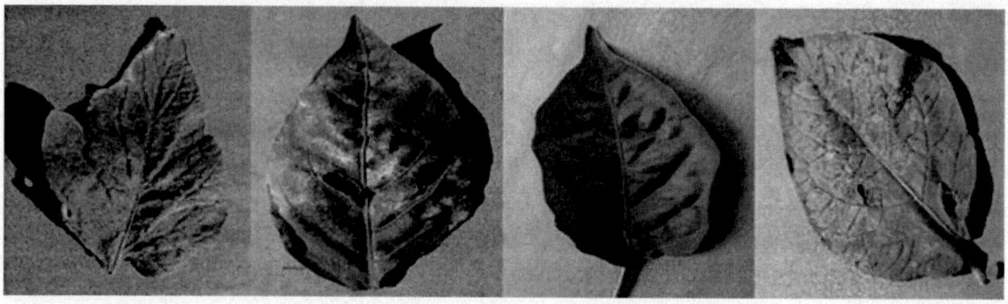

Fig. 4 Sample images from dataset

image attributes and prevents any kind of further image-related complication which impacts the quality of input images.

3.5 Image Augmentation and Feature Extraction

Image augmentation is an efficient technique that produces additional random training data from preceding training samples. It enhances the generalizability and robustness of the model. Feature extraction is the fragment of dimensionality reduction procedure in which a preliminary raw image dataset is lessened and divided into feasible categories.

3.6 Model Training

In this process, the model parameters (weights and biases) are repetitively updated so as to lessen the loss function. Prior to model training, it is essential to split the dataset. The dataset can be split randomly into two parts not necessarily of the same size. The most commonly used method is dividing 80% of data into the training part and 20% into the testing part. All in all, the end-to-end model training is attained by training data acquisition, updating the parameters, calculating the cost function, and iterating the process till the requirements are met.

3.7 Cloud Computing

The cloud is exploited for storage and processing of abundant image dataset. It also provides real-time interactive dashboards.

4 Results and Discussion

The paper presents the diverse plant leaves disease identification techniques using machine learning and deep learning models. This section confers the results related to training dataset which accommodates both augmented as well as original images. The images are converted into arrays. The number of images that can be picked from the folder can be manipulated accordingly. The size of processed images is obtained, and each image label is converted to binary label which is later saved using pickle. The ImageDataGenerator function is implemented to perform random shifts, flips, crops, and sheers generating smaller dataset with higher outcomes. The convolutional layer along with techniques like batch normalization, max pooling, and dropout is applied. The Adam optimizer provided by Keras is used to train the network. The proposed model thus plays the vital role that deep learning plays in diagnosing the plant leaves disease.

In Fig. 5 for the initial iteration, epoch value is equal to 10, followed by the initial 1e-3 learning rate which is a relatively smaller learning rate for the model to converge rather than diverging as it is a significant hyperparameter which determines the movement of minima. The batch size is 64, denoting that from the training dataset 64 sample images are obtained and utilized to estimate the direction and magnitude of the model before updating the model weights. After tuning the model with above parameters, following are the results obtained.

In Fig. 6, the third iteration is performed with epochs value 50 and learning rate 1e−3 also batch size being 32 followed by image augmentation adding hidden layers to optimize the performance along with Adam optimizer.

In Fig. 7, just by changing the batch size to 64, following are the results (It is observed theoretically that greater the value of batch size the gradients are vulnerable to lesser meaningless information, i.e., noise, resulting in finer estimation of gradient. This permits the model to achieve better minima).

The change in training and validation accuracy and training and validation loss is observed with respect to epochs of the algorithm over respective iterations. The final optimal solution obtained by the proposed algorithm is as follows: The training accuracy is 98.5%, and validation accuracy is 81.5%; further, the training loss is

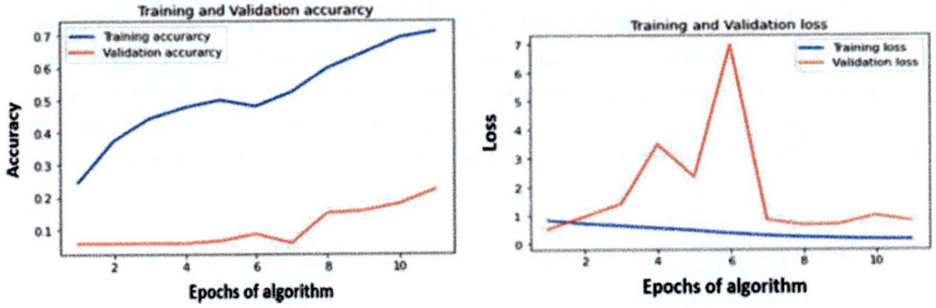

Fig. 5 Training and validation accuracy and loss with respect to epochs (iteration 1)

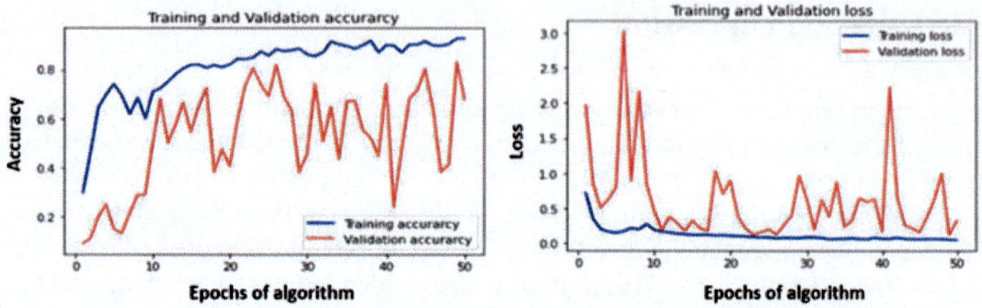

Fig. 6 Training and validation accuracy and loss with respect to epochs (iteration 3)

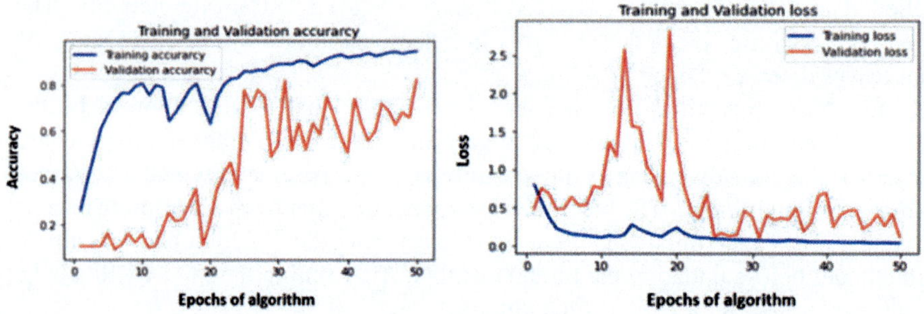

Fig. 7 Training and validation accuracy and loss with respect to epochs (iteration 4)

constant and equivalent to zero while validation loss is approximately equal to zero. Thus, the eminent model is trained and utilized for automated plant disease diagnosis with smaller validation and training time. This architecture therefore facilitates and benefits the agricultural sector.

5 Conclusion and Future Scope

Disease identification by assessing plant leaves at the initial phase is a tough job. In this paper, a DL using CNN architecture is exploited for determining and achieving highest accuracy and lowest loss. Fine-tuning the parameters has shown significant changes over each iteration. The objective of the paper to execute CNN to validate and train the model is thus accomplished. Future scope in this paper is concentrating on disease diagnosis in different regions of plants and in various phases in its disease.

References

1. Kennelly M, O'Mara J, Rivard C, Miller GL, Smith D (2012) Introduction to abiotic disorders in plants. Plant Health Instructor. https://doi.org/10.1094/PHI-I-2012-10-29-01
2. Wiesner-Hanks T, Wu H, Stewart E, DeChant C, Kaczmar N, Lipson H, Gore Michael A, Nelson Rebecca J (2019) Millimeter-level plant disease detection from aerial photographs via deep learning and crowdsourced data. Front Plant Sci 10. https://doi.org/10.3389/fpls.2019.01550
3. Arsenovic M, Karanovic M, Sladojevic S, Anderla A, Stefanovic D (2019) Solving current limitations of deep learning based approaches for plant disease detection. Symmetry 11(7):939. https://doi.org/10.3390/sym11070939
4. Akhtar A, Khanum A, Khan SA, Shaukat A (2013) Automated plant disease analysis (APDA): performance comparison of machine learning techniques. In: Proceedings of the 11th international conference on frontiers of information technology, Islamabad, Pakistan, 16–18 Dec 2013, pp 60–65
5. Bedi P, Gole P (2021) Plant disease detection using hybrid model based on convolutional autoencoder and convolutional neural network. Artif Intell Agric 5:90–101, ISSN 2589-7217. https://doi.org/10.1016/j.aiia.2021.05.002
6. Sujawat GS et al (2021) Application of artificial intelligence in detection of diseases in plants: a survey
7. Mavridou E, Vrochidou E, Papakostas GA, Pachidis T, Kaburlasos VG (2019) Machine vision systems in precision agriculture for crop farming. J Imaging 5(12):89. https://doi.org/10.3390/jimaging5120089. PMID: 34460603; PMCID: PMC8321169
8. Alzubaidi L, Zhang J, Humaidi AJ et al (2021) Review of deep learning: concepts, CNN architectures, challenges, applications, future directions. J Big Data 8:53. https://doi.org/10.1186/s40537-021-00444-8
9. Barbedo JG, Godoy CV (2015) Automatic classification of soybean diseases based on digital images of leaf symptoms. In: Embrapa Soja-Artigo em anais de congresso (ALICE), 2015
10. Phani Ratan (2020, Oct 28) What is convolutional neural network architecture?
11. git clone. https://github.com/spMohanty/PlantVillage-Dataset

Image-Based Plant Disease Detection and Classification Using Deep Convolution Neural Network

Roshani Raut, Prem Jadhav, and Abha Bodas

Abstract Agriculture is a country's economic backbone. Numerous studies have shown that plant diseases are challenging to manage because their populations vary with environmental factors. Plants can become infected with a wide range of diseases, including fungal, bacterial, and viral infections. It has been discovered that fungal-like creatures infect 85% of plants. In developing countries, farmers use a more labor-intensive and time-consuming method. Manual detection or observation with the naked eye is similarly unlikely to generate useful data. Many farmers have also been observed using pesticides to reduce the effects of disease without first recognizing the exact ailment. Farmers use pesticides in an unrestrained manner, which can have negative consequences on the plant as well as human health. As a result, through the use of several machine learning approaches, the machine learning model aids in the identification of plant diseases. In this, we used image processing methods and as well as convolutional neural networks (CNNs). This research's accuracy is 94.56%. The acquired results demonstrated that the proposed solution is usable and can be used by the farmers to identify plant diseases effectively.

Keywords Leaf disease detection · Convolutional neural networks (CNNs) · Image processing · Supervised learning · Unsupervised learning · Real-time processing · Agriculture

1 Introduction

Almost all industries rely on agricultural raw materials. Many nations rely on agriculture as their main economy. The GDP, economy, and progress of each country are dependent on agricultural expansion. In India, agriculture is the most common occupation. India is the leading producer of pulses, milk, jute, rice, and cotton. India can now meet the demands of its people while also exporting to the rest of the world. In 2019, India had a trade surplus of $8.25 billion in agriculture and forestry

R. Raut · P. Jadhav (✉) · A. Bodas (✉)
Department of Information Technology, Pimpri Chinchwad College of Engineering, Pune, India

V. K. Gunjan and J. M. Zurada (eds.), *Proceedings of 3rd International Conference on Recent Trends in Machine Learning, IoT, Smart Cities and Applications*, Lecture Notes in Networks and Systems 540, https://doi.org/10.1007/978-981-19-6088-8_63

commodities. According to the report, farmers would need to produce 70% more food as the population grows from 7 to 9 million by 2050. We can attain this figure by improving agricultural operations. Many farmers desire to adopt modern agriculture, but many are unable to do this for a variety of reasons, including a lack of knowledge about new technology, excessive technology costs, and so on. Each year, it is estimated that thirty to forty percent of crops are lost as a result of the production chain [1]. Plant disease prevention is linked to issues of sustainable agriculture and climate change [2]. According to Georgia research, approximately, $701.2 million in plant disease losses were estimated in 2010, including control expenses. The estimated crop cost was $4236.51 million, with a disease loss of 16.5% across all crops affected. The annual estimated losses in India due to nematodes are estimated to be Rs. 242.1 billion [3]. However, most illnesses manifest in some form of visual way. The primary method for identifying plant diseases in practice is a qualified expert's naked eye examination [4].

Machine learning-based techniques have proven to be effective in a wide range of image processing applications in recent years. Artificial intelligence-based learning has yielded positive results. Machine learning algorithms train the system to learn autonomously and improve results based on its own experiences [5]. Support vector machines (SVMs), artificial neural networks (ANNs), random forests, K-means method, fuzzy logic, and convolutional neural networks are examples of traditional methods [6]. Designed a system for estimating and detecting the severity of cotton plant disease. We achieved 82.5% accuracy using 40 photographs to diagnose the gray mildew disease. It has been observed that plant diseases are difficult to control. Farmers in our country use a traditional method which consumes time. The fruitful result is not possible with the help of manual detection with the naked eye. Many farmers also use pesticides unrestrictedly which harms plants as well as humans around. Using machine learning, deep learning to classify plant diseases can help farmers identify diseases and take appropriate control measures. In comparison to conventional image processing approaches, machine learning and deep learning algorithms for detecting plant diseases are more accurate and take less time.

This paper has included a literature survey in Sect. 2, followed by system architecture. In Sect. 4, we presented the findings and results of our research, followed by the work's conclusion.

2 Literature Survey

We reviewed many published studies on image processing, CNN, supervised and unsupervised learning that might assist us to understand how to classify leaf diseases based on images. Based on this study, this literature survey is composed of 3 sections. These sections are 2.1 Work related to leaf disease detection using deep learning (CNN). 2.2 Work related to leaf disease detection using supervised learning. 2.3 Work related to leaf disease detection using unsupervised learning.

2.1 Leaf Disease Detection Using Deep Learning (CNN)

In [4], authors have developed a new method for detecting plant illnesses that employs a deep CNN model that has been trained and fine-tuned to accurately match a plant's leaf database. Using deep CNN, the model was able to distinguish between damaged and healthy leaves, as well as the surroundings, using healthy leaves and backdrop images that were consistent with other classes. The classification accuracy ranged from 65 to 90%. NNE provides a more generalized understanding of learning ability by merging the findings of a large number of neural networks that have been trained. The classification accuracy using this method was 94.67%. The top-1 rate of success was 96.3% after 1 million iterations, while the top-5 rate of success was 99.99%.

In [7], 500 images were used in the dataset for disease detection and classification. The learning vector quantization (LVQ) technique and a convolutional neural network (CNN) model were used to correctly diagnose the tomato leaf disease.

A CNN model using RGB components from the Plant Village dataset's tomato leaf images was created. The learning vector quantization (LVQ) method was the primary classifier because of its topology and adaptive model. Because of its multi-layered structure, CNN is effective at analyzing visual images and can easily distinguish between the required characteristics, allowing it to recognize and categorize objects with minimal preprocessing.

In [8], there are 87,848 images with 25 different plants divided into 58 different classes. In order to create an automated plant disease detection and diagnostic system, specific CNN architectures were trained and evaluated. The LuaJIT2 programming language is used in the Torch71 machine learning computational framework was used to create the model. The best result was a success rate of 99.53%. The training set consisted of 80% of the data, while the testing set consisted of 20%.

2.2 Leaf Disease Detection Using Supervised Learning

Paper [9] focuses SVM, random forest, decision tree (DT), Naive Bayes (NB), and K-nearest neighbor (KNN) supervised machine learning techniques for maize plant disease detection. The classification techniques mentioned previously are examined and evaluated. They are compared to choose the most accurate model for plants. Disease prediction in comparison to the other classification techniques, the highest accuracy of 79.23 was achieved by the RF algorithm.

Various methods and approaches to diagnosing various types of plant leaf diseases have been proposed in the literature. SVM [10] was used to identify the leaf diseases. Tea production is critical for countries like India and Bangladesh to grow their economies. The authors focused on creating a model that can detect the tea leaf's main disease. When compared to other classification methods, this one helps the country increase its production and growth rate.

Table 1 Different algorithms used

Type of algorithms	Name of algorithms
Supervised learning	Decision tree (DT) [12], SVM [12], Naive Bayes (NB) [13], K-nearest neighbor (KNN) [14], and Random forest
Unsupervised learning	K-means clustering [15]
Deep learning	CNN [16], Learning vector quantization (LVQ) technique [17]

2.3 Work Related to Leaf Disease Detection Using Unsupervised Learning

Based on their colors, the infected leaves are divided into three clusters [11]. Cluster groups consist of yellowish, brownish, and greenish-colored leaves. Clustering is accomplished using SOM and K-means. Four different distance measuring functions are used in K-means to select the best performing function for the process. In K-means clustering, it is discovered that the cluster size of SOM corresponds to the CityBlock distance function. The F1 value indicates that the clustering is completed with 97% accuracy.

This data set can be used as a training set for the infected leaf classification process.

Table 1 shows the categorization of the algorithms used to implement machine learning in the research papers discussed in the literature survey. Therefore, we can conclude that leaf disease detection can be done using a variety of ways that include supervised learning, unsupervised learning, and also deep learning algorithms.

3 Proposed Methodology

3.1 System Architecture

Figure 1 represents the model architecture. First, we have an RGB (Red Green Black) image of dimensions (128 × 128 × 3).

Pooling Layer: This layer uses a constant factor to reduce the dimensionality of the input. This layer aims to reduce computational load while performing feature selection. There is just one output value for each of the input images that are tiled into non-overlapping sub-regions. Maxima or average, commonly known as max-pooling and avg-pooling, is the most popular options [18].

Max-pooling is widely used because it introduces the least amount of translation and distortion invariance, resulting in faster convergence and better generalization [19].

Fully Connected Layer: A multi-layer network's standard layer is the completely connected layer. The input vector is linearly combined with a weight matrix. The network either flips between convolutional and max-pooling layers, resulting in a

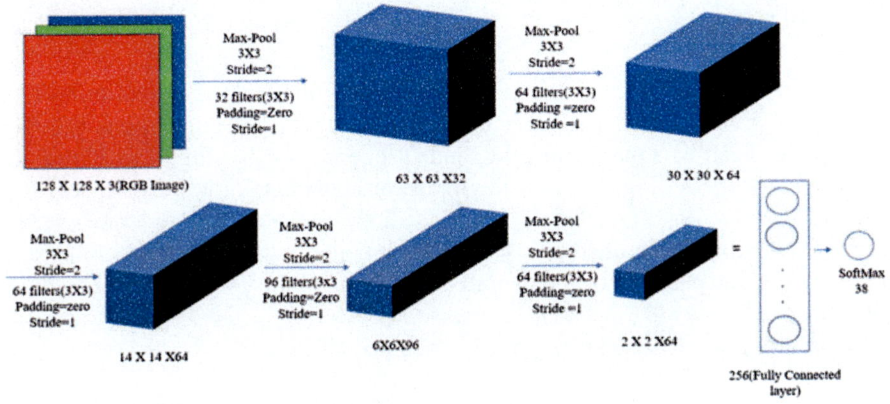

Fig. 1 Diagram for system architecture

1D feature vector (images of 11), or the images are rearranged into a 1D shape. With the same number of neurons as the classes in the classification task, the output layer is always fully connected. The SoftMax activation function is used to normalize the results, resulting in approximated posterior class probabilities [18].

In all the layers, we have used the filters of size (3 × 3), padding equal to zero, and stride equal to 1. In pooling layers, we have used pooling type as max-pooling and every time the stride used is 2 and size is (3 × 3). In the first layer, we have used 32 filters, and after using max-pool, we get a layer having 32 filters and dimensions as (63 × 63 × 32). For the next, layers, we use 64 filters each, and after we use the max-pool, we get to layer with 64 filters and dimensions as (14 × 14 × 64). For the next layer, we increase the number of filters to 96, and after using max-pool, we get the last layer with dimensions (2 × 2 × 64). Then, there is a fully connected layer having 256 neurons, and finally, we use Softmax [20] having 38 values as the total number of classes in our dataset is 38.

3.2 *Dataset Description*

The dataset that we have used was taken from Kaggle [21]. The dataset used in our model is divided into three parts:

1. Training dataset
   Our training dataset has 70,295 images belonging to 38 classes.
2. Validation dataset
   The validation dataset has 17,572 images belonging to 38 classes.
3. Test dataset

The test dataset has around 33 images.

4 Results

We ran a series of tests on healthy and diseased leaf image databases in order to classify them into the appropriate categories, such as potato late blight, potato early blight, potato healthy, soybean healthy, and raspberry healthy. Figure 2 shows a healthy apple leaf, and Fig. 3 shows the image of apple black rot, i.e., (the leaf is diseased).

The CNN model was trained on a large dataset containing about 70,000 images. On fine-tuning, the parameters and overall accuracy of 94.56% was achieved by the model.

The accuracy is calculated using Eq. 1.

$$\text{Accuracy} = \text{Number of correct outcomes}/\text{Total number of outcomes}. \quad (1)$$

Accuracy of the model is calculated by tuning various parameters like number of epochs, number of layers, and number of filters.

1. **Accuracy versus Number of Epochs**

We tested the accuracy for various epochs. The results are represented in Fig. 4.When we compared the number of epochs and accuracy initially, the accuracy was very less, i.e., it was around 16.06% after the first epoch. As the number of epochs increased initially, there was an exponential increase in accuracy, but after 90 epochs, the

Fig. 2 Apple healthy

Fig. 3 Apple black rot

accuracy increased at a steady rate. We chose 150 as our number of epochs as it gave us the best results on the test data. We could have chosen 120 as our number of epochs, but we did not use 120 as its performance was not that well on the test dataset and due to the under-fitting phenomenon [22].

2. **Accuracy versus Number of Layers**

We then tested 2 types of models, one containing 3 hidden layers and the other containing 4 hidden layers. The outcomes are represented in Fig. 5.

Even though the accuracy of the model having 3 layers is more than the accuracy of the model having 4 layers, we have selected the model with 4 layers. This is because the 3-layer model overfits the training data. Overfitting occurs when learned mappings only function well on training data and do not perform well on test data, resulting in a lack of flexibility and adaptability. The performance of the model on the testing data is not up to the mark, that is, the data which it has never seen. This is the same reason to select several epochs as 150.

3. **Accuracy versus Number of Filters**

We also manipulated the filters in the last layer to check the effects on the accuracy of our model. The details about the number of filters used in the last layer and its accuracy are given in Fig. 6.

Therefore, from Fig. 6, it is seen that accuracy is achieved more when we use 64 filters in the last layer of the model. When the number of filters is 32, the accuracy is quite good which is 93.56%, but when we use 64 filters, the accuracy increases by

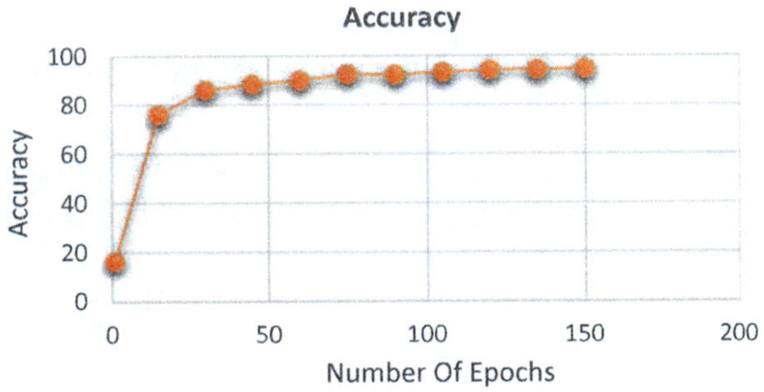

Fig. 4 Accuracy versus number of epochs

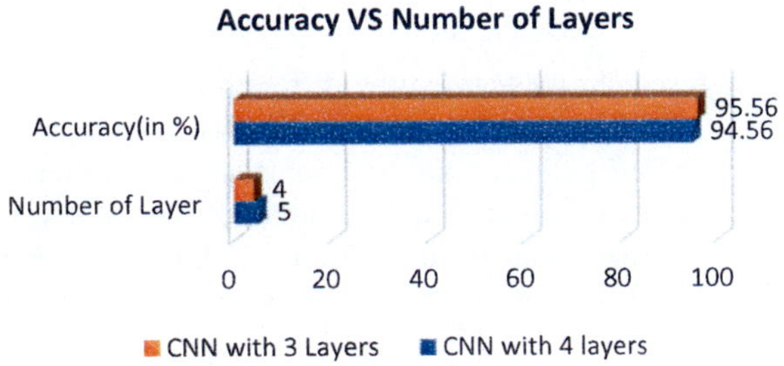

Fig. 5 Accuracy versus number of layers

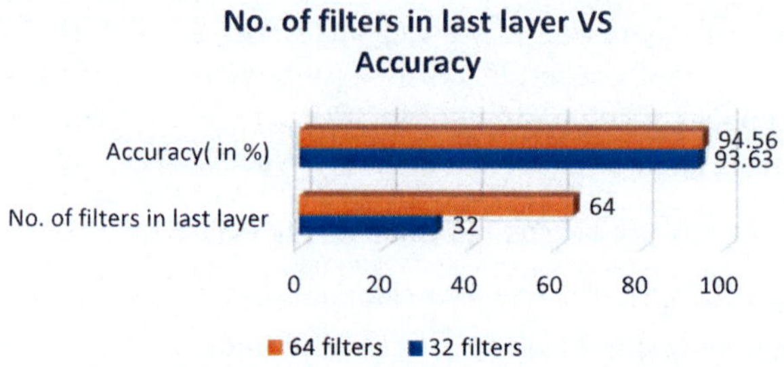

Fig. 6 Accuracy versus number of filters

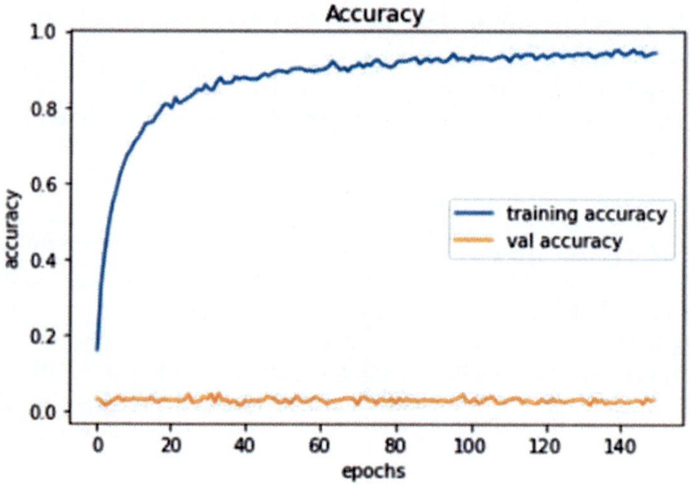

Fig. 7 Accuracy versus number of epochs (training and validation)

1%, and it fits the data perfectly without causing overfitting. Finally, 64 filters are chosen as the final value to be taken in the last layer of the model.

Figures 7 and 8 show the graphs for training accuracy and training loss, respectively. We got high accuracy from 75 iterations and above.

5 Conclusion

In this research, we learned the different methods and approaches, and algorithms used in various papers for the classification and detection of plant disease. This was done to address the serious issue of getting less yield because of disease-infested crops. Agriculture being the backbone of our country, it is necessary to ensure that we get the maximum yield per unit of land that we have available in our country. Hence, leaf disease detection is an important step to combat the diseases and improve crop yield.

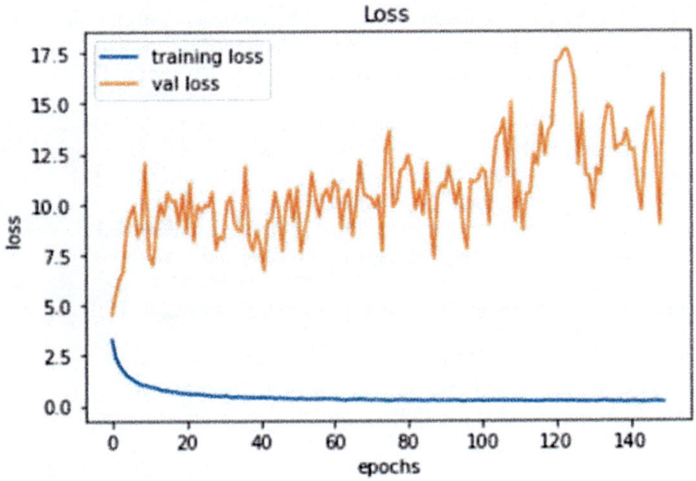

Fig. 8 Loss versus number of epochs

In this study, we developed and tested an image processing-based technique for detecting leaf diseases. We put our system to the test on a variety of diseases that affect plants. The current experimental results show that the method proposed is a valuable tool for detecting leaf diseases accurately and automatically. According to our tests, the created neural network classifier, based on statistical classification, performs well and can accurately detect and categorize the diseases with a precision of around 94.56%.

There are various opportunities for future research, like developing better segmentation techniques, choosing better feature extraction algorithms, and culling classification algorithms are all things that need to be addressed.

References

1. Flood J (2010) The importance of plant health to food security. Food Security 2(3):215–231. https://doi.org/10.1007/s12571-010-0072-5
2. Garrett KA, Dendy SP, Frank EE, Rouse MN, Travers SE (2006) Climate change effects on plant disease: genomes to ecosystems. Annu Rev Phytopathol 44:489–509. https://doi.org/10.1146/annurev.phyto.44.070505.143420
3. Dhingra G, Kumar V, Joshi HD (2018) Study of digital image processing techniques for leaf disease detection and classification. Multimedia Tools Appl 77(15):19951–20000. https://doi.org/10.1007/s11042-017-5445-8
4. Sladojevic S, Arsenovic M, Anderla A, Culibrk D, Stefanovic D (2016) Deep neural networks based recognition of plant diseases by leaf image classification. Comput Intell Neurosci 2016. https://doi.org/10.1155/2016/3289801
5. Maniyath SR et al (2018) Plant disease detection using machine learning. In: Proceedings—2018 International conference on design innovations for 3Cs compute communicate control, ICDI3C 2018, Aug 2018, pp 41–45. https://doi.org/10.1109/ICDI3C.2018.00017

6. Parikh A, Raval MS, Parmar C, Chaudhary S (2016) Disease detection and severity estimation in cotton plant from unconstrained images. In: Proceedings—3rd IEEE international conference on data science and advanced analytics, DSAA 2016, Dec 2016, pp 594–601. https://doi.org/10.1109/DSAA.2016.81
7. Sardogan M, Tuncer A, Ozen Y (2018) Plant leaf disease detection and classification based on CNN with LVQ algorithm. In: IEEE 2018 3rd International conference on computer science and engineering (UBMK)—Sarajevo, Bosnia and Herzegovina (2018.9.20–2018.9.23), pp 382–385. https://doi.org/10.1109/UBMK.2018.8566635
8. Ferentinos KP (2018) Deep learning models for plant disease detection and diagnosis. Comput Electron Agric 145:311–318. https://doi.org/10.1016/j.compag.2018.01.009
9. Panigrahi KP, Das H, Sahoo AK, Moharana SC (2020) Maize leaf disease detection and classification using machine learning algorithms. Adv Intell Syst Comput 1119:659–669. https://doi.org/10.1007/978-981-15-2414-1_66
10. Kurale NG, Vaidya MV (2018) Classification of leaf disease using texture feature and neural network classifier. In: IEEE 2018 International conference on inventive research in computing applications (ICIRCA)—Coimbatore (2018.7.11–2018.7.12), pp 1–6. https://doi.org/10.1109/ICIRCA.2018.8597434
11. ICONSTEM 2017: Digital India and smart cities. In: Proceedings: Third IEEE international conference on science, technology, engineering and management: 23rd & 24th March 2017. Jeppiaar Engineering College and Institute of Electrical and Electronics Engineers
12. Hassan S; Rafi M, Shaikh MS (2011) Comparing SVM and Naïve Bayes classifiers for text categorization with Wikitology as knowledge enrichment. In: IEEE 2011 14th International multi-topic conference (INMIC)—Karachi, Pakistan (2011.12.22–2011.12.24), pp 31–34. https://doi.org/10.1109/INMIC.2011.6151495
13. Safavian SR, Landgrebe D (1991) A survey of decision tree classifier methodology. IEEE Trans Syst Man Cybern 21(3):660–674. https://doi.org/10.1109/21.97458
14. Ali N, Neagu D, Trundle P (2019) Evaluation of k-nearest neighbour classifier performance for heterogeneous data sets. SN Appl Sci 1(12), Dec 2019. https://doi.org/10.1007/s42452-019-1356-9
15. Shi N, Liu X, Guan Y (2010) Research on k-means clustering algorithm: An improved k-means clustering algorithm. In: 3rd International symposium on intelligent information technology and security informatics, IITSI 2010, 2010, pp 63–67. https://doi.org/10.1109/IITSI.2010.74
16. Albawi S, Mohammed TA, Al-Zawi S (2017) Understanding of a convolutional neural network. In: IEEE 2017 International conference on engineering and technology (ICET)—Antalya, Turkey (2017.8.21–2017.8.23), pp 1–6. https://doi.org/10.1109/ICEngTechnol.2017.8308186
17. Kohonen T (1990) Improved versions of learning vector quantization. In: 1990 IJCNN International joint conference on neural networks, vol 1, pp 545–550 (1990.06.17–1990.06.21). https://doi.org/10.1109/ijcnn.1990.137622
18. Masci J, Meier U, Ciresan D, Schmidhuber J, Fricout G (2012) Steel defect classification with max-pooling convolutional neural networks. In: The 2012 international joint conference on neural networks (IJCNN). https://doi.org/10.1109/ijcnn.2012.6252468
19. Scherer D, Müller A, Behnke S (2010) Evaluation of pooling operations in convolutional architectures for object recognition. Lect Notes Comp Sci:92–101.https://doi.org/10.1007/978-3-642-15825-4_10
20. Wang F, Cheng J, Liu W, Liu H (2018) Additive margin Softmax for face verification. IEEE Signal Process Lett 25(7):926–930. https://doi.org/10.1109/LSP.2018.2822810
21. https://www.kaggle.com/vipoooool/new-plant-diseases-dataset
22. Zhang H, Zhang L, Jiang Y (2019) Overfitting and underfitting analysis for deep learning based end-to-end communication systems. In: 2019 11th International conference on wireless communications and signal processing (WCSP). https://doi.org/10.1109/wcsp.2019.8927876

Author Index

V. K. Gunjan and J. M. Zurada (eds.), *Proceedings of 3rd International Conference on Recent Trends in Machine Learning, IoT, Smart Cities and Applications*, Lecture Notes in Networks and Systems 540, https://doi.org/10.1007/978-981-19-6088-8

R

S

T

U

V

W

Y